RISING ABOVE THE GATHERING STORM

Energizing and Employing America for a Brighter Economic Future

Committee on Prospering in the
Global Economy of the 21st Century:
An Agenda for American Science and Technology

Committee on Science, Engineering, and Public Policy

NATIONAL ACADEMY OF SCIENCES,
NATIONAL ACADEMY OF ENGINEERING, *AND*
INSTITUTE OF MEDICINE
OF THE NATIONAL ACADEMIES

THE NATIONAL ACADEMIES PRESS
Washington, D.C.
www.nap.edu

THE NATIONAL ACADEMIES PRESS • 500 Fifth Street, N.W. • Washington, DC 20001

NOTICE: The project that is the subject of this report was approved by the Governing Board of the National Research Council, whose members are drawn from the councils of the National Academy of Sciences, the National Academy of Engineering, and the Institute of Medicine. The members of the committee responsible for the report were chosen for their special competences and with regard for appropriate balance.

Support for this project was provided by the National Academies. Any opinions, findings, conclusions, or recommendations expressed in this publication are those of the author(s) and do not necessarily reflect the views of the organizations or agencies that provided support for the project.

Library of Congress Cataloging-in-Publication Data

Rising above the gathering storm : energizing and employing America for a brighter economic future : Committee on Prospering in the Global Economy of the 21st Century : an agenda for American science and technology ; Committee on Science, Engineering, and Public Policy.

p. cm.

Includes bibliographical references and index.

ISBN 978-0-309-10039-7 (hardcover) — ISBN 978-0-309-65442-5 (pdf) 1. United States—Economic conditions—Forecasting. 2. Globalization. 3. United States—Economic policy. I. Committee on Prospering in the Global Economy of the 21st Century (U.S.) II. Committee on Science, Engineering, and Public Policy (U.S.)
HC106.83.R57 2006
331.12'0420973—dc22

2006025998

ISBN13: 978-0-309-18758-9 (paperback)

For more information about the Committee on Science, Engineering, and Public Policy, see http://www.nationalacademies.org/cosepup.

Available from the National Academies Press, 500 Fifth Street, N.W., Lockbox 285, Washington, DC 20055; (800) 624-6242 or (202) 334-3313 (in the Washington metropolitan area); Internet, http://www.nap.edu.

Printed in the United States of America

COMMITTEE ON PROSPERING IN THE GLOBAL ECONOMY OF THE 21ST CENTURY

NORMAN R. AUGUSTINE (Chair), Retired Chairman and CEO, Lockheed Martin Corporation, Bethesda, MD
CRAIG R. BARRETT, Chairman of the Board, Intel Corporation, Chandler, AZ
GAIL CASSELL, Vice President, Scientific Affairs, and Distinguished Lilly Research Scholar for Infectious Diseases, Eli Lilly and Company, Indianapolis, IN
STEVEN CHU, Director, E. O. Lawrence Berkeley National Laboratory, Berkeley, CA
ROBERT M. GATES, President, Texas A&M University, College Station, TX
NANCY S. GRASMICK, Maryland State Superintendent of Schools, Baltimore, MD
CHARLES O. HOLLIDAY, JR., Chairman of the Board and CEO, DuPont Company, Wilmington, DE
SHIRLEY ANN JACKSON, President, Rensselaer Polytechnic Institute, Troy, NY
ANITA K. JONES, Lawrence R. Quarles Professor of Engineering and Applied Science, University of Virginia, Charlottesville, VA
JOSHUA LEDERBERG, Sackler Foundation Scholar, Rockefeller University, New York, NY
RICHARD LEVIN, President, Yale University, New Haven, CT
C. D. (DAN) MOTE, JR., President, University of Maryland, College Park, MD
CHERRY MURRAY, Deputy Director for Science and Technology, Lawrence Livermore National Laboratory, Livermore, CA
PETER O'DONNELL, JR., President, O'Donnell Foundation, Dallas, TX
LEE R. RAYMOND, Chairman and CEO, Exxon Mobil Corporation, Irving, TX
ROBERT C. RICHARDSON, F. R. Newman Professor of Physics and Vice Provost for Research, Cornell University, Ithaca, NY
P. ROY VAGELOS, Retired Chairman and CEO, Merck, Whitehouse Station, NJ
CHARLES M. VEST, President Emeritus, Massachusetts Institute of Technology, Cambridge, MA
GEORGE M. WHITESIDES, Woodford L. & Ann A. Flowers University Professor, Harvard University, Cambridge, MA
RICHARD N. ZARE, Marguerite Blake Wilbur Professor in Natural Science, Stanford University, Stanford, CA

Principal Project Staff

DEBORAH D. STINE, Study Director
PETER HENDERSON, Senior Program Officer
JO L. HUSBANDS, Senior Program Officer
LAUREL L. HAAK, Program Officer
TOM ARRISON, Senior Program Officer
DAVID ATTIS, Policy Consultant
ALAN ANDERSON, Consultant Writer
STEVE OLSON, Consultant Writer
RACHEL COURTLAND, Research Associate
NEERAJ P. GORKHALY, Senior Program Assistant
JOHN B. SLANINA, Christine Mirzayan Science and Technology Policy Graduate Fellow
BENJAMIN A. NOVAK, Christine Mirzayan Science and Technology Policy Graduate Fellow
NORMAN GROSSBLATT, Senior Editor
KATE KELLY, Editor

COMMITTEE ON SCIENCE, ENGINEERING, AND PUBLIC POLICY

Preface

Ninety-nine percent of the discoveries are made
by one percent of the scientists.

Julius Axelrod, Nobel Laureate[1]

The prosperity the United States enjoys today is due in no small part to investments the nation has made in research and development at universities, corporations, and national laboratories over the last 50 years. Recently, however, corporate, government, and national scientific and technical leaders have expressed concern that pressures on the science and technology enterprise could seriously erode this past success and jeopardize future US prosperity. Reflecting this trend is the movement overseas not only of manufacturing jobs but also of jobs in administration, finance, engineering, and research.

The councils of the National Academy of Sciences and the National Academy of Engineering, at their annual joint meeting in February 2005, discussed these tensions and examined the position of the United States in today's global knowledge-discovery enterprise. Participants expressed concern that a weakening of science and technology in the United States would inevitably degrade its social and economic conditions and in particular erode the ability of its citizens to compete for high-quality jobs.

On the basis of the urgency expressed by the councils, the National Academies' Committee on Science, Engineering, and Public Policy

[1]*Proceedings of the American Philosophical Society*, Vol. 149, No. 2, June 2005.

(COSEPUP) was charged with organizing a planning meeting, which took place May 11, 2005. One of the speakers at the meeting was Senator Lamar Alexander, the former secretary of education and former president of the University of Tennessee.

Senator Alexander indicated that the Energy Subcommittee of the Senate Energy and Natural Resources Committee, which he chairs, had been given the authority by the full committee's chair, Senator Pete Domenici, to hold a series of hearings to identify specific steps that the federal government should take to ensure the preeminence of America's science and technology enterprise. Senator Alexander asked the National Academies to provide assistance in this effort by selecting a committee of experts from the scientific and technical community to assess the current situation and, where appropriate, make recommendations. The committee would be asked to identify urgent challenges and determine specific steps to ensure that the United States maintains its leadership in science and engineering to compete successfully, prosper, and be secure in the 21st century.

On May 12, 2005, the day after the planning meeting, three members of the House of Representatives who have jurisdiction over science and technology policy and funding announced that a conference would be held in fall 2005 on science, technology, innovation, and manufacturing. Appearing at a Capitol Hill press briefing to discuss the conference were representatives Frank Wolf, Sherwood Boehlert, and Vern Ehlers. Representative Boehlert said of the conference: "It can help forge a national consensus on what is needed to retain US leadership in innovation. A summit like this, with the right leaders, under the aegis of the federal government, can bring renewed attention to science and technology concerns so that we can remain the nation that the world looks to for the newest ideas and the most skilled people."

In describing the rationale for the conference, Representative Wolf recalled meeting with a group of scientists and asking them how well the United States was doing in science and innovation. None of the scientists, he reported, said that the nation was doing "okay." About 40% said that we were "in a stall," and the remaining 60% said that we were "in decline." He asked a similar question of the executive board of a prominent high-technology association, which reported that in its view the United States was "in decline."

Later, the National Academies received a bipartisan letter addressing the subject of America's competitiveness from Senators Lamar Alexander and Jeff Bingaman. The letter, dated May 27, 2005, requested that the National Academies conduct a formal study on the issue to assist in congressional deliberations. That was followed by a bipartisan letter from Representatives Sherwood Boehlert and Bart Gordon, of the House Committee on

Science, which expanded on the Senate request. In response, the National Academies initiated a study with its own funds.

To undertake the study, COSEPUP established the Committee on Prospering in the Global Economy of the 21st Century: An Agenda for American Science and Technology. The committee members included presidents of major universities, Nobel laureates, CEOs of Fortune 100 corporations, and former presidential appointees. They were asked to investigate the following questions:

- What are the top 10 actions, in priority order, that federal policymakers could take to enhance the science and technology enterprise so that the United States can successfully compete, prosper, and be secure in the global community of the 21st century?
- What implementation strategy, with several concrete steps, could be used to implement each of those actions?

This study and report were carried out with an unusual degree of urgency—only a matter of weeks elapsed from the committee's initial gathering to release of its report. The process followed the regular procedures for an independent National Research Council study, including review of the report, in this case, by 37 experts. The report relies on customary reference to the scientific literature and on consensus views and judgments of the committee members.

The committee began by assembling the recommendations of 13 issue papers summarizing past studies of topics related to the present study. It then convened five focus groups consisting of 66 experts in K–12 education, higher education, research, innovation and workforce issues, and national and homeland security and asked each group to recommend three actions it considered to be necessary for the nation to compete, prosper, and be secure in the 21st century. The committee used those suggestions and its own judgment to make its recommendations. The key thematic issues underlying these discussions were the nation's need to create jobs and need for affordable, clean, and reliable energy.

In this report, a description of the key elements of American prosperity in the 21st century is followed by an overview of how science and technology are critical to that prosperity. The report then evaluates how the United States is doing in science and technology and provides recommendations for improving our nation's prosperity. Finally, it posits the status of prosperity if the United States maintains a narrow lead (the current situation), falls behind, or emerges as the leader in a few selected fields of science and technology.

We strayed from our charge in that we present not 10 actions but 4 recommendations and 20 specific actions to implement them. The committee members deeply believe in the fundamental linkage of all the recommendations and their integrity as a coordinated set of policy actions. To emphasize one or neglect another, the members decided, would substantially weaken what should be viewed as a coherent set of high-priority actions to create jobs and enhance the nation's energy supply in an era of globalization. For example, there is little benefit in producing more researchers if there are no funds to support their research.

The committee thanks the focus-group members, who took precious personal time in midsummer to donate the expertise that would permit a highly focused, detailed examination of a question of extraordinary complexity and importance. We thank the staff of the National Academies. They quickly mobilized the knowledge resources and practical skills needed to complete this study in a rapid, thorough manner.

Norman R. Augustine
Chair, Committee on Prospering in the Global Economy of the 21st Century

CRAIG BARRETT

NANCY GRASMICK

GAIL CASSELL

CHARLES HOLLIDAY, JR.

STEVEN CHU

SHIRLEY ANN JACKSON

ROBERT GATES

ANITA K. JONES

JOSHUA LEDERBERG

ROBERT C. RICHARDSON

RICHARD LEVIN

P. ROY VAGELOS

C. D. (DAN) MOTE, JR.

CHARLES M. VEST

CHERRY MURRAY

GEORGE M. WHITESIDES

PETER O'DONNELL, JR.

RICHARD N. ZARE

LEE R. RAYMOND

Acknowledgments

This report is the product of many people. First, we thank all the focus-group members, listed in Appendix C, for contributing their time and knowledge at the focus-group session in August 2005. Second, we would like to thank all the committees and analysts at other organizations who have gone before us, producing reports and analyses on the topics discussed in this report. There are too many to mention here, but they are cited throughout the report and range from individual writers and scholars, such as Thomas Friedman and Richard Freeman, to committees and organizations, such as the Glenn Commission on K–12 education, the Council on Competitiveness, the Center for Strategic and International Studies, the Business Roundtable, the Taskforce on the Future of American Innovation, the President's Council of Advisors on Science and Technology, the National Science Board, and other National Academies committees. Without their insight and analysis, this report would not have been possible.

This report has been reviewed in draft form by persons chosen for their diverse perspectives and technical expertise in accordance with procedures approved by the National Research Council's Report Review Committee. The purpose of this independent review is to provide candid and critical comments that will assist the institution in making the published report as sound as possible and to ensure that the report meets institutional standards of objectivity, evidence, and responsiveness to the study charge. The review comments and draft manuscript remain confidential to protect the integrity of the deliberative process.

We wish to thank the following for their review of this report: Miller

Adams, Boeing Phantom Works; John Ahearne, Sigma Xi; Robert Aiken, CISCO Systems, Inc.; Bruce Alberts, University of California, San Francisco; Richard Atkinson, University of California, San Diego; William Badders, Cleveland Municipal School District; Roger Beachy, Ronald Danforth Plant Service Center; George Bugliarello, Polytechnic University; Paul Citron, Medtronic, Inc.; Michael Clegg, University of California, Irvine; W. Dale Compton, Purdue University; Robert Dynes, University of California, San Diego; Joan Ferrini-Mundy, Michigan State University; Richard Freeman, Harvard University; William Friend, Bechtel Group, Inc. (retired); Lynda Goff, University of California, Santa Cruz; William Happer, Princeton University; Robert Hauser, University of Wisconsin; Ron Hira, Rochester Institute of Technology; Dale Jorgenson, Harvard University; Thomas Keller, Medomak Valley High School, Maine; Edward Lazowska, University of Washington; W. Carl Lineberger, University of Colorado, Boulder; James Mongan, Partners Healthcare System; Gilbert Omenn, University of Michigan; Helen Quinn, Stanford Linear Accelerator Center; Mary Ann Rankin, University of Texas; Barbara Schaal, Washington University; Thomas Südhof, Howard Hughes Medical Institute; Michael Teitelbaum, Sloan Foundation; C. Michael Walton, University of Texas; Larry Welch, Institute for Defense Analyses; and Sheila Widnall, Massachusetts Institute of Technology.

Although the reviewers listed above have provided many constructive comments and suggestions, they were not asked to endorse the conclusions or recommendations, nor did they see the final draft of the report before its release. The review of this report was overseen by Floyd Bloom, Robert Frosch, and M. R. C. Greenwood, appointed by the Report Review Committee, who were responsible for making certain that an independent examination of the report was carried out in accordance with institutional procedures and that all review comments were carefully considered. Responsibility for the final content of the report rests entirely with the author committee and the institution.

Finally, we would like to thank the staff who supported this project, including Deborah Stine, study director and associate director of the Committee on Science, Engineering, and Public Policy (COSEPUP), who managed the project; program officers Peter Henderson (higher education), Jo Husbands (national security), Thomas Arrison (innovation), Laurel Haak (K–12 education), and (on loan from the Council on Competitiveness) policy consultant David Attis (research funding and management), who conducted research and analysis; Alan Anderson, Steve Olson, and research associate Rachel Courtland, the science writers and editors for this report; Rita Johnson, the managing editor for reports; Norman Grossblatt and Kate Kelly, editors; Neeraj P. Gorkhaly, senior program assistant, who coordinated and provided support throughout the project with the assistance of

Marion Ramsey and Judy Goss; science and technology policy fellows John Slanina, Benjamin Novak, and Ian Christensen who provided research and analytic support; Brian Schwartz, who compiled the bibliography; and Richard Bissell, executive director of COSEPUP and of Policy and Global Affairs. Additional thanks are extended to Rachel Marcus, Will Mason, Estelle Miller, and Francesca Moghari at the National Academies Press for their work on the production of this book.

Contents

Boxes, Figures, and Tables

BOXES

FIGURES

TABLES

Executive Summary

The United States takes deserved pride in the vitality of its economy, which forms the foundation of our high quality of life, our national security, and our hope that our children and grandchildren will inherit ever-greater opportunities. That vitality is derived in large part from the productivity of well-trained people and the steady stream of scientific and technical innovations they produce. Without high-quality, knowledge-intensive jobs and the innovative enterprises that lead to discovery and new technology, our economy will suffer and our people will face a lower standard of living. Economic studies conducted even before the information-technology revolution have shown that as much as 85% of measured growth in US income per capita was due to technological change.[1]

Today, Americans are feeling the gradual and subtle effects of globalization that challenge the economic and strategic leadership that the United States has enjoyed since World War II. A substantial portion of our workforce finds itself in direct competition for jobs with lower-wage workers around the globe, and leading-edge scientific and engineering work is being accomplished in many parts of the world. Thanks to globalization, driven by modern communications and other advances, workers in virtually every sector must now face competitors who live just a mouse-click away in Ireland, Finland, China,

[1]For example, work by Robert Solow and Moses Abramovitz published in the middle 1950s demonstrated that as much as 85% of measured growth in US income per capita during the 1890-1950 period could not be explained by increases in the capital stock or other measurable inputs. The unexplained portion, referred to alternatively as the "residual" or "the measure of ignorance," has been widely attributed to the effects of technological change.

India, or dozens of other nations whose economies are growing. This has been aptly referred to as "the Death of Distance."

CHARGE TO THE COMMITTEE

The National Academies was asked by Senator Lamar Alexander and Senator Jeff Bingaman of the Committee on Energy and Natural Resources, with endorsement by Representative Sherwood Boehlert and Representative Bart Gordon of the House Committee on Science, to respond to the following questions:

> What are the top 10 actions, in priority order, that federal policymakers could take to enhance the science and technology enterprise so that the United States can successfully compete, prosper, and be secure in the global community of the 21st century? What strategy, with several concrete steps, could be used to implement each of those actions?

The National Academies created the Committee on Prospering in the Global Economy of the 21st Century to respond to this request. The charge constitutes a challenge both daunting and exhilarating: to recommend to the nation specific steps that can best strengthen the quality of life in America—our prosperity, our health, and our security. The committee has been cautious in its analysis of information. The available information is only partly adequate for the committee's needs. In addition, the time allotted to develop the report (10 weeks from the time of the committee's first gathering to report release) limited the ability of the committee to conduct an exhaustive analysis. Even if unlimited time were available, definitive analyses on many issues are not possible given the uncertainties involved.[2]

This report reflects the consensus views and judgment of the committee members. Although the committee consists of leaders in academe, industry, and government—including several current and former industry chief executive officers, university presidents, researchers (including three Nobel prize winners), and former presidential appointees—the array of topics and policies covered is so broad that it was not possible to assemble a committee of 20 members with direct expertise in each relevant area. Because of those limitations, the committee has relied heavily on the judgment of many experts in the study's focus groups, additional consultations via e-mail and telephone with other experts, and an unusually large panel of reviewers.

---

[2]Since the prepublication version of the report was released in October, certain changes have been made to correct editorial and factual errors, add relevant examples and indicators, and ensure consistency among sections of the report. Although modifications have been made to the text, the recommendations remain unchanged, except for a few corrections, which have been footnoted.

Although other solutions are undoubtedly possible, the committee believes that its recommendations, if implemented, will help the United States achieve prosperity in the 21st century.

FINDINGS

Having reviewed trends in the United States and abroad, the committee is deeply concerned that the scientific and technological building blocks critical to our economic leadership are eroding at a time when many other nations are gathering strength. We strongly believe that a worldwide strengthening will benefit the world's economy—particularly in the creation of jobs in countries that are far less well-off than the United States. But we are worried about the future prosperity of the United States. Although many people assume that the United States will always be a world leader in science and technology, this may not continue to be the case inasmuch as great minds and ideas exist throughout the world. We fear the abruptness with which a lead in science and technology can be lost—and the difficulty of recovering a lead once lost, if indeed it can be regained at all.

The committee found that multinational companies use such criteria[3] as the following in determining where to locate their facilities and the jobs that result:

- Cost of labor (professional and general workforce).
- Availability and cost of capital.
- Availability and quality of research and innovation talent.
- Availability of qualified workforce.
- Taxation environment.
- Indirect costs (litigation, employee benefits such as healthcare, pensions, vacations).
- Quality of research universities.
- Convenience of transportation and communication (including language).
- Fraction of national research and development supported by government.

[3]D. H. Dalton, M. G. Serapio, Jr., and P. G. Yoshida. *Globalizing Industrial Research and Development.* Washington, DC: US Department of Commerce, Technology Administration, Office of Technology Policy, 1999; Grant Gross. "CEOs Defend Moving Jobs Offshore at Tech Summit." *InfoWorld*, October 9, 2003; Bruce Mehlman. 2003. Offshore Outsourcing and the Future of American Competitiveness"; Bruce Einhorn et al. "High Tech in China: Is It a Threat to Silicon Valley?" *Business Week* online, October 28, 2002; B. Callan, S. Costigan, and K. Keller. *Exporting U.S. High Tech: Facts and Fiction About the Globalization of Industrial R&D.* New York: Council on Foreign Relations, 1997.

• Legal-judicial system (business integrity, property rights, contract sanctity, patent protection).
• Current and potential growth of domestic market.
• Attractiveness as place to live for employees.
• Effectiveness of national economic system.

Although the US economy is doing well today, current trends in each of those criteria indicate that the United States may not fare as well in the future without government intervention. This nation must prepare with great urgency to preserve its strategic and economic security. Because other nations have, and probably will continue to have, the competitive advantage of a low wage structure, the United States must compete by optimizing its knowledge-based resources, particularly in science and technology, and by sustaining the most fertile environment for new and revitalized industries and the well-paying jobs they bring. We have already seen that capital, factories, and laboratories readily move wherever they are thought to have the greatest promise of return to investors.

RECOMMENDATIONS

The committee reviewed hundreds of detailed suggestions—including various calls for novel and untested mechanisms—from other committees, from its focus groups, and from its own members. The challenge is immense, and the actions needed to respond are immense as well.

The committee identified two key challenges that are tightly coupled to scientific and engineering prowess: creating high-quality jobs for Americans, and responding to the nation's need for clean, affordable, and reliable energy. To address those challenges, the committee structured its ideas according to four basic recommendations that focus on the human, financial, and knowledge capital necessary for US prosperity.

The four recommendations focus on actions in K–12 education (*10,000 Teachers, 10 Million Minds*), research (*Sowing the Seeds*), higher education (*Best and Brightest*), and economic policy (*Incentives for Innovation*) that are set forth in the following sections. Also provided are a total of 20 implementation steps for reaching the goals set forth in the recommendations.

Some actions involve changes in the law. Others require financial support that would come from reallocation of existing funds or, if necessary, from new funds. Overall, the committee believes that the investments are modest relative to the magnitude of the return the nation can expect in the creation of new high-quality jobs and in responding to its energy needs.

The committee notes that the nation is unlikely to receive some sudden "wakeup" call; rather, the problem is one that is likely to evidence itself gradually over a surprisingly short period.

10,000 TEACHERS, 10 MILLION MINDS, AND K–12 SCIENCE AND MATHEMATICS EDUCATION

Recommendation A: *Increase America's talent pool by vastly improving K–12 science and mathematics education.*

Implementation Actions

The highest priority should be assigned to the following actions and programs. All should be subjected to continuing evaluation and refinement as they are implemented.

Action A-1: Annually recruit 10,000 science and mathematics teachers by awarding 4-year scholarships and thereby educating 10 million minds. Attract 10,000 of America's brightest students to the teaching profession every year, each of whom can have an impact on 1,000 students over the course of their careers. The program would award competitive 4-year scholarships for students to obtain bachelor's degrees in the physical or life sciences, engineering, or mathematics with concurrent certification as K–12 science and mathematics teachers. The merit-based scholarships would provide up to $20,000 a year for 4 years for qualified educational expenses, including tuition and fees, and require a commitment to 5 years of service in public K–12 schools. A $10,000 annual bonus would go to participating teachers in underserved schools in inner cities and rural areas. To provide the highest-quality education for undergraduates who want to become teachers, it would be important to award matching grants, on a one-to-one basis, of $1 million a year for up to 5 years, to as many as 100 universities and colleges to encourage them to establish integrated 4-year undergraduate programs leading to bachelor's degrees in the physical and life sciences, mathematics, computer sciences, or engineering *with teacher certification*. The models for this action are the UTeach and California Teach program.

Action A-2: Strengthen the skills of 250,000 teachers through training and education programs at summer institutes, in master's programs, and in Advanced Placement (AP) and International Baccalaureate (IB) training programs. Use proven models to strengthen the skills (and compensation, which is based on education and skill level) of 250,000 *current* K–12 teachers.

- *Summer institutes:* Provide matching grants to state and regional 1- to 2-week summer institutes to upgrade the skills and state-of-the-art knowledge of as many as 50,000 practicing teachers each summer. The material covered would allow teachers to keep current with recent developments in science, mathematics, and technology and allow for the exchange of best teaching practices. The Merck Institute for Science Education is one model for this action.

• *Science and mathematics master's programs:* Provide grants to research universities to offer, over 5 years, 50,000 current middle school and high school science, mathematics, and technology teachers (with or without undergraduate science, mathematics, or engineering degrees) 2-year, part-time master's degree programs that focus on rigorous science and mathematics content and pedagogy. The model for this action is the University of Pennsylvania Science Teacher Institute.

• *AP, IB, and pre-AP or pre-IB training:* Train an additional 70,000 AP or IB and 80,000 pre-AP or pre-IB instructors to teach advanced courses in science and mathematics. Assuming satisfactory performance, teachers may receive incentive payments of $1,800 per year, as well as $100 for each student who passes an AP or IB exam in mathematics or science. There are two models for this program: the Advanced Placement Incentive Program and Laying the Foundation, a pre-AP program.

• *K–12 curriculum materials modeled on a world-class standard:* Foster high-quality teaching with world-class curricula, standards, and assessments of student learning. Convene a national panel to collect, evaluate, and develop rigorous K–12 materials that would be available free of charge as a *voluntary* national curriculum. The model for this action is the Project Lead the Way pre-engineering courseware.

Action A-3: Enlarge the pipeline of students who are prepared to enter college and graduate with a degree in science, engineering, or mathematics by increasing the number of students who pass AP and IB science and mathematics courses. Create opportunities and incentives for middle school and high school students to pursue advanced work in science and mathematics. By 2010, increase the number of students who take at least one AP or IB mathematics or science exam to 1.5 million, and set a goal of tripling the number who pass those tests to 700,000.[4] Student incentives for success would include 50% examination fee rebates and $100 mini-scholarships for each passing score on an AP or IB science or mathematics examination.

Although it is not included among the implementation actions, the committee also finds attractive the expansion of two approaches to improving K–12 science and mathematics education that are already in use:

• *Statewide specialty high schools:* Specialty secondary education can foster leaders in science, technology, and mathematics. Specialty schools immerse students in high-quality science, technology, and mathematics education; serve as a mechanism to test teaching materials; provide a training

[4]This sentence was incorrectly phrased in the original October 12, 2005, edition of the executive summary and has now been corrected.

ground for K–12 teachers; and provide the resources and staff for summer programs that introduce students to science and mathematics.

- *Inquiry-based learning:* Summer internships and research opportunities provide especially valuable laboratory experience for both middle-school and high-school students.

SOWING THE SEEDS THROUGH SCIENCE AND ENGINEERING RESEARCH

Recommendation B: ***Sustain and strengthen the nation's traditional commitment to long-term basic research that has the potential to be transformational to maintain the flow of new ideas that fuel the economy, provide security, and enhance the quality of life.***

Implementation Actions

Action B-1: Increase the federal investment in long-term basic research by 10% each year over the next 7 years through reallocation of existing funds[5] or, if necessary, through the investment of new funds. Special attention should go to the physical sciences, engineering, mathematics, and information sciences and to Department of Defense (DOD) basic-research funding. This special attention does not mean that there should be a disinvestment in such important fields as the life sciences or the social sciences. A balanced research portfolio in all fields of science and engineering research is critical to US prosperity. **Increasingly, the most significant new scientific and engineering advances are formed to cut across several disciplines.** This investment should be evaluated regularly to realign the research portfolio to satisfy emerging needs and promises—unsuccessful projects and venues of research should be replaced with research projects and venues that have greater potential.

Action B-2: Provide new research grants of $500,000 each annually, payable over 5 years, to 200 of the nation's most outstanding *early-career* researchers. The grants would be made through existing federal research agencies—the National Institutes of Health (NIH), the National Science Foundation (NSF), the Department of Energy (DOE), DOD, and the National Aeronautics and Space Administration (NASA)—to underwrite new research opportunities at universities and government laboratories.

Action B-3: Institute a National Coordination Office for Advanced Research Instrumentation and Facilities to manage a fund of $500 million in incremental funds per year over the next 5 years—through reallocation of existing funds or, if necessary, through the investment of new funds—to ensure that universities and government laboratories create and maintain

[5]The funds may come from anywhere in government, not just other research funds.

the facilities, instrumentation, and equipment needed for leading-edge scientific discovery and technological development. Universities and national laboratories would compete annually for these funds.

Action B-4: Allocate at least 8% of the budgets of federal research agencies to discretionary funding that would be managed by technical program managers in the agencies and be focused on catalyzing high-risk, high-payoff research of the type that often suffers in today's increasingly risk-averse environment.

Action B-5: Create in the Department of Energy an organization like the Defense Advanced Research Projects Agency (DARPA) called the Advanced Research Projects Agency-Energy (ARPA-E).[6] The director of ARPA-E would report to the under secretary for science and would be charged with sponsoring specific research and development programs to meet the nation's long-term energy challenges. The new agency would support creative "out-of-the-box" transformational generic energy research that industry by itself cannot or will not support and in which risk may be high but success would provide dramatic benefits for the nation. This would accelerate the process by which knowledge obtained through research is transformed to create jobs and address environmental, energy, and security issues. ARPA-E would be based on the historically successful DARPA model and would be designed as a lean and agile organization with a great deal of independence that can start and stop targeted programs on the basis of performance and do so in a timely manner. The agency would itself perform no research or transitional effort but would fund such work conducted by universities, startups, established firms, and others. Its staff would turn over approximately every 4 years. Although the agency would be focused on specific energy issues, it is expected that its work (like that of DARPA or NIH) will have important spinoff benefits, including aiding in the education of the next generation of researchers. Funding for ARPA-E would start at $300 million the first year and increase to $1 billion per year over 5-6 years, at which point the program's effectiveness would be evaluated and any appropriate actions taken.

Action B-6: Institute a Presidential Innovation Award to stimulate scientific and engineering advances in the national interest. Existing presidential awards recognize lifetime achievements or promising young scholars, but the proposed new awards would identify and recognize persons who develop unique scientific and engineering innovations in the national interest at the time they occur.

---

[6]One committee member, Lee Raymond, does not support this action item. He does not believe that ARPA-E is necessary, because energy research is already well funded by the federal government, along with formidable funding by the private sector. Also, ARPA-E would, in his view, put the federal government into the business of picking "winning energy technologies"—a role best left to the private sector.

BEST AND BRIGHTEST IN SCIENCE AND ENGINEERING HIGHER EDUCATION

Recommendation C: *Make the United States the most attractive setting in which to study and perform research so that we can develop, recruit, and retain the best and brightest students, scientists, and engineers from within the United States and throughout the world.*

Implementation Actions

Action C-1: Increase the number and proportion of US citizens who earn bachelor's degrees in the physical sciences, the life sciences, engineering, and mathematics by providing 25,000 new 4-year competitive undergraduate scholarships each year to US citizens attending US institutions. The Undergraduate Scholar Awards in Science, Technology, Engineering, and Mathematics (USA-STEM) would be distributed to states on the basis of the size of their congressional delegations and awarded on the basis of national examinations. An award would provide up to $20,000 annually for tuition and fees.

Action C-2: Increase the number of US citizens pursuing graduate study in "areas of national need" by funding 5,000 new graduate fellowships each year. NSF should administer the program and draw on the advice of other federal research agencies to define national needs. The focus on national needs is important both to ensure an adequate supply of doctoral scientists and engineers and to ensure that there are appropriate employment opportunities for students once they receive their degrees. Portable fellowships would provide a stipend of $30,000[7] annually directly to students, who would choose where to pursue graduate studies instead of being required to follow faculty research grants, and up to $20,000 annually for tuition and fees.

Action C-3: Provide a federal tax credit to encourage employers to make continuing education available (either internally or through colleges and universities) to practicing scientists and engineers. These incentives would promote career-long learning to keep the workforce productive in an environment of rapidly evolving scientific and engineering discoveries and technological advances and would allow for retraining to meet new demands of the job market.

Action C-4: Continue to improve visa processing for international students and scholars to provide less complex procedures and continue to make improvements on such issues as visa categories and duration, travel for

[7]An incorrect number was provided for the graduate student stipend in the original October 12, 2005, edition of the executive summary.

scientific meetings, the technology alert list, reciprocity agreements, and changes in status.

Action C-5: Provide a 1-year automatic visa extension to international students who receive doctorates or the equivalent in science, technology, engineering, mathematics, or other fields of national need at qualified US institutions to remain in the United States to seek employment. If these students are offered jobs by US-based employers and pass a security screening test, they should be provided automatic work permits and expedited residence status. If students are unable to obtain employment within 1 year, their visas would expire.

Action C-6: Institute a new skills-based, preferential immigration option. Doctoral-level education and science and engineering skills would substantially raise an applicant's chances and priority in obtaining US citizenship. In the interim, the number of H-1B visas should be increased by 10,000, and the additional visas should be available for industry to hire science and engineering applicants with doctorates from US universities.[8]

Action C-7: Reform the current system of "deemed exports." The new system should provide international students and researchers engaged in fundamental research in the United States with access to information and research equipment in US industrial, academic, and national laboratories comparable with the access provided to US citizens and permanent residents in a similar status. It would, of course, exclude information and facilities restricted under national-security regulations. In addition, the effect of deemed-exports[9] regulations on the education and fundamental research work of international students and scholars should be limited by removing from the deemed-exports technology list all technology items (information and equipment) that are available for purchase on the overseas open market from foreign or US companies or that have manuals that are available in the public domain, in libraries, over the Internet, or from manufacturers.

---

[8]Since the report was released, the committee has learned that the Consolidated Appropriations Act of 2005, signed into law on December 8, 2004, exempts individuals that have received a master's or higher education degree from a US university from the statutory cap (up to 20,000). The bill also raised the H-1B fee and allocated funds to train American workers. The committee believes that this provision is sufficient to respond to its recommendation—even though the 10,000 additional visas recommended is specifically for science and engineering doctoral candidates from US universities, which is a narrower subgroup.

[9]The controls governed by the Export Administration Act and its implementing regulations extend to the transfer of technology. Technology includes "specific information necessary for the 'development,' 'production,' or 'use' of a product." Providing information that is subject to export controls—for example, about some kinds of computer hardware—to a foreign national within the United States may be "deemed" an export, and that transfer requires an export license. The primary responsibility for administering controls on deemed exports lies with the Department of Commerce, but other agencies have regulatory authority as well.

INCENTIVES FOR INNOVATION

Recommendation D: ***Ensure that the United States is the premier place in the world to innovate; invest in downstream activities such as manufacturing and marketing; and create high-paying jobs based on innovation by such actions as modernizing the patent system, realigning tax policies to encourage innovation, and ensuring affordable broadband access.***

Implementation Actions

Action D-1: Enhance intellectual-property protection for the 21st-century global economy to ensure that systems for protecting patents and other forms of intellectual property underlie the emerging knowledge economy but allow research to enhance innovation. The patent system requires reform of four specific kinds:

- Provide the US Patent and Trademark Office with sufficient resources to make intellectual-property protection more timely, predictable, and effective.
- Reconfigure the US patent system by switching to a "first-inventor-to-file" system and by instituting administrative review *after* a patent is granted. Those reforms would bring the US system into alignment with patent systems in Europe and Japan.
- Shield research uses of patented inventions from infringement liability. One recent court decision could jeopardize the long-assumed ability of academic researchers to use patented inventions for research.
- Change intellectual-property laws that act as barriers to innovation in specific industries, such as those related to data exclusivity (in pharmaceuticals) and those that increase the volume and unpredictability of litigation (especially in information-technology industries).

Action D-2: Enact a stronger research and development tax credit to encourage private investment in innovation. The current Research and Experimentation Tax Credit goes to companies that *increase* their research and development spending above a base amount calculated from their spending in prior years. Congress and the Administration should make the credit permanent,[10] and it should be increased from 20 to 40% of the qualifying increase so that the US tax credit is competitive with those of other countries. The credit should be extended to companies that have consistently spent large amounts on research and development so that they will

[10]The current R&D tax credit expires in December 2005.

not be subject to the current de facto penalties for having previously invested in research and development.

Action D-3: Provide tax incentives for US-based innovation. Many policies and programs affect innovation and the nation's ability to profit from it. It was not possible for the committee to conduct an exhaustive examination, but alternatives to current economic policies should be examined and, if deemed beneficial to the United States, pursued. These alternatives could include changes in overall corporate tax rates and special tax provisions providing incentives for the purchase of high-technology research and manufacturing equipment, treatment of capital gains, and incentives for long-term investments in innovation. The Council of Economic Advisers and the Congressional Budget Office should conduct a comprehensive analysis to examine how the United States compares with other nations as a location for innovation and related activities with a view to ensuring that the United States is one of the most attractive places in the world for long-term innovation-related investment and the jobs resulting from that investment. From a tax standpoint, that is not now the case.

Action D-4: Ensure ubiquitous broadband Internet access. Several nations are well ahead of the United States in providing broadband access for home, school, and business. That capability can be expected to do as much to drive innovation, the economy, and job creation in the 21st century as did access to the telephone, interstate highways, and air travel in the 20th century. Congress and the administration should take action—mainly in the regulatory arena and in spectrum management—to ensure widespread affordable broadband access in the very near future.

CONCLUSION

The committee believes that its recommendations and the actions proposed to implement them merit serious consideration if we are to ensure that our nation continues to enjoy the jobs, security, and high standard of living that this and previous generations worked so hard to create. Although the committee was asked only to recommend actions that can be taken by the federal government, it is clear that related actions at the state and local levels are equally important for US prosperity, as are actions taken by each American family. The United States faces an enormous challenge because of the disparity it faces in labor costs. Science and technology provide the opportunity to overcome that disparity by creating scientists and engineers with the ability to create entire new industries—much as has been done in the past.

It is easy to be complacent about US competitiveness and preeminence in science and technology. We have led the world for decades, and we continue to do so in many research fields today. But the world is changing rapidly, and our advantages are no longer unique. Some will argue that this is a problem for market forces to resolve—but that is exactly the concern. Market forces are *already at work* moving jobs to countries with less costly, often better educated, highly motivated workforces and friendlier tax policies.

Without a renewed effort to bolster the foundations of our competitiveness, we can expect to lose our privileged position. For the first time in generations, the nation's children could face poorer prospects than their parents and grandparents did. We owe our current prosperity, security, and good health to the investments of past generations, and we are obliged to renew those commitments in education, research, and innovation policies to ensure that the American people continue to benefit from the remarkable opportunities provided by the rapid development of the global economy and its not inconsiderable underpinning in science and technology.

SOME COMPETITIVENESS INDICATORS

US Economy

- The United States is today a net importer of *high-technology* products. Its trade balance in high-technology manufactured goods shifted from *plus* $54 billion in 1990 to *negative* $50 billion in 2001.[1]
- In one recent period, low-wage employers, such as Wal-Mart (now the nation's largest employer) and McDonald's, created 44% of the new jobs while high-wage employers created only 29% of the new jobs.[2]
- The United States is one of the few countries in which industry plays a major role in providing healthcare for its employees and their families. Starbucks spends more on healthcare than on coffee. General Motors spends more on healthcare than on steel.[3]
- US scheduled airlines currently outsource portions of their aircraft maintenance to China and El Salvador.[4]
- IBM recently sold its personal computer business to an entity in China.[5]
- Ford and General Motors both have junk bond ratings.[6]
- It has been estimated that within a decade nearly 80% of the world's middle-income consumers would live in nations outside the currently industrialized world. China alone could have 595 million middle-income consumers and 82 million upper-middle-income consumers. The total population of the United States is currently 300 million[7] and it is projected to be 315 million in a decade.
- Some economists estimate that about half of US economic growth since World War II has been the result of technological innovation.[8]
- In 2005, American investors put more new money in foreign stock funds than in domestic stock portfolios.[9]

Comparative Economics

- Chemical companies closed 70 facilities in the United States in 2004 and tagged 40 more for shutdown. Of 120 chemical plants being built around the world with price tags of $1 billion or more, one is in the United States and 50 are in China. No new refineries have been built in the United States since 1976.[10]
- The United States is said to have 7 million illegal immigrants,[11] but under the law the number of visas set aside for "highly qualified foreign workers," many of whom contribute significantly to the nation's innovations, dropped to 65,000 a year from its 195,000 peak.[12]
- When asked in spring 2005 what is the most attractive place in the world in which to "lead a good life", respondents in only 1 (India) of the 16 countries polled indicated the United States.[13]

- A company can hire nine factory workers in Mexico for the cost of one in America. A company can hire eight young professional engineers in India for the cost of one in America.[14]
- The share of leading-edge semiconductor manufacturing capacity owned or partly owned by US companies today is half what it was as recently as 2001.[15]
- During 2004, China overtook the United States to become the leading exporter of information-technology products, according to the Organisation for Economic Co-operation and Development (OECD).[16]
- The United States ranks only 12th among OECD countries in the number of broadband connections per 100 inhabitants.[17]

K–12 Education

- Fewer than one-third of US 8th-grade students performed at or above a level called "proficient" in mathematics; "proficiency" was considered the ability to exhibit competence with challenging subject matter. Alarmingly, about one-fifth of the 4th graders and one-third of the 8th graders lacked the competence to perform even basic mathematical computations.[18]
- In 1999, 69% of US 5-8th-grade students received instruction from a mathematics teacher who did not hold a degree or certification in mathematics.[19]
- In 2000, 93% of students in grades 5-8 were taught physical science by a teacher lacking a major or certification in the physical sciences (chemistry, geology, general science, or physics).[20]
- In 1995 (the most recent data available), US 12th graders performed below the international average for 21 countries on a test of general knowledge in mathematics and science.[21]
- US 15-year-olds ranked 27th out of 39 countries that participated in a 2003 administration of the Program for International Student Assessment (PISA) examination, which assessed students' ability to apply mathematical concepts to real-world problems.[22]
- According to a recent survey, 86% of US voters believe that the United States must increase the number of workers with a background in science and mathematics or America's ability to compete in the global economy will be diminished.[23]
- American youth spend more time watching television[24] than in school.[25]
- Because the United States does not have a set of national curricula, changing K–12 education is challenging, given that there are almost 15,000 school systems in the United States and the average district has only about six schools.[26]

Higher Education

• In South Korea, 38% of all undergraduates receive their degrees in natural science or engineering. In France, the figure is 47%, in China, 50%, and in Singapore, 67%. In the United States, the corresponding figure is 15%.[27]

• Some 34% of doctoral degrees in natural sciences (including the physical, biological, earth, ocean, and atmospheric sciences) and 56% of engineering PhDs in the United States are awarded to foreign-born students.[28]

• In the US science and technology workforce in 2000, 38% of PhDs were foreign-born.[29]

• Estimates of the number of engineers, computer scientists, and information-technology students who obtain 2-, 3-, or 4-year degrees vary. One estimate is that in 2004, China graduated about 350,000 engineers, computer scientists, and information technologists with 4-year degrees, while the United States graduated about 140,000. China also graduated about 290,000 with 3-year degrees in these same fields, while the US graduated about 85,000 with 2- or 3-year degrees.[30] Over the past 3 years alone, both China[31] and India[32] have doubled their production of 3- and 4-year degrees in these fields, while the United States[33] production of engineers is stagnant and the rate of production of computer scientists and information technologists doubled.

• About one-third of US students intending to major in engineering switch majors before graduating.[34]

• The proportion of bachelor's degrees in physics to total degrees awarded was twice as high the year before Sputnik, deemed a time of dangerous educational neglect, as in 2004.[35]

• More S&P 500 CEOs obtained their undergraduate degrees in engineering than in any other field.[36]

Research

• In 2001 (the most recent year for which data are available), US industry spent more on tort litigation than on research.[37]

• In 2005, only four American companies ranked among the top 10 corporate recipients of patents granted by the *United States* Patent and Trademark Office.[38]

• Beginning in 2007, the most capable high-energy particle accelerator on Earth will, for the first time, reside outside the United States.[39]

• Federal funding of research in the physical sciences, as a percentage of gross domestic product (GDP), was 45% less in fiscal year (FY) 2004 than in FY 1976.[40] The amount invested annually by the US federal government in research in the physical sciences, mathematics, and engineering combined equals the annual increase in US healthcare costs incurred every six weeks.[41]

PERSPECTIVES

- "If you can solve the education problem, you don't have to do anything else. If you don't solve it, nothing else is going to matter all that much." —Alan Greenspan, outgoing Federal Reserve Board chairman[42]
- "We go where the smart people are. Now our business operations are two-thirds in the U.S. and one-third overseas. But that ratio will flip over the next ten years." —Intel Corporation spokesman Howard High[43]
- "If we don't step up to the challenge of finding and supporting the best teachers, we'll undermine everything else we are trying to do to improve our schools." —Louis V. Gerstner, Jr., Former Chairman, IBM[44]
- "If you want good manufacturing jobs, one thing you could do is graduate more engineers. We had more sports exercise majors graduate than electrical engineering grads last year." —Jeffrey R. Immelt, Chairman and Chief Executive Office, General Electric[45]
- "If I take the revenue in January and look again in December of that year 90% of my December revenue comes from products which were not there in January." —Craig Barrett, Chairman of Intel Corporation[46]
- "When I compare our high schools to what I see when I'm traveling abroad, I am terrified for our workforce of tomorrow." —Bill Gates, Chairman and Chief Software Architect of Microsoft Corporation[47]
- "Where once nations measured their strength by the size of their armies and arsenals, in the world of the future knowledge will matter most." —President Bill Clinton[48]
- "Science and technology have never been more essential to the defense of the nation and the health of our economy." —President George W. Bush[49]

NOTES FOR *SOME COMPETITIVENESS INDICATORS AND PERSPECTIVES*

[1]For 2001, the dollar value of high-technology imports was $561 billion; the value of high-technology exports was $511 billion. See National Science Board. *Science and Engineering Indicators 2004*. NSB 04-01. Arlington, VA: National Science Foundation, 2004. Appendix Table 6-01. Page A6-5 provides the export numbers for 1990 and 2001 and page A6-6 has the import numbers.

[2]S. Roach. *More Jobs, Worse Work*. New York Times, July 22, 2004.

[3]C. Noon. "Starbuck's Schultz Bemoans Health Care Costs." *Forbes.com,* September 19, 2005. Available at: http://www.forbes.com/; R. Scherer. "Rising Benefits Burden." *Christian Science Monitor,* June 9, 2005. Available at: http://www.csmonitor.com/.

[4]S. K. Goo. *Airlines Outsource Upkeep*. Washington Post, August 21, 2005. Available at: http://www.washingtonpost.com/wp-dyn/content/article/2005/08/20/AR200508 2000979.html; S. K. Goo. *Two-Way Traffic in Airplane Repair*. Washington Post, June 1, 2004. Available at: http://www.washingtonpost.com/.

[5]M. Kanellos. "IBM Sells PC Group to Lenovo." *News.com,* December 8, 2004. Available at: http://news.com.com/IBM+sells+PC+group+to+Lenovo/2100-1042_3-5482284.html.

[6]See http://www.nytimes.com/.

[7]In China, P. A. Laudicina. *World Out of Balance: Navigating Global Risks to Seize Competitive Advantage*. New York: McGraw-Hill, 2005. P. 76. For the United States, see US Census Bureau. "US Population Clock." Available at: http://www.census.gov. For current population and for the projected population, see Population Projections Program, Population Division, US Census Bureau. "Population Projections of the United States by Age, Sex, Race, Hispanic Origin, and Nativity: 1999 to 2100." Washington, DC, January 13, 2000. Available at: http://www.census.gov/population/www/projections/natsum-T3.html.

[8]M. J. Boskin and L. J. Lau. Capital, Technology, and Economic Growth. In N. Rosenberg, R. Landau, and D. C. Mowery, eds. *Technology and the Wealth of Nations*. Stanford, CA: Stanford University Press, 1992.

[9]P. J. Lim. *Looking Ahead Means Looking Abroad*. New York Times, January 8, 2006.

[10]M. Arndt. "No Longer the Lab of the World: U.S. Chemical Plants are Closing in Droves as Production Heads Abroad." *BusinessWeek,* May 2, 2005. Available at: http://www.businessweek.com/ and http://www.usnews.com/usnews/.

[11]As of 2000, the unauthorized resident population in the United States was 7 million. See US Citizenship and Immigration Services. "Executive Summary: Estimates of the Unauthorized Immigrant Population Residing in the United States: 1990 to 2000." January 31, 2003. Available at: http://uscis.gov/graphics/shared/statistics/publications/2000ExecSumm.pdf.

[12]Section 214(g) of the Immigration and Nationality Act sets an annual limit on the number of aliens that can receive H-1B status in a fiscal year. For FY 2000 the limit was set at 115,000. The American Competitiveness in the Twenty-First Century Act increased the annual limit to 195,000 for 2001, 2002, and 2003. After that date the cap reverts back to 65,000. H-1B visas allow employers to have access to highly educated foreign professionals who have experience in specialized fields and who have at least a bachelor's degree or the equivalent. The cap does not apply to educational institutions. In November 2004, Congress created an exemption for 20,000 foreign nationals earning advanced degrees from US universities. See Immigration and Nationality Act, Section 101(a)(15)(h)(1)(b). See US Citizenship and Immigration Services. "USCIS Announces Update Regarding New H-1B Exemptions." July 12, 2005. Available at: http://uscis.gov/ and US Citizenship and Immigration Services. "Questions and Answers: Changes to the H-1B Program." November 21, 2000. Available at: http://uscis.gov.

[13]Pew Research Center. "U.S. Image Up Slightly, But Still Negative, American Character Gets Mixed Reviews." Washington, DC: Pew Global Attitudes Project, 2005. Available at: http://pewglobal.org/reports/display.php?ReportID=247.

The interview asked nearly 17,000 people the question: "Suppose a young person who wanted to leave this country asked you to recommend where to go to lead a good life—what country would you recommend?" Except for respondents in India, Poland, and Canada, no more than one-tenth of the people in the other nations said they would recommend the United States. Canada and Australia won the popularity contest.

[14]US Bureau of Labor Statistics. "International Comparisons of Hourly Compensation Costs for Production Workers in Manufacturing, 2004." November 18, 2005. Available at: ftp://ftp.bls.gov/.

Starting annual salaries for U.S. class of 2005 engineering graduates averaged $49,567 to $53,813 depending upon specialty. Fresh engineering graduates might receive starting salaries of $4,300 to $8,000 per year at leading Indian companies. See S. Hargreaves. 2005. "College majors that boost your paycheck." CNN-Money.com, August 17. http://money.cnn.com/2005/08/11/pf/college/starting_salaries/index.htm.

M. Kanellos. 2005. "India's Tech Renaissance." CNET News. June 28. http://news.cnet.com/Indias-renaissance-U.S.-style-labor-pains—page-2/2009-1041_3-5752021-2.html?tag=st.num

[15]Semiconductor Industry Association. "Choosing to Compete." December 12, 2005. Available at: http://www.sia-online.org/.

[16]Organisation for Economic Co-operation and Development. "China Overtakes U.S. as World's Leading Exporter of Information Technology Goods." December 12, 2005. Available at: http://www.oecd.org/. The main categories included in OECD's definition of ICT (information and communications technology) goods are electronic components, computers and related equipment, audio and video equipment, and telecommunication equipment.

[17]Organisation for Economic Co-operation and Development. "OECD Broadband Statistics, June 2005." October 20, 2005. Available at: http://www.oecd.org/.

[18]National Center for Education Statistics. 2006. "The Nation's Report Card: Mathematics 2005." Available at: http://nces.ed.gov/nationsreportcard/pdf/main2005/2006453.pdf.

[19]National Science Board. *Science and Engineering Indicators 2004*. NSB 04-01. Arlington, VA: National Science Foundation, 2004. Chapter 1.

[20]National Center for Education Statistics. Schools and Staffing Survey, 2004. "Qualifications of the Public School Teacher Workforce: Prevalence of Out-of-Field Teaching 1987-88 to 1999-2000 (Revised)." 2004. P. 10. Available at: http://nces.ed.gov/pubs2002/2002603.pdf.

[21]National Center for Education Statistics. "Highlights from TIMSS." 2004. Available at: http://nces.ed.gov/pubs99/1999081.pdf.

[22]National Center for Education Statistics. "International Outcomes of Learning in Mathematics Literacy and Problem Solving: PISA 2003 Results from the U.S. Perspective." 2005. Pp. 15 and 29. Available at: http://nces.ed.gov/pubs2005/2005003.pdf.

[23]The Business Roundtable. "Innovation and U.S. Competitiveness: Addressing the Talent Gap. Public Opinion Research." January 12, 2006. Available at: http://www.businessroundtable.org/pdf/20060112Two-pager.pdf.

[24]American Academy of Pediatrics. "Television—How it Affects Children." Available at: http://www.aap.org/. The American Academy of Pediatrics reports, "Children in the United States watch about four hours of TV every day"; this works out to be 1,460 hours per year.

[25]National Center for Education Statistics. 2005. "The Condition of Education." Table 26-2, Average Number of Instructional Hours per Year Spent in Public School, by Age or Grade of Student and Country: 2000 and 2001. Available at: http://nces.ed.gov/. NCES reports that in 2000 US 15-year-olds spent 990 hours in school, during the same year 4th graders spent 1,040 hours.

[26]National Center for Education Statistic. "Public Elementary and Secondary Students, Staff, Schools, and School Districts: School Year 2003-04." 2006. Available at: http://nces.ed.gov/.

[27]Analysis conducted by the Association of American Universities. 2006. "National Defense Education and Innovation Initiative." Based on data in National Science Board. *Science and Engineering Indicators 2004*. NSB 04-01. Arlington, VA: National Science Foundation, 2004.

Appendix Table 2-33. For countries with both short and long degrees, the ratios are calculated with both short and long degrees as the numerator.

[28]National Science Board. *Science and Engineering Indicators 2004*. NSB 04-01. Arlington, VA: National Science Foundation, 2004. Chapter 2, Figure 2-23.

[29]National Science Board. A companion to *Science and Engineering Indicators 2004*. NSB 04-01. Arlington, VA: National Science Foundation, 2004.

[30]G. Gereffi and V. Wadhwa. 2005. "Framing the Engineering Outsourcing Debate: Placing the United States on a Level Playing Field with China and India." Available at: http://memp.pratt.duke.edu/downloads/duke_outsourcing_2005.pdf.

[31]Ministry of Science and Technology (MOST). *Chinese Statistical Yearbook 2004*. People's Republic of China: National Bureau of Statistics of China, 2004. Chapter 21, Table 21-11. Available at: http://www.stats.gov.cn/english/statisticaldata/yearlydata/yb2004-e/indexeh.htm. The extent to which engineering degrees from China are comparable to those from the United States is uncertain.

[32]National Association of Software and Service Companies. *Strategic Review 2005*. National Association of Software and Service Companies, India, 2005. Chapter 6, Sustaining the India Advantage. Available at: http://www.nasscom.org/strategic2005.asp.

[33]National Center for Education Statistics. *Digest of Education Statistics 2004*. Washington, DC: Institute of Education Sciences, Department of Education, 2004. Table 250. Available at: http://nces.ed.gov/.

[34]M. Boylan. Assessing Changes in Student Interest in Engineering Careers Over the Last Decade. CASEE, National Academy of Engineering, 2004. Available at: http://www.nae.edu/; C. Adelman. *Women and Men on the Engineering Path: A Model for Analysis of Undergraduate Careers*. Washington, DC: US Department of Education, 1998. Available at: http://www.ed.gov. According to this Department of Education analysis, the majority of students who switch from engineering majors complete a major in business or other non-science and engineering fields.

[35]American Physical Society. 2007. APS News. August/September.

[36]S. Stuart. "2004 CEO Study: A Statistical Snapshot of Leading CEOs." 2005. Available at: http://content.spencerstuart.com/sswebsite/pdf/lib/Statistical_Snapshot_of_Leading_CEOs_relB3.pdf#search='ceo%20educational%20background'.

[37]US research and development spending in 2001 was $273.6 billion, of which industry performed $194 billion and funded about $184 billion. National Science Board. *Science and Engineering Indicators 2004*. NSB 04-01. Arlington, VA: National Science Foundation, 2004. One estimate of tort litigation costs in the United States was $205 billion in 2001.

Towers Perrin. 2006. U.S. Tort Costs and Cross-Border Perspectives: 2005 Update. Available at: www.towersperrin.com.

J. A. Leonard. 2003. "How Structural Costs Imposed on U.S. Manufacturers Harm Workers and Threaten Competitiveness." Prepared for the Manufacturing Institute of the National Association of Manufacturers. Available at: http://www.nam.org/.

[38]US Patent and Trademark Office. "USPTO Annual List of Top 10 Organizations Receiving Most U.S. Patents." January 10, 2006. Available at: http://www.uspto.gov/web/offices/com/speeches/06-03.htm.

[39]CERN. Internet Homepage. Available at: http://public.web.cern.ch/Public/Welcome.html.

[40]American Association for the Advancement of Science. "Trends in Federal Research by Discipline, FY 1976-2004." October 2004. Available at: http://www.aaas.org/.

[41]Centers for Medicare and Medicaid Services. "National Heath Expenditures." 2005. Available at: http://www.cms.hhs.gov/NationalHealthExpendData/downloads/tables.pdf.

[42]US Department of Education, Office of the Secretary. *Meeting the Challenge of a Changing World: Strengthening Education for the 21st Century*. Washington, DC: US Department of Education, 2006.

[46]C. Barrett. Comments at public briefing on the release of Rising Above the Gathering Storm report. October 12, 2005. Available at: http://www.nationalacademies.org/morenews/20051012.html.

[47]B. Gates. Speech to the National Education Summit on High Schools. February 26, 2005. Available at: http://www.gatesfoundation.org/MediaCenter/Speeches/BillgSpeeches/BGSpeechNGA-050226.htm.

[48]W. J. Clinton. Commencement address at Morgan State University in Baltimore, Maryland. In *1997 Public Papers of the Presidents of the United States, Books I and II*. Washington, DC: Government Printing Office, May 18, 1997. Available at: http://www.gpoaccess.gov/pubpapers/wjclinton.html.

[49]Remarks by President George W. Bush in meeting with high-tech leaders. March 28, 2001. Available at: http://www.whitehouse.gov/.

1

A Disturbing Mosaic[1]

In *The World Is Flat: A Brief History of the Twenty-First Century*,[2] Thomas Friedman asserts that the international economic playing field is now "more level" than it has ever been.[3] The causes of this "flattening" include easier access to information technology and rising technical competences abroad that have made it possible for US companies to locate call centers in India, coordinate the complex supply chains and work flows that enable manufacturing in China, and conduct "back office" service functions abroad. It is not uncommon for radiologists in India, for example, to read x-ray pictures of patients in US hospitals. Architects in the United States have their drawings made in Brazil. Software is written for US firms in Bangalore. Ireland has successfully put into place a set of policies to attract companies and their research activities, as has Finland. The European Union is actively pursuing policies to enhance the innovation environment, as are Singapore, China, Japan, South Korea, Taiwan, and many other countries.

Friedman argues that, despite the dangers, a flat world is on balance a *good thing*—economically and geopolitically. Lower costs benefit consumers and shareholders in developed countries, and the rising middle class in

---

[1]Major portions of this chapter were adapted from an article of the same name by Wm. A. Wulf, president of the National Academy of Engineering in the fall 2005 issue of *The Bridge*, a journal of the National Academies.

[2]T. L. Friedman. *The World Is Flat: A Brief History of the Twenty-First Century*. New York: Farrar, Straus, and Giroux, 2005.

[3]An alternative point of view is presented in Box 1-1.

BOX 1-1
Another Point of View: The World Is Not Flat[a]

Some believe that although the world is certainly a more competitive place, it is not "flat." It is more competitive because access to knowledge is easier than ever before, but the rise of scientific competence and the apparent flight of high-technology jobs abroad is no more likely to dislodge the United States from its science and technology leadership than were previous challenges from the Soviet Union in the 1950s and 1960s or from Japan in the 1980s.

For example, Americans are alarmed to read of the large numbers of well-educated, English-speaking young people in India vying with US workers for jobs via the Internet. In fact, only about 6% of Indian students make it to college; of those who do, only two-thirds graduate. Just a small fraction of India's citizenry can read English; of these, a smaller fraction can speak it well enough to be understood by Americans. In China, where the numbers of engineers and other technically trained people are rising, government skepticism about the Internet and aspects of free markets is likely to hinder the advance of national power.

China and India indeed have low wage structures, but the United States has many other advantages. These include a better science and technology infrastructure, stronger venture-capital markets, an ability to attract talent from around the world, and a culture of inventiveness. Comparative advantage shifts from place to place over time and always has; the earth cannot really be flattened. The US response to competition must include proper retraining of those who are disadvantaged and adaptive institutional and policy responses that make the best use of opportunities that arise.

---

[a]This box was adapted from J. Bhagwati. *The World Is Not Flat.* Wall Street Journal, August 4, 2005. P. A12.

India and China will become consumers of those countries' products as well as ours. That same rising middle class will have a stake in the "frictionless" flow of international commerce—and hence in stability, peace, and the rule of law. Such a desirable state, writes Friedman, will not be achieved without problems, and whether global flatness is good for a particular country depends on whether that country is prepared to compete on the global playing field, which is as rough and tumble as it is level.

Friedman asks rhetorically whether his own country is proving its readiness by "investing in our future and preparing our children the way we need to for the race ahead." Friedman's answer, not surprisingly, is no.

This report addresses the possibility that our lack of preparation will reduce the ability of the United States to compete in such a world. Many underlying issues are technical; some are not. Some are "political"—not in the sense of partisan politics, but in the sense of "bringing the rest of the body politic along." Scientists and engineers often avoid such discussions, but the stakes are too high to keep silent any longer.

Friedman's term *quiet crisis*, which others have called a "creeping crisis," is reminiscent of the folk tale about boiling a frog. If a frog is dropped into boiling water, it will immediately jump out and survive. But a frog placed in cool water that is heated slowly until it boils won't respond until it is too late.

Our crisis is not the result of a one-dimensional change; it is more than a simple increase in water temperature. And we have no single awakening event, such as Sputnik. The United States is instead facing problems that are developing slowly but surely, each like a tile in a mosaic. None by itself seems sufficient to provoke action. But the collection of problems reveals a disturbing picture—a recurring pattern of abundant short-term thinking and insufficient long-term investment. Our collective reaction thus far seems to presuppose that the citizens of the United States and their children are *entitled* to a better quality of life than others, and that all Americans need do is circle the wagons to defend that entitlement. Such a presupposition does not reflect reality and neither recognizes the dangers nor seizes the opportunities of current circumstances. Furthermore, it won't work.

In 2001, the Hart–Rudman Commission on national security, which foresaw large-scale terrorism in America and proposed the establishment of a cabinet-level Homeland Security organization before the terrorist attacks of 9/11, put the matter this way:[4]

> The inadequacies of our system of research and education pose a greater threat to U.S. national security over the next quarter century than any potential conventional war that we might imagine.

President George W. Bush has said

> "Science and technology have never been more essential to the defense of the nation and the health of our economy."[5]

[4]US Commission on National Security. *Road Map for National Security: Imperative for Change*. Washington, DC: US Commission on National Security, 2001.

[5]Remarks by the President in a meeting with high-tech leaders, March 28, 2001.

A letter from the leadership of the National Science Foundation to the President's Council of Advisors on Science and Technology put the case even more bluntly:[6]

> Civilization is on the brink of a new industrial order. The big winners in the increasingly fierce global scramble for supremacy will not be those who simply make commodities faster and cheaper than the competition. They will be those who develop talent, techniques and tools so advanced that there is no competition.

This chapter addresses the relevant issues in three related clusters. Later chapters examine each cluster in more detail and recommend ways to address the problems that are identified.

CLUSTER 1: TILTED JOBS IN A GLOBAL ECONOMY

Is the world flat, or is it tilted? Many people who once had jobs in the textile, furniture, apparel, automotive, and other manufacturing industries might be forgiven for saying that world is decidedly slanted. They watched their jobs run downhill to countries where the workforce earns far lower wages. The movement of jobs has accelerated sharply in the past 5 years, surprising many employers and employees and disrupting the lives of those who have been underbid by "hungry," skilled job-seekers abroad.

Large companies use various criteria in making a decision to relocate administrative, production, or research and development (R&D) facilities, and they often have a number of options. Some reasons cited for relocations in past studies include capitalizing on:

- Foreign R&D personnel (scientists, engineers, and programmers)[7] who are highly skilled and eager to work.[8]
- New science and technology in fresh environments.[9]
- Technological developments abroad.[10]
- Joint and cooperative research products.[11]

---

[6]The President's Council of Advisors on Science and Technology. "Sustaining the Nation's Innovation Ecosystems." Report on Information Technology Manufacturing and Competitiveness, January 2004.

[7]D. H. Dalton, M. G. Serapio, Jr., and P. G. Yoshida. *Globalizing Industrial Research and Development*. Washington, DC: US Department of Commerce, Technology Administration, Office of Technology Policy, 1999.

[8]G. Gross. "CEOs Defend Moving Jobs Offshore at Tech Summit." *InfoWorld*, October 9, 2003.

[9]Dalton, 1999.

[10]Ibid.

[11]Ibid.

- Proximity to offshore manufacturing.[12]
- Lower costs of conducting R&D, particularly labor costs.[13]
- Reduced labor costs associated with employing foreign workers.[14]
- Proximity to growing markets.
- US regulation and R&D climates, including strict regulatory regimes, high risks of legal liability, and technology transfer limitations.[15]
- High-technology centers with skilled personnel, world-class R&D infrastructure, vibrant research cultures, government incentives, and intellectual-property protection.[16]
- Lower corporate tax rates and special tax incentives.
- Increasingly high-quality research universities.

The global forces that affect employment have swirled into the service sector, once thought secure from international competition. First, there was outsourcing, which allows employers to reassign some jobs by contracting them to specialty firms that can do the jobs better or more cheaply. At first, jobs were outsourced within the United States, but "offshoring" soon sent jobs overseas, beyond the reach of US workers. That practice has become especially controversial, and there has been an outcry for measures to protect those jobs for the domestic market. In some states, legislation has been proposed to curb outsourcing through such initiatives as Opportunity Indiana, the Keep Jobs in Colorado Act, and the American Jobs Act of Wisconsin.[17]

Offshoring has become established, however, and it is merely one logical outcome of a flatter world. Furthermore, protectionist measures have historically proved counterproductive. For several years, US companies that outsource information-technology jobs have all but ordered their contractors to send some portion of the work overseas to gain hiring flexibility, cut employment costs—by 40% in some cases[18]—and cut overhead costs for

[12]B. Mehlman, Assistant Secretary for Technology Policy, US Department of Commerce. "Offshore Outsourcing and the Future of American Competitiveness." Speech to Business Roundtable Working Group presented on July 31, 2003. Available at: http://www.technology.gov/Speeches/BPM_2003-Outsourcing.pdf.

[13]Dalton, 1999.

[14]See, for example, "High Tech in China: Is It a Threat to Silicon Valley?" *Business Week* online, October 28, 2002.

[15]B. Callan, S. Costigan, and K. Keller. *Exporting U.S. High Tech: Facts and Fiction About the Globalization of Industrial R&D*. New York: Council on Foreign Relations, 1997.

[16]Dalton, 1999.

[17]D. C. Sharma and M. Yamamoto. "How India is Handling International Backlash." *CNET news.com*, May 6, 2004.

[18]The Gartner Group, an organization that analyzes the information-technology sector, estimates that companies can achieve cost savings of 25-30% through successful outsourcing. But Gartner also warns that offshoring could produce lower savings than estimated if backup service and other costs are not considered.

the home company.[19] Employers also hire offshore workers to gain access to better-trained workers or those with specialized skills, to move the workforce closer to manufacturing or production facilities, or to gain access to desirable markets.[20] In India, US companies can hire insurance-claims processors, medical transcriptionists, accountants, engineers, computer scientists, and other English-speaking workers for, on average, about one-fifth the salaries those employees would earn here. Because about three-fourths of all US jobs are now in the service sector,[21] millions of US employees are at risk of losing their jobs to overseas workers.[22]

Offshoring also could place downward pressure on wages at home.[23] Fewer than a million jobs have been sent overseas so far,[24] but even that number could be broadly affecting the economy as displaced workers seek jobs held by others or are forced to accept lower wages to keep their existing jobs.

Because offshoring of service-sector jobs is a recent phenomenon, few analysts offer predictions about its long-term effects on the US economy. The classical view of free trade, as articulated nearly two centuries ago by British economist David Ricardo, states that if a nation specializes in making a product in which it has a comparative cost advantage and if it trades with another nation for a product in which that nation has a similar cost advantage, both countries will be better off than if they had each made both products themselves.[25] But does that theory hold in a world where not only goods but many services are tradable as well? Will wages merely fall worldwide as more knowledge workers enter the jobs arena?

Most economists believe that Ricardo is still correct—that there will be gains for all such nations. They acknowledge that there might be a transition phase in which wages for lower-skilled workers in a rich country like the United States will fall. Some say that there is, however, no reason to

---

[19]J. King. "Its Itinerary: Offshore Outsourcing Is Inevitable." *Computerworld*, September 15, 2003.

[20]R. Hira, Rochester Institute of Technology, presentation to Committee on Science, Engineering, and Public Policy, Workshop on International Students and Postdoctoral Scholars, National Academies, July 2004.

[21]G. Colvin. "Can Americans Compete? Is America the World's 97-lb. Weakling?" *Fortune*, July 25, 2005.

[22]Forrester Research, a technology and market research company, estimates that 3.3 million white-collar jobs could be sent offshore by 2015. Tom Pohlman. "Topic Overview, Outsourcing, Q3 2005." September 12, 2005. Available at: http://www.forrester.com/Research/Document/0,7211,37613,00.html.

[23]R. Freeman. *It's a Flat World, After All.* New York Times, April 3, 2005. Section 6, Column 1, Magazine Desk, P. 33.

[24]Colvin, 2005.

[25]"Biography of David Ricardo." *The Concise Encyclopedia of Economics*. Available at: http://www.econlib.org/library/Enc/bios/Ricardo.html.

believe that wages for highly skilled workers will fall in either the short run or the long run.[26] Economist Paul Romer[27] argues that technological change continues to increase the demand for workers with high levels of education.[28] As a result, wages for US workers with at least a college education continue to rise faster than wages for other workers. The low wages for highly skilled workers seen in such countries as China and India are not a sign that the worldwide supply of highly skilled workers is so large that worldwide wages are now falling or are about to fall, says Romer. In those economies, wages for skilled workers are low because these workers were previously cut off from the deep and rapidly growing pool of technological knowledge that existed outside their borders. As they have opened up their economies so that this knowledge can now flow in, wages for highly skilled workers have grown rapidly.

With the collapse of the high-technology bubble, some highly skilled workers in the United States have experienced a fall in their wages from the values that prevailed at the peak. Moreover, at every level of education, there is wide variation in compensation and career paths. Some engineers and scientists, even now, are unemployed or underemployed, just as some physicians, MBAs, and lawyers are unemployed or underemployed. It would be a mistake, according to Romer, for public policy to limit the training of new physicians only because some of them end up with careers that are not as lucrative or rewarding as they had hoped. In the same way, public-policy decisions about the supply of scientists and engineers should not be guided by an attempt to provide a guaranteed high level of income for every recipient of an advanced degree. It is also important that scientists and engineers tend, through innovation, to create new jobs not only for themselves but also for workers throughout the economy.

Some economists believe that there might be a transition phase in some fields during which wages fall, but they assert that there is no reason to believe that such a dip would be permanent, because the global economic pie keeps growing.[29]

It has also been argued that in a period of tectonic change such as the one that the global community is now undergoing, there will inevitably be nations and individuals that are winners or losers. It is the view of this committee that the determining factors in such outcomes are the extent of a nation's commitment to get out and compete in the global marketplace.

---

[26]Friedman, 2005, p. 227.

[27]E-mail communication from P. Romer to D. Stine, September 22, 2005.

[28]D. Autor, L. Katz, and M. Kearney. *Trends in U.S. Wage Inequality: Re-Assessing the Revisionists*. Working Paper 11627. Washington, DC: National Bureau of Economic Research, 2005, for a recent summary of the evidence on this point, see http://www.nber.org.

[29]Friedman, 2005, p. 227.

New generations of US scientists and engineers, assisted by progressive government policies, could lead the way to US leadership in the new, flatter world—as long as US workers remain among the best educated, hardest-working, best trained, and most productive in the world.

That, of course, is the challenge.

CLUSTER 2: DISINVESTMENT IN THE FUTURE

The most effective way for the United States to meet the challenges of a flatter world would be to draw heavily and quickly on its investments in human capital. We need people who have been prepared for the kinds of knowledge-intensive occupations in which the nation must excel. Yet the United States has for a number of decades fallen short in making the kinds of investments that will be essential in a global economy.

Loss of Human Capital

An educated, innovative, motivated workforce—human capital—is the most precious resource of any country in this new, flat world. Yet there is widespread concern about our K–12 science and mathematics education system, the foundation of that human capital in today's global economy. A recent Gallup poll[30] asked respondents, "Overall, how satisfied are you with the quality of education students receive in kindergarten through grade twelve in the United States today—would you say you are completely satisfied, somewhat satisfied, somewhat dissatisfied or completely dissatisfied?" More than 50% were either "completely dissatisfied" or "somewhat dissatisfied" with our schooling. According to the poll results, the critical required change would be to produce better educated, higher-quality teachers.[31] This committee shares that view, particularly in connection with education in science and mathematics. By far the highest leverage to be found in our education system resides with teachers, if for no other reason than that they influence such a large number of future workers.

Students in the United States are not keeping up with their counterparts in other countries. In 2003 the Organisation for Economic Co-operation and Development's (OECD's) Programme for International Student Assessment[32] measured the performance of 15-year-olds in 49 industrialized coun-

[30]Gallup poll, August 8-11, 2005, ± 3% margin of error, sample size = 1,001. As found at: http://www.gallup.com/ on September 14, 2005.

[31]Gallup poll, August 9-11, 2004, ± 3% margin of error, sample size = 1,017. As found at: http://www.gallup.com/ on September 14, 2005.

[32]Organization for Economic Co-operation and Development. "Program for International Student Assessment." Available at: http://www.pisa.oecd.org.

tries. It found that US students scored in the middle or in the bottom half of the group in three important ways: our students placed 16th in reading, 19th in science literacy, and 24th in mathematics.[33] In 1996 (the most recent data available), US 12th graders performed below the international average of 21 countries on a test of general knowledge in mathematics and science.[34]

After secondary school, fewer US students pursue science and engineering degrees than is the case of students in other countries. About 6% of our undergraduates major in engineering; that percentage is the second lowest among developed countries. Engineering students make up about 12% of undergraduates in most of Europe, 20% in Singapore, and more than 40% in China. Students throughout much of the world see careers in science and engineering as the path to a better future.

Higher Education as a Private Good

Our culture has always considered higher education a public good—or at least we have seemed to do so. We have agreed as a society that educated citizens benefit the whole society; that the benefit accrues to us all and not just to those who receive the education. That was a primary reason for the creation in the 1860s of the land-grant college system; it is why early in the 20th century universal primary and secondary schooling was supported; it is why a system of superior state universities was created and generously supported and scholarships were given to needy students; and it is why the Serviceman's Readjustment Act of 1944—the GI Bill—was established and why the National Defense Education Act was passed in 1958 shortly after the launch of Sputnik.

Now, however, funding for state universities is dwindling, tuition is rising, and students are borrowing more than they receive in grants. These seem to be indications that our society increasingly sees higher education as a private good, of value only to the individual receiving it. A disturbing aspect of that change is its consequences for low-income students. College has been a traditional path for upward mobility—and this has been particularly true in the field of engineering for students who were first in their family to attend college. The acceptance of higher education as a personal benefit rather than a public good, the growth of costly private K–12 schooling, and the shift of the cost burden to individuals have made it increasingly difficult for low-income students to advance beyond high school. In the

---

[33]The report included results from 49 countries, available at: http://www.pisa.oecd.org/dataoecd/1/63/34002454.pdf.

[34]National Science Board. *Science and Engineering Indicators 2004*. NSB 04-01. Arlington, VA: National Science Foundation, 2004. Chapter 1.

long run, the nation as a whole will suffer from the lack of new talent that could have been discovered and nurtured in affordable, accessible, high-quality public schools, colleges, and universities.

Trends in Corporate Research

The US research structure that evolved after World War II was a self-reinforcing triangle of industry, academe, and government. Two sides of that triangle—industrial research and government investment in R&D as a fraction of gross domenstic product (GDP) have changed dramatically. Some of the most important fundamental research in the 20th century was accomplished in corporate laboratories—Bell Labs, GE Research, IBM Research, Xerox PARC, and others. Since that time, the corporate research structure has been significantly eroded. One reason might be the challenge of capturing the results of research investments within one company or even a single nation on a long-term basis. The companies and nation can, however, capture high-technology discoveries at least for the near term (5-10 years) and enhance the importance of innovation in jobs.[35] For example, the United States has successfully capitalized on research in monoclonal antibodies, network systems, and speech recognition. As a result, corporate funding of certain applied research has been enhanced at such companies as Google and Intel and at many biotechnology companies. Nonetheless, the increasing pressure on corporations for short-term results has made investments in research highly problematic.

Funding for Research in the Physical Sciences and Engineering

Although support for research in the life sciences increased sharply in the 1990s and produced remarkable results, funding for research in most physical sciences, mathematics, and engineering has declined or remained relatively flat—in real purchasing power—for several decades. Even to those whose principal interest is in health or healthcare, that seems short-sighted: Many medical devices and procedures—such as endoscopic surgery, "smart" pacemakers, kidney dialysis, and magnetic resonance imaging—are the result of R&D in the physical sciences, engineering, and mathematics. The need is to strengthen investment in the latter areas while not disinvesting in those areas of the health sciences that are producing promising results. Many believe that federal funding agencies—perhaps influenced by the stagnation of funding levels in the physical sciences, mathematics, and engineering—have become increasingly risk-averse and focused on

[35]NAS/NAE/IOM. *Capitalizing on Investments in Science and Technology*. Washington, DC: National Academy Press, 1999.

short-term results. For example, even the generally highly effective Defense Advanced Research Projects Agency (DARPA) has been criticized in this regard in congressional testimony.[36]

Widespread, if anecdotal, evidence shows that even the National Science Foundation and the National Institutes of Health (NIH) have changed their approach in this regard. A recent National Academies study[37] revealed that the average age at which a principal investigator receives his or her *first* grant is 42 years—partly because of requirements for evidence of an extensive "track record" to reduce risk to the grant-makers.[38] But reducing the risk for individual research projects *increases* the likelihood that breakthrough, "disruptive" technologies will not be found—the kinds of discoveries that often yield huge returns. History also suggests that young researchers make disproportionately important discoveries. The NIH roadmap[39] established in fiscal year (FY) 2004, recognizes this concern, but the amount of funds devoted to long-term, high-payoff, high-risk research remains very limited.

CLUSTER 3: REACTIONS TO 9/11

Three other pieces in the mosaic also appear to provide short-term security but little long-term benefit. These relate to the events of 9/11, which profoundly changed our world and made it necessary to re-examine national security issues in an entirely new context. This re-examination led to changes in visa policies, export controls, and the treatment of "sensitive but unclassified" information. There appears today to be a need to better balance security concerns with the benefits of an open, creative society.

New Visa Policies

Much has been written about new immigration and visa policies for students and researchers. Although there have been improvements in the last

---

[36]See US Congress House of Representatives Committee on Science. Available at: http://www.house.gov/science/hearings/full05/may12/. The current director of DARPA, however, points out that DARPA's job has always been to mine fundamental research, looking for those ideas whose time has come to move on to applied developmental research.

[37]National Research Council. *Bridges to Independence: Fostering the Independence of New Investigators in Biomedical Research*. Washington, DC: The National Academies Press, 2004.

[38]Other observers note that part of the reason for this is the length of the biomedical PhD and postdoctoral period and the difficulty of young biomedical researchers in finding initial tenure-track positions, for which many institutions require principal-investigator status on an NIH grant proposal. These trends, which are occurring in spite of the recent doubling of the NIH grants budget, suggest an imbalance between demand for and supply of recent PhDs.

[39]The purpose of the roadmap was to identify major opportunities and gaps in biomedical research that no single NIH institute could tackle alone but that the agency as a whole must address to make the biggest impact on the progress of medical research.

several months (at this writing, the average time to process a student visa is less than 2 weeks), there is still concern about response times in particular cases. Some promising students wait a year or more for visas; some senior scholars are subjected to long and sometimes demeaning review processes. Those cases, not the shorter *average* processing time, are emphasized in the international press. The United States is portrayed less as a welcoming land of opportunity than as a place that is hostile to foreigners.

Immigration procedures implemented since 9/11 have discouraged students from applying to US programs, prevented international research leaders from organizing conferences here, and dampened international collaboration. As a result, we are damaging the image of our country in the eyes of much of the world. Although there are recent signs of improvement, the matter remains a concern.

This committee is generally not privy to whatever evidence lies in the government's library of classified information, but it is important to recognize that our nation's borders have been crossed by more than 10 million people who are still residing illegally in the United States. Set against this background, a way is needed to quickly, legally, and safely admit to our shores the relatively small numbers of highly talented people who possess the skills needed to make major contributions to our nation's future competitiveness and well-being.

Some observers are also concerned that encouraging international students to come to the United States will ultimately fill jobs that could be occupied by American citizens. Others worry that such visitors will reduce the compensation that scientists and engineers receive—diminishing the desire of Americans to enter those professions. Studies show, however, that the financial impact is minimal, especially at the PhD level. Furthermore, scientists and engineers tend to be creators of new jobs and not simply consumers of a fixed set of existing jobs. If Americans make up a larger percentage of a graduating class, a larger percentage of Americans will be hired by corporations. In the end, the United States needs the smartest people, wherever they come from throughout the world. The United States will be more prosperous if those people live and work in the United States rather than elsewhere. History has emphatically proven this point.

The Use of Export Controls

Export controls were first instituted in the United States in 1949 to keep weapons technology out of the hands of potential adversaries. They have since been used, on occasion, as an economic tool against competitors.

The export of controlled technology requires a license from the Department of Commerce or from the Department of State. Since 1994, the disclosure of information regarding a controlled technology to some foreign nationals—even when the disclosure takes place inside the United States, a practice sometimes called "deemed export"—has been considered the same as the export of the technology itself and thus requires an export license.

Some recent reports[40] suggest that implementation of the rules that govern deemed exports should be tightened even further—for example, by altering or eliminating the exemption for basic research and by broadening the definition of "access" to controlled technology.

The academic research community is deeply concerned that a literal interpretation of these suggestions could prevent foreign graduate students from participating in US-based research and would require an impossibly complex system of enforcement. Given that 55% of the doctoral students in engineering in the United States are foreign-born and that many of these students currently remain in the United States after receiving their degrees, the effect could be to drastically reduce our talent pool.

The United States is not the world's only country capable of performing research; China and India, for example, have recognized the value of research universities to their economic development and are investing heavily in them. By putting up overly stringent barriers to the exchange of information about basic research, we isolate ourselves and impede our own progress. At the same time, the information we are protecting often is available elsewhere.

The current fear that foreign students in our universities pose a security risk must be balanced against the great advantages of having them here. It is, of course, prudent to control entry to our nation, but as those controls become excessively burdensome they can unintentionally harm us. In this regard, it should be noted that Albert Einstein, Edward Teller, Enrico Fermi, and many other immigrants enabled the United States to develop the atomic bomb and bring World War II to an earlier conclusion than would otherwise have been the case. In addition, immigrant scientists and engineers have contributed to US economic growth throughout the nation's history by founding or cofounding new technology-based companies. Examples include Andrew Carnegie (US Steel, born in Scotland), Alexander Graham Bell (AT&T, born in Scotland), Herbert Henry Dow (Dow Chemical, born in Canada), Henry Timken (Timken Company, born in Germany), Andrew Grove (Intel, born in Hungary), Davod Lam (Lam Research, born in China), Vinod Khosla (Sun Microsystems, born in India), and Sergey Brin (Google, born in Russia).

---

[40]Reports from the inspectors general of the US Departments of Commerce, Defense, and State. As an example, see Bureau of Industry and Security, Office of Inspections and Program Evaluations. "Deemed Export Controls May Not Stop the Transfer of Sensitive Technology to Foreign Nationals in the U.S." Final Inspection Report No. IPE-16176-March 2004.

Similarly, it has been noted that

- Many students from abroad stay here after their education is complete and contribute greatly to our economy.
- Foreign students who do return home often are our best ambassadors.
- The United States benefits economically from open trade, and our security is reinforced by rising living standards in developing countries.
- The quality of life in the United States has been improved as a result of shared scientific results. Some foreign-born students do return home to work as competitors, but others join in international collaborations that help us move faster in the development and adaptation of new technology and thereby create new jobs.

Yet, Section 214b of the Immigration and Nationality Act requires applicants for student or exchange visas to provide convincing evidence that they plan to return to their home countries—a challenging requirement.

Sensitive but Unclassified Information

Since 9/11, the amount of information designated sensitive but unclassified (SBU) by the US government has presented a problem that is less publicized than visas or deemed exports but is a complicating factor in academic research. The SBU category, as currently applied, is inconsistent with the philosophy of building high fences around small places associated with the traditional protection of scientific and technical information. There are no laws, no common definitions, and no limits on who can declare information "SBU," nor are there provisions for review and disclosure after a specific period. There is little doubt that the United States would profit from a serious discussion about what kinds of information should be classified, but such a discussion is not occurring.

THE PUBLIC RECOGNIZES THE CHALLENGES

Does the public truly see the challenge to our prosperity? In recent months, polls have indicated persistent concern not only about the war in Iraq and issues of terrorism but also, and nearly equally, about jobs and the economy. One CBS-*New York Times* poll showed security leading economic issues by only 1%;[41] another[42] showed that our economy and job security

[41]CBS News-*New York Times* poll, June 10-15, 2005; of 1,111 adults polled nationwide, 19% found the war in Iraq the most important problem, 18% cited the economy and jobs. Available at: http://www.cbsnews.com/htdocs/CBSNews_polls/bush616.pdf.

[42]ABC News-*Washington Post* poll, June 2-5, 2005; of 1,002 adults polled nationwide, 30% rated the economy and jobs of highest concern, 24% rated Iraq of highest concern.

are of slightly greater concern to respondents than are issues of national security and terrorism. On the eve of the 2004 presidential election, the Gallup organization asked respondents what issues concerned them most. Terrorism was first, ranked "extremely important" by 45% of respondents; next came the economy (39%), health care (33%), and education (32%).[43] Only 35% say that now is a good time to find a high-quality job; 61% say that it is not.[44] Polls, of course, only provide a snapshot of America's thinking, but presumably one can conclude that Americans are generally worried about jobs—if not for themselves then for their children and grandchildren.

Investors are worried, too. According to a Gallup poll, 83% percent of US investors say job outsourcing to foreign countries is currently hurting the investment climate "a lot" (61%) or "a little" (22%). The numbers who are worried about outsourcing are second only to the numbers who are worried about the price of energy, according to a July 2005 Gallup poll on investor concerns.[45]

DISCOVERY AND APPLICATION: KEYS TO COMPETITIVENESS AND PROSPERITY

A common denominator of the concerns expressed by many citizens is the need for and use of knowledge. Well-paying jobs, accessible healthcare, and high-quality education require the discovery, application, and dissemination of information and techniques. Our economy depends on the knowledge that fuels the growth of business and plants the seeds of new industries, which in turn provides rewarding employment for commensurately educated workers. Chapter 2 explains that US prosperity since World War II has depended heavily on the excellence of its "knowledge institutions": high-technology industries, federal R&D agencies, and research universities that are generally acknowledged to be the best in the world.

The innovation model in place for a half-century has been so successful in the United States that other nations are now beginning to emulate it. The governments of Finland, Korea, Ireland, Canada, and Singapore have mapped and implemented strategies to increase the knowledge base of students and researchers, strengthen research institutions, and promote exports of high-technology products—activities in which the United States has in the past

[43]D. Jacob, Gallup chief economist, in "More Americans See Threat, Not Opportunity, in Foreign Trade: Most Investors See Outsourcing as Harmful." Available at: http://www.gallup.com/poll/content/default.aspx?ci=14338.

[44]F. Newport, Gallup poll editor-in-chief, in "Bush Approval, Economy, Election 2008, Iraq, John Roberts, Civil Rights." August 9, 2005. Available at: http://www.gallup.com/poll/content/?ci=17758&pg=1.

[45]Gallup poll, June 24-26, 2005, ± 3% margin of error, sample size = 1,009. As found at: http://www.gallup.com/poll/content/?ci=17605&pg=1 on September 14, 2005.

excelled.[46] China formally adopted a pro-R&D policy in the middle of the 1990s and has been moving rapidly to raise government spending on basic research, to reform old structures in a fashion that supports a market economy, and to build indigenous capacity in science and technology.[47]

The United States is now part of a connected, competitive world in which many nations are empowering their indigenous "brainware" and building new and effective performance partnerships—and they are doing so with remarkable focus, vigor, and determination. The United States must match that tempo if it hopes to maintain the degree of prosperity it has enjoyed in the past.

ACTION NOW

Indeed, if we are to provide prosperity and a secure environment for our children and grandchildren, we cannot be complacent. The gradual change in England's standing in the world since the 1800s and the sudden change in Russia's standing since the end of the Cold War are but two examples that illustrate how dramatically power can shift. Simply maintaining the status quo is insufficient when other nations push ahead with desire, energy, and commitment.

Today, we see in the example of Ireland how quickly a determined nation can rise from relative hunger to burgeoning prosperity. In the 1980s, Ireland's unemployment rate was 18%, and during that decade 1% of the population—mostly young people—left the country, largely to find jobs.[48] In response, a coalition of government, academic institutions, labor unions, farmers, and others forged an ambitious and sometimes painful plan of tax and spending cuts and aggressively courted foreign investors and skilled scientists and engineers. Today, Ireland is, on a per capita basis, one of Europe's wealthiest countries.[49] In 1990, Ireland's per capita GDP of $12,891 (in current US dollars) ranked it 23rd of the 30 OECD member countries. By 2002, Ireland's per capita GDP had grown to $32,646, making it 4th highest among OECD member countries.[50] Ireland's unemploy-

[46]Organisation for Economic Co-operation and Development. "Main Science & Technology Indicators, 2005." Available at: http://www.oecd.org/document/26/0,2340,en_2649_34451_1901082_1_1_1_1,00.html.

[47]"China's Science and Technology Policy for the Twenty-First Century—A View from the Top." Report from the US Embassy, Beijing, November 1996.

[48]W. C. Harris, director general, Science Foundation Ireland, personal communication, August 15, 2005.

[49]T. Friedman. *The End of the Rainbow*. New York Times, June 29, 2005.

[50]Organisation for Economic Co-operation and Development. "OECD Factbook 2005." Available at: http://puck.sourceoecd.org/vl=2095292/cl=23/nw=1/rpsv/factbook/.

ment rate (as a percentage of the total labor force) was 13.4% in 1990. By 1993, it had risen to 15.6%. By 2004 the unemployment rate declined to 4.5%.[51] Since 1995, Ireland's economic growth has averaged 7.9%. Over the same time period, economic growth averaged 2% in Europe and 3.3% in the United States.[52]

History is the story of people mobilizing intellectual and practical talents to meet demanding challenges. World War II saw us rise to the military challenge, quickly developing nuclear weapons and other military capabilities. After the launch of Sputnik[53] in 1957, we accepted the challenge of the space race, landed 12 Americans on the moon, and fortified our science and technology capacity.

Today's challenge is economic—no Pearl Harbor, Sputnik, or 9/11 will stir quick action. It is time to shore up the basics, the building blocks without which our leadership will surely decline. For a century, many in the United States took for granted that most great inventions would be homegrown—such as electric power, the telephone, the automobile, and the airplane—and would be commercialized here as well. But we are less certain today who will create the next generation of innovations, or even what they will be. We know that we need a more secure Internet, more-efficient transportation, new cures for disease, and clean, affordable, and reliable sources of energy. But who will dream them up, who will get the jobs they create, and who will profit from them? If our children and grandchildren are to enjoy the prosperity that our forebears earned for us, our nation must quickly invigorate the knowledge institutions that have served it so well in the past and create new ones to serve in the future.

CONCLUSION

A few of the tiles in the mosaic are apparent; many other problems could be added to the list. The three clusters discussed in this chapter share a common characteristic: short-term responses to perceived problems can give the *appearance* of gain but often bring real, long-term losses.

---

[51]Ibid.

[52]R. Samuelson. "The World Is Still Round." *Newsweek*, July 25, 2005.

[53]The fall 1957 launch of Sputnik I, the first artificial satellite, caused many in the United States to believe that we were quickly falling behind the USSR in science education and research. That concern led to major policy reforms in education, civilian and military research, and federal support for researchers. Within a year, the National Aeronautics and Space Administration and DARPA were founded. In that era, science and technology became a major focus of the public, and a presidential science adviser was appointed.

This report emphasizes the need for world-class science and engineering—not simply as an end in itself but as the principal means of creating new jobs for our citizenry as a whole as it seeks to prosper in the global marketplace of the 21st century. We must help those who lose their jobs; they need financial assistance and retraining. It might even be appropriate to protect some selected jobs for a very short time. But in the end, the country will be strengthened only by learning to compete in this new, flat world.

2

Why Are Science and Technology Critical to America's Prosperity in the 21st Century?

Since the Industrial Revolution, the growth of economies throughout the world has been driven largely by the pursuit of scientific understanding, the application of engineering solutions, and continual technological innovation.[1] Today, much of everyday life in the United States and other industrialized nations, as evidenced in transportation, communication, agriculture, education, health, defense, and jobs, is the product of investments in research and in the education of scientists and engineers.[2] One need only think about how different our daily lives would be without the technological innovations of the last century or so.

The products of the scientific, engineering, and health communities are, in fact, easily visible—the work-saving conveniences in our homes; medical help summoned in emergencies; the vast infrastructure of electric power, communication, sanitation, transportation, and safe drinking water we take for granted.[3] To many of us, that universe of products and

---

[1]Another point of view is provided in Box 2-1.

[2]S. W. Popper and C. S. Wagner. *New Foundations for Growth: The U.S. Innovation System Today and Tomorrow*. Santa Monica, CA: RAND Corporation, 2002. The authors state: "The transformation of the U.S. economy over the past 20 years has made it clear that innovations based on scientific and technological advances have become a major contributor to our national well being." P. ix.

[3]One study argues that "there has been more material progress in the United States in the 20th century than there was in the entire world in all the previous centuries combined," and most of the examples cited have their basis in scientific and engineering research. S. Moore and J. L. Simon. "The Greatest Century That Ever Was: 25 Miraculous Trends of the Last 100 Years." *Policy Analysis* No. 364. Washington, DC: Cato Institute, December 15, 1999.

BOX 2-1
Another Point of View: Science, Technology, and Society

For all the practical devices and wonders that science and technology have brought to society, it has also created its share of problems. Researchers have had to reapply their skills to create solutions to unintended consequences of many innovations, including finding a replacement for chlorofluorocarbon-based refrigerants, eliminating lead emissions from gasoline-powered automobiles, reducing topsoil erosion caused by large-scale farming, researching safer insecticides to replace DDT, and engineering new waste-treatment schemes to reduce hazardous chemical effluents from coal power plants and chemical refineries.

services defines modern life, freeing most of us from the harsh manual labor, infectious diseases, and threats to life and property that our forebears routinely faced. Now, few families know the suffering caused by smallpox, tuberculosis (TB), polio, diphtheria, cholera, typhoid, or whooping cough. All those diseases have been greatly suppressed or eliminated by vaccines (Figure 2-1).

We enjoy and rely on world travel, inexpensive and nutritious food, easy digital access to the arts and entertainment, laptop computers, graphite tennis rackets, hip replacements, and quartz watches. Box 2-2 lists a few examples of how completely we depend on scientific research and its application—from the mighty to the mundane.

Science and engineering have changed the very nature of work. At the beginning of the 20th century, 38% of the labor force was needed for farm work, which was hard and often dangerous. By 2000, research in plant and animal genetics, nutrition, and husbandry together with innovation in machinery had transformed farm life. Over the last half-century, yields per acre have increased about 2.5 times,[4] and overall output per person-hour has increased fully 10-fold for common crops, such as wheat and corn (Figure 2-2). Those advances have reduced the farm labor force to less than 3% of the population.

Similarly, the maintenance of a house a century ago without today's labor-saving devices left little time for outside enjoyment or work to produce additional income.

The visible products of research, however, are made possible by a large

[4]National Research Council. *Frontiers in Agricultural Research: Food, Health, Environment, and Communities*. Washington, DC: The National Academies Press, 2003.

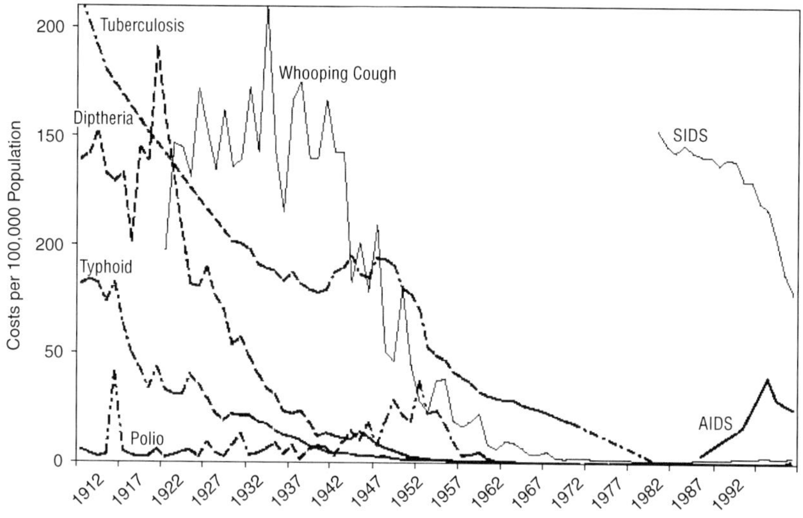

FIGURE 2-1 Incidence of selected diseases in the United States throughout the 20th century. The 20th century saw dramatic reductions in disease incidence in the United States.
NOTES: Sudden Infant Death Syndrome (SIDS) rate is per 100,000 live births. AIDS definition was substantially expanded in 1985, 1987, and 1993. TB rate prior to 1930 is estimated as 1.3 times the mortality rate.
SOURCES: S. Moore, J. L. Simon, and the CATO Institute. "The Greatest Century That Ever Was: 25 Miraculous Trends of the Past 100 Years." *Policy Analysis* No. 364, December 15, 1999. Pp. 1-32. Based on *Historical Statistic of the United States,* Series B 149, B 291, B 299-300, B 303; *Health, United States, 1999,* Table 53; and American SIDS Institute. Available at: http://www.sids.org/.

enterprise mostly hidden from public view—fundamental and applied research, an intensively trained workforce, and a national infrastructure that provides risk capital to support the nation's science and engineering innovation enterprise. All that activity, and its sustaining public support, fuels the steady flow of knowledge and provides the mechanism for converting information into the products and services that create jobs and improve the quality of modern life. Maintaining that vast and complex enterprise during an age of competition and globalization is challenging, but it is essential to the future of the United States.

ENSURING ECONOMIC WELL-BEING

Knowledge acquired and applied by scientists and engineers provides the tools and systems that characterize modern culture and the raw materials

BOX 2-2
Twenty Great Engineering Achievements of the 20th Century

Electricity: steam turbine generators; long-distance, high-voltage transmission lines; pulverized coal; large-scale electric grids
Automotive: machine tools, assembly line, self-starting ignition, balloon tire, safety-glass windshield, electronic fuel injection and ignition, airbags, antilock brakes, fuel cells
Aeronautics: aerodynamic wing and fuselage design, metal alloys and composite materials, stressed-skin construction, jet propulsion, fly-by-wire control systems, collision warning systems, Doppler weather radar
Water supply and distribution: chlorination, wastewater treatment, dams, reservoirs, storage tanks, tunnel-boring equipment, computerized contaminant detection, desalination, large-scale distillation, portable ultraviolet devices
Electronics: triodes, semiconductors, transistors, molecular-beam epitaxy, integrated circuits, digital-to-optical recording (CD-ROM), microprocessors, ceramic chip carriers
Radio and television: alternators, triodes, cathode-ray tubes, super heterodyne circuits, AM/FM, videocassette recorders, flat-screen technology, cable and high-definition television, telecommunication satellites
Agriculture: tractors, power takeoff, rubber tires, diesel engines, combine, corn-head attachments, hay balers, spindle pickers, self-propelled irrigation systems, conservation tillage, global-positioning technology
Computers: electromechanical relays; Boolean operations; stored programs; programming languages; magnetic tape; software, supercomputers, minicomputers, and personal computers; operating systems; the mouse; the Internet
Telephony: automated switchboards, dial calling, touch-tone, loading coils, signal amplifiers, frequency multiplexing, coaxial cables, microwave signal transmission, switching technology, digital systems, optical-fiber signal transmission, cordless telephones, cellular telephones, voice-over-Internet protocols
Air conditioning and refrigeration: humidity-control technology, refrigerant technology, centrifugal compressors, automatic temperature control, frost-free cooling, roof-mounted cooling devices, flash-freezing
Highways: concrete, tar, road location, grading, drainage, soil science, signage, traffic control, traffic lights, bridges, crash barriers
Aerospace: rockets, guidance systems, space docking, lightweight materials for vehicles and spacesuits, solar power cells, rechargeable batteries, satellites, freeze-dried food, Velcro
Internet: packet-switching, ARPANET, e-mail, networking services, transparent peering of networks, standard communication protocols, TCP/IP, World Wide Web, hypertext, web browsers

Imaging: diagnostic x-rays, color photography, holography, digital photography, cameras, camcorders, compact disks, microprocessor etching, electron microscopy, positron-emission tomography, computed axial tomography, magnetic-resonance imaging, sonar, radar, sonography, reflecting telescopes, radiotelescopes, photodiodes, charge-coupled devices
Household appliances: gas ranges, electric ranges, oven thermostats, nickel-chrome resistors, toasters, hot plates, electric irons, electric motors, rotary fans, vacuum cleaners, washing machines, sewing machines, refrigerators, dishwashers, can openers, cavity magnetrons, microwave ovens
Health technology: electrocardiography; heart–lung machines; pacemakers; kidney dialysis; artificial hearts; prosthetic limbs; synthetic heart valves, eye lenses, replacement joints; manufacturing techniques and systems design for large-scale drug delivery; operating microscopy; fiber-optic endoscopy; laparoscopy; radiologic catheters; robotic surgery
Petroleum and petrochemical technology: thermal-cracking oil refining; leaded gasoline; catalytic cracking; oil byproduct compounds; synthetic rubber; coal tar distillation byproduct compounds, plastics, polyvinyl chloride, polyethylene, synthetic fibers; drilling technologies; drill bits; pipelines; seismic siting; catalytic converters; pollution-control devices
Lasers and fiber optics: maser, laser, pulsed-beam laser, compact-disk players, barcode scanners, surgical lasers, fiber optic communication
Nuclear technology: nuclear fission, nuclear reactors, electric-power generation, radioisotopes, radiation therapy, food irradiation
High-performance materials: steel alloys, aluminum alloys, titanium superalloys; synthetic polymers, Bakelite, Plexiglas; synthetic rubbers, neoprene, nylon; polyethylene, polyester, Saran Wrap, Dacron, Lycra spandex fiber, Kevlar; cement, concrete; synthetic diamonds; superconductors; fiberglass, graphite composites, Kevlar composites, aluminum composites

SOURCE: G. Constable and B. Somerville. *A Century of Innovation: Twenty Engineering Achievements That Transformed Our Lives.* Washington, DC: Joseph Henry Press, 2003.

for economic growth and well-being. The knowledge density of modern economies has steadily increased, and the ability of a society to produce, select, adapt, and commercialize knowledge is critical for sustained economic growth and improved quality of life.[5,6] Robert Solow demonstrated that pro-

[5]L. B. Holm-Nielsen. Promoting Science and Technology for Development: The World Bank's Millennium Science Initiative. Paper delivered on April 30, 2002, to the First International Senior Fellows meeting, The Wellcome Trust, London, UK.

[6]The Organisation for Economic Co-operation and Development (OECD) concludes that "underlying long-term growth rates in OECD economies depend on maintaining and expanding the knowledge base." OECD. *Technology, Productivity, and Job Creation: Best Policy Practices*. Paris: OECD, 1998. P. 4.

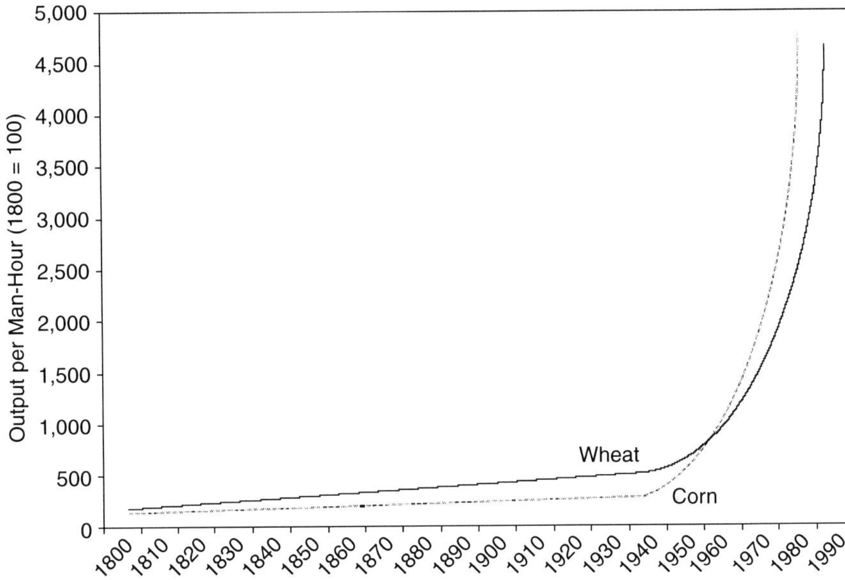

FIGURE 2-2 US farm labor productivity from 1800 to 2000. There was a 100-fold increase in US farm labor output, much of it brought about by advancements in science and technology.
SOURCE: S. Moore, J. L. Simon, and the CATO Institute. "The Greatest Century That Ever Was: 25 Miraculous Trends of the Past 100 Years." *Policy Analysis* No. 364, December 15, 1999. Pp. 1-32.

ductivity depends on more than labor and capital.[7] Intangible qualities—research and development (R&D), or the acquisition and application of knowledge—are crucial.[8] The earlier national commitment to make a substantial public investment in R&D was based partly on that assertion (Figure 2-3).

Since Solow's pioneering work, the economic value of investing in science and technology has been thoroughly investigated. Published estimates of return on investment (ROI) for publicly funded R&D range from 20 to 67% (Table 2-1). Although most early studies focused on agriculture, recent work shows high rates of return for academic science research in the

[7]R. M. Solow. "Technical Change and the Aggregate Production Function." *The Review of Economics and Statistics* 39(1957):312-320; R. M. Solow. Investment and Technical Progress. In Arrow, Karlin & Suppes, eds. *Mathematical Models in Social Sciences*, 1960. For more on Solow's work, see http://nobelprize.org/economics/laureates/1987/index.html.

[8]Solow, 1957.

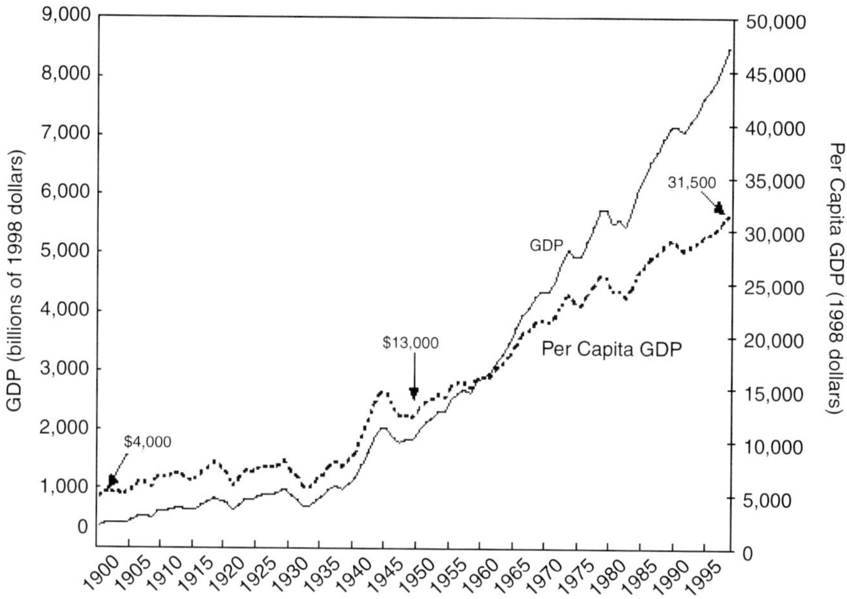

FIGURE 2-3 Gross domestic product during the 20th century. In the 20th century, US per capita gross domestic product (GDP) rose almost 7-fold.
SOURCE: S. Moore, J. L. Simon, and the CATO Institute. "The Greatest Century That Ever Was: 25 Miraculous Trends of fhe Past 100 Years." *Policy Analysis*, No. 364, December 15, 1999. Pp. 1-32.

aggregate (28%),[9] and slightly higher rates of return for pharmaceutical products in particular (30%).[10] Modern agriculture continues to respond, and the average return on investment for public funding of agricultural research for member countries of the Organisation for Economic Co-operation and Development (OECD) is estimated at 45%.[11]

Starting in the middle 1990s, investments in computers and information technology started to show payoffs in US productivity. The economy grew faster and employment rose more than had seemed possible without

---

[9]E. Mansfield. "Academic Research and Industrial Innovation." *Research Policy* 20(1991): 1-12.

[10]A. Scott, G. Steyn, A. Geuna, S. Brusoni, and W. E. Steinmeuller. "The Economic Returns of Basic Research and the Benefits of University-Industry Relationships." *Science and Technology Policy Research*. Brighton: University of Sussex, 2001. Available at: http://www.sussex.ac.uk/spru/documents/review_for_ost_final.pdf.

[11]R. E. Evenson. Economic Impacts of Agricultural Research and Extension. In B. L. Gardner and G. C. Rausser, eds. *Handbook of Agricultural Economics Vol. 1*. Rotterdam: Elsevier, 2001. Pp. 573-628.

TABLE 2-1 Annual Rate of Return on Public R&D Investment

Studies	Subject	Rate of Return to Public R&D (percent)
Griliches (1958)	Hybrid corn	20-40
Peterson (1967)	Poultry	21-25
Schmitz-Seckler (1979)	Tomato harvester	37-46
Griliches (1968)	Agriculture research	35-40
Evenson (1968)	Agriculture research	28-47
Davis (1979)	Agriculture research	37
Evebsib (1979)	Agriculture research	45
Davis and Peterson (1981)	Agriculture research	37
Mansfield (1991)	All academic science research	28
Huffman and Evenson (1993)	Agricultural research	43-67
Cockburn and Henderson (2000)	Pharmaceuticals	30+

SOURCE: A. Scott, G. Steyn, A. Geuna, S. Brusoni, W. E. Steinmeuller. "The Economic Returns of Basic Research and the Benefits of University-Industry Relationships." *Science and Technology Policy Research*. Brighton: University of Sussex, 2001. Available at: http://www.sussex.ac.uk/spru/documents/review_for_ost_final.pdf.

fueling inflation. Policy-makers previously focused almost entirely on changes in demand as the determinant of inflation, but the surge in productivity showed that changes on the supply side of the economy could be just as important and in some cases even more important.[12] Such data serve to sustain the US commitment to invest substantial public funds in science and engineering.[13]

Of equal interest are studies of the rate of return on *private* investments in R&D.[14] The return on investment to the nation is generally higher than is the return to individual investors (Table 2-2).[15] One reason is that knowledge tends to spill over to other people and other businesses, so research results diffuse to the advantage of those who are prepared to apply them.

[12]E. L. Andrews. *The Doctrine Was Not to Have One; Greenspan Will Leave No Road Map to His Successor*. New York Times, August 26, 2005. P. C1.

[13]US Congress House of Representatives Committee on Science. *Unlocking Our Future: Toward a New National Science Policy* (the "Ehlers Report"). Washington, DC: US Congress, 1998. The report notes that "the growth of economies throughout the world since the industrial revolution began has been driven by continual technological innovation through the pursuit of scientific understanding and application of engineering solutions." P. 1.

[14]Council of Economic Advisors. *Supporting Research and Development to Promote Economic Growth: The Federal Government's Role*. Washington, DC: White House, October 1995.

[15]Center for Strategic and International Studies. *Global Innovation/National Competitiveness*. Washington, DC: CSIS, 1996.

TABLE 2-2 Annual Rate of Return on Private R&D Investment

	Estimated Rate of Return %	
Researcher	Private	Social
Nadiri (1993)	20-30	50
Mansfield (1977)	25	56
Terleckyj (1974)	29	48-78
Sveikauskas (1981)	7-25	50
Goto-Suzuki (1989)	26	80
Bernstein-Nadiri (1988)	10-27	11-111
Scherer (1982, 1984)	29-43	64-147
Bernstein-Nadiri (1991)	15-28	20-110

SOURCE: Center for Strategic and International Studies. *Global Innovation/National Competitiveness*. Washington, DC: CSIS, 1996.

Those "social rates of return"[16] on investments in R&D are reported to range from 20 to 100%, with an average of nearly 50%.[17] As a single example, in recent years, graduates from one US university have founded 4,000 companies, created 1.1 million jobs worldwide, and generated annual sales of $232 billion.[18]

Although return-on-investment data vary from study to study, most economists agree that federal investment in research pays substantial economic dividends. For example, Table 2-3 shows the large number of jobs and revenues created by information-technology manufacturing and services—an industry that did not exist until the recent past. The value of public and private investment in research is so important that it has been

[16]"Social rate of return" is defined in C. I. Jones and J. C. Williams. "Measuring the Social Return to R&D." Working Paper 97002. Stanford University Department of Economics, 1997. Available at: http://www.econ.stanford.edu/faculty/workp/swp97002.pdf#search='R&D%20social%20rate%20of%20return. They state, "One can think of knowledge as an 'asset' purchased by society, held for a short period of time to reap a dividend, and then sold. The return can then be thought of as a sum of a dividend and a capital gain (or loss). . . . The dividend associated with an additional idea consists of two components. First, the additional knowledge directly raises the productivity of capital and labor in the economy. Second, the additional knowledge changes the productivity of future R&D investment because of either knowledge spillovers or because subsequent ideas are more difficult to discover." Pp. 6-8.

[17]M. I. Nadiri. "Innovations and Technological Spillovers." *Economic Research Reports*, RR 93-31. New York: C. V. Starr Center for Applied Economics, New York University Department of Economics, August 1993. Nadiri adds, "The channels of diffusion of the spillovers vary considerably and their effects on productivity growth are sizeable. These results suggest a substantial underinvestment in R&D activity."

[18]W. M. Ayers. *MIT: The Impact of Innovation*. Boston, MA: Bank Boston, 2002. Available at: http://web.mit.edu/newsoffice/founders/Founders2.pdf.

TABLE 2-3 Sales and Employment in the Information Technology (IT) Industry, 2000

	NAICS Code	Sales Revenues ($ billions)	Number of Jobs (1,000)
IT Manufacturing			
Computer and peripheral equipment	3341	110.0	190
Communications equipment	3342	119.3	291
Software	5112	88.6	331
Semiconductors and other electronic components	3344	168.5	621
IT Services			
Data processing services	5142	42.9	296
Telecommunications services	5133	354.2	1,165

SOURCE: National Research Council. *Impact of Basic Research on Industrial Performance.* Washington, DC: The National Academies Press, 2003.

described as "fuel for industry."[19] The economic contribution of science and technology can be understood by examining revenue and employment figures from technology- and service-based industries, but the largest economic influence is in the productivity gains that follow the adoption of new products and technologies.[20]

CREATING NEW INDUSTRIES

The power of research is demonstrated not only by single innovations but by the ability to create entire new industries—some of them the nation's most powerful economic drivers.

Basic research on the molecular mechanisms of DNA has produced a new field, molecular biology, and recombinant-DNA technology, or gene splicing, which in turn has led to new health therapies and the enormous growth of the biotechnology industry. The potential of those developments for health and healthcare is only beginning to be realized.

Studies of the interaction of light with atoms led to the prediction of stimulated emission of coherent radiation. That, together with the quest for a device to produce high-frequency microwaves, led to the development of

[19]Council of Economic Advisers. *Economic Report of the President.* Washington, DC: US Government Printing Office, 1995.

[20]D. J. Wilson. "Is Embodied Technological Change the Result of Upstream R&D? Industry-Level Evidence." *Review of Economic Dynamics* 5(2)(2002):342-362.

the laser, a ubiquitous device with uses ranging from surgery, precise machining, and nuclear fusion to sewer alignment, laser pointers, and CD and DVD players.

Enormous economic gains can be traced to research in harnessing electricity, which grew out of basic research (such as that conducted by Michael Faraday and James Maxwell) and applied research (such as that by Thomas Edison and George Westinghouse). Furthermore, today's semiconductor integrated circuits can be traced to the development of transistors and integrated circuits, which began with basic research into the structure of the atom and the development of quantum mechanics by Paul Dirac, Wolfgang Pauli, Werner Heisenberg, and Erwin Schrodinger[21] and was realized through the applied research of Robert Noyce and Jack Kilby.

In virtually all those examples, the original researchers did not—or could not—foresee the consequences of the work they were performing, let alone its economic implications. The fundamental research typically was driven by the desire to answer a specific question about nature or about an application of technology. The greatest influence of such work often is removed from its genesis,[22] but the genius of the US research enterprise has been its ability to afford its best minds the opportunity to pursue fundamental questions (Figures 2-4, 2-5, 2-6).

PROMOTING PUBLIC HEALTH

One straightforward way to view the practical application of research is to compare US life expectancy (Figure 2-7) in 1900 (47.3 years)[23] with that in 1999 (77 years).[24] Our cancer and heart-disease survival rates have improved (Figure 2-8), and accidental-death rates and infant and maternal mortality (Figure 2-9) have fallen dramatically since the early 20th century.[25]

Improvements in the nation's health are, of course, attributable to many factors, some as straightforward as the engineering of safe drinking-water supplies. Also responsible are the large-scale production, delivery, and storage

---

[21]J. I. Friedman. "Will Innovation Flourish in the Future?" *Industrial Physicist* 8(6)(December 2002/January 2003):22-25.

[22]See, for example, National Research Council. *Evolving the High Performance Computing and Communications Initiative to Support the Nation's Information Infrastructure.* Washington, DC: National Academy Press, 1995.

[23]US Census Bureau. "Historical Statistics of the United States, Colonial Times to 1970." Part 1, Series B 107-15. P. 55.

[24]US Census Bureau. *Statistical Abstract of the United States: 2000.* P. 84. Table 116.

[25]F. Hobbs and N. Stoops. *Demographic Trends in the 20th Century.* CENSR-4. Washington, DC: US Census Bureau, November 2004.

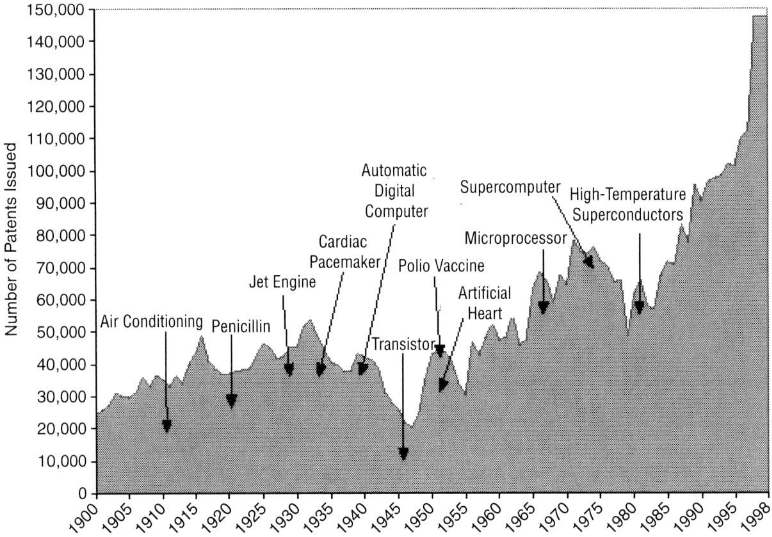

FIGURE 2-4 Number of patents granted by the United States in the 20th century with examples of critical technologies.
SOURCE: S. Moore, J. L. Simon, and the CATO Institute. "The Greatest Century That Ever Was: 25 Miraculous Trends of the Past 100 Years." *Policy Analysis* No. 364, December 15, 1999. Pp. 1-32.

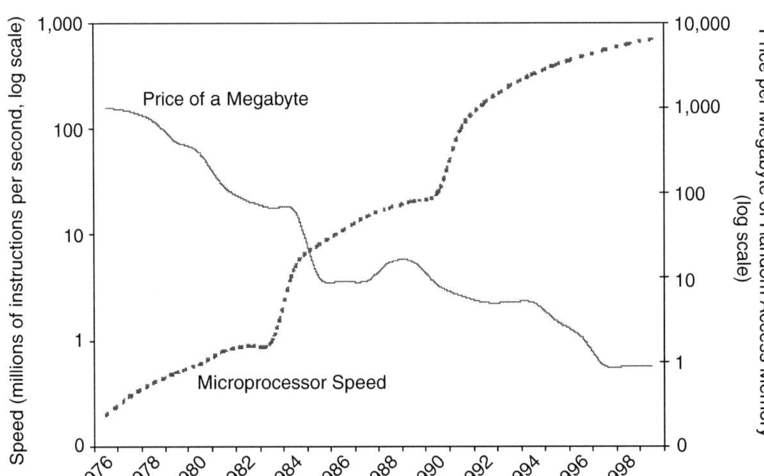

FIGURE 2-5 Megabyte prices and microprocessor speeds, 1976-2000. Moore's law maintained: megabyte prices decrease as microprocessor speeds increase.
SOURCE: S. Moore, J. L. Simon, and the CATO Institute. "The Greatest Century That Ever Was: 25 Miraculous Trends of the Past 100 Years." *Policy Analysis* No. 364, December 15, 1999. Pp. 1-32.

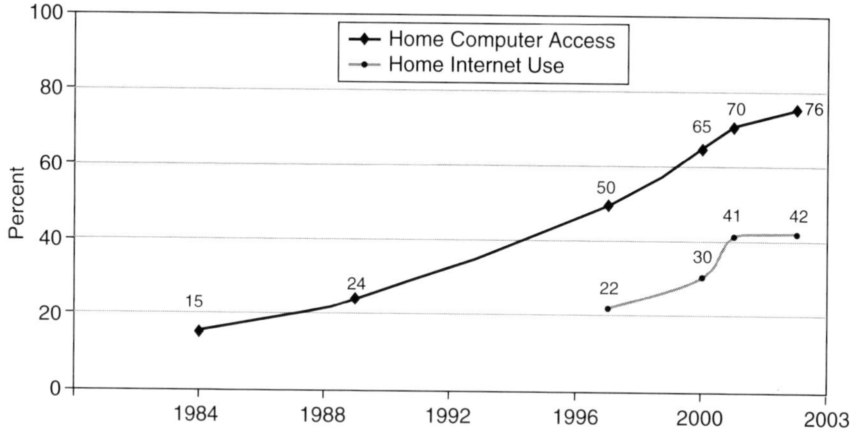

FIGURE 2-6 Percentage of children ages 3 to 17 who have access to a home computer and who use the Internet at home, selected years, 1984-2001. Many US children have access to and use computers and the Internet.
SOURCE: Child Trends Data Bank. Available at: http://www.childtrendsdatabank.org/figures/78-Figure-2.gif.

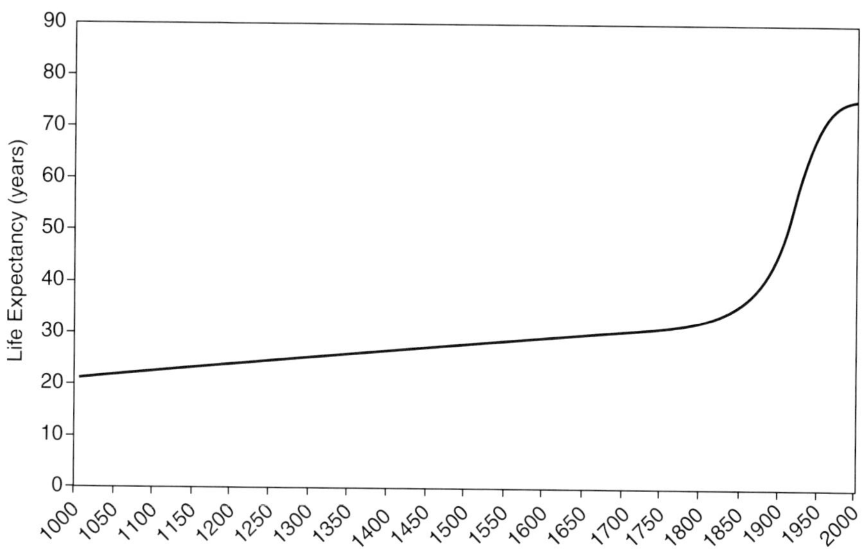

FIGURE 2-7A Life expectancy at birth, 1000-2000. Life expectancy has increased, particularly in the last century.
SOURCE: S. Moore, J. L. Simon, and the CATO Institute. "The Greatest Century That Ever Was: 25 Miraculous Trends of the Past 100 Years." *Policy Analysis* No. 364, December 15, 1999. Pp. 1-32.

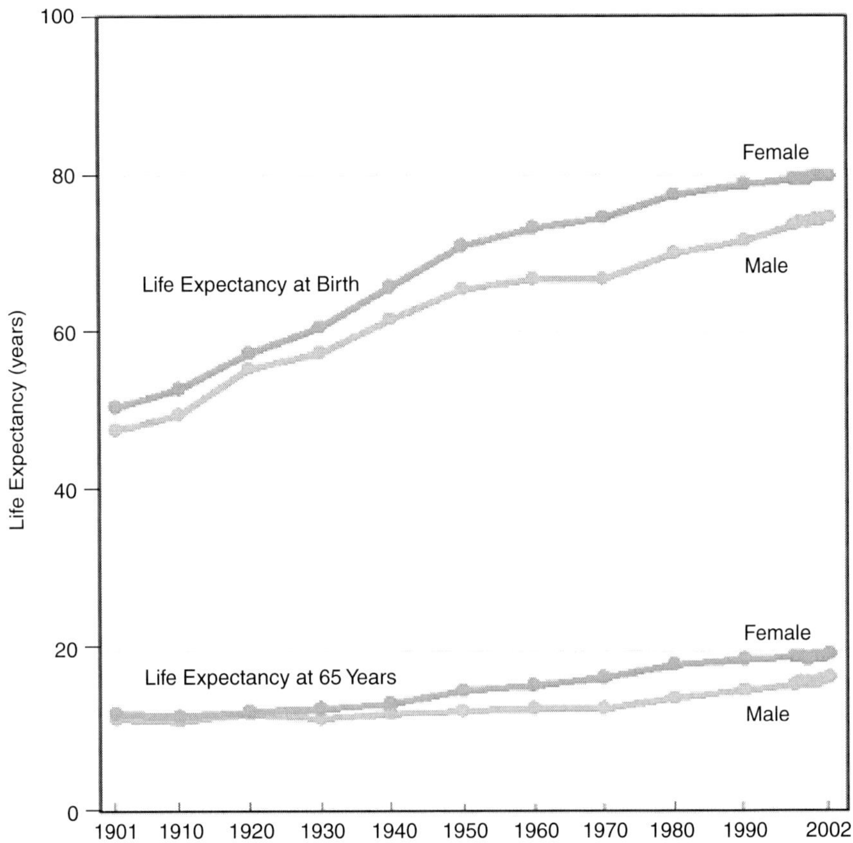

FIGURE 2-7B Life expectancy at birth and at 65 years of age, by sex, in the United States, 1901-2002. Life expectancy has increased in the United States, particularly in the last century.
SOURCE: Center for Disease Control and Prevention, National Center for Health Statistics, National Vital Statistic System.

of nutritious foods and advances in diagnosis, pharmaceuticals, medical devices, and treatment methods.[26]

Medical research also has brought economic benefit. The development of lithium as a mental-health treatment, for example, saves $9 billion in health costs each year. Hip-fracture prevention in postmenopausal women at risk for osteoporosis saves $333 million annually. Treatment for

[26]National Academy of Engineering. *A Century of Innovation*. Washington, DC: The National Academies Press, 2003.

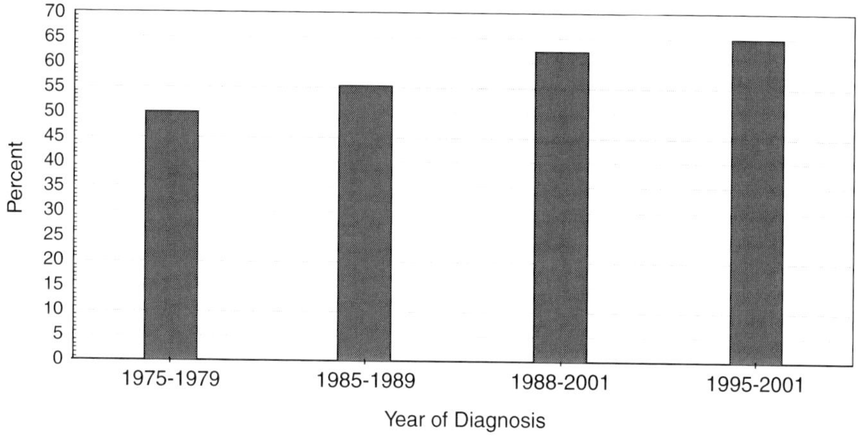

FIGURE 2-8A Five-year relative cancer survival rates for all ages, 1975-1979, 1985-1989, 1988-2001, and 1995-2001.
SOURCE: Surveillance, Epidemiology, and End Results (SEER) Program (www.seer.cancer.gov) SEER*Stat Database: Incidence—SEER 9 Regs Public-Use, November 2004 Sub (1973-2002), National Cancer Institute, DCCPS, Surveillance Research Program, Cancer Statistics Branch, released April 2005, based on the November 2004 submission.

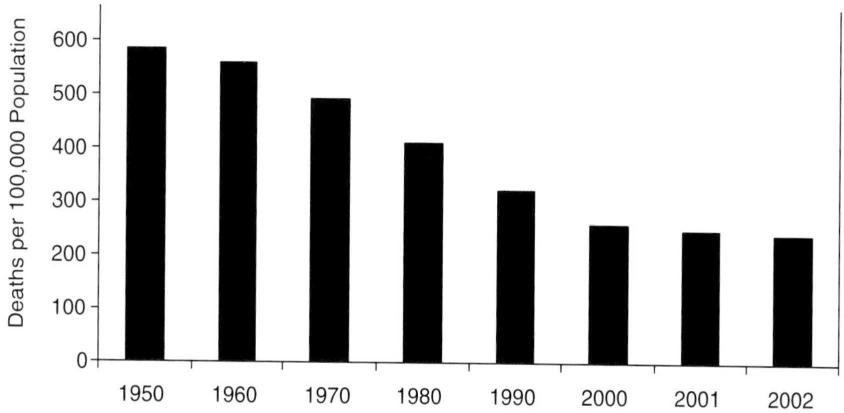

FIGURE 2-8B Heart disease mortality, 1950-2002.
SOURCE: National Center for Health Statistics. *Health, United States, 2005*. Table 29. Available at: http://www.cdc.gov/nchs/data/hus/hus05.pdf.

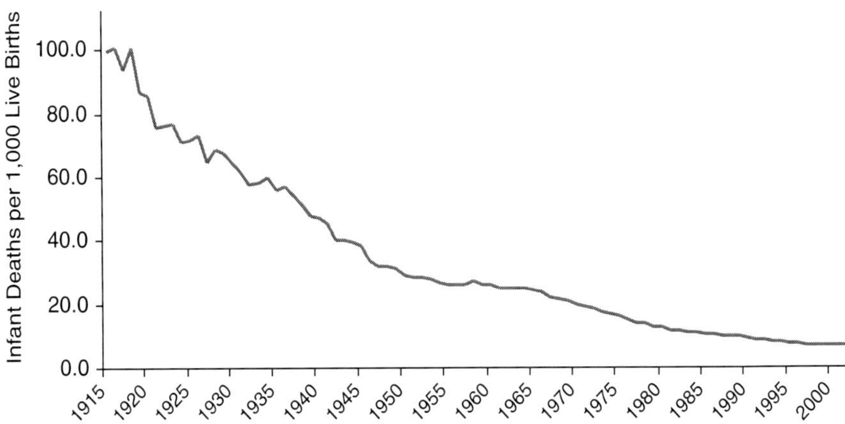

FIGURE 2-9A Infant mortality, 1915-2000.
SOURCE: National Center for Health Statistics. *National Vital Statistics Reports* (53)5:Table 11. Available at: http://www.cdc.gov/nchs/products/pubs/pubd/nvsr/53/53-21.htm.

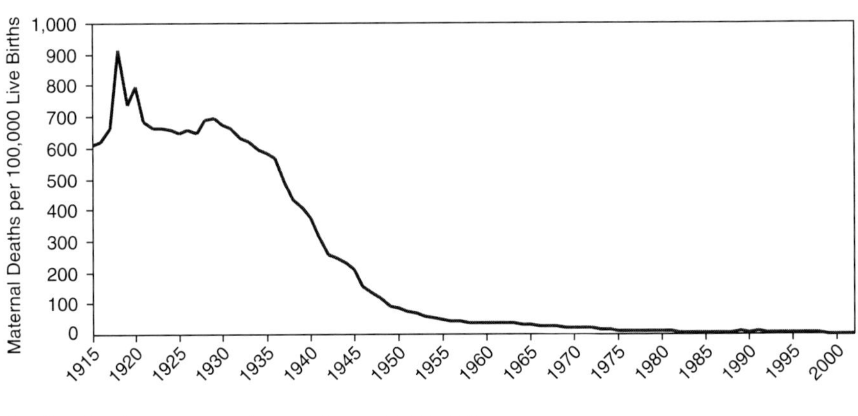

FIGURE 2-9B Maternal mortality, 1915-2000.
SOURCE: National Center for Health Statistics: *National Vital Statistics Reports* (53)5:Table 11. Available at: http://www.cdc.gov/nchs/products/pubs/pubd/nvsr/53/53-21.htm.

testicular cancer has resulted in a 91% remission rate and annual savings of $166 million.[27]

CARING FOR THE ENVIRONMENT

Advances in our understanding of the environment have led to better systems to promote human health and the health of our planet. Weather satellites, global positioning systems, and airborne-particle measurement technologies also have helped us to monitor and mitigate unexpected environmental problems. Unfortunately, some of these problems have been the consequence of unexpected side-effects of technological advances. Fortunately, in many cases additional technological understanding was able to overcome unintended consequences without forfeiting the underlying benefits.

Water Quality

Early in the 20th century, when indoor plumbing was rare, wastewater often was dumped directly into streets and rivers. Waterborne diseases—cholera, typhoid fever, dysentery, and diarrhea—were rampant and among the leading causes of death in the United States. Research and engineering for modern sewage treatment and consequent improvements in water quality have dramatically affected public and environmental health. Water-pollution controls have mitigated declines in wildlife populations, and research into wetlands and riparian habitats has informed the process of engineering water supplies for our population.

Automobiles and Gasoline

In the 1920s, engineers discovered that adding lead to gasoline caused it to burn more smoothly and improved the efficiency of engines. However, they did not predict the explosive growth of the automobile industry. The widespread use of leaded gasoline resulted in harmful concentrations of lead in the air,[28] and by the 1970s the danger was apparent. New formulations developed by petrochemical researchers not requiring the use of lead

---

[27]W. D. Nordhaus. *The Health of Nations: The Contribution of Improved Health and Living Standards.* New York: Albert and Mary Lasker Foundation, 1999. Available at: http://www.laskerfoundation.org/reports/pdf/economic.pdf; L. E. Rosenberg. "Exceptional Returns: The Economic Value of America's Investment in Medical Research." *Research Enterprise* 177(2000):368-371.

[28]US Congress House of Representatives Committee on Science, 1998, p. 38.

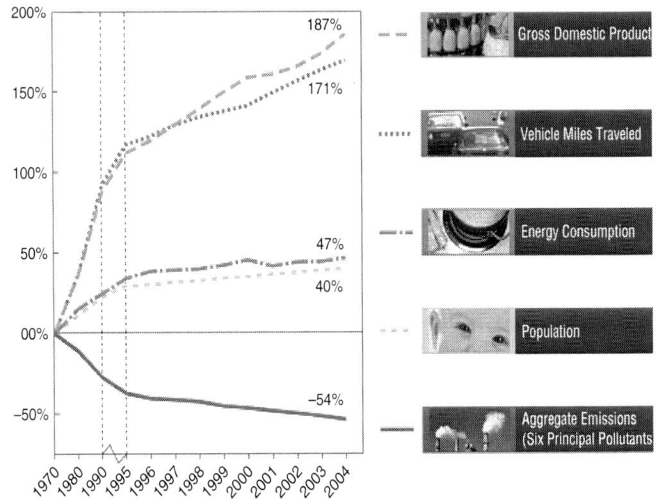

FIGURE 2-10 Comparison of growth areas and air pollution emissions, 1970-2004. US air quality has improved despite increases in gross domestic product, vehicle miles traveled, and energy consumption since the 1970s.
SOURCE: US Environmental Protection Agency. *Air Emissions Trends—Continued Progress Through 2004*. Available at: http://www.epa.gov/airtrends/2005/econ-emissions.html.

have resulted in vastly reduced emissions and improved air quality (Figure 2-10). Parallel advances in petroleum refining and the adoption and improvement of catalytic converters increased engine efficiency and removed harmful byproducts from the combustion process. Those achievements have reduced overall automobile emissions by 31%, and carbon monoxide emissions per automobile are 85% lower than in the 1970s.[29]

Refrigeration

In the early 1920s, scientists began working on nontoxic, nonflammable replacements for ammonia and other toxic refrigerants then in use. In 1928, Frigidaire synthesized the world's first chlorofluorocarbon (CFC), trademarked as Freon. By the 1970s, however, it had become clear that CFCs contribute to losses in the atmosphere's protective layer of ozone. In

[29]National Energy Policy Development Group. *National Energy Policy*. Washington, DC: US Government Printing Office, May 2001.

1974, scientists identified a chain reaction that begins with CFCs and sunlight and ends with the production of chlorine atoms. A single chlorine atom can destroy as many as 100,000 ozone molecules. The consequences could be long-lasting and severe, including increased cancer rates and global warming.[30]

In 1987, the Montreal Protocol began a global phase-out of CFC production. That in turn provided the market force that fueled the development of new, non-CFC refrigerants. Although the results of CFC use provide an example of the unintended negative consequences of technology, the response demonstrates the influence of science in diagnosing problems and providing effective solutions.

Agricultural Mechanization

Advances in agriculture have vastly increased farm productivity and food production. The food supply for the world's population of more than 6 billion people comes from a land area that is 80% of what was used to feed 2.5 billion people in 1950. However, injudicious application of mechanization also led to increased soil erosion. Since 1950, 20% of the world's topsoil has been lost—much of it in developing countries. Urban sprawl, desertification, and over-fertilization have reduced the amount of arable land by 20%.[31] Such improvements as conservation tillage, which includes the use of sweep plows to undercut wheat stalks but leave roots in place, have greatly reduced soil erosion caused by traditional plowing and have promoted the conservation of soil moisture and nutrients. Advances in agricultural biotechnology have further reduced soil erosion and water contamination because they have reduced the need for tilling and for use of pesticides.

IMPROVING THE STANDARD OF LIVING

Improvements attributable to declining mortality and better environmental monitoring are compounded by gains made possible by other advances in technology. The result has been a general enhancement in the quality of life in the United States as viewed by most observers.

---

[30]National Academy of Sciences. *Ozone Depletion, Beyond Discovery Series*. Washington, DC: National Academy Press, April 1996.

[31]P. Raven. "Biodiversity and Our Common Future." *Bulletin of the American Academy of Arts & Sciences* 58(2005):20-24.

Electrification and Household Appliances

Advances in technology in the 20th century resulted in changes at home and in the workplace. In 1900, less than 10% of the nation was electrified; now virtually every home in the United States is wired (Figure 2-11).[32] Most of us give little thought to the vast array of electrical appliances that surround us.

Transportation

As workers left farms to move to cities, transportation systems developed to get them to work and home again. Advances in highway construction in turn fueled the automotive industry. In 1900, one-fourth of US households had a horse, and many in urban areas relied on trolleys and trams to get to work and market. Today, more than 90% of US households own at least one car (Figure 2-12). Improvements in refrigeration put a refrigerator in virtually every home, and the ability to ship food across the country made it possible to keep those refrigerators stocked. The increasing speed, safety, and reliability of aircraft spawned yet another global industry that spans commercial airline service and overnight package delivery.

Communication

At the beginning of the 20th century slightly more than 1 million telephones were in use in the United States. The dramatic increase in telephone calls per capita over the following decades was made possible by advances in cable bundling, fiber optics, touch-tone dialing, and cordless communication (Figure 2-13). Cellular-telephone technology and voice-over-Internet protocols have added even more communication options. At the beginning of the 21st century, there were more than 300 million telephone communication devices and cellular telephone lines in the United States.

Radio and television revolutionized the mass media, but the Internet has provided altogether new ways of communicating. Interoperability between systems makes it possible to use one device to communicate by telephone, over the Internet, in pictures, in voice, and in text. The "persistent presence" that those devices make possible and the eventual widespread availability of wireless and broadband services will spawn another revolution in communication. At the same time, new R&D will be needed to

[32]US Department of Labor. *Report on the American Workforce, 2001*. Washington, DC: US Department of Labor, 2001. Available at: http://www.bls.gov/opub/rtaw/pdf/rtaw2001.pdf.

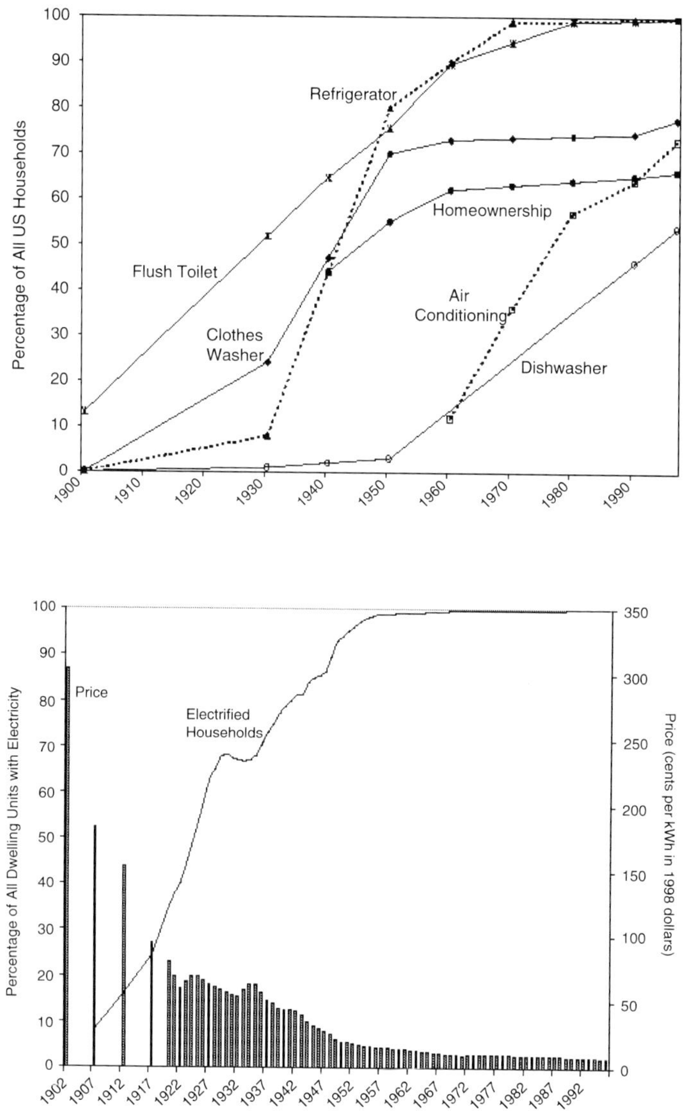

FIGURE 2-11 Improvement in US housing and electrification of US homes during the 20th century. The number of US homes with electricity, plumbing, refrigeration, and basic appliances soared in the middle of the 20th century.
SOURCE: S. Moore, J. L. Simon, and the CATO Institute. "The Greatest Century That Ever Was: 25 Miraculous Trends of the Past 100 Years." *Policy Analysis* No. 364, December 15, 1999. Pp. 1-32.

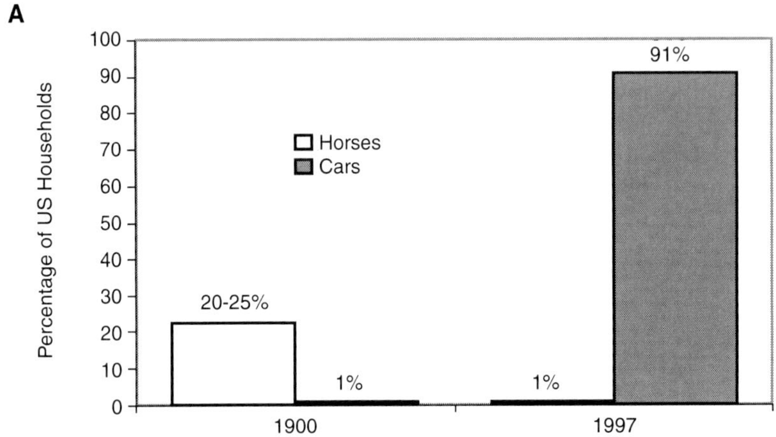

FIGURE 2-12A Ground transportation: horses to horsepower, 1900 and 1997.
SOURCE: S. Moore, J. L. Simon, and the CATO Institute. "The Greatest Century That Ever Was: 25 Miraculous Trends of the Past 100 Years." *Policy Analysis* No. 364, December 15, 1999. Pp. 1-32.

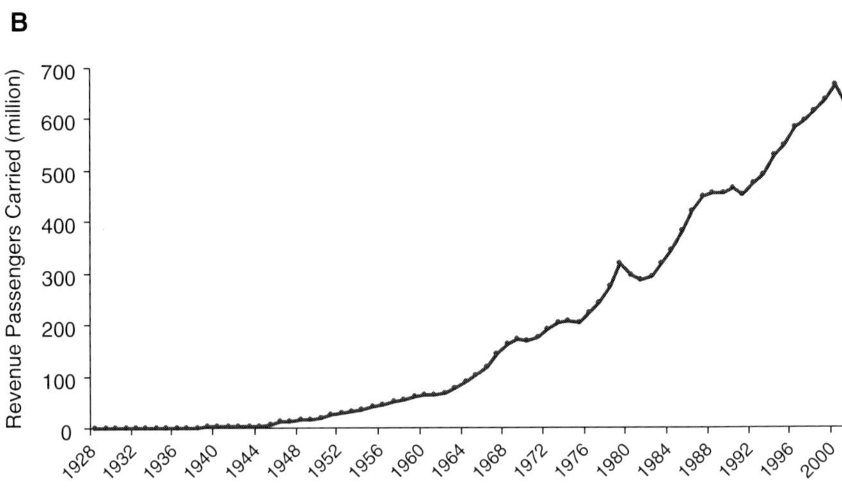

FIGURE 2-12B Air travel, United States, 1928-2002.
SOURCE: US Census Bureau. "Statistical Abstract of the United States." Available at: http://www.census.gov/statab/hist/HS-41.pdf.

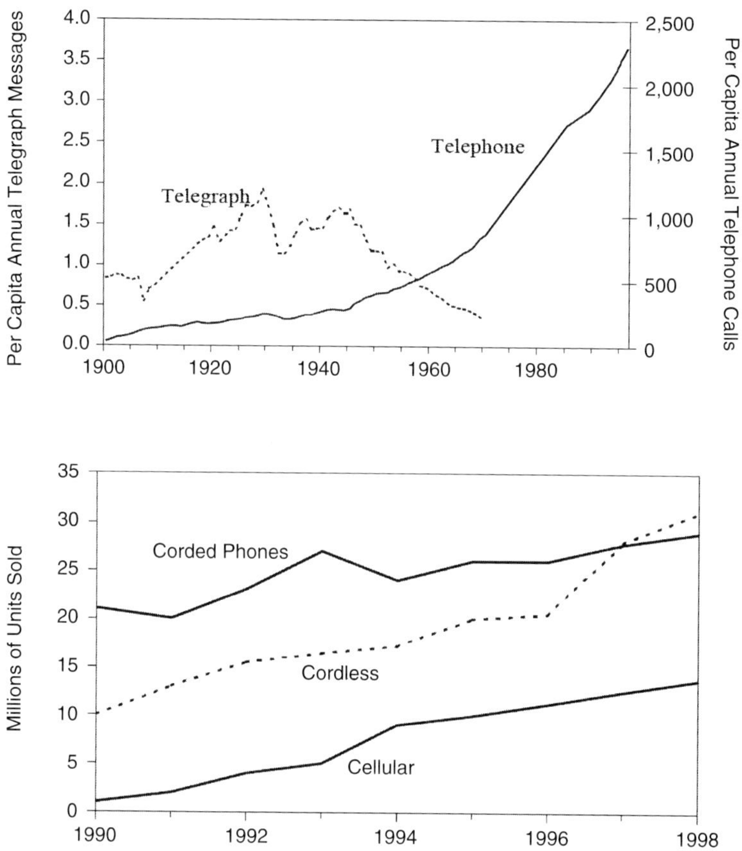

FIGURE 2-13 Modern communication, 1900-1998. More telephones than ever are used to make more calls per capita, thanks to enormous technological advances in a host of disciplines.
SOURCE: S. Moore, J. L. Simon, and the CATO Institute. "The Greatest Century That Ever Was: 25 Miraculous Trends of the Past 100 Years." *Policy Analysis* No. 364, December 15, 1999. Pp. 1-32.

reduce the energy demands of the new devices and their sensor-net support infrastructures.

Disaster Mitigation

Structural design, electrification, transportation, and communication come together in coordinating responses to natural disasters. Earthquake engineering and related technologies now make possible quake-resistant skyscrapers in high-risk zones. The 1989 Loma Prieta earthquake in central California caused 60 deaths and more than $6 billion in property damage, but occupants of the 49-story Transamerica Pyramid building in San Francisco were unharmed, as was the building itself, even though its top swayed from side to side by more than 1 foot for more than a minute.[33] In December 1988, an earthquake in Georgia in the former USSR of the same magnitude as Loma Prieta led to the deaths of 22,000 people—illustrating the impact of the better engineered building protection available in California.

A US Geological Survey radio system increases safety for cleanup crews during aftershocks. After Loma Prieta, workers in Oakland were given almost a half hour notice of aftershocks 50 miles away, thanks to the speed differential between radio and seismic waves.[34]

Weather prediction, enabled by satellites and advances in imaging technology, has helped mitigate losses from hurricanes. Early-warning systems for tornadoes and tsunamis offer another avenue for reducing the effects of natural disasters—but only when coupled with effective on-the-ground dissemination. As is the case for many technologies, this last step of getting a product implemented, especially in underserved areas or developing countries, can be the most difficult. Furthermore, as hurricane Katrina in New Orleans demonstrated, early warning is not enough—sound structural design and a coordinated human response are also essential.

Energy Conservation

The last century saw demonstrations of the influence of technology in every facet of our lives. It also revealed the urgent need to use resources wisely. Resource reduction and recycling are expanding across the United States. Many communities, spurred by advances in recycling technologies, have instituted trash-reduction programs. Industries are producing increasingly energy-efficient products, from refrigerators to automobiles. Today's cars use about 60% of the gasoline per mile driven that was used in 1972. With the advent of hybrid automobiles, further gains are now being realized. Similarly, refrigerators today require one-third of the electricity that they needed 30 years ago. In the 1990s, manufacturing output in the United States expanded by 41%, but industrial consumption of

---

[33]US Geological Survey. *Building Safer Structures. Fact Sheet 167-95.* Reston, VA: USGS, June 1998. Available at: http://quake.wr.usgs.gov/prepare/factsheets/SaferStructures/SaferStructures.pdf.

[34]US Geological Survey. *Speeding Earthquake Disaster Relief. Fact Sheet 097-95.* Reston, VA: USGS, June 1998. Available at: http://quake.wr.usgs.gov/prepare/factsheets/Mitigation/Mitigation.pdf.

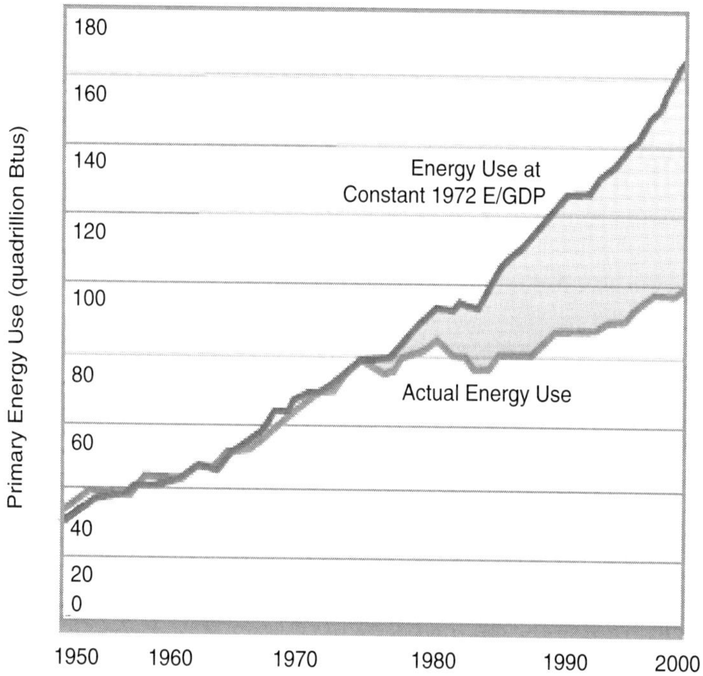

FIGURE 2-14 US primary energy use, 1950-2000. The efficiency of energy use has improved substantially over the last 3 decades.
SOURCE: National Energy Policy Development Group. *National Energy Policy*. Washington, DC: US Government Printing Office, May 2001.

electricity grew by only 11%. The introduction and use of energy-efficient products have enabled the US economy to grow by 126% since 1973 while energy use has increased by only 30% (Figure 2-14).[35] Those improvements in efficiency are the result of work in a broad spectrum of science and engineering fields.

UNDERSTANDING HOW PEOPLE LEARN

Today, an extraordinary scientific effort is being devoted to the mind and the brain, the processes of thinking and learning, the neural processes that occur during thought and learning, and the development of competence. The

[35]National Energy Policy Development Group. *National Energy Policy*. Washington, DC: US Government Printing Office, May 2001.

revolution in the study of the mind that has occurred in recent decades has important implications for education.[36] A new theory of learning now coming into focus will lead to very different approaches to the design of curriculum, teaching, and assessment from those generally found in schools today.

Research in the social sciences has increased understanding of the nature of competent performance and the principles of knowledge organization that underlie people's abilities to solve problems in a wide variety of fields, including mathematics, science, literature, social studies, and history. It has also uncovered important principles for structuring learning experiences that enable people to use what they have learned in new settings. Collaborative studies of the design and evaluation of learning environments being conducted by cognitive and developmental psychologists and educators are yielding new knowledge about the nature of learning and teaching in a variety of settings.

SECURING THE HOMELAND

Scientific and engineering research demonstrated its essential role in the nation's defense during World War II. Research led to the rapid development and deployment of the atomic bomb, radar and sonar detectors, nylon that revolutionized parachute use, and penicillin that saved battlefield lives. Throughout the Cold War the United States relied on a technological edge to offset the larger forces of its adversaries and thus generously supported basic research. The US military continues to depend on new and emerging technologies to respond to the diffuse and uncertain threats that characterize the 21st century and to provide the men and women in uniform with the best possible equipment and support.[37]

Just as Vannevar Bush described a tight linkage between research and security,[38] the Hart–Rudman Commission a half-century later argued that security can be achieved only by funding more basic research in a variety of fields.[39] In the wake of the 9/11 attacks and the anthrax mailings, it is clear that innovation capacity and homeland security are also tightly coupled.

---

[36]National Research Council. *How People Learn: Brain, Mind, Experience, and School: Expanded Edition*. Washington, DC: National Academy Press, 2000.

[37]Joint Chiefs of Staff. *Joint Vision 2020*. Washington, DC: Department of Defense, 2000; Department of Defense. *Quadrennial Defense Review Report*. Washington, DC: Department of Defense, 2001.

[38]V. Bush. *Science: The Endless Frontier*. Washington, DC: US Government Printing Office, 1945.

[39]US Commission on National Security. *Road Map for National Security: Imperative for Change*. Washington, DC: US Commission on National Security, 2001.

There can be no security without the economic vitality created by innovation, just as there can be no economic vitality without a secure environment in which to live and work.[40] Investment in R&D for homeland security has grown rapidly; however, most of it has been in the form of development of new technologies to meet immediate needs.

Human capacity is as important as research funding. As part of its comprehensive overview of how science and technology could contribute to countering terrorism, for example, the National Research Council recommended a human-resources development program similar to the post-Sputnik National Defense Education Act (NDEA) of 1958.[41] A Department of Defense proposal to create and fund a new NDEA is currently being examined in Congress.[42]

CONCLUSION

The science and technology research community and the industries that rely on that research are critical to the quality of life in the United States. Only by continuing investment in advancing technology—through the education of our children, the development of the science and engineering workforce, and the provision of an environment conducive to the transformation of research results into practical applications—can the full innovative capacity of the United States be harnessed and the full promise of a high quality of life realized.

---

[40]Council on Competitiveness. *Innovate America*. Washington, DC: Council on Competitiveness, 2004. P. 19.

[41]National Research Council. *Making the Nation Safer: The Role of Science and Technology in Countering Terrorism*. Washington, DC: The National Academies Press, 2002.

[42]See H.R. 1815, National Defense Authorization Act for Fiscal Year 2006, Sec. 1105. Science, Mathematics, and Research for Transformation (SMART) Defense Education Program—National Defense Education Act (NDEA), Phase I. Introduced to the House of Representatives on April 26, 2005; referred to Senate committee on June 6, 2005; status as of July 26, 2005: received in the Senate and read twice and referred to the Committee on Armed Services.

3

How Is America Doing Now in Science and Technology?

By most available criteria, the United States is still the undisputed leader in the performance of basic and applied research (see Box 3-1). In addition, many international comparisons put the United States as a leader in applying research and innovation to improve economic performance. In the latest IMD International *World Competitiveness Yearbook,* the United States ranks first in economic competitiveness, followed by Hong Kong and Singapore.[1] The survey compares economic performance, government efficiency, business efficiency, and infrastructure. Larger economies are further behind, with Zhejiang (China's wealthiest province), Japan, the United Kingdom, and Germany ranked 20 though 23, respectively.[2] An extensive review by the Organisation for Economic Co-operation and Development (OECD) concludes that since World War II, US leadership in science and engineering has driven its dominant strategic position, economic advantages, and quality of life.[3]

---

[1]IMD International. *World Competitiveness Yearbook. 2005.* Lausanne, Switzerland: IMD International, 2005. The United States leads the world (with a score of 100), followed in order by Hong Kong (93), Singapore, Iceland, Canada, Finland, Denmark, Switzerland, Australia, and Luxembourg (80).

[2]Mainland China ranks 31st.

[3]Organization for Economic Co-operation and Development. "Science, Technology and Industry Scoreboard, 2003, R&D Database." Available at: http://www1.oecd.org/publications/e-book/92-2003-04-1-7294/. The scoreboard uses four indicators in its ranking: the creation and diffusion of knowledge; the information economy; the global integration of economic activity; and productivity and economic structure. In the United States, investment in knowledge—the sum of investment in research and development (R&D), software, and higher education—amounted to almost 7% of GDP in 2000, well above the share for the European Union or Japan.

BOX 3-1
Pasteur's Quadrant

The writers of this report, like many others, faced a semantic question in the discussions of different kinds of research. *Basic research,* presumably pursued for the sake of fundamental understanding but without thought of use, generally is distinguished from *applied research,* which is pursued to convert basic understanding into practical use. This view, called the "linear model" is shown here:

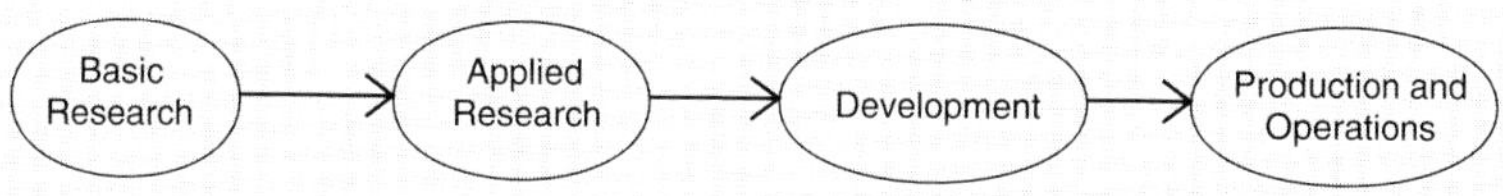

But that classification quickly breaks down in the real world because "basic" discoveries often emerge from "applied" or even "developmental" activities. In his 1997 book, *Pasteur's Quadrant,*[a] Donald Stokes responded to that complexity with a more nuanced classification that describes research according to intention. He distinguishes four types:

- Pure basic research, performed with the goal of fundamental understanding (such as Bohr's work on atomic structure).
- Use-inspired basic research, to pursue fundamental understanding but motivated by a question of use (such as Pasteur's work on the biologic bases of fermentation and disease).
- Pure applied research, motivated by use but not seeking fundamental understanding (such as that leading to Edison's inventions).
- Applied research that is not motivated by a practical goal (such as plant taxonomy).

In Stokes's argument, research is better depicted as a box than as a line:

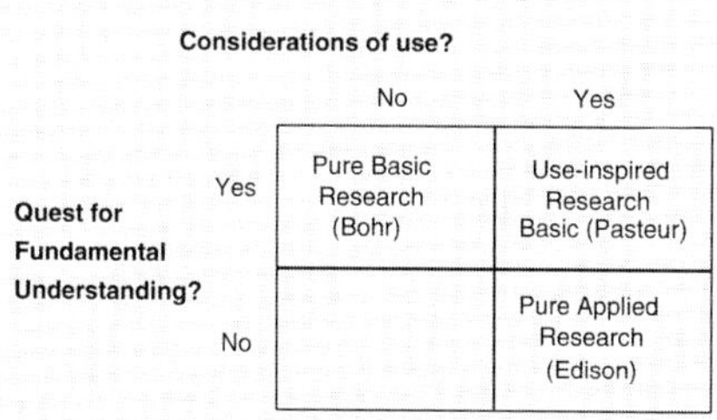

In contrast to the basic–applied dichotomy, Stokes's taxonomy explicitly recognizes research that is simultaneously inspired by a use but that also seeks fundamental knowledge, which he calls "Pasteur's Quadrant."

[a]D. Stokes. *Pasteur's Quadrant.* Washington, DC: Brookings Institution Press, 1997.

Researchers in the United States lead the world in the volume of articles published and in the frequency with which those papers are cited by others.[4] US-based authors were listed on one-third of all scientific articles worldwide in 2001.[5] Those publication data are significant because they reflect original research productivity and because the professional reputations, job prospects, and career advancement of researchers depend on their ability to publish significant findings in the open peer-reviewed literature.

The United States also excels in higher education and training. A recent comparison concluded that 38 of the world's 50 leading research institutions—those that draw the greatest interest of science and technology students—are in the United States.[6] Since World War II, the United States has been the destination of choice for science and engineering graduate students and for postdoctoral scholars choosing to study abroad. Our nation—about 6% percent of the world's population—has for decades produced more than 20% of the world's doctorates in science and engineering.[7]

Because of globalization in the fields of science and engineering, however, it is difficult to compare research leadership among countries. Research teams commonly include members from several nations, and industries have dispersed many activities, including research, across the globe.

SCIENCE AND ENGINEERING ADVANTAGE

The strength of science and engineering in the United States rests on many advantages: the diversity, quality, and stability of its research and teaching institutions; the strong tradition of public and private investment in research and advanced education; the quality of academic personnel; the prevalence of English as the language of science and engineering; the availability of venture capital; a relatively open society in which talented people of any background or nationality have opportunities to succeed; the US custom, unmatched in other countries, of providing positions for postdoctoral scholars;[8] and the strength of the US peer-review and free-

[4] D. A. King. "The Scientific Impact of Nations." *Nature* 430(6997)(July 15, 2004):311-316.

[5] National Science Board. *Science and Engineering Indicators 2004*. NSB 04-01. Arlington, VA: National Science Foundation, 2004. Chapter 5.

[6] Shanghai's Jiao Tong University Institute of Higher Education. "Academic Ranking of World Universities." 2004. Available at: http://ed.sjtu.edu.cn/rank/2004/2004Main.htm. The ranking emphasizes prizes, publications, and citations attributed to faculty and staff, as well as the size of institutions. *The Times* Higher Education Supplement citation has provided similar results in comparing universities worldwide.

[7] National Science Board. *Science and Engineering Indicators 2004*. NSB 04-01. Arlington, VA: National Science Foundation, 2004. P. 2-36.

[8] The National Academies. *Policy Implications of International Graduate Students and Postdoctoral Scholars*. Washington, DC: The National Academies Press, 2005. P. 81.

enterprise systems in weeding out noncompetitive academic and business pursuits.

In addition to such tangible advantages, US leadership might also be attributed to many favorable public policy priorities: research activities funded by public and private sources that have led to new industries, products, and jobs; an economic climate that encourages investment in technology-based companies; an outward-looking international economic policy; and support for lifelong learning.[9]

However, things are changing, as noted in *Innovate America,* a 2004 report from the Council on Competitiveness:[10]

- Innovation is diffusing at an ever-increasing rate. It took 55 years for automobile use to spread to a quarter of the US population, 35 years for the telephone, 22 years for the radio, 16 years for the personal computer, 13 years for the cell phone, and just 7 years for the World Wide Web once the Internet had matured (through technology and policy developments) to the point of takeoff.
- Innovation is increasingly multidisciplinary and technologically complex, arising from the intersection of different fields and spheres of activity.
- Innovation is collaborative. It requires active cooperation and communication among scientists and engineers and between creators and users.
- Innovation is creative. Workers and consumers demand ever more new ideas, technologies, and content.
- Innovation is global. Advances come from centers of excellence around the world and are prompted by the demands of billions of customers.

Central to the strength of US innovation is our tradition of public funding for science and engineering research. Graduate education in the United States is supported mainly by federal grants from the National Science Foundation (NSF) and the National Institutes of Health (NIH) to faculty researchers, buttressed by a smaller volume of federally funded fellowships. One study reported that 73% of applicants for US patents said that publicly funded research formed part or all of the foundation for their innovations.[11] Much of the nation's research in engineering and the physical sciences is performed in federal laboratories, part of whose mission is to assist the commercialization of new technology.

---

[9]K. H. Hughes. "Facing the Global Competitiveness Challenge." *Issues in Science and Technology* 21(4)(Summer 2005):72-78.

[10]Council on Competitiveness. *Innovate America.* Washington, DC: Council on Competitiveness, 2004. P. 6.

[11]M. I. Nadiri. *Innovations and Technical Spillovers.* Working Paper 4423. Cambridge, MA: National Bureau of Economic Research, 1993.

OTHER NATIONS ARE FOLLOWING OUR LEAD—AND CATCHING UP[12]

It is no surprise that as the value of research becomes more widely understood, other nations are strengthening their own programs and institutions. If imitation is flattery, we can take pride in watching as other nations eagerly adopt major components of the US innovation model.[13] Their strategies include the willingness to increase public support for research universities, to enhance protections for intellectual property rights, to promote venture capital activity, to fund incubation centers for new businesses, and to expand opportunities for innovative small companies.[14]

Many nations have made research a high priority. To position the European Union (EU) as the most competitive knowledge-based economy in the world and enhance its attractiveness to researchers worldwide, EU leaders are urging that, by 2010, member nations spend 3% of gross domestic product (GDP) on research and development (R&D).[15] In 2000, R&D as a percentage of GDP was 2.72 in the United States, 2.98 in Japan, 2.49 in Germany, 2.18 in France, and 1.85 in the United Kingdom.[16]

Many nations also are investing more aggressively in higher education and increasing their public investments in R&D (Figure 3-1). Those investments are stimulating growth in the number of research universities in those countries; the number of researchers; the number of papers listed in the *Science Citation Index;* the number of patents awarded; and the number of doctoral degrees granted (Table 3-1, Figures 3-2, 3-3, 3-4).[17]

China is emulating the US system as well. The Chinese Science Foundation is modeled after our National Science Foundation, and peer review methodology and startup packages for junior faculty are patterned on US practices. In China, national spending in the past few years for all R&D activities rose 500%, from $14 billion in 1991 to $65 billion in 2002. US

[12]For another point of view, see Box 3-2.

[13]Council on Competitiveness. *Innovate America.* Washington, DC: Council on Competitiveness, 2004. P. 6.

[14]K. H. Hughes. "Facing the Global Competitiveness Challenge." *Issues in Science and Technology* 21(4)(Summer 2005):72-78. See also M. Enserink. "France Hatches 67 California Wannabes." *Science* 309(2005):547.

[15]R. M. May. "Raising Europe's Game." *Nature* 430(2004):831; P. Busquin. "Investing in People." *Science* 303(2004):145.

[16]National Science Board. *Science and Engineering Indicators 2004.* NSB 04-01. Arlington, VA: National Science Foundation, 2004. Appendix Table 4-43.

[17]D. Hicks. 2004. "Asian Countries Strengthen Their Research." *Issues in Science and Technology* 20(4)(Summer 2004):75-78. The author notes that the number of doctoral degrees awarded in China has increased 50-fold since 1986.

BOX 3-2
Another Point of View: US Competitiveness

"Americans are having another Sputnik moment," writes Robert J. Samuelson, "one of those periodic alarms about some foreign technological and economic menace. It was the Soviets in the 1950s and early 1960s, the Germans and Japanese in the 1970s and 1980s, and now it's the Chinese and Indians."[a] Sputnik moments come when the nation worries about its scientific and technological superiority and its ability to compete globally. And, according to Samuelson, the nation tends to be overly concerned.

Sputnik led to the theory of a "missile gap that turned out to be a myth. The competitiveness crisis of the 1980s suggested that Japan would surge ahead of us because they were better savers, innovators, workers, and managers. But in 2004, per capita US income averaged $38,324 compared to $26,937 for Germany and $29,193 for Japan."

Similarly, Samuelson argues that our current fears are unfounded, another "illusion" in which "a few selective happenings" are transformed into a "full blown theory of economic inferiority or superiority." He argues that low wages and rising skills in China and India could cost us some jobs, but that US gains and losses in response to the rising economic power of those countries will tend to balance out.

Samuelson indicates that he believes "the apparent American deficit in scientists and engineers is also exaggerated." He notes that only about one-third of our science and engineering graduates work in science and engineering occupations and that if there were a shortage, salaries for those jobs would increase and scientists and engineers would return to them. Of greater importance, Samuelson concludes, is that the United States must continue to draw on the strengths that overcome its weaknesses: "ambitiousness; openness to change (even unpleasant change); competition; hard work; and a willingness to take and reward risk."

[a]R. J. Samuelson. *Sputnik Scare, Updated.* Washington Post, August 26, 2005. P. A27.

R&D spending increased 140%, from $177 billion to $245 billion, in the same period.[18]

The rapid rise of South Korea as a major science and engineering power has been fueled by the establishment of the Korea Science Founda-

[18]Organisation of Economic Co-operation and Development. *Science, Technology and Industry Outlook 2004*. Paris: OECD, 2004. P. 190. The United States spends significantly more than China on R&D in gross terms and in percentage of R&D. However, if China's US$65 billion in R&D spending were adjusted based on purchasing power parity, it would approach US$300 billion.

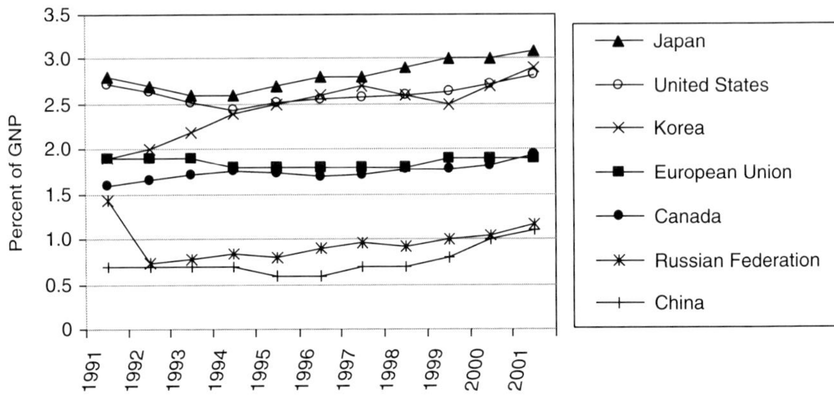

FIGURE 3-1 R&D expenditures as a percentage of GNP, 1991-2002. These expenditures are beginning to rise worldwide.
SOURCE: Organisation for Economic Co-operation and Development. *Main Science and Engineering Indicators*. Paris: OECD, 2005.

TABLE 3-1 Publications and Citations in the United States and European Union per Capita and per University Researcher, 1997-2001

	United States	European Union
Publications	1,265,608	1,347,985
Publications/population	4.64	3.60
Publications/researcher	6.80	4.30
Researchers/population	0.68	0.84
Citations	10,850,549	8,628,152
Citations/population	39.75	23.03
Citations/researcher	58.33	27.52
Top 1% publications	23,723	14,099
Top 1% publications/population	0.09	0.04
Top 1% publications/researcher	0.13	0.04

NOTES: Number of publications, citations, and top 1% publications refer to 1997-2001. Population (measured in thousands) and number of university researchers (measured in full-time equivalents) refer to 1999. Each cited paper is allocated once to every author. European Union totals are adjusted to account for duplications by removing papers with multiple EU national authorship to give an accurate net total.
SOURCE: G. Dosi, P. Llerena, and M. S. Labini. "Evaluating and Comparing the Innovation Performance of the United States and the European Union." Expert report prepared for the Trend Chart Policy Workshop. June 29, 2005. Available at: http://trendchart.cordis.lu/scoreboards/scoreboard2005/pdf/EIS%202005%20EU%20versus%20US.pdf.

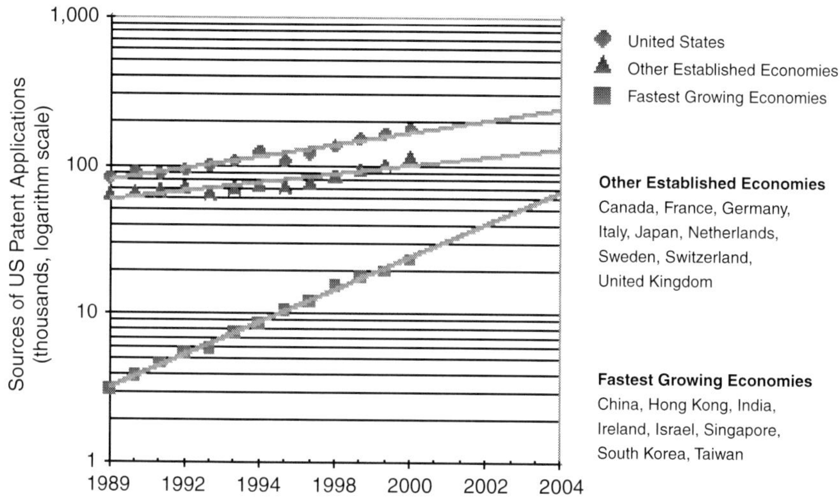

FIGURE 3-2 US patent applications, by country of applicant, 1989-2004.
SOURCE: Task Force on the Future of American Innovation based on data from National Science Foundation. *Science and Engineering Indicators 2004*. Arlington, VA: APS Office and Public Affairs, 2004.

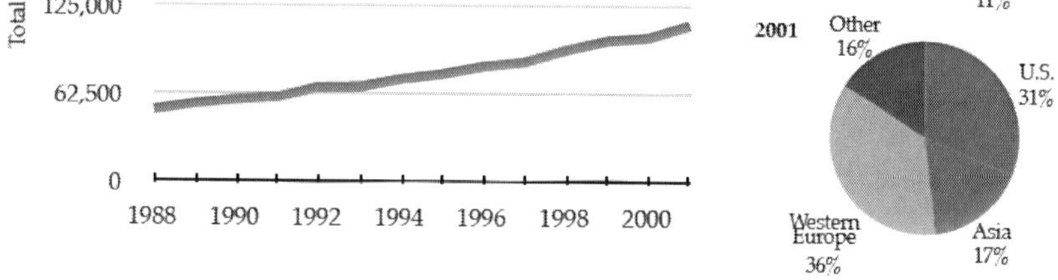

FIGURE 3-3 Total science and engineering articles with international coauthors, 1988-2001.
NOTE: Internationally coauthored articles were counted more than once so each country represented on the author list was included. So if an article was written by authors from the United States and Switzerland, it would be included in the count for both countries.
SOURCES: Task Force on the Future of American Innovation based on data from National Science Foundation. *Science and Engineering Indicators 2004*. Arlington, VA: APS Office and Public Affairs, 2004.

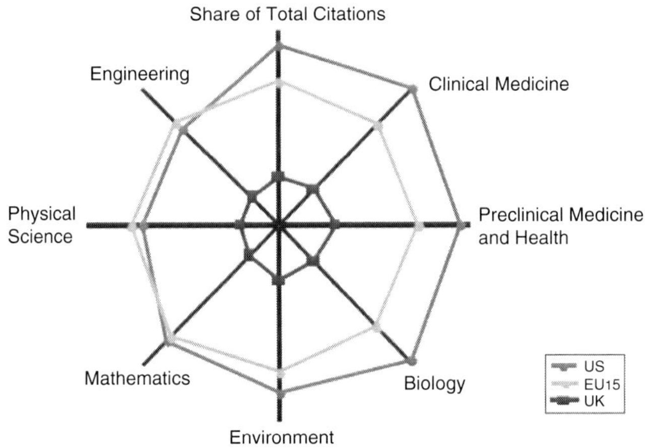

FIGURE 3-4 Disciplinary strengths in the United States, the 15 European Union nations in the comparator group (EU15), and the United Kingdom.
NOTE: The distance from the origin to the data point is proportional to citation share.
SOURCE: D. A. King. "The Scientific Impact of Nations." *Nature* 430(2004):311-316. Data are from citations in ISI Thompson.

tion—funded primarily by the national sports lottery—to enhance public understanding, knowledge, and acceptance of science and engineering throughout the nation.[19] Similarly, the government uses contests and prizes specifically to stimulate the scientific enterprise and public appreciation of scientific knowledge.

Other nations also are spending more on higher education and providing incentives for students to study science and engineering. To attract the best graduate students from around the world, universities in Japan, Switzerland, and elsewhere are offering science and engineering courses in English. In the 1990s, both China and Japan increased the number of students pursuing science and engineering degrees, and there was steady growth in South Korea.[20]

Some consequences of this new global science and engineering activity are already apparent—not only in manufacturing but also in services. India's software services exports rose from essentially zero in 1993 to about $10 billion in 2002.[21] In broader terms, the US share of global

[19]Korean Ministry of Science and Engineering (MOST). Available at: http://www.most.go.kr/most/english/link_2.jsp.

[20]National Science Board. *Science and Engineering Indicators 2004*. NSB 04-01. Arlington, VA: National Science Foundation, 2004. P. 2-35.

[21]S. S. Athreye. "The Indian Software Industry." Carnegie Mellon Software Industry Center Working Paper 03-04. Pittsburgh, PA: Carnegie Mellon University, October 2003.

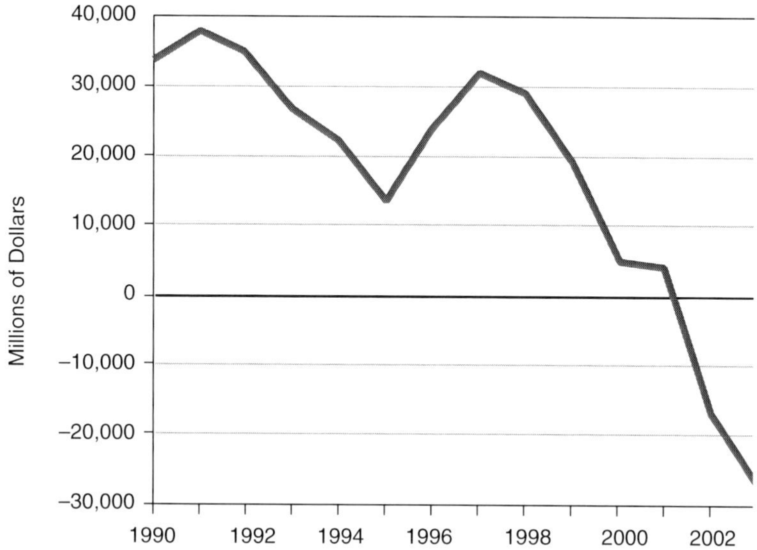

FIGURE 3-5 United States trade balance for high-technology products, in millions of dollars, 1990-2003.
SOURCE: Task Force on the Future of American Innovation based on data from US Census Bureau Foreign Trade Statistics, *U.S. International Trade in Goods and Services.* Compiled by the American Psychological Society Office of Public Affairs.

exports has fallen in the past 20 years from 30 to 17%, while the share for emerging countries in Asia grew from 7 to 27%.[22] The United States now has a negative trade balance even for high-technology products (Figure 3-5). That deficit raises concern about our competitive ability in important areas of technology.[23]

Although US scientists and engineers still lead the world in publishing results, new trends emerge from close examination of the data. From 1988 to 2001, world publishing in science and engineering increased by almost 40%,[24] but most of that increase came from Western Europe, Japan, and several emerging East Asian nations (South Korea, China, Singapore, and Taiwan). US publication in science and engineering has remained essen-

[22]For 2004, the dollar value of high-technology imports was $560 billion; the value of high-technology exports was $511 billion.

[23]D. R. Francis. "U.S. Runs a High-Tech Trade Gap." *Christian Science Monitor* 96(131) (June 2, 2004):1-1.

[24]National Science Board. *Science and Engineering Indicators 2004.* NSB 04-01. Arlington, VA: National Science Foundation, 2004. Chapter 5.

tially constant since 1992.[25] Since 1997, researchers in the 15 EU countries have published more papers than have their US counterparts, and the gap in citations between the United States and other countries has narrowed steadily.[26] The global increase in the production of scientific knowledge eventually benefits all countries. Yet trends in publication could be a troubling bellwether about our competitive position in the global science community.

INTERNATIONAL COMPETITION FOR TALENT

The graduate education of our scientists and engineers largely follows an apprenticeship model. Graduate students and postdoctoral scholars gain direct experience under the guidance of veteran researchers. The important link between graduate education and research that has been forged through a combination of research assistantships, fellowships, and traineeships has been tremendously beneficial to students and researchers and is a critical component of our success in the last half-century.

One measure of other nations' successful adaptation of the US model is doctoral production, which increased rapidly around the world but most notably in China and South Korea (Figure 3-6). In South Korea, doctorate production rose from 128 in 1975 to 2,865 in 2001. In China, doctorate production was essentially zero until 1985, but 15 years later, 7,304 doctorates were conferred. In 1975, the United States conferred 59% of the world's doctoral degrees in science and engineering; by 2001, our share had fallen to 41%. China's 2001 portion was 12%.[27]

Another challenge for US research institutions is to attract the overseas students on whose talents the nation depends. The US research enterprise, especially at the graduate and postdoctoral levels, has benefited from the work of foreign visitors and immigrants. They came first from Europe, fleeing fascism, and more recently they have come from China, India, and the former Soviet Union, seeking better education and more economic opportunity. International students account for nearly half the US doctorates awarded in engineering and computer science[28] (Figure 3-7). Similarly, more than 35% of US engineering and computer science university faculty are foreign-born.[29] According to US Census data from 2000,

---

[25]Ibid., Table 5-30.

[26]D. A. King. "The Scientific Impact of Nations." *Nature* 430(6997)(July 15, 2004):311-316.

[27]National Science Board. *Science and Engineering Indicators 2004*. NSB 04-01. Arlington, VA: National Science Foundation, 2004. Appendix Table 2-38.

[28]National Science Board. *Science and Engineering Indicators 2004*. NSB 04-01. Arlington, VA: National Science Foundation, 2004.

[29]Ibid.

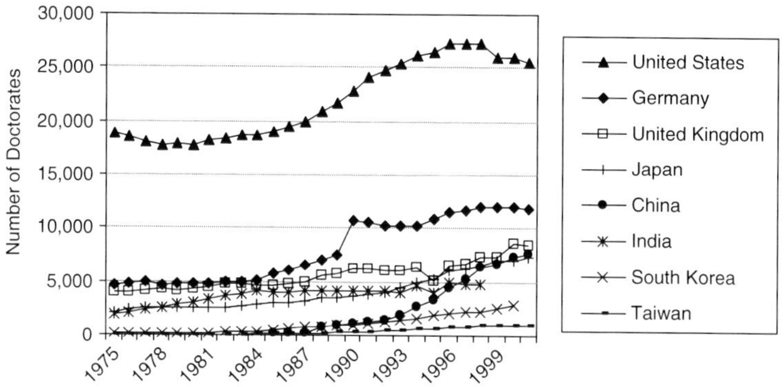

FIGURE 3-6 Science and engineering doctorate production for selected countries, 1975-2001. US doctorate production in science and engineering is decreasing; European Union and Asian production are rising but are still well below US levels. SOURCE: Based on National Science Board. *Science and Engineering Indicators 2004*. NSB 04-01. Arlington, VA: National Science Foundation, 2004. Appendix Tables 2-38 and 2-39.

the proportion of doctoral-level employees in the science and engineering research labor force is about equivalent to the percentage of doctorates produced by US universities.

Many nations are seeking to reap the benefits of advanced education, including strong positive effects on GDP growth. They are working harder to attract international students and to encourage the movement of skilled personnel into their countries.[30]

- China implemented an "opening-up" policy in 1978 and began to send large numbers of students and scholars abroad to gain the skills they need to bolster that country's economic and social development.
- India liberalized its economy in 1991 and started encouraging students to go abroad for advanced education and training. Since 2001, the Indian government has been providing money ($5 billion in fiscal year 2005) for "soft loans," which require no collateral, to students who wish to travel abroad for their education. In 2002, India surpassed China as the largest exporter of graduate students to the United States.[31]

---

[30]Conference Board of Canada. *The Economic Implications of International Education for Canada and Nine Comparator Countries: A Comparison of International Education Activities and Economic Performance*. Ottawa: Department of Foreign Affairs and International Trade, 1999.

[31]Institute for International Education. *Open Doors Report on International Educational Exchange*. New York: Institute for Internal Education, 2004.

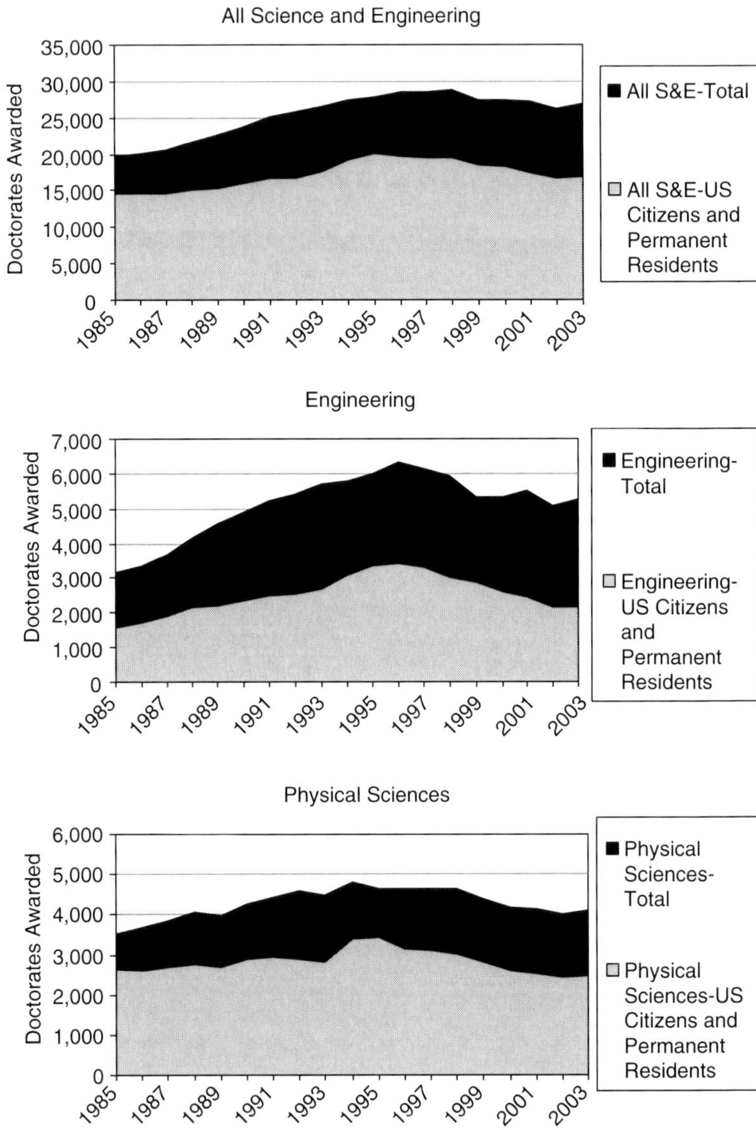

FIGURE 3-7 Doctorates awarded by US institutions, by field and citizenship status, 1985-2003. US citizens and permanent residents earn about 62% of the doctorates in all fields of science and engineering (S&E), about 60% in the physical sciences, and 41% of those awarded in engineering and the combined fields of mathematics and computer sciences (CS).
SOURCE: National Science Foundation. *Survey of Earned Graduates*. Arlington, VA: National Science Foundation, 2005.

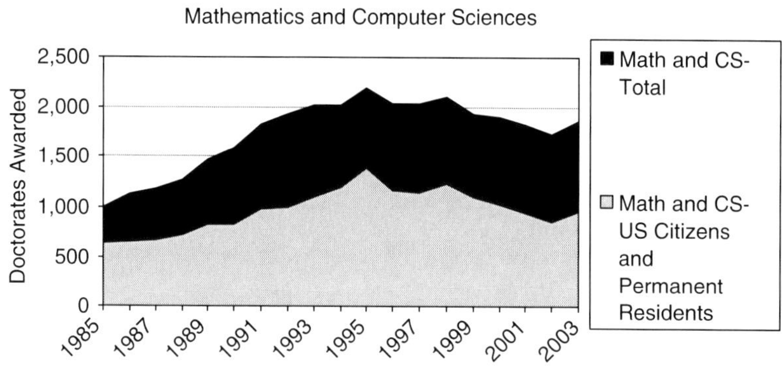
Mathematics and Computer Sciences
Doctorates Awarded
2,500
2,000
1,500
1,000
500
0
1985
1987
1989
1991
1993
1995
1997
1999
2001
2003
Math and CS-Total
Math and CS-US Citizens and Permanent Residents

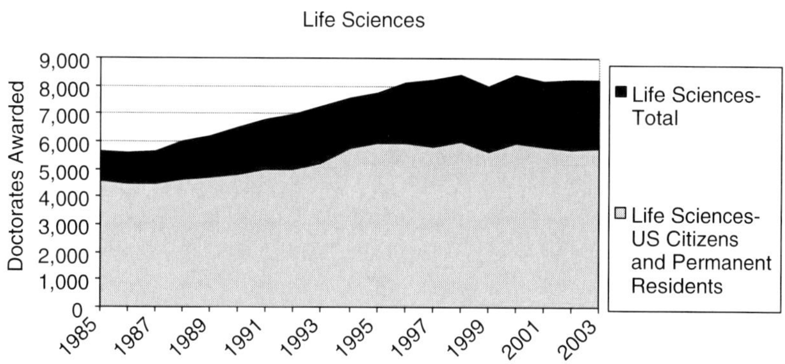
Life Sciences
Doctorates Awarded
9,000
8,000
7,000
6,000
5,000
4,000
3,000
2,000
1,000
0
1985
1987
1989
1991
1993
1995
1997
1999
2001
2003
Life Sciences-Total
Life Sciences-US Citizens and Permanent Residents

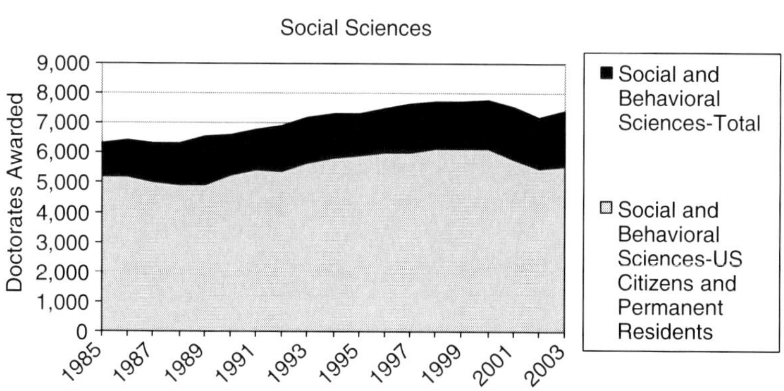
Social Sciences
Doctorates Awarded
9,000
8,000
7,000
6,000
5,000
4,000
3,000
2,000
1,000
0
1985
1987
1989
1991
1993
1995
1997
1999
2001
2003
Social and Behavioral Sciences-Total
Social and Behavioral Sciences-US Citizens and Permanent Residents

- The United Kingdom's points-based Highly Skilled Migrant Programme, which began in the mid-1990s, has increased the number of work permits issued to skilled workers.
- The Irish government permits relatively easy immigration of skilled workers in information technology and biotechnology through intra-company transfers from non-Irish to Irish locations.
- Several EU countries and the EU itself have programs that facilitate networking among students and researchers working abroad, providing contact information, collaborative possibilities, and funding and job opportunities in the EU. The German Academic Exchange Service has launched GAIN (German Academic International Network); the Italian Ministry of Foreign Affairs has launched DAVINCI, an Internet database that tracks the work of Italian researchers overseas; and the EU has its Researcher's Mobility Portal.
- Nigeria and other oil-producing nations use petroleum profits to support the overseas education of thousands of students.

In addition to sending students abroad for training, emerging economic powers, notably India and China, have lured their skilled scientists and engineers to return home by coupling education-abroad programs with strategic investments in the science and engineering infrastructure—in essence sending students away to gain skills and providing jobs to draw them back.[32]

The global competition for talent was already under way when the events of September 11, 2001, disrupted US travel and immigration plans of many international graduate students, postdoctoral researchers, and visiting scholars. The intervening years have seen security-related changes in federal visa and immigration policy that, although intended to restrict the illegal movements of only a few, have had a wider effect on many foreign-born graduate students and postdoctoral scholars who either were already in the United States or were contemplating studying here. Many potential visitors who in the past might have found the United States welcoming them for scientific meetings and sabbaticals now look elsewhere or stay home.[33] Much of this is to our detriment: Hosting international meetings and visiting researchers is essential to staying at the forefront of international science.

The flow of graduate students and postdoctoral researchers is unlikely to be curtailed permanently, at least as long as the world sees the United

---

[32]R. A. Mashelkar. "India's R&D: Reaching for the Top." *Science* 307(2005):1415-1417; L. Auriol. "Why Do We Need Indicators on Careers of Doctorate Holders?" Workshop on User Needs for Indicators on Careers of Doctorate Holders. OECD: Paris, September 27, 2004. Available at: http://www.olis.oecd.org/olis/2004doc.nsf.

[33]The National Academies. *Policy Implications of International Graduate Students and Postdoctoral Scholars*. Washington, DC: The National Academies Press, 2005. P. 61.

TABLE 3-2 Change in Applications, Admissions, and Enrollment of International Graduate Students, 2003-2005

	Total	Engineering	Life Sciences	Physical Sciences
Applications	–28% (–5%)	–36% (–7%)	–24% (–1%)	–26% (–3%)
Admissions	–18%	–24%	–19%	–17%
Enrollment	–6%	–8%	–10%	+6%

NOTES: There have been large declines in applications and admissions and a more moderate decrease in enrollment. The admissions data for the 2005 academic year are shown in parentheses.
SOURCES: H. Brown and M. Doulis. *Findings from the 2005 CGS International Graduate Survey I*. Washington, DC: Council of Graduate Schools, 2005; H. Brown. *Council of Graduate Schools Finds Decline in New International Graduate Student Enrollment for the Third Consecutive Year*. Washington, DC: Council of Graduate Schools, November 4, 2004.

States as the best place for science and engineering education, training, and technology-based employment (Table 3-2). If that perception shifts, and if international students find equally attractive educational and professional opportunities in other countries, including their own, the difficulty of visiting the United States could gain decisive importance.[34]

STRAINS ON RESEARCH IN THE PRIVATE SECTOR

A large fraction of all those with doctorates in science and engineering in the United States—more than half in some fields—find employment in industry (Figure 3-8). There they make major contributions to innovation and economic growth. US industry has traditionally excelled at innovation and at capitalizing on the results of research.[35] For decades after World War II, corporate central research laboratories paid off in fledgling technologies that grew into products or techniques of profound consequence. Researchers at Bell Laboratories pursued lines of groundbreaking research that resulted in the transistor and the laser, which revolutionized the electronics industry and led to several Nobel prizes.[36]

---

[34]Ibid., p. 79.

[35]S. W. Popper and C. S. Wagner. *New Foundations for Growth: The US Innovation System Today and Tomorrow*. Arlington, VA: RAND, January 2002. The authors note the following advantages of industry: rapid responses, flexibility and adaptability, efficiency, fast entry and exit, smooth capital flows, and mobility.

[36]US Congress House of Representatives Committee on Science. *Unlocking Our Future: Toward a New National Science Policy* ("the Ehlers Report"). Washington, DC: US Congress, 1998. P. 38. Available at: http://www.house.gov/science/science_policy_report.htm.

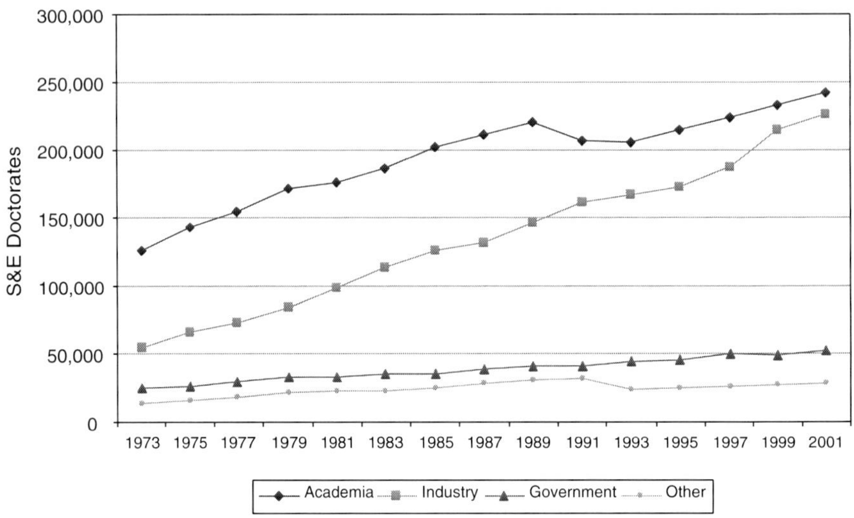

FIGURE 3-8 US S&E doctorates, by employment sector, 1973-2001. The majority of people with science and engineering doctorates obtain nonacademic jobs. About equal numbers work in academic and industrial settings, and about 15% work in government or other sectors.
SOURCE: National Science Foundation. *Survey of Doctoral Recipients*. Arlington, VA: National Science Foundation, 2004.

Although industry-funded R&D has increased steadily overall (Figure 3-9A), that new money has gone overwhelmingly to activities that are near-term and incremental rather than to long-term or discovery-oriented research, and R&D as a share of gross domestic product has declined (Figure 3-9B). Several explanations are offered for industry's turn away from fundamental research. First, the Bell Laboratories model was supported by funding from a monopoly that now is dismantled and no longer relevant to the organization of science and engineering research in the United States. Second, Wall Street analysts increasingly focus on quarterly financial results and assign little value to long-term (and therefore risky) research investments or to social returns. Third, companies cannot always fully capture a return that justifies long-term research with results that often spill over to other researchers, sometimes including those of competitors. Fourth, private-sector research is more fragmented across national boundaries in the era of globalization. Capital follows opportunity with little attention to geopolitical borders—this may lead more multinational companies to pursue opportunities outside the United States.

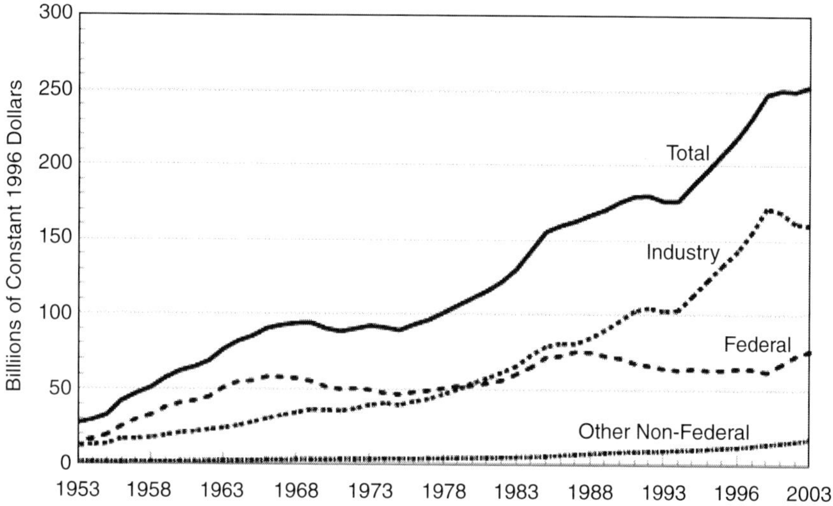

FIGURE 3-9A US R&D funding, by source of funds, 1953-2003.
SOURCE: NSF Division of Science Resources Statistics. *National Patterns of Research Development Resources,* annual series. Appendix Tables B-2 and B-22. Available at: http://www.nsf.gov/statistics/nsf05308/secta.htm.

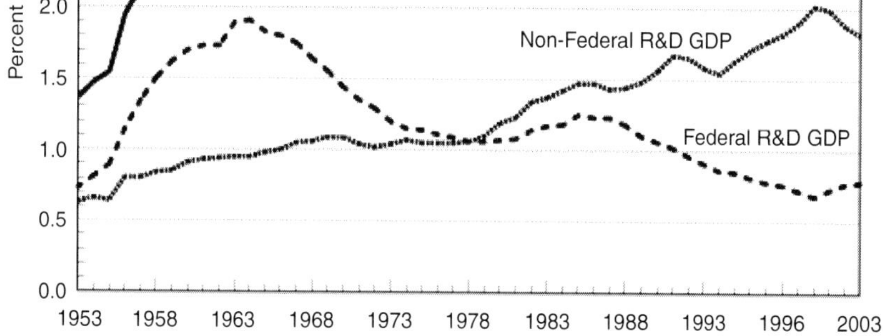

FIGURE 3-9B R&D shares of US gross domestic product, 1953-2003.
SOURCE: NSF Division of Science Resources Statistics. *National Patterns of Research Development Resources,* annual series. Appendix Table B-9. Available at: http://www.nsf.gov/statistics/nsf05308/sectd.htm.

The National Science Board[37] has made the following observations:

- Two-thirds of the R&D performed overseas in 2000 by US-owned companies ($13.2 billion of $19.8 billion) was conducted in six countries: the United Kingdom, Germany, Canada, Japan, France, and Sweden. At the same time, emerging markets—such as those in Singapore, Israel, Ireland, and China—were increasingly attracting R&D activities by subsidiaries of US companies. In 2000, each of those emerging markets reached US-owned R&D expenditures of $500 million or more, considerably more than in 1994.
- Three manufacturing sectors dominated overseas R&D activity by US-owned companies: transportation equipment, computer and electronic products, and chemicals and pharmaceuticals. The same industries accounted for most foreign-owned R&D in the United States, implying a high degree of R&D globalization in those industries.

As some large companies reduce their investment in basic research, smaller research-based enterprises often assume risk as the only way to break into a competitive market. Those startup companies commonly rely on the initial capital provided by their investors to finance early research, coupled with the granting of potential future financial gains in the form of stock options to compensate employees. If the money runs out, they can seldom interest venture capital firms until they have grown considerably larger. Many of those companies thus expire before reaching commercialization.[38]

The overall amount of venture capital invested also has collapsed since the stock market decline of 2000, sinking in 2002 to one-fifth the amount invested in 2000[39] (Figure 3-10). Venture capital investments in US companies have since stabilized at around $20 billion in 2003 and 2004,[40] just one-fifth of their 2000 peak but well above 1998 funding. Led by a resurgence in late-stage financing, total venture capital investment rose 10.5% to $20.9 billion in 2004, according to the MoneyTree Survey by PricewaterhouseCoopers, Thomson Venture Economics, and the National Venture

[37]National Science Board. *Science and Engineering Indicators 2004*. NSB 04-01. Arlington, VA: National Science Foundation, 2004. P. 4-65.

[38]National Research Council. Board on Science, Technology, and Economic Policy. *The Small Business Innovation Research Program: An Assessment of the Department of Defense Fast Track Initiative*. Washington, DC: National Academy Press, 2000. Available at: http://books.nap.edu/catalog/9985.html; US Congress House of Representatives Committee on Science. *Unlocking Our Future: Toward a New National Science Policy* (the "Ehlers Report"). Washington, DC: US Congress, 1998. P. 39.

[39]National Science Board. *Science and Engineering Indicators 2004*. NSB 04-01. Arlington, VA: National Science Foundation, 2004. Appendix Table 6-15.

[40]National Venture Capital Association. Available at: http://www.nvca.org/ffax.html.

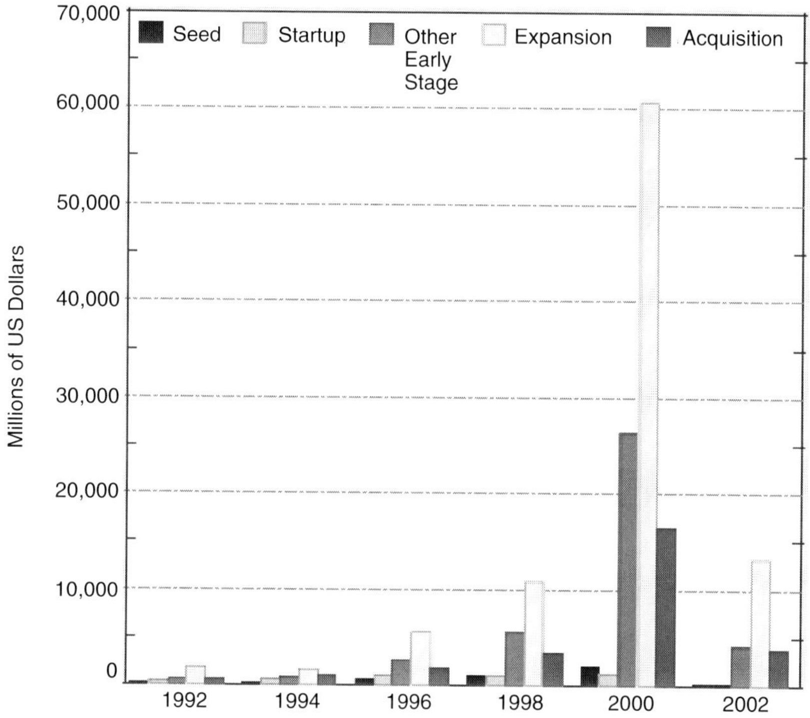

FIGURE 3-10 US venture capital disbursements, by stage of financing, 1992-2002. Venture capital funding is returning to pre-2000 levels.
SOURCES: Thompson Venture Economics, special tabulations, June 2003. See National Science Board. *Science and Engineering Indicators 2004*. NSB 04-01. Arlington, VA: National Science Foundation, 2004. Appendix Table 6-16.

Capital Association (NVCA).[41] With stock values rising, the climate for initial public offerings and acquisitions has improved, attracting capital from investors considering exit opportunities.

Another positive sign is a recent increase in capital raised by venture funds, suggesting an improving attitude toward risk taking. According to NVCA and Thomson Venture Economics,[42] venture funds raised \$17.6 billion in 2004, more than in the prior 2 years combined (albeit at just one-sixth their 2000 peak). There is a strong funding pipeline to support ven-

[41]PricewaterhouseCoopers. "MoneyTree Survey." Available at: http://www.pwcmoneytree.com/moneytree/index.jsp. Accessed December 20, 2005.

[42]Ibid.

ture capital investments in 2005, especially early-stage investments with particular emphasis on biotechnology.

In addition to private venture capital, small companies can obtain federal tax incentives and other help through the research and experimentation (R&E) tax credit (Table 3-3) and the federal Small Business Innovation Research (SBIR) program and Advanced Technology Program[43] (Table 3-4).

The US workforce faces the additional pressure of competing with workers in nations with lower wage structures. A US company can hire five chemists in China or at least that many engineers (depending on the field) in India for the cost of one employee of equivalent training in the United States.[44] The upshot has been the growing trend of corporations moving work offshore because of wage disparities (Figure 3-11). Wage differences at the factory and clerical levels are even more pronounced.

A recent McKinsey and Company study[45] reported that the supply of young professionals (university graduates with up to 7 years of experience) in low-wage countries vastly outstrips the supply in high-wage countries. There were 33 million people in that category in 28 low-wage countries, and 15 million in 8 high-wage countries, including 7.7 million in the United States.[46] With opportunities to study or work abroad or to work at home for a multinational corporation, workers in low-wage countries increasingly will be in direct competition with workers from developed nations.

The same study estimates, however, that only 13% of the potential talent supply in low-wage nations is suited to work for multinational corporations because these individuals lack language skills, because of low-quality domestic education systems, and because of a lack of cultural fit. For the United States to compete, then, its workers can and must bring to the workplace not only technical skills and knowledge but other valuable skills, including knowledge of other cultures, the ability to interact comfortably with diverse clientele, and the motivation to apply their skills. US workers also must be able to communicate effectively orally and in writing, lead teams, manage projects, and solve problems. Although much of our education system is working to teach those skills, there is much to do to prepare

---

[43]The other program is the Manufacturing Technology Program in the Department of Defense.

[44]The Web site http://www.payscale.com/about.asp tracks and compares pay scales in many countries. R. Hira, of the University of Rochester, calculates average salaries for engineers in the United States and India as $70,000 and $13,580, respectively.

[45]McKinsey and Company. *The Emerging Global Labor Market: Part II—The Supply of Offshore Talent in Services*. New York: McKinsey and Company, June 2005.

[46]Ibid.

TABLE 3-3 R&E Tax Claims and US Corporate Tax Returns, 1990-2001

	R&E Tax Credit Claims		
Year	Current Dollars (millions)	2000 Constant Dollars (millions)	Returns
1990	1,547	1,896	8,699
1991	1,585	1,877	9,001
1992	1,515	1,754	7,750
1993	1,857	2,101	9,933
1994	2,423	2,684	9,150
1995	1,422	1,544	7,877
1996	2,134	2,274	9,709
1997	4,398	4,609	10,668
1998	5,208	5,399	9,849
1999	5,281	5,396	10,019
2000	7,079	7,079	10,495
2001	6,356	6,207	10,388

NOTES: Data exclude IRS forms 1120S (S corporations), 1120-REIT (Real Estate Investment Trusts), and 1120-RIC (Regulated Investment Companies). Constant dollars based on calendar year 2000 GDP price deflator. The R&E credit is designed to stimulate company R&D over time by reducing after-tax costs. Companies that qualify may deduct or subtract from corporate income taxes an amount equal to 20% of qualified research expenses above a base amount. For established companies, that amount depends on historical expenses over a statutory base period relative to gross receipts; startups follow other provisions.
SOURCE: US Internal Revenue Service, Statistics of Income program, unpublished tabulations.

US students for work in a more competitive global economy—as well as to provide the rudimentary skills needed in any economy.

RESTRAINTS ON PUBLIC FUNDING

Public financial support is the backbone of America's research establishment. In the 1960s and 1970s, university researchers could look to a dozen or so federal sources for grant support, including NSF, NIH, predecessors of the Office of Science in the Department of Energy (DOE),[47] the Department of Defense (DOD), the National Aeronautics and Space Administration, and the Department of Agriculture. Funding from those sources, combined with private money, provided flexibility and generosity unmatched in any other nation. Large numbers of today's senior scientists and engineers owe their ability to pursue their professions to grants from those federal agencies.

[47]The Department of Energy Office of Science began as a component of the Atomic Energy Commission.

TABLE 3-4 Federally and Privately Funded Early-Stage Venture Capital in Millions of Dollars, 1990-2002

Year	Federal SBIR	Federal ATP	Private Early-Stage Venture Capital
1990	461	46	1,148
1991	483	93	826
1992	508	48	1,186
1993	698	60	2,100
1994	718	309	1,581
1995	835	414	2,143
1996	916	19	2,658
1997	1,107	162	3,373
1998	1,067	235	4,700
1999	1,097	110	10,995
2000	1,190	144	20,260
2001	1,294	164	764
2002	NA	156	1,813

NOTES: Federally funded sources include SBIR and ATP. ATP, Advanced Technology Program; NA, not available; SBIR, Small Business Innovation Research. Data reflect disbursements funded publicly through federal SBIR and ATP and privately through US venture capital funds.
SOURCE: National Science Board. *Science and Engineering Indicators 2004*. NSB 04-01. Arlington, VA: National Science Foundation, 2004. P. 6-31.

Several trends cast doubt on our continuing commitment to the above strategy. The first accompanied the end of the Cold War, when reductions in military funding had the perhaps unintentional effect of cutting basic and applied DOD research budgets. The portion of funding DOD devoted to basic research (the "6.1 account") declined from 3.3% in fiscal year (FY) 1994 to about 1.9% in FY 2005[48] (Figure 3-12). Military research funding has gradually shifted from basic and applied research toward the more immediate needs of the combat forces.

Public funding for science and engineering rose through the 1990s, but virtually all of the increase went to biomedical research at NIH. Federal spending on the physical sciences remained roughly flat, and increases for mathematics and engineering only slightly surpassed inflation (Figure 3-13). Funding for important areas of the life sciences—plant science, ecology, environmental research—supported by agencies other than NIH also has leveled off. The lack of new funding for research in the physical sci-

[48]National Science Board. *Science and Engineering Indicators 2004*. NSB 04-01. Arlington, VA: National Science Foundation, 2004.

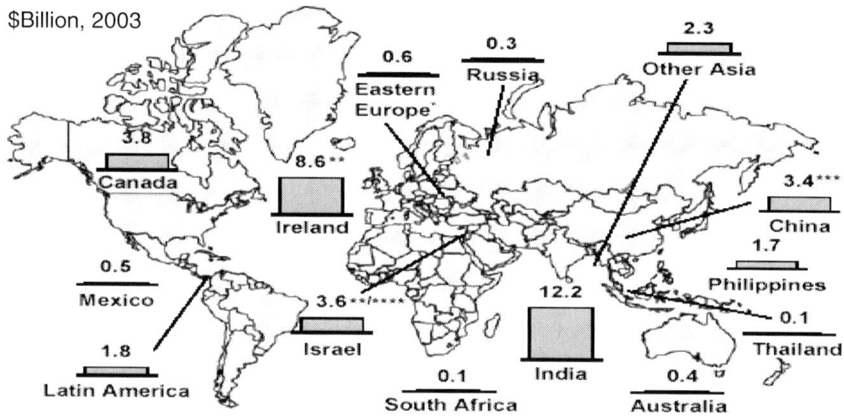

*Includes Poland, Romania, Hungary, Ukraine, and Czech Republic.
**Primarily composed of MNC captives.
***Estimate, based on total Chinese BPO and IT services revenue (7.0) minus domestic demand for IT services (4.4).
****Estimate, based on 2001 market size of 3.0 and assumed growth rate of 20% p.a.

FIGURE 3-11 Offshored services market size, in billions of dollars, 2003.
NOTE: Offshored services market size includes Business Process Outsourcing and Information Technology, Captive and Outsourced.
SOURCE: Based on Software Associations; US country commercial reports; press articles; Gartner; IDC; Country government Web sites; Ministry of Information Technology for various countries; Enterprise Ireland; NASSCOM; McKinsey Global Institute analysis. McKinsey and Company. *The Emerging Global Labor Market: Part II—The Supply of Offshore Talent in Services*. New York: McKinsey and Company, June 2005.

ences, mathematics, and engineering raises concern about the overall health of the science and engineering research enterprise, including that of the health sciences. Yet, these are disciplines that lead to innovation across the spectrum of modern life.[49]

Figure 3-9B shows that total R&D as a percentage of GDP bottomed out in the late 1970s at around 2.1%, then rebounded to about 2.6%. That rate of investment has stayed relatively constant since the early 1980s. Federal R&D as a percentage of GDP peaked in the early 1960s and has fallen since then.

---

[49]The National Academies. *Observations on the President's Fiscal Year 2003 Federal Science and Technology Budget*. Washington, DC: The National Academies Press, 2002. Pp. 14-16.

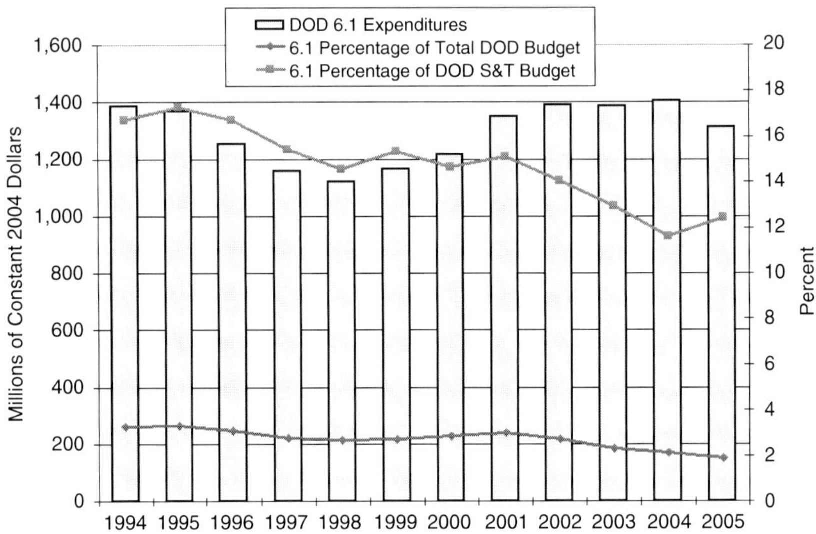

FIGURE 3-12 Department of Defense (DOD) 6.1 expenditures, in millions of constant 2004 dollars, 1994-2005.
SOURCE: National Science Board. *Science and Engineering Indicators 2004*. NSB 04-01. Arlington, VA: National Science Foundation, 2004.

EXPANDED MISSION FOR FEDERAL LABORATORIES

Among the nation's most significant investments in R&D are some 700 laboratories funded directly by the federal government, about 100 of which are considered significant contributors to the national innovation system.[50] Work performed by the government's own laboratories accounts for about 35% of the total federal R&D investment.[51] The largest and best known of these laboratories are run by DOD and DOE. NIH also has an extensive research facility in Maryland. The DOE laboratories focus mainly on national security research, as at Lawrence Livermore National Laboratory, or more broadly on scientific and engineering research, as at Oak Ridge National Laboratory or Argonne National Laboratory.

The national laboratories could potentially fill the gap left when the

[50]In contrast, there are approximately 14,000 industrial laboratories with about 1,000 that are considered to be substantive contributors to national innovation according to M. Crow and B. Bozeman. *Limited by Design: R&D Laboratories and the U.S. National Innovation System*. New York: Columbia University, 1998.

[51]Ibid., pp. 5-6.

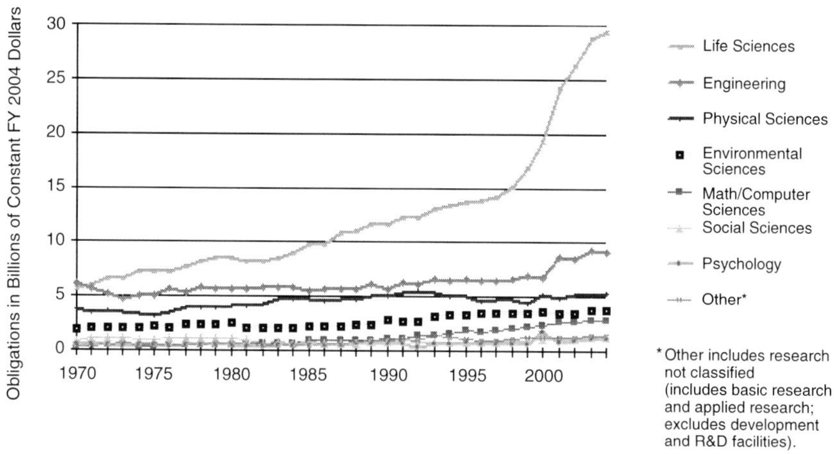

FIGURE 3-13 Trends in federal research funding by discipline, obligations in billions of constant FY 2004 dollars, FY 1970-FY 2004.
NOTE: Life sciences—split into NIH support for biomedical research and all other agencies' support for life sciences.
SOURCE: American Association for the Advancement of Science analysis based on National Science Foundation. *Federal Funds for Research and Development: Fiscal Years 2002, 2003, 2004*. FY 2003 and FY 2004 data are preliminary. Constant-dollar conversions based on OMB's GDP deflector.

large corporate R&D laboratories reduced their commitment to high-risk, long-term research in favor of short-term R&D work, often conducted in overseas laboratories close to their manufacturing plants and to potential markets for their products. The payoff for the US economy from the old corporate R&D system was huge. Today, that work is difficult for business to justify: Its profitability is best measured in hindsight, after many years of sustained investment, and the probability for the success of any single research project often is small.

Nonetheless, it was that type of corporate research which provided the disruptive technologies and technical leaps that fueled US economic leadership in the 20th century. If properly managed and adequately funded, the large multidisciplinary DOE laboratories could assist in filling the void left by the shift in corporate R&D emphasis. The result would be a stable, world-class science and engineering workforce focused both on high-risk, long-term basic research and on applied research for technology development. The national laboratories now offer the right mix of basic scientific inquiry and practical application. They often promote collaboration with research universities and with large teams of applied scientists and engineers, and the enterprise has demonstrated an early ability to translate prototypes into commercial products. National defense-homeland security and new technologies for clean, affordable, and reliable energy are particularly appropriate areas of inquiry for the national laboratory system.

EDUCATIONAL CHALLENGES

The danger exists that Americans may not know enough about science, technology, or mathematics to significantly contribute to, or fully benefit from, the knowledge-based society that is already taking shape around us. Moreover, most of us do not have enough understanding of the importance of those skills to encourage our children to study those subjects—both for their career opportunities and for their general benefit. Other nations have learned from our history, however, and they are boosting their investments in science and engineering education because doing so pays immense economic and social dividends.

The rise of new international competitors in science and engineering is forcing the United States to ask whether its education system can meet the demands of the 21st century. The nation faces several areas of challenge: K–12 student preparation in science and mathematics, limited undergraduate interest in science and engineering majors, significant student attrition among science and engineering undergraduate and graduate students, and science and engineering education that in some instances inadequately prepares students to work outside universities.

K–12 Performance

Education in science, mathematics, and technology has become a focus of intense concern within the business and academic communities. The domestic and world economies depend more and more on science and engineering. But our primary and secondary schools do not seem able to produce enough students with the interest, motivation, knowledge, and skills they will need to compete and prosper in the emerging world.

Although there was steady improvement in mathematics test scores from 1990 through 2005, only 36% of 4th-grade students and 30% of 8th-grade students who took the 2005 National Assessment of Educational Progress (NAEP) performed at or above the "proficient" level in mathematics (Figure 3-14). (Proficiency was demonstrated by competence with "challenging subject matter".)[52] The results of the science 2000 NAEP test were

[52]Educational Programs. Available at: http://nces.ed.gov/pubsearch/pubsinfo.asp?pubid=2005451. Accessed December 20, 2005; J. S. Braswell, G. S. Dion, M. C. Daane, and Y. Jin. *The Nation's Report Card*. NCES 2005451. Washington, DC: US Department of Education, 2004. Based on National Assessment of Educational Progress.

similar. Only 29% of 4th-grade students, 32% of 8th-grade students, and 18% of 12th-grade students performed at or above the proficient level (Figure 3-15). Without fundamental knowledge and skills, the majority of students scoring below this level—particularly those below the basic level—lack the foundation for good jobs and full participation in society.

Our 4th-grade students perform as well in mathematics and science as do their peers in other nations, but in the most recent assessment (1999) 12th graders were almost last among students who participated in the Trends in International Mathematics and Science Study. Of the 20 nations assessed in advanced mathematics and physics, none scored significantly *lower* than did the United States in either subject. The relative standing of US high school students in those areas has been attributed both to inadequate quality of teaching and to a weak curriculum.

There has, however, been some arguably good news about student achievement. Our 8th graders did better on an international assessment of mathematics and science in 2003 than the same age group did in 1995. Unfortunately, in both cases they ranked poorly in comparison with students from other nations. The achievement gap that separates African American and Hispanic students from white students narrowed during that period. However, a recent assessment by the OECD Programme for International Student Assessment revealed that US 15-year-olds are near the bottom worldwide in their ability to solve practical problems that require mathematical understanding. Test results for the last 30 years show that although scores of US 9- and 13-year-olds have improved, scores of 17-year-olds have remained stagnant.[53]

One key to improving student success in science and mathematics is to increase interest in those subjects, but that is difficult because mathematics and science teachers are, as a group, largely ill-prepared. Furthermore, many adults with whom students come in contact seemingly take pride in "never understanding" or "never liking" mathematics. Analyses of the teacher pool indicate that an increasing number do not major or minor in the discipline they teach, although there is growing pressure from the No Child Left Behind Act for states to hire more highly qualified teachers (see Table 5-1). About 30% of high school mathematics students and 60% of those enrolled in physical sciences have teachers who either did not major in the

[53]The Programme for International Student Assessment (PISA) Web site is available at: http://www.pisa.oecd.org. PISA, a survey every 3 years (2000, 2003, 2006, etc.) of 15-year-olds in the principal industrialized countries, assesses to what degree students near the end of compulsory education have acquired some of the knowledge and skills that are essential for full participation in society.

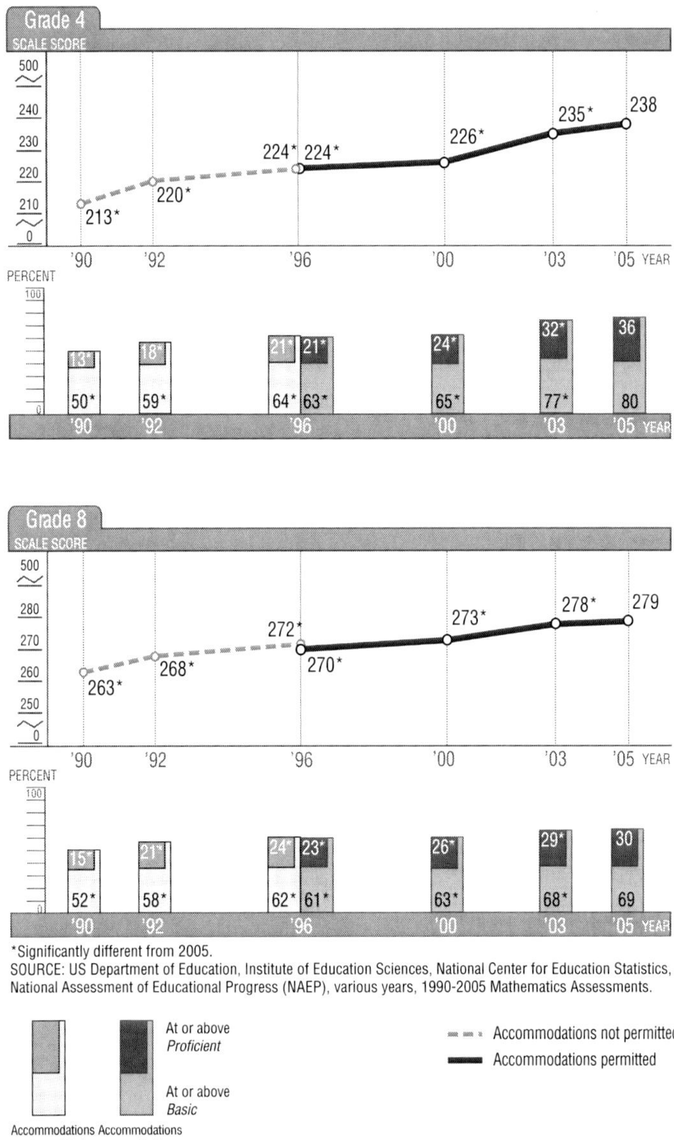

FIGURE 3-14 Average scale NAEP scores and achievement-level results in mathematics, grades 4 and 8: various years, 1990-2005.
SOURCE: National Center for Education Statistics. Available at: http://nces.ed.gov/nationsreportcard/.

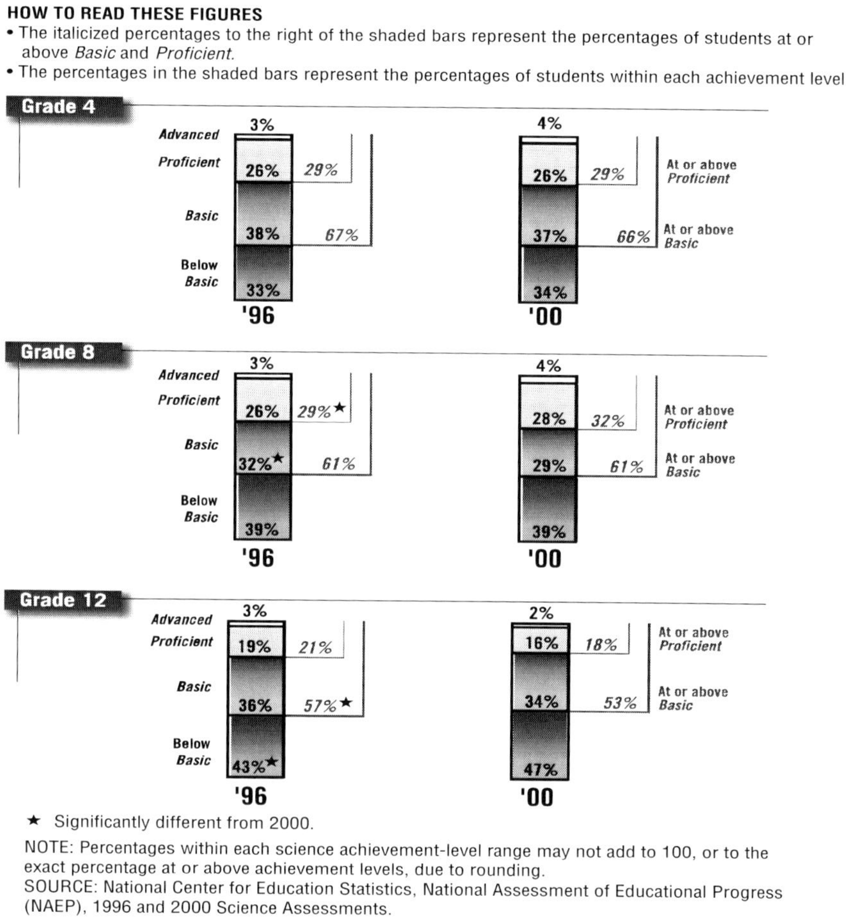

FIGURE 3-15 Percentage of students within and at or above achievement levels in science, grades 4, 8, and 12, 1996 and 2000.
SOURCE: National Center for Education Statistics. Available at: http://nces.ed.gov/nationsreportcard/.

subject in college or are not certified to teach it. The situation is worse for low-income students: 70% of their middle school mathematics teachers majored in some other subject in college.

Meanwhile, an examination of curricula reveals that middle school mathematics and science courses lack focus, cover too many topics, repeat material, and are implemented inconsistently. That could be changing, at least in part because of new science and mathematics teaching and learning standards that emphasize inquiry and detailed study of fewer topics.

Another major challenge—and opportunity—has been the diversity of the student population and the large variation in quality of education between schools and districts, particularly between suburban, urban, and rural schools. Some schools produce students who consistently score at the top of national and international tests; while others consistently score at the bottom. Furthermore, accelerated mathematics and science courses are less frequently offered in rural and city schools than in suburban ones. How to achieve an equitable distribution of funding and high-quality teaching should be a top-priority issue for the United States. It is an issue that is exacerbated by the existence of almost 15,000 school districts, each containing an average of six schools.

Student Interest in Science and Engineering Careers

The United States ranks 16 of 17 nations in the proportion of 24-year-olds who earn degrees in natural sciences or engineering as opposed to other majors (Figure 3-16A) and 20 of 24 nations when looking at all 24-year-olds (Figure 3-16B).[54] The number of bachelor's degrees awarded in the United States fluctuates greatly (see Figure 3-17).

About 30% of students entering college in the United States (more than 95% of them US citizens or permanent residents) intend to major in science or engineering. That proportion has remained fairly constant over the past 20 years. However, undergraduate programs in those disciplines report the lowest retention rates among all academic disciplines, and very few students transfer into these fields from others. Throughout the 1990s, fewer than half of undergraduate students who entered college intending to earn a science or engineering major completed a degree in one of those subjects.[55] Undergraduates who opt out of those programs by switching majors are

[54]National Science Board. *Science and Engineering Indicators 2004*. NSB 04-01. Arlington, VA: National Science Foundation, 2004. Appendix Table 2-23 places the following countries ahead of the United States: Finland (13.2), Hungary (11.9), France (11.2), Taiwan (11.1), South Korea (10.9), United Kingdom (10.7), Sweden (9.5), Australia (9.3), Ireland (8.5), Russia (8.5), Spain (8.1), Japan (8.0), New Zealand (8.0), Netherlands (6.8), Canada (6.7), Lithuania (6.7), Switzerland (6.5), Germany (6.4), Latvia (6.4), Slovakia (6.3), Georgia (5.9), Italy (5.9), and Israel (5.8).

[55]L. K. Berkner, S. Cuccaro-Alamin, and A. C. McCormick. *Descriptive Summary of 1989-90 Beginning Postsecondary Students: 5 Years Later with an Essay on Postsecondary Persistence and Attainment*. NCES 96155. Washington, DC: National Center for Education Statistics, 1996; T. Smith. *The Retention and Graduation Rates of 1993-1999 Entering Science, Mathematics, Engineering, and Technology Majors in 175 Colleges and Universities*. Norman, OK: Center for Institutional Data Exchange and Analysis, University of Oklahoma, 2001.

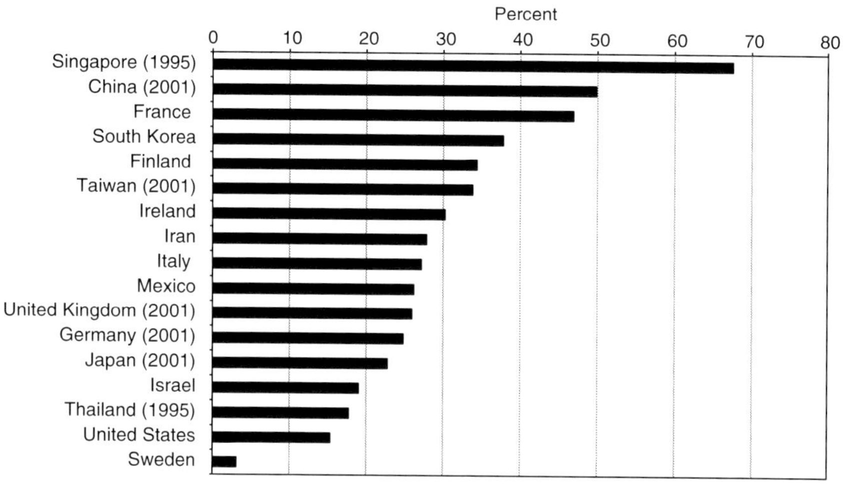

FIGURE 3-16A Percentage of 24-year-olds with first university degrees in the natural sciences or engineering, relative to all first university degree recipients, in 2000 or most recent year available.
SOURCE: Analysis conducted by the Association of American Universities. 2006. *National Defense Education and Innovation Initiative* based on data from Appendix Table 2-35 in National Science Board. *Science and Engineering Indicators 2004*. NSB 04-01. Arlington, VA: National Science Foundation, 2004.

often among the most highly qualified college entrants,[56] and they are disproportionately women and students of color. The implication is that potential science or engineering majors become discouraged well before they can join the workforce.[57]

---

[56]S. Tobias. *They're Not Dumb, They're Different. Stalking the Second Tier*. Tucson, AZ: Research Corporation, 1990; E. Seymour and N. Hewitt. *Talking About Leaving: Why Undergraduates Leave the Sciences*. Boulder, CO: Westview Press, 1997; M. W. Ohland, G. Zhang, B. Thorndyke, and T. J. Anderson. Grade-Point Average, Changes of Major, and Majors Selected by Students Leaving Engineering. 34th ASEE/IEEE Frontiers in Education Conference. Session T1G:12-17, 2004.

[57]M. F. Fox and P. Stephan. "Careers of Young Scientists: Preferences, Prospects, and Reality by Gender and Field." *Social Studies of Science* 31(2001):109-122; D. L. Tan. *Majors in Science, Technology, Engineering, and Mathematics: Gender and Ethnic Differences in Persistence and Graduation*. Norman, OK: University of Oklahoma, 2002. Available at: http://www.ou.edu/education/csar/literature/tan_paper3.pdf; Building Engineering and Science Talent (BEST). *The Talent Imperative: Diversifying America's Science and Engineering Workforce*. San Diego: BEST, 2004; G. D. Heyman, B. Martyna, and S. Bhatia. "Gender and Achievement-related Beliefs Among Engineering Students." *Journal of Women and Minorities in Science and Engineering* 8(2002):33-45.

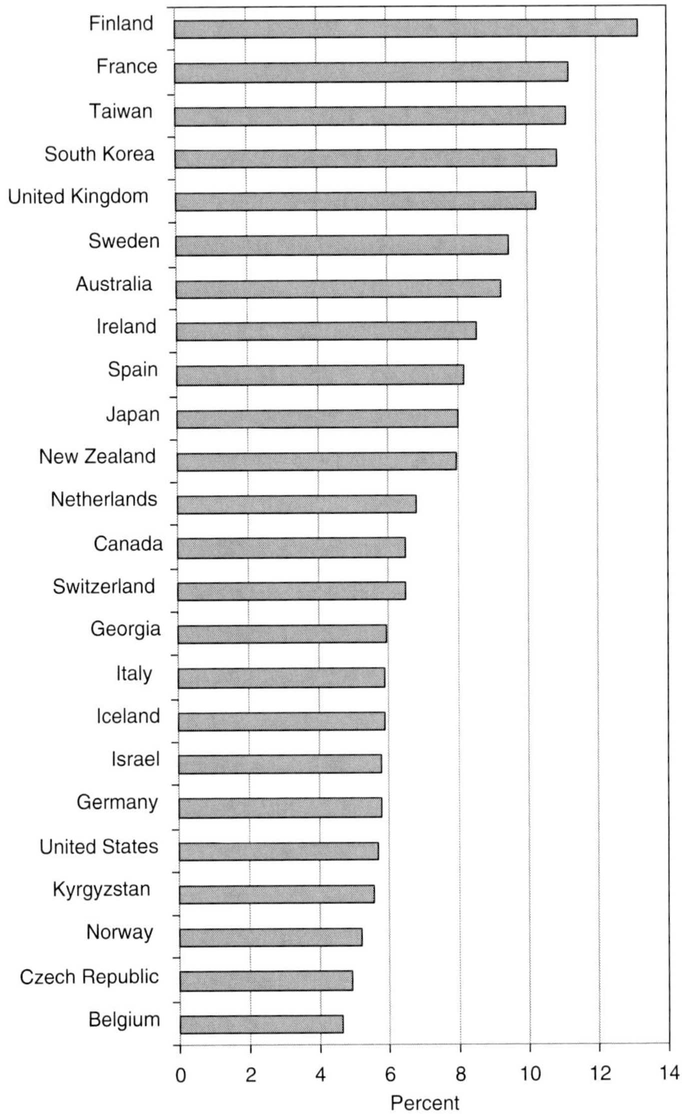

FIGURE 3-16B Percentage of 24-year-olds with first university degrees in the natural sciences or engineering relative to all 24-year-olds, in 2000 or most recent year available.

NOTE: Natural sciences and engineering include the physical, biological, agricultural, computer, and mathematical sciences and engineering.

SOURCE: National Science Board. *Science and Engineering Indicators 2004*. NSB 04-01. Arlington, VA: National Science Foundation, 2004.

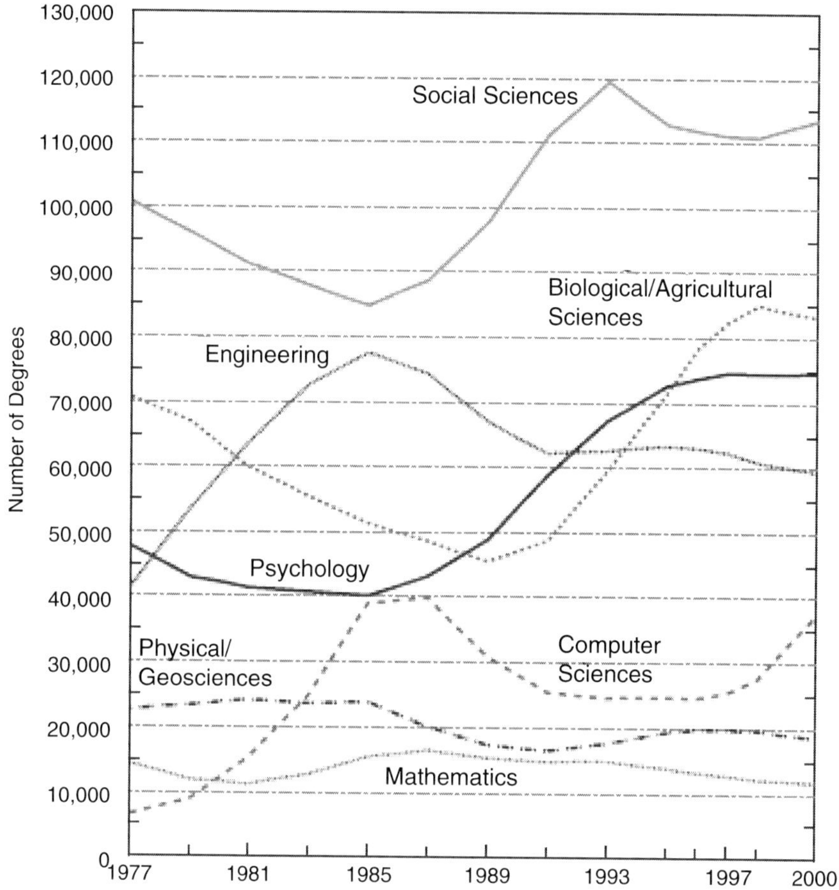

FIGURE 3-17 Science and engineering bachelor's degrees, by field: selected years, 1977-2000.

NOTES: Geosciences include earth, atmosphere, and ocean sciences. Degree production for many science, technology, engineering, and mathematics fields increased and computer science decreased in 2001. See graphs in the *Attracting the Most Able US Students to Science and Engineering* paper located in Appendix D.

SOURCE: National Science Board. *Science and Engineering Indicators 2004*. NSB 04-01. Arlington, VA: National Science Foundation, 2004. Appendix Table 2-23.

Graduate school enrollments in science and engineering in the United States have been relatively stable since 1993, at 22-26% of the total enrollment. More women and under represented minorities participate than has been the case in the past, but a relative decline in the enrollment of US whites and males in the late 1990s has been reversed only since 2001.[58] Indeed, for the past 15 years, growth in the number of doctorates awarded is attributable primarily to the increased number of international students. Attrition is generally lower in the doctoral programs than among undergraduates in science, technology, engineering, and mathematics, but doctoral programs in the sciences nonetheless report dropout rates from 24 to 67%, depending on the discipline.[59] If the primary objective is to maintain excellence, a major challenge is to determine how to continue to attract the best international students and still encourage the best domestic students to enter the programs—and to remain in them.

Student interest in research careers is dampened by several factors. First, there are important prerequisites for science and engineering study. Students who choose not to or are unable to finish algebra 1 before 9th-grade—which is needed for them to proceed in high school to geometry, algebra 2, trigonometry, and precalculus—effectively shut themselves out of careers in the sciences. In contrast, the decision to pursue a career in law or business typically can wait until the junior or senior year of college, when students begin to commit to postgraduate entrance examinations.

Science and engineering education has a unique hierarchical nature that requires academic preparation for advanced study to begin in middle school. Only recently have US schools begun to *require* algebra in the 8th-grade curriculum. The good news is that more schools are now offering integrated science curricula and more districts are working to coordinate curricula for grades 7–12.[60]

For those students who do wish to pursue science and engineering, there are further challenges. Introductory science courses can function as "gatekeepers" that intentionally foster competition and encourage the best stu-

---

[58]National Science Foundation. *Graduate Enrollment Increases in Science and Engineering Fields, Especially in Engineering and Computer Sciences*. NSF 03-315. Arlington, VA: National Science Foundation, 2003.

[59]Council of Graduate Schools. "Ph.D. Completion and Attrition: Policy, Numbers, Leadership, and Next Steps." 2004. The Council of Graduate Schools' PhD Completion Project's goal is to improve completion and attrition rates of doctoral candidates. This 3-year project had provided funding to 21 major universities to create intervention strategies and pilot projects and to evaluate the impact of these projects on doctoral completion rates and attrition patterns.

[60]National Research Council. *Learning and Understanding: Improving Advanced Study of Mathematics and Science in US High Schools*. Washington, DC: National Academy Press, 2002.

dents to continue, but in so doing they also can discourage highly qualified students who could succeed if they were given enough support in the early days of their undergraduate experience.

Beyond the prospect of difficult and lengthy undergraduate and graduate study and postdoctoral requirements, career prospects can be tenuous. At a general level, news about companies that send jobs overseas can foster doubt about the domestic science and engineering job market. Graduate students are sometimes discouraged by a perceived mismatch between education and employment prospects in the academic sector. The number of tenured academic positions is decreasing, and an increasing majority of those with doctorates in science or engineering now work outside of academia. Doctoral training, however, still typically assumes students will work in universities and often does not prepare graduates for other careers.[61] Finally, it is harder to stay current in science and engineering than it is to keep up with developments in many other fields. Addressing the issues of effective lifelong training, time-to-degree, attractive career options, and appropriate type and amount of financial support are all critical to recruiting and retaining students at all levels.

Where are the top US students going, if not into science and engineering? They do not appear to be headed in large numbers to law school or medical school, where enrollments also have been flat or declining. Some seem attracted to MBA programs, which grew by about one-third during the 1990s. In the 1990s, many science and engineering graduates entered the workforce directly after college, lured by the booming economy. Then, as the bubble deflated in the early part of the present decade, some returned to graduate school. A larger portion of the current crop of science and engineering graduates seems to be interested in graduate school.[62] In 2003, enrollment in graduate science and engineering programs reached an all-time high, gaining 4% over 2002 and 9% over 1993, the previous peak year. Increasingly, the new graduate students are US citizens or permanent residents—67% in 2003 compared with 60% in 2000[63]—and their prospects seem good: In 2001, the share of top US citizen scorers on the Gradu-

---

[61]NAS/NAE/IOM. *Reshaping Graduate Education.* Washington, DC: National Academy Press, 1995; National Research Council. *Assessing Research-Doctorate Programs: A Methodology Study.* Washington, DC: The National Academies Press, 2003.

[62]W. Zumeta and J. S. Raveling. The Best and the Brightest for Science: Is There a Problem Here? In M. P. Feldman and A. N. Link, eds. *Innovation Policy in the Knowledge-Based Economy.* Boston: Klewer Academic Publishers, 2001. Pp. 121-161.

[63]National Science Foundation. *Graduate Enrollment in Science and Engineering Programs Up in 2003, but Declines for First-Time Foreign Students.* NSF 05-317. Arlington, VA: National Science Foundation, 2005.

ate Record Exam quantitative scale (above 750) heading to graduate school in the natural sciences and engineering was 31% percent higher than in 1998. That group had declined by 21% in the previous 6 years.[64]

There is still ample reason for concern about the future. A number of analysts expect to see a leveling off of the number of US-born students in graduate programs. If the number of foreign-born graduate students decreases as well, absent some substantive intervention, the nation could have difficulty meeting its need for scientists and engineers.

BALANCING SECURITY AND OPENNESS

Science thrives on the open exchange of information, on collaboration, and on the opportunity to build on previous work. The United States gained and maintained its preeminence in science and engineering in part by embracing the values of openness and by welcoming students and researchers from all parts of the world to America's shores. Openness has never been unqualified, of course, and the nation actively seeks to prevent its adversaries from acquiring scientific information and technology that could be used to do us harm. Scientists and engineers are citizens too, and those communities recognize both their responsibility and their opportunity to help protect the United States, as they have in the past. This has been done by harnessing the best science and engineering to help counter terrorism and other national security threats, even though that could mean accepting some limitations on research and its dissemination.[65]

But now concerns are growing that some measures put in place in the wake of September 11, 2001, seeking to increase homeland security, will be ineffective at best and could in fact hamper US economic competitiveness and prosperity.[66] New visa restrictions have had the unintended consequence of discouraging talented foreign students and scholars from coming here to work, study, or participate in international collaborations. Fortunately, the federal agencies responsible for these restrictions have recently implemented changes.[67] Of principal concern now are other forms of disincentive:

---

[64]W. Zumeta and J. S. Raveling. "The Market for PhD Scientists: Discouraging the Best and Brightest? Discouraging All?" AAAS Symposium, February 16, 2004. Press release available at: http://www.eurekalert.org/pub_releases/2004-02/uow-rsl021304.php.

[65]See, for example, National Research Council. *Making the Nation Safer: The Role of Science and Technology in Countering Terrorism*. Washington, DC: The National Academies Press, 2002.

[66]Letter from the Presidents of the National Academies to Secretary of Commerce Carlos Gutierrez, June 24, 2005. Available at: http://www.nationalacademies.org/morenews/20050624.html.

[67]The National Academies. *Policy Implications of International Graduate Students and Postdoctoral Scholars*. Washington, DC: The National Academies Press, 2005. Pp. 56-57.

• Expansion of the restrictions on "deemed exports," the passing of technical information to foreigners in the United States that requires a formal export license, is expected to cover a much wider range of university and industry settings.[68] Companies that rely on the international members of their R&D teams and university laboratories staffed by foreign graduate students and scholars could find their work significantly hampered by the new restrictions.

• Expanded or new categories of "sensitive but unclassified" information could restrict publication or other forms of dissemination. The new rules have been proposed or implemented even though many of the lists of what is to be controlled are sufficiently vague or obsolete that it could be difficult to ascertain compliance.[69] The result could be to force researchers to err on the side of caution and thus substantially impede the flow of scientific information.

Both approaches could undermine the protections for fundamental research established in National Security Decision Directive 189 (NSDD-189), the Reagan Administration's 1985 executive order declaring that publicly funded research, such as that conducted in universities and laboratories, should "to the maximum extent possible" be unrestricted.[70] Where restriction is considered necessary, the control mechanism should be formal classification: "No restrictions may be placed upon the conduct or reporting of federally-funded fundamental research that has not received national security classification, except as provided in applicable U.S. statutes." The NSDD-189 policy remains in force and has been reaffirmed by senior officials of the current administration, but it appears to be at odds with other policy developments and some recent practices.

---

[68]In 2000, Congress mandated annual reports by the Office of Inspector General (IG) on the transfer of militarily sensitive technology to countries and entities of concern; the 2004 reports focused on deemed exports. The individual agency IG reports and a joint interagency report concluded that enforcement of deemed-export regulations had been ineffective; most of the agency reports recommended particular regulatory remedies.

[69]Center for Strategic and International Studies. *Security Controls on Scientific Information and the Conduct of Scientific Research*. Washington, DC: CSIS, June 2005.

[70]Fundamental research is defined as "basic and applied research in science and engineering, the results of which ordinarily are published and shared broadly within the scientific community, as distinguished from proprietary research and from industrial development, design, production and product utilization, the results of which ordinarily are restricted for proprietary or national security reasons." National Security Decision Directive 189, September 21, 1985. Available at: http://www.aau.edu/research/ITAR-NSDD189.html.

CONCLUSION

Although the United States continues to possess the world's strongest science and engineering enterprise, its position is jeopardized both by evolving weakness at home and by growing strength abroad.[71] Because our economic, military, and cultural well-being depends on continued science and engineering leadership, the nation faces a compelling call to action. The United States has responded energetically to challenges of such magnitude in the past:

- Early in the 20th century, we determined to provide free education to all, ensuring a populace that was ready for the economic growth that followed World War II.
- The GI Bill eased the return of World War II veterans to civilian life and established postsecondary education as the fuel for the postwar economy.
- The Soviet space program spurred a national commitment to science education and research. The positive effects are seen to this day—for example, in much of our system of graduate education.
- The decline of the US semiconductor manufacturing industry in the middle 1980s was met with SEMATECH, the government–industry consortium credited by many with stimulating the resurgence of that industry.

Today's challenges are even more diffuse and more complex than many of the challenges we have confronted in our past. Research, innovation, and economic competition are worldwide, and the nation's attention, unlike that of many competitors, is not focused on the importance of its science and engineering enterprise. If the United States is to retain its edge in the technology-based industries that generate innovation, quality jobs, and high wages, we must act to broker a new, collaborative understanding among the sectors that sustain our knowledge-based economy—industry, academe, and government—and we must do so promptly.

[71]Note that some do not believe this is the case. See Box 3-2.

4

Method

The charge to the Committee on Prospering in the Global Economy of the 21st Century constitutes a challenge both daunting and exhilarating: To recommend to the nation specific steps that can best strengthen the quality of life in America—our prosperity, our health, our security. This chapter is an overview of the committee's methods for arriving at its recommendations and for identifying the specific steps it proposes for their implementation. Chapters 5-8 identify the committee's list of action items. Appendix E is an overview of the committee's investment cost of its proposed actions and programs. Appendix F provides the rationale for the K–12 programs proposed in Chapter 5.

Despite a demanding schedule for completion of the study, members reviewed literature and case studies, studied the results of other expert panels, and convened focus groups with expertise in K–12 education, higher education, research, innovation and workforce issues, and national and homeland security to arrive at a slate of recommendations.

The focus groups, involving over 66 individual experts, were asked to identify, within their issue areas, the three recommendations they believed were of the highest urgency. The results became raw material for the committee's discussion of recommendations. The committee later met numerous times via conference call to refine its recommendations as it consulted with additional experts. Final coordination involved extensive e-mail interactions as the committee sought to avail itself of the technology that is pervading modern decision-making and making the world "flat," in the words of Thomas Friedman (see Chapter 1).

REVIEW OF LITERATURE AND PAST COMMITTEE RECOMMENDATIONS

Before meeting in person, the committee requested a compilation of the results of past studies on the topics it was likely to address. Appendix D provides these background papers on topics such as science, mathematics, and technology education; research funding and productivity; the environment for innovation; and science and technology issues in national and homeland security.

The committee used those documents as a means to review the work of many other groups. Some were individual writers and scholars[1] and others were blue ribbon groups, such as the one chaired by former Senator John Glenn, which produced the report *Before It's Too Late*[2] for the National Commission on Mathematics and Science Teaching for the 21st Century and others at the Council on Competitiveness,[3] Center for Strategic and International Studies,[4] Business Roundtable,[5] Taskforce on the Future of American Innovation,[6] President's Council of Advisors on Science and Technology,[7] National Science Board,[8] and other National Academies committees, such as those which produced *A Patent System for the 21st Century,*[9] *Policy Implications of International Graduate Students and Postdoctoral Scholars in the United States,*[10] and *Advanced Research Instrumentation and Facili-*

---

[1]R. B. Freeman. *Does Globalization of the Scientific/Engineering Workforce Threaten US Economic Leadership?* NBER Working Paper 11457. Cambridge, MA: National Bureau of Economic Research, 2005.

[2]*Before It's Too Late: A Report to the Nation from the National Commission on Mathematics and Science Teaching for the 21st Century.* Glenn Commission Report. Washington, DC: US Department of Education, 2000.

[3]Council on Competitiveness. *Innovate America.* Washington, DC: Council on Competitiveness, 2004.

[4]Center for Strategic and International Studies. *Global Innovation/National Competitiveness.* Washington, DC: Center for Strategic and International Studies, 1996.

[5]Business Roundtable. *Tapping America's Potential.* Washington, DC: Business Roundtable, 2005.

[6]Task Force on the Future of American Innovation. *The Knowledge Economy: Is America Losing Its Competitive Edge?* Washington, DC: Task Force on the Future of American Innovation, 2005.

[7]The President's Council of Advisors on Science and Technology. *Sustaining the Nation's Innovation Ecosystems.* Report on Information Technology Manufacturing and Competitiveness, January 2004.

[8]National Science Board. *Science and Engineering Indicators 2004.* NSB 04-01. Arlington, VA: National Science Foundation, 2004.

[9]National Research Council. *A Patent System for the 21st Century.* Washington, DC: The National Academies Press, 2004.

[10]The National Academies. *Policy Implications of International Graduate Students and Postdoctoral Scholars in the United States.* Washington, DC: The National Academies Press, 2005.

ties.[11] Others were the committee and analyst at other organizations who have gone before us producing reports focusing on the topics discussed in this report. There are too many to mention here, but they are cited throughout the report and range from individual scholars to the Glenn Commission on K–12 education, the Council on Competitiveness, the President's Council of Advisors on Science and Technology, the National Science Board, and other National Academies committees. Such work and the reaction to it once published were invaluable to the committee's deliberations.

The committee decided to provide a "box" in each chapter containing alternative points of view as captured in a review of existing reports, studies, reviewer comments, and informal consultations with experts and policy-makers.

The committee examined numerous case studies to gain a better understanding of which policies had the most potential to influence national prosperity. For example, many of the recommendations on K–12 and higher education rely on extrapolating successful state or local programs to the national level. The committee also reviewed existing federal programs for higher education and research policy that work well in one place and could potentially be applicable to other parts of the federal infrastructure. The committee also studied other nations' experiences in implementing policy changes to encourage innovation.

FOCUS GROUPS

The focus groups (Appendix C) convened experts in five broad areas—K–12 education, higher education, science and technology research policy, innovation and workforce issues, and homeland security. Group members were asked to identify ways the United States can successfully compete, prosper, and be secure in the global community of the 21st century.

Their contributions were compiled with the results of the literature search and with recommendations gathered during committee interviews. More than *150 concrete recommendations and implementation steps* were identified and discussed at a weekend focus group session in Washington, DC. Each focus group, following its own discussions, presented its top three proposed recommendations to the committee members and to other focus-group participants.

COMMITTEE DISCUSSION AND ANALYSIS

The committee itself met over that same weekend and then in weekly conference calls. Using the focus-group recommendations as a starting point,

[11]NAS/NAE/IOM. *Advanced Research Instrumentation and Facilities*. Washington, DC: The National Academies Press, 2006.

the committee developed four key recommendations (labeled A through D in this report), which it ranked, and 20 actions to implement them. It assigned ratings of either *most urgent* or *urgent* to each of the four recommendations. They are summarized here. Specific implementing actions are discussed in later sections of this report.

Most Urgent

10,000 Teachers, 10 Million Minds, and K–12 Science and Mathematics Education. Increase America's talent pool by vastly improving K–12 science and mathematics education.

Sowing the Seeds Through Science and Engineering Research. Sustain and strengthen the nation's traditional commitment to long-term basic research that has the potential to be transformational to maintain the flow of new ideas that fuel the economy, provide security, and enhance the quality of life.

Urgent

Best and Brightest in Science and Engineering Higher Education. Make the United States the most attractive setting in which to study and perform research so that we can develop, recruit, and retain the best and brightest students, scientists, and engineers from within the United States and throughout the world.

Incentives for Innovation. Ensure that the United States is the premier place in the world to innovate; invest in downstream activities such as manufacturing and marketing; and create high-paying jobs that are based on innovation by modernizing the patent system, realigning tax policies to encourage innovation and the location of resulting facilities in the United States, and ensuring affordable broadband access.

Unless the nation has the science and engineering experts and the resources to generate new ideas, and unless it encourages the transition of those ideas through policies that enhance the innovation environment, we will not continue to prosper in an age of globalization. Each recommendation represents one element of an interdependent system essential for US prosperity.

Some of the committee's proposed actions and programs involve changes in the law. Some require substantial investment. Funding would ideally come from reallocation of existing funds, but if necessary, via new funds. The committee believes the investments are small relative to the return the nation can expect in the creation of new high-quality jobs, inasmuch as economic studies show that the social rate of return on federal and private investment in research is often 30% or more (Tables 2-1 and 2-2). The committee fully recognizes the extant demands on the federal budget, but it believes that few problems facing the nation have more profound implications for America than the one addressed herein and, thus believes, that the investment it entails should be given high priority.

CAUTIONS

The committee has been cautious in its analysis of information. However, the available information is, in some instances, insufficient for the committee's needs. In addition, the limited timeframe to develop the report (10 weeks from the time of the committee's meeting to report release) is inadequate to conduct an independent analysis. Even if unlimited time were available, definitive analysis of many issues is simply not possible given the uncertainties involved.

The recommendations in this report rely heavily on the experience, consensus views, and judgments of the committee members. Although the committee consists of leaders from academe, industry, and government—including several current and former industry chief executive officers, university presidents, researchers (including three Nobel prize winners), and former presidential appointees—the array of topics and policies covered in this study is so broad that it was impossible to assemble a committee of 20 members with directly relevant expertise in each. The committee has therefore relied heavily on the judgments of experts in the study's focus groups, additional consultations with other experts, and the panel of 37 expert reviewers.

The recommendations herein should be subjected to continuing evaluation and refinement. In particular, the committee encourages regular evaluations to determine the efficacy of its policy recommendations in reaching the nation's goals. If the proposals prove successful, more investment may be warranted. If not, programs should be modified or dropped from the portfolio.

CONCLUSION

The committee's recommendations are the fundamental actions the nation should take if it is to prosper in the 21st century. Just as "reading, writing, and arithmetic" are essential for any student to succeed—regardless of career—"education, research, and innovation" are essential if the nation is to succeed in providing jobs for its citizenry.

5

What Actions Should America Take in K–12 Science and Mathematics Education to Remain Prosperous in the 21st Century?

10,000 TEACHERS, 10 MILLION MINDS

Recommendation A: *Increase America's talent pool by vastly improving K–12 science and mathematics education.*

The US system of public education must lay the foundation for developing a workforce that is literate in mathematics and science, among other subjects. It is the creative intellectual energy of our workforce that will drive successful innovation and create jobs for all citizens.

In 1944, during the final phases of a global war, President Franklin D. Roosevelt asked Vannevar Bush, his White House director of scientific research, to study areas of public policy having to do with science. The president observed, "New frontiers of the mind are before us, and if they are pioneered with the same vision, boldness and drive with which we have waged this war, we can create a fuller and more fruitful employment and a fuller and more fruitful life." In the intervening years, our country appears to have lost sight of the importance of scientific literacy for our citizens, and it has become increasingly reliant on international students and workers to fuel our knowledge economy.

The lack of a natural constituency for science causes short- and long-term damage. Without basic scientific literacy, adults cannot participate effectively in a world increasingly shaped by science and technology. Without a flourishing scientific and engineering community, young people are not motivated to dream of "what can be," and they will have no motivation to become the next generation of scientists and engineers who can address persistent national problems, including national and homeland security, healthcare, the provision of energy, the preservation of the environment, and the growth of the economy, including the creation of jobs.

Laying a foundation for a scientifically literate workforce begins with developing outstanding K–12 teachers in science and mathematics.[1] A highly qualified corps of teachers is a critical component of the No Child Left Behind initiative.[2] Improvements in student achievement are solidly linked to teacher excellence, the hallmarks of which are thorough knowledge of content, solid pedagogical skills, motivational abilities, and career-long opportunities for continuing education.[3] Excellent teachers inspire young people to develop analytical and problem-solving skills, the ability to interpret information and communicate what they learn, and ultimately to master conceptual understanding. Simply stated, teachers are the key to improving student performance.

Today there is such a shortage of highly qualified K–12 teachers that many of the nation's 15,000 school districts[4] have hired uncertified or underqualified teachers. Moreover, middle and high school mathematics and science teachers are more likely than not to teach outside their own fields of study (Table 5-1). A US high school student has a 70% likelihood of being taught English by a teacher with a degree in English but about a 40% chance of studying chemistry with a teacher who was a chemistry major.

These problems are compounded by chronic shortages in the teaching workforce. About two-thirds of the nation's K–12 teachers are expected to retire or leave the profession over the coming decade, so the nation's schools will need to fill between 1.7 million and 2.7 million positions[5] during that

---

[1]See, for example, The Glenn Commission. *Before It's Too Late: A Report to the Nation from the National Commission on Mathematics and Science Teaching for the 21st Century.* Washington, DC: US Department of Education, 2000.

[2]No Child Left Behind Act of 2001. Pub. L. No. 107-110, signed by President George W. Bush on January 8, 2001, 107th Congress.

[3]National Research Council. *Learning and Understanding: Improving Advanced Study of Mathematics and Science in U.S. Schools.* Washington, DC: National Academy Press, 2002.

[4]National Center for Education Statistic. 2006. "Public Elementary and Secondary Students, Staff, Schools, and School Districts: School Year 2003–04." Available at: http://nces.ed.gov/pubs2006/2006307.pdf.

[5]National Center for Education Statistics. *Predicting the Need for Newly Hired Teachers in the United States to 2008-09.* NCES 1999-026. Washington, DC: US Government Printing Office, 1999. Available at: http://nces.ed.gov/pubs99/1999026.pdf. According to the Bureau of Labor Statistics, job opportunities for K–12 teachers over the next 10 years will vary from good to excellent, depending on the locality, grade level, and subject taught. Most job openings will be attributable to the expected retirement of a large number of teachers. In addition, relatively high rates of turnover, especially among beginning teachers employed in poor, urban schools, also will lead to numerous job openings for teachers. Competition for qualified teachers among some localities will likely continue, with schools luring teachers from other states and districts with bonuses and higher pay. See http://stats.bls.gov/oco/ocos069.htm#emply.

TABLE 5-1 Students in US Public Schools Taught by Teachers with No Major or Certification in the Subject Taught, 1999-2000

Discipline	Grades 5–8	Grades 9–12
English	58%	30%
Mathematics	69%	31%
Physical science	93%	63%
Biology–life sciences	—	45%
Chemistry	—	61%
Physics	—	67%
Physical education	19%	19%

SOURCE: National Center for Education Statistics. *Qualifications of the Public School Teacher Workforce: Prevalence of Out-of-Field Teaching 1987-1988 to 1999-2000.* Washington, DC: US Department of Education, 2003.

period, about 200,000 of them in secondary science and mathematics classrooms.[6]

We need to recruit, educate, and retain excellent K–12 teachers who fundamentally understand biology, chemistry, physics, engineering, and mathematics. The critical lack of technically trained people in the United States can be traced directly to poor K–12 mathematics and science instruction. Few factors are more important than this if the United States is to compete successfully in the 21st century.

The Committee on Prospering in the 21st Century recommends a package of K–12 programs that is based on tested models, including financial incentives for teachers and students and high standards for, and measurable achievement by, teachers, students, and administrators. The programs will create broad-based academic leadership for K–12 mathematics and science, and they will provide for rigorous curricula. Support for the action items in this recommendation should have the highest priority for the federal government as it addresses America's ability to compete for quality jobs in the future.

The strengths of the proposed actions derive from their focus on *teachers*—those who are entering the profession and those who currently teach science and mathematics—*and on the students they will teach.* The recommendations cover the spectrum of K–12 teachers, and several programs are recommended to tailor education for different populations. Each recommendation has specific, measurable objectives. At the same time, we must emphasize the need for research and evaluation to serve as a foundation for

[6]National Research Council. *Attracting Science and Mathematics PhDs to Secondary School Education.* Washington, DC: National Academy Press, 2000. Available at: http://www.nap.edu/catalog/9955.html.

change in K–12 mathematics and science education. In particular, a better understanding of what actions can be taken to excite children about science, mathematics, and technology would be useful in designing future educational programs.

The first two action items focus on K–12 teacher education and professional development. They are designed to give new K–12 science, mathematics, and technology teachers a solid science, mathematics, and technology foundation; provide continuing professional development for current teachers and for those entering the profession from technology-sector jobs so they gain mastery in science and mathematics and the means to teach those subjects; and provide continuing education for current teachers in grades 6–12 so they can teach vertically aligned advanced science and mathematics courses.[7] One fortunate spinoff of enhanced education of K–12 teachers is that salaries—in many school districts—are tied to teacher educational achievements.

ACTION A-1: 10,000 TEACHERS FOR 10 MILLION MINDS

Annually recruit 10,000 science and mathematics teachers by awarding 4-year scholarships and thereby educating 10 million minds. Our public education system must attract at least 10,000 of our best college graduates to the teaching profession each year. A competitive federal scholarship program will allow bright, motivated students to earn bachelors' degrees in science, engineering, and mathematics with concurrent certification as K–12 mathematics and science teachers.

Students could enter the program at any of several points and would receive annual scholarships of up to $20,000 per year for tuition and qualified educational expenses. Awards would be given on the basis of academic merit.[8] Each scholarship would carry a 5-year postgraduate commitment to teach in a public school.[9]

---

[7]"Vertically aligned curricula" use sequenced materials over several years. An example is pre-algebra followed by algebra, geometry, trigonometry, pre-calculus, and calculus. The systematic approach to education reform emphasizes that teachers, school and district administrative personnel, and parents work together to align their efforts. See, for example, Southwest Education Development Laboratory. "Alignment in SEDL's Working Systemically Model, 2004 Progress Report to Schools and Districts." Available at: http://www.sedl.org/rel/resources/ws-report-summary04.pdf.

[8]Teacher education programs would be 4 years in duration with multiple entry points. A first-year student entering the program would be eligible for a 4-year scholarship, while students entering in their second or later undergraduate years would be eligible for fewer years of support.

[9]If the scholarship recipients do not fulfill the 5-year service requirement, they would be obligated to repay a prorated portion of their scholarship.

To provide the highest quality education for students who want to become teachers, it is important to award competitive matching grants of $1 million per year, to be matched on a one-for-one basis, for 5 years to help 100 universities and colleges establish integrated 4-year undergraduate programs that lead to bachelors' degrees in physical and life sciences, mathematics, computer science, and engineering *with teacher certification.*[10] To qualify, science, technology, engineering, and mathematics (STEM) departments would collaborate with colleges of education to develop teacher education and certification programs with in-depth content education and subject-specific education in pedagogy. STEM departments also would offer high-quality research experiences and thorough training in the use of educational technologies. Colleges or universities without education departments or schools could collaborate with such departments in nearby colleges or universities.

A well-prepared corps of teachers is central to the development of a literate student population.[11] The National Center for Teaching and America's Future unequivocally shows the positive effect of better teaching on student achievement.[12] The Center for the Study of Teaching[13] reported that the most consistent and powerful predictor of student achievement in science and mathematics was the presence of teachers who were fully certified and had at least a bachelor's degree in the subjects taught. Teachers with content expertise, like experts in all fields, understand the structure of their disciplines and have cognitive "roadmaps" for the work they assign, the assessments they use to gauge student progress, and the questions they ask in the classroom.[14] The investment in educating those teachers is money well spent because they are likely to prepare internationally competitive students.

---

[10]The institutional awards would be matching grants awarded competitively to applicants who had identified partners, such as universities, industries, or philanthropic foundations, to contribute additional resources. Public-public and public-private consortia would be encouraged. Institutions that demonstrate success would be eligible for competitive renewals.

[11]National Research Council. *Attracting PhDs to K–12 Education: A Demonstration Program for Science, Mathematics, and Technology.* Washington, DC: The National Academies Press, 2002.

[12]National Center for Teaching and America's Future. *Doing What Matters Most: Teaching for America's Future.* New York: NCTAF, 1996. See also H. C. Hill, B. Rowan, and D. L. Ball. "Effects of Teachers' Mathematical Knowledge for Teaching on Student Achievement." *American Educational Research Journal* 42(2)(2005):371-406.

[13]L. Darling-Hammond. *Teacher Quality and Student Achievement: A Review of State Policy Evidence.* New York: Center for the Study of Teaching and Policy, 1999. Available at: http://depts.washington.edu/ctpmail/Publications/PDF_versions/LDH_1999.pdf.

[14]National Research Council. *How People Learn: Brain, Mind, Experience, and School: Expanded Edition.* Washington, DC: National Academy Press, 2000. Available at: http://books.nap.edu/catalog/6160.html.

Some of the nation's top research universities are leading the way to prepare a cadre of highly skilled teachers. Two in particular have developed innovative programs that combine undergraduate degrees in science, technology, engineering, or mathematics with pedagogy education and teacher certification.

UTeach, a program in the College of Natural Sciences, headed by the Dean of Natural Sciences at the University of Texas (UT) at Austin, recruits from among the 25% of undergraduate science and mathematics students who express a serious desire to teach. As a result of this program, UT-Austin has been able to increase the number of science and math teachers it graduates who have both degrees in a science or mathematics as well as teacher certification.

Program enrollees have SAT scores above the average for the university's College of Natural Sciences, have higher grade point averages, and are retained in the degree program at more than twice the rate of other students in that college (Figure 5-1). UTeach has a 26% minority enrollment, compared with 16% universitywide.

Each year the program graduates about 70 students who have teaching certification and bachelors' degrees in chemistry, physics, computer science, biology, or mathematics. Students receive strong practical education and continuing mentoring, especially in the critical first few years in the classroom, as that increases effectiveness and promotes professional retention as teachers. As also shown in Figure 5-1, UTeach graduates have deep disciplinary grounding, they know how to engage students in scientific inquiry, and they know how to use new technology to improve student achievement. The UTeach experience shows that an effective scholarship program must be coupled with a teacher education program that is interesting and attractive to students. The program's most effective tools are the field experience courses for first-year students and the use of master teachers as their supervisors.

Starting with the current academic year, the 10-campus University of California (UC) system offers its California Teach program, which, by 2010, should graduate a thousand highly qualified science and mathematics teachers each year.[15] California Teach provides every STEM student in the university with an opportunity to complete the STEM major and pedagogical training in a 4-year program. Early in the program, students work as paid classroom assistants in elementary and middle schools, supervised by mentor teachers. Students enroll in seminars taught by master teachers and participate in 10-week summer institutes to help them develop methods for

[15]Even more teachers may come from a similar program being conducted by the California state university system.

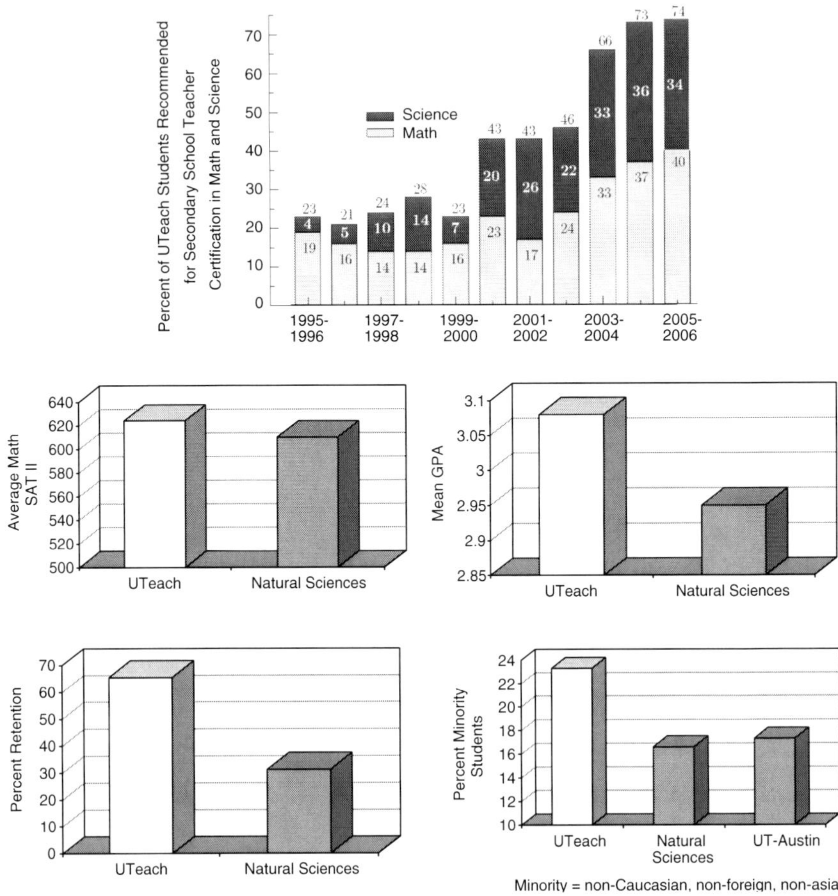

FIGURE 5-1 UTeach minority enrollment, quality of undergraduate students in the certification recommendations program, student retention, and performance compared with all students in the UT-Austin College of Natural Sciences.
SOURCE: Information based on e-mail from M. Marder of UTeach to D. Stine dated February 2, 2006.

teaching in a specific discipline. Students from throughout the university system in the California Teach program who satisfactorily complete their courses through the junior year participate in subject-area institutes. UC-San Diego, for example, might host a high school chemistry institute that would be open to students and faculty from all campuses.

At each institute, students and faculty (those from UC, those who are visiting, and master secondary school teachers) collaborate to develop case study videos of teaching methods and approaches that will be archived by the University of California television system for use by students and faculty in subsequent institutes and by teachers in the field. Students develop the portfolios that eventually will be required of teachers to become certified by a national board. Students who complete the institutes receive $5,000 scholarships.

Both the UTeach and California Teach programs provide a continuum of pre- and in-service teacher education and professional development and established cohorts and relationships that are crucial for retaining the most talented individuals in the profession. California Teach also will provide the nation with a large-scale experiment to show which elements of teacher preparation are most effective. Replicating the strong points of such programs around the country will transform the quality of our science and mathematics teaching.[16]

ACTION A-2: A QUARTER OF A MILLION TEACHERS INSPIRING YOUNG MINDS EVERY DAY

Strengthen the skills of 250,000 teachers through training and education programs at summer institutes, in master's programs, and in Advanced Placement (AP) and International Baccalaureate (IB) training programs. Excellent professional development models exist to strengthen the skills of the 250,000 current mathematics and science teachers, but they reach too few in the profession. The four-part program recommended by the committee consists of (1) summer institutes, (2) master's degree programs in science and mathematics, (3) training for advanced placement and International Baccalaureate teachers, and (4) development of a voluntary national K–12 science and mathematics curriculum.

We need to reach all K–12 science and mathematics teachers and provide them with high-quality continuing professional development opportunities—specifically those that emphasize rigorous content education. High-quality, content-driven professional development has a significant effect on student performance, particularly when augmented with classroom practice, year-long mentoring, and high-quality curricular materials.[17]

---

[16]The National Academies has also published a report on demonstration programs for PhD K–12 teacher programs: National Research Council. *Attracting PhDs to K–12 Education: A Demonstration Program for Science, Mathematics, and Technology*. Washington, DC: The National Academies Press, 2002.

[17]D. K. Cohen and H. C. Hill. "Instructional Policy and Classroom Performance: The Mathematics Reform in California." *Teachers College Record* 102(2)(2000):294-343; W. H. Schmidt, C. McKnight, R. T. Houang, and D. E. Wiley. "The Heinz 57 Curriculum: When More May Be Less." Paper presented at the 2005 annual meeting of the American Education

About 10% of the nation's 3 million K–12 teachers provide instruction in science and mathematics in middle and high schools.[18] The No Child Left Behind Act requires all of them to participate regularly in professional development, and in most states professional development already is required to maintain teaching credentials. Funding for continuing education now comes from the No Child Left Behind appropriation and from the states.

As the number of programs has ballooned, many teachers report that they are "buried in opportunities" for continuing education. They also complain that it is difficult to know which programs are worthwhile and which are irrelevant and disconnected. The object of this implementation action is to identify outstanding programs that improve content knowledge and pedagogical skills, especially for those who enter the profession from other careers. Over 5 years, these programs could reach all teachers of middle and high school mathematics and science. Furthermore, as these teachers become more qualified, they can be provided increased financial rewards without confronting the historical culture that largely dismisses the concept of pay-for-performance.

Action A-2 Part 1: Summer Institutes

In the first implementation action, the committee recommends a summer education program for 50,000 classroom teachers each year. Matching grants would be provided on a one-for-one basis to state and regional summer institutes to develop and provide 1- to 2-week sessions. The expected federal investment per participant is about $1,200 per week, excluding participant stipends, which would be covered by local school districts.

Summer institutes for secondary school teachers of science and mathematics have existed in various forms at least since the 1950s, often with corporate sponsors.[19] The National Science Foundation (NSF) started funding teacher institutes in 1953, when shortages of adequately trained person-

---

Research Association, Montreal, Quebec; National Research Council. *Educating Teachers of Science, Mathematics, and Technology: New Practices for a New Millennium.* Washington, DC: National Academy Press, 2001; National Research Council. *Improving Teacher Preparation and Credentialing Consistent with the National Science Education Standards: Report of a Symposium.* Washington, DC: National Academy Press, 1997.

[18]In 1999-2000, the latest year for which we have figures, of the total number of public K–12 teachers, 191,000 taught science (including biology, physics, and chemistry) and 160,000 taught mathematics.

[19]Summer institutes at Union College in Schenectady and at the Case Institute of Technology in Cleveland were supported by the General Electric Company, institutes at the University of Minnesota were supported by the Ford Foundation, and institutes at the University of Tennessee were supported by the Martin Marietta Corporation.

nel in scientific and technical fields became increasingly evident.[20] In 2004, the NSF Math and Science Partnership began making awards under a new program, Teacher Institutes for the 21st Century.[21]

There is a particularly strong need for elementary and middle school teachers to have a deeper education in science and mathematics.[22] Many school children are systematically discouraged from learning science and mathematics because of their teachers' lack of preparation, or in some cases, because of their teachers' disdain for science and mathematics. In many school systems, no science at all is taught before middle school.

Teachers who are not required to teach science have little reason to increase their knowledge and skills through professional development. No Child Left Behind requirements, however, will expand testing to the sciences in 2007. Elementary school teachers thus need training now in many areas of science; they need to see the relationships between mathematics and the sciences; and, most important if they are to excite young minds, they need the ability to integrate information across disciplines. In short, *all* teachers need to be scientifically literate and preferably excited about science.

The Merck Institute for Science Education (MISE)[23] is an in-service professional development program for K–6 teachers established in 1993 with a 10-year commitment from Merck & Company. An intensive 3-year course combines multiple-year summer institutes in inquiry-based science instruction that is tied to state and national standards with in-classroom follow-up and reinforcement from September to June. MISE also provides curriculum materials and training in their use. The current participants are K–6 teachers in New Jersey and Pennsylvania public schools. In all, about 4,000 teachers have participated in the program. Analysis by an external evaluator indicates that students of teachers who participated in MISE pro-

---

[20]Funding for institutes for the continuing education of high school science teachers began to decline in number in the late 1960s, when the shortages of technical personnel including science teachers, began to decline. After a leveling period during the 1970s, National Science Foundation support for teacher institutes was discontinued in 1982. Support for the teacher institute programs was resumed the following year following several national reports detailing the severe problems facing science teaching and with growing recognition of the shortage of qualified science teachers.

[21]These awards are directed to disciplinary faculty of higher education institutions to work with experienced teachers of mathematics and the sciences to deepen teachers' content knowledge and instructional skills so they may become school-based intellectual leaders in their fields.

[22]National Research Council. *Science for All Children: A Guide to Improving Elementary Science Education in Your School District*. Washington, DC: National Academy Press, 1997.

[23]"Merck Institute for Science Education (MISE)." Available at: http://www.mise.org/mise/index.jsp.

fessional development programs for at least 3 years substantially outperformed those whose teachers participated for a year or less.[24]

Local MISE programs have made science a priority in each district. New science frameworks and instructional materials developed by MISE have been adopted by all of the participating districts. Added benefits are seen in improvements in hiring and recruitment of teachers and administrators, increased expenditures for instructional materials, changes in how teachers are observed and evaluated in the classroom, augmented instructional support services, development of new districtwide science assessments, and the leveraging of significant additional external resources for science education programs. MISE also has helped to lead the way in the creation of statewide science content standards and professional development standards.

Similar to MISE in its focus on K–6 science education is the Washington State Leadership and Assistance for Science Education Reform (LASER) program,[25] which began in 1999 with a strategic planning institute to coordinate standards, curricula, and evaluation. Six more institutes have convened since then, and now 131 school districts, enrolling more than 60% of Washington's students, are at various stages of implementing an inquiry-based science program.[26]

In 2005, achievement in the 5th-grade science portion of the Washington Assessment of Student Learning (WASL) was measured and correlated with teacher participation in LASER. Primary among the findings was a significant relationship between professional development among teachers and the percentage of students meeting the science standard on the 2004 test (Figure 5-2). LASER teachers' classroom practices changed incrementally until they had more than 80 hours of professional development; at that point, more dramatic shifts to inquiry-based methods were observed.

---

[24]Consortium for Policy Research in Education. 2002. *A Report on the Eighth Year of the Merck Institute for Science Education.* Philadelphia, PA: CPRE, University of Pennsylvania, 2002. Available at: http://www.mise.org/pdf/cpre2000_2001.pdf. When MISE was created in 1995, there were no districtwide or state assessments in science in Pennsylvania or New Jersey, where MISE programs were based. The absence of assessment often meant that less attention was given to science in elementary classrooms, and it meant that there was no easy way to measure the impact of MISE's work on student learning. MISE has been exploring the use of performance tasks for districtwide assessment. For the past two years, performance tasks drawn from the Third International Mathematics and Science Study (TIMSS) have been administered in grades 3 and 7 in all four districts. This has been a collaborative project involving MISE staff, central office staff, and many interested teachers.

[25]"Washington State Leadership and Assistance for Science Education Reform (LASER) Program." Available at: http://www.wastatelaser.org.

[26]"Inquiry" is a set of interrelated processes by which scientists and students pose questions about the natural world and investigate knowledge. Using an inquiry-based approach students learn science in a way that reflects how science actually works. See National Research Council. *National Science Education Standards.* Washington, DC: National Academy Press, 1995.

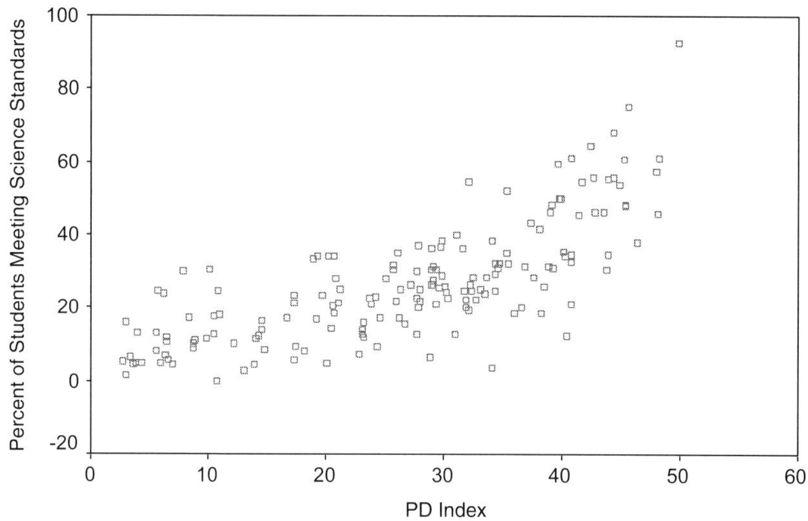

FIGURE 5-2 Professional development index relative to percent of students meeting science standards. Professional development of teachers increases student achievement in science. The scatter plot shows the PD index (total professional development hours per 100 students provided over a 3-year period to the teachers of 5th graders who took the WASL in spring 2004) compared with the percentage of students who met the WASL standards. Each box represents a school. There is a gradual increase in the percentage of students meeting the standard as the PD index increases. The data suggest the rate of increase accelerated after teachers received a critical amount of professional development, although the exact point at which that change occurred cannot be determined without access to classroom-level aggregates and the ability to track the professional development of the teachers of individual students. The relationship between professional development and student achievement holds even after adjustments for the influence of percentage of students eligible for free and reduced-price lunches and for the percentage of Asian students.
SOURCE: D. Schatz, D. Weaver, and P. D. Finch. *Washington State LASER—Evaluation Results*. WSTA Journal (July 22, 2005).

The system of national laboratories also can be tapped for continuing education of K–12 teachers. The Laboratory Science Teacher Professional Development program was designed by the Office of Science in the Department of Energy (DOE) to create a cadre of outstanding middle and high school science and mathematics teachers who will serve as leaders in their local and regional teaching communities.[27] Through this 3-year program, teachers establish long-term relationships with DOE mentor scientists and

[27]US Department of Energy, Office of Science, Office of Workforce Development for Teachers and Scientists. "Laboratory Science Teacher Professional Development Program: About LSTPD." Available at: http://www.scied.science.doe.gov/scied/LSTPD/about.htm.

with teaching colleagues. Teachers are expected to spend at least 4 weeks at one of the DOE laboratories during the first year and at least 2 weeks at one of the laboratories for each of 2 years after that. If such a program were used to train two teachers from each of the 15,000 school districts in the country over a 10-year period, about 3,200 teachers each year would be brought into the 17 DOE laboratories, eventually reaching a 3-year steady state of 9,600 teachers. The Science and Mathematics Education Task Force of the Secretary of Energy Advisory Board is currently reviewing such a proposal.[28]

The National Aeronautics and Space Administration (NASA) also has an educational program whose focus is to "inspire and motivate students to pursue careers in science, technology, engineering, and mathematics." It supports education in schools and also participates in informal education and public outreach efforts. NASA's programs focus on increasing elementary and secondary education participation in NASA programs; enhancing higher education capability in science, technology, engineering, and mathematics disciplines; increasing participation by underrepresented and underserved communities; expanding e-education; and expanding NASA's participation with the informal-education community. Among its activities for teachers and students are summer academies at its flight centers and workshops.[29]

Action A-2 Part 2: Science and Mathematics Master's Programs

The second element of this implementation action would, through part-time 2-year master's degree programs granted by the colleges of science and engineering (working with the colleges of education) at the nation's research universities, enhance the education and skills of current middle and high school science, mathematics, and technology teachers as well as those with science, mathematics, and engineering degrees who decide to pursue teaching either upon graduation or later in their career.

The master's in science education programs (identified for each specific field) would take place over three full-time summers plus alternate weekends during the academic year in science, mathematics, and technology education for current teachers. Over the course of 5 years, it would enhance the education and skills of 50,000 current science, mathematics, and technology teachers nationwide and qualify them for higher pay under existing rules in nearly all school districts.

---

[28]US Department of Energy. "Secretary of Energy Advisory Board: Subcommittees." Available at: http://www.seab.energy.gov/sub/committees.htm.

[29]NASA. "Overview: NASA Education Programs." February 1, 2004. Available at: http://education.nasa.gov/edprograms/overview/index.html.

To implement this action, the committee recommends that the federal government provide 100 to 125 academic research universities (2 or more per state) the ability to offer four to five programs in mathematics, biology, chemistry, physics, engineering, computer science, or integrated science for a total of 500 competitive institutional grants nationwide. The programs would focus on content education and pedagogy and would each provide in-classroom training and continuous evaluation for approximately 20 in-service middle and high school teachers and career changers.[30]

The program's master teachers[31] would provide leadership in their own districts for all the programs included in this recommendation. They would be mentors for new college graduates teaching in their schools and for the many very able current teachers who would welcome the opportunity to upgrade their skills through summer institutes or education to become AP or IB teachers or pre-AP–IB teachers. Teachers who complete the program would receive federally funded incentive stipends of $10,000 annually for up to 5 years provided that they remain in the classroom and engage in teaching leadership activities.[32] Once the 5-year limit has been reached, teachers can pursue national certification for which many states offer a financial bonus.

Students learn best from teachers who have strong content knowledge and pedagogical skills.[33] Unfortunately, it is uncertain what science and mathematics preparation, beyond the basics, is the best training for teachers. Nonetheless, it is known that teachers need to stay current with their disciplines. Master's degree programs, particularly those emphasizing content knowledge, keep teachers updated and provide working teachers the skills to teach for the future.

The Science Teacher Institute in the University of Pennsylvania's School of Arts and Sciences and Graduate School of Education[34] is a rigorous pro-

---

[30]An example of such a program is Math for America's Newton fellowship program in New York City. In this 5-year program, new and mid-career scientists, engineers, and mathematics receive a stipend to pursue a master's level teaching program, obtain a teaching certificate, begin teaching, and are mentored, coached, and provided support as they begin their teaching career. See http://www.mathforamerica.org.

[31]This program may be even more effective if such master teachers would be nationally board certified, and would then become a national pool of teacher leaders.

[32]Such master teachers should also be eligible for some release time from classroom teaching to engage in leadership activities.

[33]National Research Council. *Learning and Understanding: Improving Advanced Study of Mathematics and Science in U.S. Schools*. Washington, DC: National Academy Press, 2002; M. Cochran-Smith and K. M. Zeichner. *Studying Teacher Education*. Washington, DC: American Educational Research Association, 2005; M. Allen. 2003. *Eight Questions on Teacher Preparation: What Does the Research Say?* Washington, DC: Education Commission of the States, 2003. Available at: http://www.ecs.org/tpreport/.

[34]"Science Teacher Institute." Available at: http://www.sas.upenn.edu/PennSTI/.

gram that trains middle and high school science teachers. Eighty percent of the education is in a participant's scientific discipline and 20% percent is in pedagogy, emphasizing the secondary-classroom applications of inquiry-based instruction. At the end of 2 years (three summers and alternate Saturdays during the school year), teachers graduate with master's of science degrees in chemistry education or integrated science education. Those teachers have demonstrated a major influence in their schools.[35] They mentor other teachers, update the schools' curricula, and recruit students into demanding science courses. They are the "teachers of teachers" who provide the academic leadership so urgently needed in school districts across the country.

An additional 50,000 of those truly outstanding teachers could inspire and support students and other teachers to work harder at mathematics and science. Our recommendation would provide the funding and structure to reach about one-sixth of the nation's science and mathematics teachers—about three teachers in each of the more than 15,000 school districts in the nation.

Action A-2 Part 3: Advanced Placement, International Baccalaureate, and Pre-AP/IB Education

The third implementation action for the K–12 educational recommendation is a program to train an additional 70,000 AP and IB teachers and 80,000 pre-AP/IB teachers of mathematics and science (at present, the AP program serves many more students than does the IB program). Teachers from schools where there are few or no AP or IB courses would receive priority for this program. The model for this recommendation is the College Board's AP program, which has wide acceptance in secondary and higher education. It also could be implemented in schools certified by the International Baccalaureate Organization. So long as they demonstrate satisfactory performance, AP and IB teachers would receive incentives to attend professional development seminars and to tutor and prepare students outside regular classroom hours. Under the proposed program, their development fees would be paid, and they would receive a bonus for each student who passed an AP or IB examination in mathematics or science. Implementation in each state would require the creation of a non-profit organization staffed by talented master teachers who would help local schools manage the program and enforce high standards.

---

[35]C. Blasie and G. Palladino. "Implementing the Professional Development Standards: A Research Department's Innovative Masters Degree Program for High School Chemistry Teachers." *Journal of Chemical Education* 82(4)(2005):567-570.

The model for this recommendation is the Dallas-based AP Incentive Program (APIP),[36] which offers financial incentives to prepare teachers to teach demanding courses to ever-increasing numbers of secondary school students. To serve as large a percentage of students as possible, APIP has been coupled with a pre-AP program, Laying the Foundation, which begins in the 6th grade to help students prepare for 11th-and 12th-grade AP and IB examinations. Teachers use vertically aligned lessons based on national standards and final, comprehensive end-of-course examinations to measure mastery of essential concepts. The process continues through middle and high schools to ensure that graduating seniors are prepared for college work.

The foundation for each program is intensive, 4-year professional development, focused on content, delivered by the College Board and by master teachers in local school districts. Assuming satisfactory performance, AP/IB teachers can, under the proposed program, receive annual incentive payments of $1,800 and pre-AP teachers receive annual incentive payments of $1,000. AP/IB teachers also receive a $100 bonus for each student who passes an AP examination in mathematics or science. Pre-AP teachers receive a $25 bonus for each students who passes the end-of-course examination.

To reach currently underserved areas or populations of students with specific learning needs, it might be useful to consider implementing online learning. The University of California College Prep program (UCCP) makes AP courses available to students who enroll individually or as part of a school group. In either case, they have online access to teachers and tutors. The more than 5,000 students currently enrolled are taught by certified teachers and tutored by paid university undergraduates and graduate students.

---

[36]APIP is part of a statewide initiative to raise educational standards. See Texas Education Agency. *Advanced Placement and International Baccalaureate Examination Results in Texas, 2001-2002*. Doc. No. GE03 601 08. Austin, TX: TEA, 2003. In 2001, the Texas Legislature enacted the Gold Performance Acknowledgement (GPA) system to acknowledge districts and campuses for high performance on indicators not used to determine accountability ratings (TEC, §39.0721, 2001). Included is an AP/IB indicator that measures the percentage of non-special-education students who take an AP or IB examination and the combined percentage of non-special-education examinees at or above the criterion score on at least one AP or IB examination (TEC §39.0721, 2001). The percentage of examinations with high scores on AP or IB was kept as a report-only performance indicator (TEA, 2002). GPA acknowledgment is given when non-special-education 11th- and 12th-graders take at least one AP or IB examination, represent 15% or more of the non-special-education in 11th- and 12th-grade students, and 50% or more of those examinees have at least one score of 3 or above on an AP examination or 4 or above on an IB examination.

Action A-2 Part 4: K–12 Curricular Materials Modeled on World-Class Standards

The fourth part of the K–12 recommendation asks the Department of Education to convene a national panel to collect K–12 science and mathematics teaching materials that have been proven effective or develop new ones where no effective models exist. All materials would be made available online, free of charge, as a voluntary national curriculum that would provide an effective high standard for K–12 teachers.

High-quality teaching is grounded in careful vertical alignment of curricula, assessments, and student achievement standards. Efforts to directly evaluate curricular quality have often foundered in the past,[37] but the need still exists. Excellent resources for the development of K–12 science, technology, and mathematics curricular materials include the National Science Education Standards,[38] Project 2061,[39] and numerous Web-based compendia, including the National Science Digital Library.[40] Gateway to Educational Materials (GEM), sponsored by the US Department of Education, is a collaborative effort to collect materials and provide them free to educators. The GEM Web site offers more than 20,000 educational resources, catalogued by type and grade level. Although GEM can be cumbersome to use, it has been lauded as an exemplary effort. GEM also has made it clear that teacher education programs need to add a technology component.[41]

Project Lead the Way (PLTW) is a national program with partners in public schools, colleges and universities, and the private sector.[42] The project

---

[37]Math and Science Expert Panel. *Exemplary Promising Mathematics Programs.* Washington, DC: US Department of Education, 1999; National Research Council. *On Evaluating Curricular Effectiveness: Judging the Quality of K–12 Mathematics Evaluations.* Washington, DC: The National Academies Press, 2004.

[38]National Research Council. *National Science Education Standards.* Washington, DC: National Academy Press, 1996; National Council of Teachers of Mathematics. *Principles and Standards for School Mathematics.* Washington, DC: NCTM, 2000. Available at: http://standards.nctm.org.

[39]Project 2061, sponsored by the American Association for the Advancement of Science, is an initiative to reform K–12 education nationwide so that all high school graduates are science literate. In the first stage of its work, Project 2061 published *Science for All Americans*, which outlines what all students should know and be able to do in science, mathematics, and technology after 13 years of schooling. See F. J. Rutherford and A. Ahlgren. *Science for All Americans.* Washington, DC: AAAS, October 1990. Available at: http://www.project2061.org/default_flash.htm.

[40]The "National Digital Science Library." See: http://nsdl.org.

[41]For example, see M. A. Fitzgerald and J. McClendon. 2002. "The Gateway to Educational Materials: An Evaluation Study, Year 3." A technical report submitted to the US Department of Education, October 10, 2002. Available at: http://www.geminfo.org/Evaluation/Fitzgerald_02.10.pdf.

[42]PLTW is now offered in 45 states and the District of Columbia. See http://www.pltw.org/aindex.htm.

has developed a 4-year sequence of courses that, when combined with college preparatory mathematics and science, introduces students to the scope, rigor, and discipline of engineering and engineering technology. PLTW also has developed a middle school technology curriculum, Gateway to Technology. Students participating in PLTW courses are better prepared for college engineering programs than those exposed only to the more traditional curricula.

Comprehensive teacher education is a critical component of PLTW, and the curriculum uses cutting-edge technology and software that require specialized education. Continuing education supports teachers as they implement the program and provides for continuous improvement of skills.

ACTION A-3: ENLARGE THE PIPELINE

Enlarge the pipeline of students who are prepared to enter college and graduate with a degree in science, engineering, or mathematics by increasing the number of students who pass AP and IB science and mathematics courses. The competitiveness of US knowledge industries will be purchased largely in the K–12 classroom: We must invest in our students' mathematics and science education. A new generation of bright, well-trained scientists and engineers will transform our future only if we begin in the 6th grade to significantly enlarge the pipeline and prepare students to engage in advanced coursework in mathematics and science.

The "other side" of the classroom equation, of course, is the students,[43] our innovators of the future.[44] Despite expressing an interest in the subjects, many US students avoid rigorous high school work in mathematics and science.[45] All US students should be held to high expectations, and rigorous coursework should be available to all students. Particular attention should be paid to increasing the participation of those students in groups that are underrepresented in science, technology, and mathematics education, training, and employment.

The first goal of the proposed action is to have 1,500,000 students taking at least one AP or IB mathematics or science examination by 2010, an increase to 23% from 6.5% of juniors and seniors who took at least one AP or IB mathematics or science examination in 2004. We also must in-

[43]National Research Council. *Engaging Schools: Fostering High-School Students' Motivation to Learn.* Washington, DC: The National Academies Press, 2004.

[44]K. Hunter. "Education Key to Jobs, Microsoft CEO Says." *Stateline.org*, August 17, 2005.

[45]T. Lewin. *Many Going to College Are Not Ready, Report Says.* New York Times, August 17, 2005. Among those who took the 2005 American College Testing (ACT), only 51% achieved the benchmark in reading, 26% in science, and 41% in mathematics; the figure for English was 68%.

crease the number of students who pass those examinations from 230,000 in 2004 to at least 700,000 by 2010. AP and IB programs would be voluntary and open to all and would give students a head start by providing them with college-level courses taught by outstanding high school teachers.[46] The result will be better prepared undergraduates who will have a better chance of completing their bachelor's degrees in science, engineering, and mathematics.[47] Table 5-2 shows that a student who passes an AP examination has a better chance overall—regardless of ethnicity—of completing a bachelor's degree within 6 years. Students would be eligible for a 50% examination fee rebate and a $100 mini-scholarship for each passing score on an AP or IB mathematics or science examination.

This action is built on standards, testing, and incentives to achieve excellence in science and mathematics. The APIP program has been successful across gender, ethnicity, and economic groups. The program proposed herein would give students the further background they need to study science, engineering, and mathematics as undergraduates.

Such advanced coursework can provide the foundation for students to be internationally competitive in the fields of focus. For example, US students who passed AP calculus in 2000 were administered the 1995 Trends in International Mathematics and Science Study (TIMSS) test.[48] Their scores were significantly higher than the average 1995 US score, and they were higher

---

[46]One researcher estimates that each year 25,000 interested and adequately prepared students in the United States are told they cannot take AP or IB courses. He further speculates that another 75,000 or more students who could do well elect not to take them because no one encourages them to do so. See J. Mathews. *Class Struggle: What's Wrong (and Right) with America's Best Public High Schools.* New York: Times Books, 1998. Limiting access to advanced study occurs in all kinds of educational settings, including the most competitive high schools in America—schools with adequate resources, qualified teachers, and well-prepared students. Those schools, while typically advocating college preparation for everyone, create layers of curricular differentiation, such that only a select group of students are allowed entrance into certain AP and honors courses; other students are placed in less vigorous courses. See P. Attewell. "The Winner Take-All High School: Organizational Adaptations to Educational Stratification." *Sociology of Education* 74(4)(2001):267-296. For a larger discussion of access to advanced coursework, see National Research Council. 2002. *Learning and Understanding: Improving Advanced Study of Mathematics and Science in U.S. Schools.* Washington, DC: National Academy Press, 2002.

[47]Academic opportunities such as AP and IB programs benefit students in several ways. High school students who participate in AP and IB courses and associated examinations are exposed to college-level academic content and are challenged to complete more rigorous coursework. Students with qualifying examination scores are provided the opportunity to earn college credit or advanced placement, depending on the college or university they attend. Texas Education Agency. *Advanced Placement and International Baccalaureate Examination Result in Texas 2003-2004.* Document no. GE05 601 11. Austin, TX, 2005. P. 6.

[48]See Chapter 3 or Appendix D for more detailed discussion of the exam. Available at: http://nces.ed.gov/timss/.

TABLE 5-2 Six-Year Graduation Rate of Students Who Passed AP Examinations and Students Who Did Not Take AP Examinations

Ethnicity	Passed AP Examination	Did Not Take AP Examination
White	72%	30%
Hispanic	62%	15%
Blacks	60%	17%

NOTES: Data are for all students graduating from Texas public high schools in 1998 and enrolling in a Texas public college or university (88,961 students). AP examinations were given in the core subjects of English, mathematics, science, and social studies to students in grades 10–12. The percentage shown is the proportion of students who obtained bachelor's degrees or higher within 6 years of secondary-school graduation.
SOURCE: National Center for Educational Accountability at: http://www.nc4ea.org.

than the 1995 average scores of the students from all 14 participating countries. Similarly, US students who passed AP physics in 2000 outperformed the 1995 US national TIMSS average and exceeded the 1995 scores for all participating countries except Norway (Table 5-3). It is clear that engaging K–12 students in challenging courses taught by qualified teachers will enhance their educational experiences and may increase the number of students who enter college and complete higher education degrees.

Data from the Texas APIP demonstrate that combining incentives and teacher education can increase student participation (Figure 5-3), and APIP has increased academic performance for minority students in high school. The Dallas school district is the nation's 12th largest. It has a 93% minority enrollment, and 81% of its students come from low-income households. Yet Dallas students achieve outstanding AP results. African American and Hispanic students pass AP examinations in mathematics, science, and English at a rate four times higher than the national average for minority students, and female students pass the examinations at twice the national rate.[49]

EFFECTIVE CONTINUING PROGRAMS

The committee proposed expansion of two additional approaches to improving K–12 science and mathematics education that are already in use:

• **Statewide Specialty High Schools.** An effective way to increase student achievement in science and mathematics is to provide an intensive

[49]Passing rate is calculated as number of students passing exam per 1,000 junior and senior high school students in the Dallas Independent School District compared with all of Texas and all of the United States.

TABLE 5-3 Achievement of US AP Calculus and Physics Students Who Participated in the Trends in International Mathematics and Science Study (TIMSS) in 2000 Compared with Average International Scores from 1995

Mathematics		Physics	
	Average Score		Average Score
US AP calculus students scoring 3, 4, or 5	**596**	Norway	581
US AP calculus students	**573**	**US AP physics students scoring 3, 4, or 5**	**577**
France	557	Sweden	573
Russian Federation	542	Russian Federation	545
Switzerland	533	**US AP physics students**	**529**
Australia	525	Germany	522
Cyprus	518	Australia	518
Lithuania	516	*International Average*	*501*
Greece	513	Cyprus	494
Sweden	512	Latvia	488
Canada	509	Switzerland	488
International Average	*501*	Greece	486
Italy	474	Canada	485
Czech Republic	469	France	466
Germany	465	Czech Republic	451
United States	**442**	Austria	435
Austria	436	**United States**	**423**

NOTE: Advanced placement scores on a 5-point scale; 3 is considered a passing score by the College Board, the organization that administers the courses, and colleges and universities generally require a score of 3, 4, or 5 to qualify for course credit.
SOURCE: E. J. Gonzalez, K. M. O'Connor, and J. A. Miles. *How Well Do Advanced Placement Students Perform on the TIMSS Advanced Mathematics and Physics Tests?* International Study Center, Lynch School of Education, Boston College, June 2001. Available at: http://www.timss.org.

learning experience for high-performing students.[50] These schools immerse students in high-quality science and mathematics education, serve as testing grounds for curricula and materials, provide in-classroom educational opportunities for K–12 teachers, and have the resources and staff for summer programs to introduce students to science and mathematics. One model is the North Carolina School of Science and Mathematics (NCSSM), which opened in 1980. NCSSM enrolls juniors and seniors from most of North Carolina's 100 counties. NCSSM's unique living and learning experience

[50]K. Powell. "Science Education: Hothouse High." *Nature* 435(June 16, 2005):874-875.

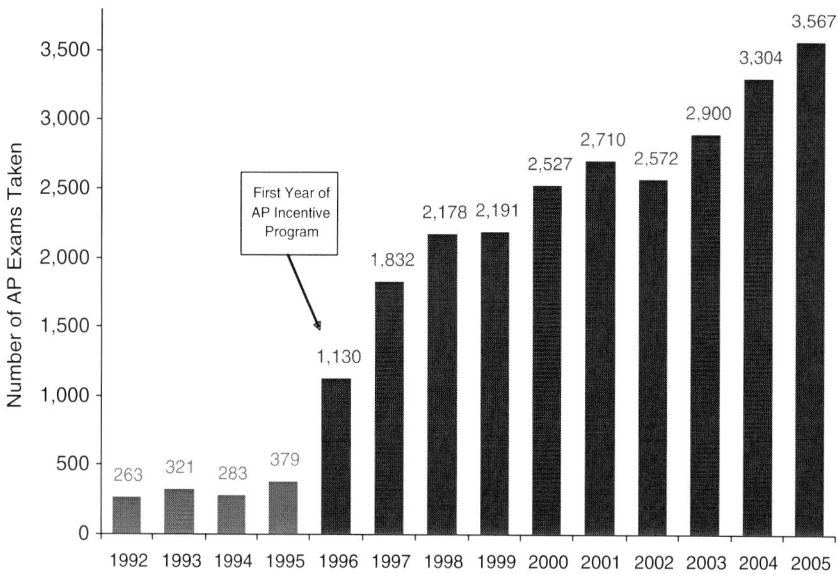

FIGURE 5-3 The number of AP examinations in mathematics, science, and English taken in APIP schools in the Dallas Independent School District (DISD). The number of AP examinations taken has increased more than 9-fold over 10 years.
SOURCE: Advanced Placement Strategies. 2005. The 2004 results are based on updated data received from the Dallas Independent School District for AP examinations in mathematics, science, and English.

made it the model for 16 similar schools around the world. It is the first school of its kind in the nation—a public, residential high school where students study a specialized science and mathematics curriculum. At NCSSM, teachers come for a "sabbatical year," and the school has a structure and the personnel it needs to offer summer institutes for outstanding students.

• **Inquiry-Based Learning.** Summer research programs stimulate student interest and achievement in science, mathematics, and technology. Programs that involve several institutions or public–private partnerships should be encouraged, as should those designed to stimulate low-income and minority student participation.

CONCLUSION

Public education is potentially our country's most valuable asset, yet our system has too long ignored the development of critical teaching and workforce skills.

BOX 5-1
Another Point of View: K–12 Education

Some of those who provided comments to the committee questioned the ability of K–12 reform based on the existing US educational model to produce effective, long-lasting improvements in the way our children learn. The United States currently spends more per student than all but one other country (Switzerland),[a] but it is losing ground in educational performance. Its relatively low student achievement through high school clearly shows that the system is inefficient, and dedicating additional funding to this system is not a guarantee of success. In fact, the biggest concerns involve disparate quality among K–12 institutions and the difficulty of measuring success.

Some question whether K–12 education in the United States really suffers from low student achievement. International comparisons might serve merely to highlight the huge funding inequities among US school districts.[b] American scholastic achievement, unlike that in most other Western nations, varies widely from school to school and even from state to state. Eighth graders in high-achieving states score even in mathematics with students in the highest-achieving foreign countries. Some in other states score, on the average, about even with schoolchildren in scarcely developed nations. In the United States, many more suburban school districts can provide smaller classes, better-paid teachers, and more computers than can the schools for most urban and rural children. The underprivileged groups struggle with gross overcrowding, decayed buildings, and inadequate funding even for basic instruction. Standardized test scores generally reflect the disparate distribution of resources.

The committee has examined a number of educational programs that have been demonstrated to work, identified core program components—strong content knowledge, practical pedagogical training, ongoing mentoring and education, and incentives—and recommended that programs be implemented as one would implement a research program: with built-in benchmarks, evaluations, and ongoing education—with the expectation that no one program will fit every situation.

Thorough education in science, mathematics, and technology will start students on the path to high-technology jobs in our knowledge economy. To develop an innovative workforce, we must begin now to improve public education in science and mathematics.[51]

[51]For another point of view on K–12 education reform, see Box 5-1.

Some commentators also argue that in industrialized countries there is no correlation between school achievement and economic success but that educational reforms often are the least controversial way of planning social improvement.[c] School changes are less threatening than are direct structural changes, which can involve confronting the whole organization of industry and government. Reforming education, it is claimed, is easier and less expensive than examining and correcting the societal problems that affect our schools directly—economic weaknesses, wealth and income inequality, an aging population, the prevalence of violence and drug abuse, and the restructuring of work.

Because there is not a well-developed literature on the effectiveness of K–12 learning and teaching interventions, it is challenging to recommend programs with high confidence. For example, some have argued that the International Baccalaureate program has established neither teacher qualifications nor standards for faculties and that the Advanced Placement curriculum needs better quality control.[d] Others have suggested that summer teacher-education programs are merely vehicles for textbook companies; others argue that any teacher-education programis worthless unless there is a strong in-classroom, continuing mentoring component.

---

[a]Organisation for Economic Co-operation and Development. *Education at a Glance 2005.* Paris: OECD, 2005. Available at: http://www.oecd.org/dataoecd/41/13/35341210.pdf.

[b]D. C. Berliner and B. J. Biddle. *The Manufactured Crisis: Myths, Fraud, and the Attack on America's Public Schools.* New York: Addison-Wesley, 1995.

[c]Ibid.

[d]National Research Council. *Learning and Understanding: Improving Advanced Study of Mathematics and Science in U.S. Schools.* Washington, DC: National Academy Press, 2002.

Virtually all quality jobs in the global economy will require certain mathematical and scientific skills. The committee's objectives are to ensure that all students will gain these necessary skills and have the opportunity to become part of a cadre of world-class scientists and engineers who can create the new products that will in turn broadly enhance the nation's standard of living. In short, our goal in producing highly qualified scientists and engineers is to ensure that, through their innovativeness, high-quality jobs are available to all Americans.

When fully implemented, the committee's recommendations will produce the academic achievement in science, mathematics, and technology that every student should exhibit and will afford numerous opportunities for further learning. Excellent teachers, increasing numbers of students meeting high academic standards, and measurable results will become the academic reality.

6

What Actions Should America Take in Science and Engineering Research to Remain Prosperous in the 21st Century?

SOWING THE SEEDS

Recommendation B: *Sustain and strengthen the nation's traditional commitment to long-term basic research that has the potential to be transformational to maintain the flow of new ideas that fuel the economy, provide security, and enhance the quality of life.*

Flat or declining research budgets for federal agencies and programs hamper long-term basic and high-risk research, funding for early-career researchers, and investments in infrastructure. Yet all of those activities are critical for attracting and retaining the best and brightest students in science and engineering and producing important research results. These factors are the seeds of innovation for the applied research and development on which our national prosperity depends.

The Committee on Prospering in the Global Economy of the 21st Century has identified a series of actions that will help restore the national investment in research in mathematics, the physical sciences, and engineering. The proposals concern basic-research funding, grants for researchers early in their careers, support for high-risk research with a high potential for payoff, the creation of a new research agency within the US Department of Energy (DOE), and the establishment of prizes and awards for breakthrough work in science and engineering.

ACTION B-1: FUNDING FOR BASIC RESEARCH

The United States must ensure that an adequate portion of the federal research investment addresses long-term challenges across all fields, with

the goal of creating new technologies. The federal government should increase our investment in long-term basic research—ideally through reallocation of existing funds,[1] but if necessary via new funds—by 10% annually over the next 7 years. It should place special emphasis on research in the physical sciences, engineering, mathematics, and information sciences and basic research conducted by the Department of Defense (DOD). This special attention does not mean that there should be a disinvestment in such important fields as the life sciences (which have seen substantial growth in recent years) or the social sciences. A balanced research portfolio in all fields of science and engineering research is critical to US prosperity. Increasingly, the most significant new scientific and engineering advances are formed to cut across several disciplines. Investments should be evaluated regularly to reprioritize the research portfolio—dropping unsuccessful programs or venues and redirecting funds to areas that appear more promising.

The United States currently spends more on research and development (R&D) than the rest of the G7 countries combined. At first glance (see Box 6-1), it might seem questionable to argue that the United States should invest more than it already does in R&D. Furthermore, federal spending on nondefense research nearly doubled, after inflation, from slightly more than $30 billion in fiscal year (FY) 1976 to roughly $55 billion in FY 2004.[2]

However, the committee believes that the commitment to *basic* research, particularly in the physical sciences, mathematics, and engineering, is inadequate. In 1965, the federal government funded more than 60% of all US R&D; by 2002 that share had fallen below 30%. During the same period, there was an extraordinary increase in corporate R&D spending: IBM, for example, now spends more than $5 billion annually[3]—more than the entire federal budget for physical sciences research. Corporate R&D has thus become the linchpin of the US R&D enterprise, but it cannot replace federal investment in R&D, because corporations fund relatively little basic research—for several reasons: basic research typically offers greater benefits to society than to its sponsor; it is almost by definition risky and shareholder pressure for short-term results discourages long-term, speculative investment by industry.

Although federal funding of R&D as a whole has increased in dollar terms, its share of the gross domestic product (GDP) dipped from 1.25% in 1985 to about 0.78% in 2003 (Figure 6-1). Furthermore, in recent years much of the federal research budget has been shifted to the life sciences. From 1998 to 2003, funding for the National Institutes of Health (NIH)

---

[1]The funds could come from anywhere in an agency, not just other research funds.

[2]P. N. Spotts. "Pulling the Plug on Science?" *Christian Science Monitor*, April 14, 2005.

[3]"Corporate R&D Scorecard." *Technology Review*, September 2005. Pp. 56-61.

BOX 6-1
Another Point of View: Research Funding

The committee heard commentary from several respondents who believe that current R&D funding is robust and that significant additional federal funding for research is unjustified. Their arguments include the following:

- Overall, research and development spending in the United States is high by international standards and continues to increase. Total R&D spending (government and industry) has remained remarkably consistent as a percentage of the gross domestic product, indicating that R&D spending has kept pace with the relatively rapid growth of the US economy. The fraction of the US federal domestic *discretionary* budget devoted to science has remained practically constant for the last 30 years.
- Annual nondefense research spending by the federal government has nearly doubled in real terms since 1976 and exceeds $56 billion per year—more than that in the rest of the G-7 countries combined. Government funding of overall basic research is increasing in real dollars and holding its own as a percentage of GDP.
- Additional federal funds should not be committed without better programmatic justification and improved processes to ensure that such funds are used effectively. Increases in federal R&D funding should be based on specific demonstrated needs rather than on a somewhat arbitrary decision to increase funds by a given percentage.

Some critics also worry about the challenges of implementing a rapid increase in research funding. For example, they say that doubling the NIH budget was a precipitous move. It takes time to recruit new staff and expand laboratory space, and by the time capacity has expanded, the pace of budget increases has\ve slowed and researchers have difficulty in readjusting. Others fear that reallocating additional funds to basic research will draw resources away from the commercialization efforts that are a critical part of the innovation system.

doubled; funding for the physical sciences, engineering, and mathematics has remained relatively flat for 15 years (Figure 6-2).

The case of the National Science Foundation (NSF) illustrates the trends. Despite the *authorization* in 2002 to double NSF's budget over a 5-year period, its funding has actually decreased in recent years.[4] This af-

[4]American Association for the Advancement of Science. "Historical Data on Federal R&D, FY 1976-2006." March 22, 2005. Available at: http://www.aaas.org/spp/rd/hist06p2.pdf.

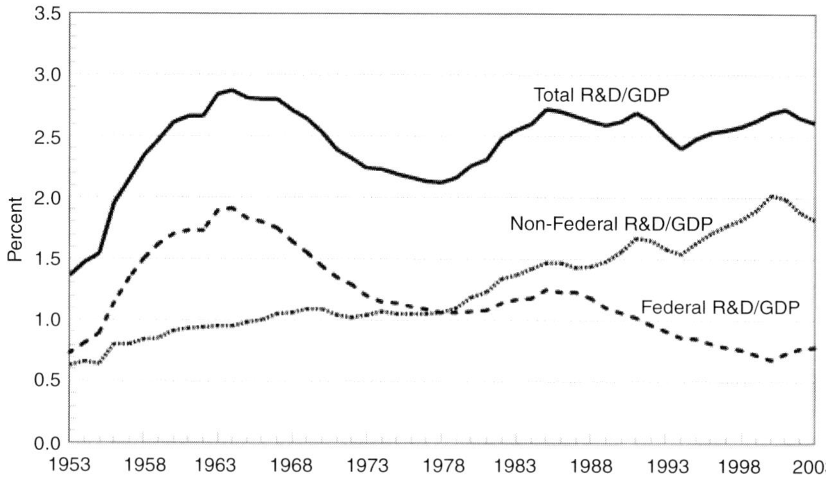

FIGURE 6-1 Research and development shares of US gross domestic product, 1953-2003.
SOURCE: NSF Division of Science Resources Statistics. "National Patterns of Research Development Resources," annual series. Appendix Table B-9. Available at: http://www.nsf.gov/statistics/nsf05308/sectd.htm.

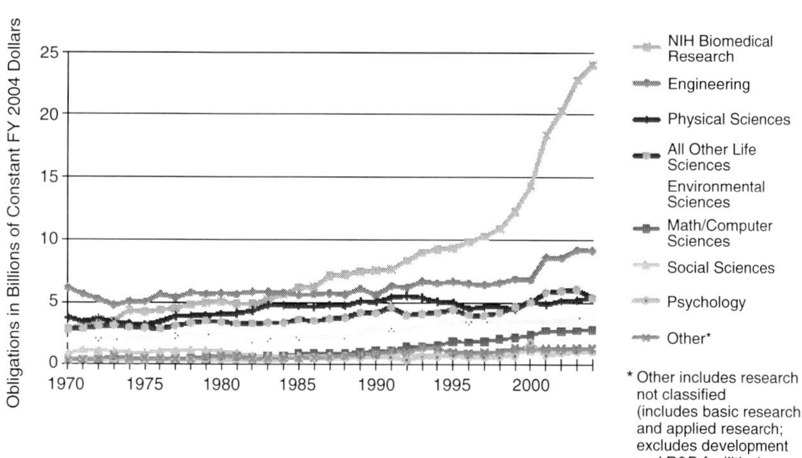

FIGURE 6-2 Trends in federal research funding by discipline, obligations in billions of constant FY 2004 dollars, FY 1970-FY 2004. Trends in federal research funding show the life sciences increasing rapidly in the late 1990s; funding for research in mathematics, computer sciences, the physical sciences, and engineering remained relatively steady.
SOURCE: American Association for the Advancement of Science. "Trends in Federal Research by Discipline, FY 1970-2004." Available at: http://www.aaas.org/spp/rd/discip04.pdf.

fects both the number and the grant size of researcher proposals funded. In 2004, for example, only 24% of all proposals to NSF were funded, the lowest proportion in 15 years.[5]

Ultimately, increases in research funding must be justified by the results that can be expected rather than by the establishment of overall budget targets. But there is a great deal of evidence today that agencies do not support high-potential research because funding will not allow it. Furthermore, because of lack of funds, NSF in 2004 declined to support $2.1 billion in proposals that its independent external reviewers rated as very good or excellent.[6]

The DOD research picture is particularly troubling in this regard. As the US Senate Committee on Armed Services has noted, "investment in basic research has remained stagnant and is too focused on near-term demands."[7] A 2005 National Research Council panel's assessment is similar: "In real terms the resources provided for Department of Defense basic research have declined substantially over the past decade."[8] Reductions in funding for basic research at DOD—in the "6.1 programs"—have a particularly large influence outside the department. For example, DOD funds 40% of the engineering research performed at universities, including more than half of all research in electrical and mechanical engineering, and 17% of basic research in mathematics and computer science.[9]

The importance of DOD basic research is illustrated by its products—in defense areas these include night vision; stealth technology; near-real-time delivery of battlefield information; navigation, communication, and weather satellites; and precision munitions. But the investments pay off for civilian applications too. The Internet, communications and weather satellites, global positioning technology, the standards that became JPEG, and even the search technologies used by Google all had origins in DOD basic research. John Deutch and William Perry point out that "the [Department of Defense] technology base program has also had a major effect on American industry. Indeed, it is the primary reason that the United States leads the world today in information technology."[10]

---

[5]National Science Board. *Report of the National Science Board on the National Science Foundation's Merit Review Process Fiscal Year 2004*. NSB 05-12. Arlington, VA: National Science Board, March 2005. P. 7.

[6]Ibid., pp. 5, 21.

[7]The Senate Armed Services Committee. Report 108-046 accompanying S.1050, National Defense Authorization Act for FY 2004.

[8]National Research Council. *Assessment of Department of Defense Basic Research*. Washington, DC: The National Academies Press, 2005. P. 4.

[9]Ibid., p. 21.

[10]J. M. Deutch and W. J. Perry. *Research Worth Fighting For*. New York Times, April 13, 2005. P. 19.

There is also a significant federal R&D budget for homeland security. For FY 2006 the total is nearly $4.4 billion across all agencies. The Department of Homeland Security itself has a $1.5 billion R&D budget, but only a small portion—$112 million—is earmarked for basic research. The rest will be devoted to applied research ($399 million), development ($746 million), and facilities and equipment ($210 million).[11]

Business organizations, trade associations, military commissions, bipartisan groups of senators and representatives, and scientific and academic groups have all reiterated the critical importance of increased R&D investment across our economic, military, and intellectual landscape (Table 6-1). After reviewing the proposals provided in the table and other related materials, the committee concluded that a 10% annual increase over a 7-year period would be appropriate. This achieves the doubling that was in principle part of the NSF Authorization Act of 2002 but would expand it to other agencies, albeit over a longer period. The committee believes that this rate of growth strikes an appropriate balance between the urgency of the issue being addressed and the ability of the research community to apply new funds efficiently.

The committee is recommending special attention to the physical sciences, engineering, mathematics, and the information sciences and to DOD basic research to restore balance to the nation's research portfolio in fields that are essential to the generation of both ideas and skilled people for the nation's economy and national and homeland security. Most assuredly, this does not mean that there should be a *disinvestment* in such important fields as the life sciences or the social sciences. A balanced research portfolio in all fields of science and engineering research is critical to US prosperity.

As indicated in the National Academies report *Science, Technology, and the Federal Government: National Goals for a New Era,* the United States needs to be among the world leaders in all fields of research so that it can

- Bring the best available knowledge to bear on problems related to national objectives even if that knowledge appears unexpectedly in a field not traditionally linked to that objective.
- Quickly recognize, extend, and use important research results that occur elsewhere.

---

[11]American Association for the Advancement of Science. *R&D Funding Update March 4, 2005—Homeland Security R&D in the FY 2006 Budget.* Available at: http://www.aaas.org/spp/rd/hs06.htm1.

TABLE 6-1 Specific Recommendations for Federal Research Funding

Source	Report	Recommendation
Rep. Frank Wolf (R-Virginia), chair, Subcommittee on Commerce, Justice, Science, and Related Agencies	Letter to President George W. Bush, May 2005	Triple federal basic R&D over the next decade
US Congress and President Bush	NSF Authorization Act of 2002, passed by Congress; signed by the President	Double the NSF budget over 5 years to reach $9.8 million by FY 2007
US Commission on National Security in the 21st Century (Hart–Rudman)	*Road Map for National Security: Imperative for Change, The Phase III Report,* 2001	Double the federal R&D budget by 2010
Defense of Defense	*Quadrennial Defense Review Report*, 2001	Allocate at least 3% of the total DOD budget for defense science and technology
President's Council of Advisors on Science and Technology (PCAST)	*Assessing the US R&D Investment,* January 2003	Target the physical sciences and engineering to bring them "collectively to parity with the life sciences over the next 4 budget cycles"
Coalition of 15 industry associations, including US Chamber of Commerce, National Association of Manufacturers, and Business Roundtable	*Tapping America's Potential: The Education for Innovation Initiative,* 2005	Increase R&D spending, particularly for basic research in the physical sciences and engineering, at NSF, NIST, DOD, and DOE by at least 7% annually
167 Members of Congress	Letter to Rep. Wolf, chair, Subcommittee on Commerce, Justice, Science, and Related Agencies, May 4, 2005	Increase NSF budget to $6.1 billion in FY 2006, 6% above the FY 2005 request
68 Senators	Letter to Sen. Pete Domenici (R-New Mexico), chair, Energy and Water Development Subcommittee	Increase funding for DOE Office of Science by an inflation-adjusted 3.2% over FY 2005 appropriation, a 7% increase over the Bush administration's FY 2006 request

TABLE 6-1 continued

Source	Report	Recommendation
Council on Competitiveness	*Innovate America*, 2004	Allocate at least 3% of the total DOD budget for defense science and technology; direct at least 20% of that amount to long-term, basic research; intensify support for the physical sciences and engineering
National Science Board	*Fulfilling the Promise: A Report to Congress on the Budgetary and Programmatic Expansion of the National Science Foundation*, NSB 2004-15	Fund NSF annually at $18.7 billion, including about $12.5 billion for R&D

NOTES: NSF, National Science Foundation; DOD, Department of Defense; NIST, National Institute of Standards and Technology; DOE, Department of Energy.

- Prepare students in American colleges and universities to become leaders who can extend the frontiers of knowledge and apply new concepts.
- Attract the brightest young students both domestically and internationally.[12]

ACTION B-2: EARLY-CAREER RESEARCHERS

The federal government should establish a program to provide 200 new research grants each year at $500,000 each, payable over 5 years, to support the work of outstanding early-career researchers. The grants would be funded by federal agencies (NIH, NSF, DOD, DOE, and the National Aeronautics and Space Administration [NASA]) to underwrite new research opportunities at universities and government laboratories.

About 50,000 people hold postdoctoral appointments in the United States.[13] Those early-career researchers are particularly important because they often are the forefront innovators. A report in the journal *Science* states

[12]NAS/NAE/IOM. *Science, Technology, and the Federal Government: National Goals for a New Era*. Washington, DC: National Academy Press, 1993.

[13]National Science Foundation. "WebCASPAR, Integrated Science and Engineering Data System." Available at: http://www.casper.nsf.gov.

that postdoctoral scholars (those who had completed doctorates but who had not yet obtained long-term research positions) comprised 43% of the first authors on the research articles it published in 1999.[14] However, as funding processes have become more conservative and as money becomes tighter, it has become more difficult for junior researchers to find support for new or independent research. In 2002, the median age at which investigators received a first NIH grant was 42 years, up from about 35 years in 1981.[15] At NSF, the percentage of first-time applicants who received grant funding fell from 25% in 2000 to 17% in 2004.[16]

There is a wide divergence among fields in the use of postdoctoral researchers and in the percentages heading toward industry rather than academe. Recent trends suggest that more students are opting for postgraduate study and that the duration of postdoctoral appointments is increasing, particularly in the life sciences.[17] But new researchers face challenges across a range of fields.

The problem is particularly acute in the biomedical sciences. In 1980, investigators under the age of 40 received more than half of the competitive research awards; by 2003, fewer than 17% of those awards went to researchers under 40.[18] Both the percentage and the number of awards made to new investigators—regardless of age—have declined for several years; new investigators received fewer than 4% of NIH research awards in 2002.[19] One conclusion is that academic biomedical researchers are spending long periods at the beginning of their careers unable to set their own research directions or establish their independence. New investigators thus have diminished freedom to risk the pursuit of independent research, and they continue instead with their postdoctoral work or with otherwise conservative research projects.[20]

Postdoctoral salaries are relatively low,[21] although several federal programs support early-career researchers in tenure-track or equivalent posi-

---

[14]G. Vogel. "A Day in the Life of a Topflight Lab." *Science* 285(1999):1531-1532.

[15]National Research Council. *Bridges to Independence: Fostering the Independence of New Investigators in Biomedical Research*. Washington, DC: The National Academies Press, 2005. P. 37.

[16]National Science Board, March 2005.

[17]National Research Council. *Bridges to Independence: Fostering the Independence of New Investigators in Biomedical Research*. Washington, DC: The National Academies Press, 2005. P. 43.

[18]Ibid., p. 43.

[19]Ibid., p. 1.

[20]Ibid., p. 1.

[21]A Sigma Xi survey found that the median postdoctoral salary was $38,000—below that of all bachelor's degree recipients ($45,000). See G. Davis. "Doctors Without Orders." *American Scientist* 93(3, Supplement)(May–June 2005).

tions. The NSF Faculty Early Career Development Program makes 350-400 awards annually, ranging from $400,000 to nearly $1 million over 5 years, to support career research and education.[22] Corresponding DOD programs include the Office of Defense Programs' Early Career Scientist and Engineer Award and the Navy Young Investigator Program. The Presidential Early Career Award for Scientists and Engineers (PECASE) is the highest national honor for investigators in the early stages of their careers. In 2005, there were 58 PECASE awards that each provided funding of $100,000 annually for 5 years (Table 6-2). Still, that group is a tiny fraction of the postdoctoral research population.

In making its recommendation, the committee decided to use the PECASE awards as a model for the magnitude and duration of awards. In determining the number of awards, the committee considered the number of awards in other award programs and the overall reasonableness of the extent of the program.

ACTION B-3: ADVANCED RESEARCH INSTRUMENTATION AND FACILITIES

The federal government should establish a National Coordination Office for Advanced Research Instrumentation and Facilities to manage a fund of $500 million per year over the next 5 years—ideally through reallocation of existing funds, but if necessary via new funds—for construction and maintenance of research facilities, including the instrumentation, supplies, and other physical resources researchers need. Universities and the government's national laboratories would compete annually for the funds.

Advanced research instrumentation and facilities (ARIF) are critical to successful research that benefits society. For example, eight Nobel prizes in physics were awarded in the last 20 years to the inventors of new instrument technology, including the electron and scanning tunneling microscopes, laser and neutron spectroscopy, particle detectors, and the integrated circuit.[23] Five Nobel prizes in chemistry were awarded for successive generations of mass-spectrometry instruments and applications.

Advanced research instrumentation and facilities[24] are defined as instrumentation and facilities housing closely related or interacting instruments and includes networks of sensors, databases, and cyberinfrastructure.

---

[22]J. Tornow, National Science Foundation, personal communication, August 2005.

[23]National Science Board. *Science and Engineering Infrastructure for the 21st Century: The Role of the National Science Foundation.* Arlington, VA: National Science Foundation, 2003. P. 1.

[24]NAS/NAE/IOM. *Advanced Research Instrumentation and Facilities.* Washington, DC: The National Academies Press, 2006.

TABLE 6-2 Annual Number of PECASE Awards, by Agency, 2005

Agency	Awards
National Science Foundation	20
National Institutes of Health	12
Department of Energy	9
Department of Defense	6
Department of Commerce	4
Department of Agriculture	3
National Aeronautics and Space Administration	2
Department of Veterans Affairs	2
TOTAL	58

ARIF are distinguished from other types of instrumentation by their expense and in that they are commonly acquired by large-scale centers or research programs rather than individual investigators. The acquisition of ARIF by an academic institution often requires a substantial institutional commitment and depends on high-level decision-making at both the institution and federal agencies. ARIF at academic institutions are often managed by institution administration. Furthermore, the advanced nature of ARIF often requires expert technical staff for its operation and maintenance.

A recent National Academies committee[25] found that there is a critical gap in federal programs for ARIF. Although federal research agencies research do have instrumentation programs, few allow proposals for instrumentation when the capital cost is greater than $2 million. No federal research agency has an agencywide ARIF program.

In addition, the ARIF committee found that instrumentation programs are inadequately supported. Few provide funds for continuing technical support and maintenance. The programs tend to support instrumentation for specific research fields and rarely consider broader scientific needs. The shortfalls in funding for instrumentation have built up cumulatively and are met by temporary programs that address short-term issues but rarely long-term problems. The instrumentation programs are poorly integrated across (or even within) agencies. The ad hoc ARIF programs are neither well organized nor visible to most investigators, and they do not adequately match the research community's increasing need for ARIF.

When budgets for basic research are stagnant, it is particularly difficult to maintain crucial investments in instrumentation, and facilities. The Na-

[25]Ibid.

tional Science Board (NSB) reports that over the last decade funding for the US academic research instrumentation and facilities has not kept pace with funding in the rest of the world.[26] Nations that are relative newcomers to science and technology research—South Korea, China, and some European nations, for example—are investing heavily in instrumentation and facilities that serve as a major attraction to scientists from throughout the world. NSB recommends increasing the share of the NSF budget devoted to such tools from the current 22 to 27%.

NSB also cites reports by other organizations that point to major deficiencies in federal research infrastructure including instrumentation and facilities.[27] These organizations include:

- The National Science and Technology Council, which in 1995 stated that $8.7 billion would be needed just to rectify then-current infrastructure deficits.[28]
- NSF, which estimated in 1998 that it would cost $11.4 billion to construct, repair, or renovate US academic research facilities.[29]
- NIH, which in 2001 estimated health research infrastructure needs at $5.6 billion.[30]
- NASA, which reported a $900 million construction backlog in 2001 and said that $2 billion more would be needed to revitalize and modernize the aerospace research infrastructure.[31]
- The DOE Office of Science, which reported that in 2001 more than 60% of its laboratory space was more than 30 years old and identified more than $2 billion in capital investments it needed for the next decade.[32]
- NSF directorates, which, when surveyed in FY 2001, estimated additional infrastructure needs of $18 billion through 2010.[33]

---

[26]Ibid., p. 2.

[27]Ibid., pp. 18-19.

[28]National Science and Technology Council. *Final Report on Academic Research Infrastructure: A Federal Plan for Renewal.* Washington, DC: White House Office of Science and Technology Policy, March 17, 1995.

[29]National Science Foundation, Division of Science Resources Statistics. *Science and Engineering Research Facilities at Colleges and Universities, 1998.* NSF-01-301. Arlington, VA: National Science Foundation, October 2000.

[30]National Institutes of Health, Working Group on Construction of Research Facilities. *A Report to the Advisory Committee of the Director, National Institutes of Health.* Bethesda, MD: National Institutes of Health, July 6, 2001.

[31]Jefferson Morris. "NASA Considering Closing, Consolidating Centers as Part of Restructuring Effort." *Aerospace Daily* 200(1)(October 17, 2001).

[32]US Department of Energy. *Infrastructure Frontier: A Quick Look Survey of the Office of Science Laboratory Infrastructure.* Washington, DC: US Department of Energy, April 2001.

[33]National Science Board. *Science and Engineering Infrastructure for the 21st Century: The Role of the National Science Foundation.* Arlington, VA: National Science Foundation, 2003. P. 19.

- A blue ribbon panel convened by NSF, which estimated that $850 million more per year is needed for cyber infrastructure.[34]

One contributor to infrastructure deficits has been the imposition by the federal government in 1991 of a 26% cap on reimbursement to universities for "administrative costs," including funding for construction, maintenance, and operation of research facilities. Universities have in most cases been unable to increase their spending on infrastructure and have had to shift funds from other nongovernment sources to cover their investments in this area.[35]

NSB concludes that researchers are less productive than they could be and somewhat more likely to take positions abroad where resources are increasingly available. It is also important to note that the federal government alone has the ability to fund this type of research infrastructure. Industry has little incentive to do so, and state governments and universities do not have the resources. If the federal government fails to maintain the national research infrastructure, this infrastructure will continue to decay.

The committee used the 2001 estimates to determine the advanced research instrumentation and facilities needs of the nation. The recommendation would fund only a portion of that built-up demand, but the committee believes the proposed amount would be sufficient to at least keep the research enterprise moving forward.

The National Academies committee that developed the report on ARIF recommended that the White House Office of Science and Technology Policy (OSTP) enhance federal research agency coordination and cooperation with respect to ARIF. Federal agencies could work together to develop joint solicitations, invite researchers from diverse disciplines to present opportunities for ARIF that would be useful to many fields to multiple agencies, simultaneously, seek out and identify best practices, and discuss the appropriate balance of funding among people, tools, and ideas, which could become part of the regular White House Office of Management and Budget-OSTP budget memorandum.

Therefore, in terms of the management of this fund, this committee believes that the best model is that of a national coordination office such as the National Coordination Office for Networking and Information Technology Research and Development (NCO/NITRD).[36] The National Coor-

---

[34]Report of the National Science Foundation Advisory Panel on Cyberinfrastructure. *Revolutionizing Science and Engineering Through Cyberinfrastructure*. Arlington, VA: National Science Foundation, February 2003.

[35]Council on Governmental Relations. *Report of the Working Group on the Cost of Doing Business*. Washington, DC: Council on Governmental Relations, June 2, 2003.

[36]See http://www.nitrd.gov/.

dination Office director reports to the director of the OSTP through the assistant director for technology. Twelve agencies participate, with each agency retaining its own funds, but, through the National Coordination Office, agencies are able to work together on technical and budget planning.

The other example using the National Coordination Office is the National Nanotechnology Initiative (NNI),[37] which coordinates the multiagency efforts in nanoscale science, engineering, and technology and is managed similarly. Twenty-three federal agencies participate in the National Nanotechnology Initiative, 11 of which have an R&D budget for nanotechnology. Other federal organizations contribute with studies, applications of results, and other collaborations. A third comparable program is the global climate change program. Again, the funding remains within each agency but supports a coordinated research effort.

Federal managers will probably be in the best position to determine the management of the proposed National Coordination Office for research infrastructure, but one model might be a design analogous to the management of the major research instrumentation (MRI) program of NSF. In that program, all proposals for instrumentation are submitted to a central source—the Office of Integrative Activities (OIA). This office then distributes the proposals throughout NSF for review. Proposal evaluations are then collected and prioritized, and funding decisions are made. The funding remains in the different divisions of NSF, but funds are also pooled to support the instrument based on the relationship to that office's mission. A similar mechanism could be used at the interagency level with the National Coordination Office acting in a similar fashion to NSF's Office of Integrative Activities.

ACTION B-4: HIGH-RISK RESEARCH

At least 8% of the budgets of federal research agencies should be set aside for discretionary funding managed by technical program managers in those agencies to catalyze high-risk, high-payoff research.

An important subset of basic research is the high-risk or transformative research that involves the new theories, methods, or tools that are often developed by new investigators—the group demonstrably most likely to generate radical discoveries or new technologies. These opportunities are generally first identified at the working level, not by research planning staffs. Today, there is anecdotal evidence that several barriers have reduced the national capacity for high-risk, high-payoff work:

---

[37]See http://www.nano.gov.

- Flat or declining funding in many disciplines makes it harder to justify risky or unorthodox projects.
- The peer review system tends to favor established investigators who use well-known methods.
- Industry, university, and federal laboratories are under pressure to produce short-term results—especially DOD, which once was the nation's largest source of basic-research funding.
- Increased public scrutiny of government R&D spending makes it harder to justify non-peer-reviewed awards, and peer reviewers tend to place confidence in older, established researchers.
- High-risk, high-potential projects are prone to failure, and government oversight and media and public scrutiny make those projects increasingly untenable to those responsible for the work.

A National Research Council study indicates that the Department of Defense's budgets for basic research have declined and that "there has been a trend within DOD for reduced attention to unfettered exploration in its basic research program."[38] The Defense Advanced Research Projects Agency (DARPA) was created in part because of this consideration (see Box 6-2).[39]

Defense Advanced Research Projects Agency managers, unlike program managers at NSF or NIH, for example, were encouraged to fund promising work for long periods in highly flexible programs—in other words, to take risks.[40] The National Institutes of Health and National Science Foundation recently acknowledged that their peer review systems today tend to screen out risky projects, and both organizations are working to reverse this trend.

In 2004, the National Institutes of Health awarded its first Director's Pioneer Award to foster high-risk research by investigators in the early to middle stages of their careers. Similarly, in 1990 the National Science Foundation started a program called Small Grants for Exploratory Research (SGER), which allows program officers to make grants without formal external review. Small Grants Exploratory Research awards are for "preliminary work on untested and novel ideas; ventures into emerging research; and potentially transformative ideas."[41] At $29.5 million, however, the total SGER budget for 2004 was just 0.5% of NSF's operating budget for

[38]National Research Council. *Assessment of Department of Defense Basic Research*. Washington, DC: The National Academies Press, 2005. P. 2.

[39]*It's Time to Sound the Alarm Over Shift from Basic, University Projects*. Editorial. San Jose Mercury News, April 17, 2005.

[40]National Research Council. *Assessment of Department of Defense Basic Research*. Washington, DC: The National Academies Press, 2005. P. 2.

[41]National Science Board. *Report of the National Science Board on the National Science Foundation's Merit Review Process Fiscal Year 2004*. NSB 05-12. Arlington, VA: National Science Foundation, March 2005. P. 27.

BOX 6-2
DARPA

The Defense Advanced Research Projects Agency (DARPA) was established with a budget of $500 million in 1958 following the launch of Sputnik to turn innovative technology into military capabilities. The agency is highly regarded for its work on the Internet, high-speed microelectronics, stealth and satellite technologies, unmanned vehicles, and new materials.[a]

DARPA's FY 2005 budget is $3.1 billion. In terms of personnel, it is a small, relatively nonhierarchical organization that uses highly flexible contracting and hiring practices that are atypical of the federal government as a whole. Its workforce of 220 includes 120 technical staffers, and it can hire quickly from the academic world and industry at wages that are substantially higher than those elsewhere in the government. Researchers, as intended, typically stay with DARPA only for a few years. Lawrence Dubois says that DARPA puts the following questions to its principal investigators, individual project leaders, and program managers:[b]

- What are you trying to accomplish?
- How is it done today and what are the limitations? What is truly new in your approach that will remove current limitations and improve performance? By how much? A factor of 10? 100? More? If successful, what difference will it make and to whom?
- What are the midterm exams, final exams, or full-scale applications required to prove your hypothesis? When will they be done?
- What is DARPA's exit strategy? Who will take the technologies you develop and turn them into new capabilities or real products?
- How much will it cost?

Dubois quotes a former DARPA program manager who describes the agency this way:[c]

> Program management at DARPA is a very proactive activity. It can be likened to playing a game of multidimensional chess. As a chess player, one always knows what the goal is, but there are many ways to reach checkmate. Like a program manager, a chess player starts out with many different pieces (independent research groups) in different geographic locations (squares on the board) and with different useful capabilities (fundamental and applied research or experiment and theory, for example). One uses this team to mount a coordinated attack (in one case to solve key technical problems and for another to defeat one's opponent). One of the challenges in both cases is that the target is continually moving. The DARPA program manager has to deal

continued

BOX 6-2 Continued

with both emerging technologies and constantly changing customer demand, whereas the chess player has to contend with his or her opponent's king and surrounding players always moving. Thus, both face changing obstacles and opportunities. The proactive player typically wins the chess game, and it is the proactive program manager who is usually most successful at DARPA.

---

[a]L. H. Dubois. DARPA's Approach to Innovation and Its Reflection in Industry. In *Reducing the Time from Basic Research to Innovation in the Chemical Sciences: A Workshop Report to the Chemical Sciences Roundtable*. Washington, DC: The National Academies Press, 2003. Chapter 4.

[b]Ibid.

[c]Ibid.

research and education. In 2004, the National Science Board convened a Task Force on Transformative Research to consider how to adapt NSF processes to encourage more funding of high-risk, potentially high-payoff research.

Several accounts indicate that although program managers might have the authority to fund at least some high-risk research, they often lack incentives do so. Partly for this reason, the percentage of effort represented by such pursuits is often quite small—1 to 3% being common. The committee believes that additional discretionary funding will enhance the transformational nature of research without requiring additional funding. Some committee members thought 5% was sufficient, others 10%. Thus, 8% seemed a reasonable compromise and is reflected in the committee's recommended action. The degree to which such a program will be successful depends heavily on the quality and coverage of the program staff.

ACTION B-5: USE DARPA AS A MODEL FOR ENERGY RESEARCH

The federal government should create a DARPA-like organization within the Department of Energy called the Advanced Research Projects Agency-Energy (ARPA-E) that reports to the under secretary for science and is charged with sponsoring specific R&D programs to meet the nation's long-term energy challenges.[42]

---

[42]One committee member, Lee Raymond, shares the alternative point of view on this recommendation as summarized in Box 6-3.

BOX 6-3
Another Point of View: ARPA-E

Energy issues are potentially some of the most profound challenges to our future prosperity and security, and science and technology will be critical in addressing them. But not everyone believes that a federal program like the proposed ARPA-E would be an effective mechanism for developing bold new energy technologies. This box summarizes some of the views the committee heard about ARPA-E from those who disagree with its utility.

Some believe that such applied energy research is already well funded by the private sector—by large energy companies and, increasingly, by venture capital firms—and that the federal government should fund only basic research. They argue that there is no shortage of long-term research funding in energy, including that sponsored by the federal government. DOE is the largest individual government supporter of basic research in the physical sciences, providing more than 40% of associated federal funding. DOE provides funding and support to researchers in academe, other government agencies, nonprofit institutions, and industry. The government spends substantial sums annually on research, including $2.8 billion on basic research and on numerous technologies. Given the major investment DOE is already making in energy research, it is argued that if additional federal research is desired in a particular field of energy, it should be accomplished by reallocating and optimizing the use of funds currently being invested.

It is therefore argued that no additional federal involvement in energy research is necessary, and given the concerns about the apparent shortage in scientific and technical talent, any short-term increase in federally directed research might crowd out more productive private-sector research. Furthermore, some believe that industry and venture capital investors will already fund the things that have a reasonable probability of commercial utility (the invisible hand of the free markets at work), and what is not funded by existing sources is not worthy of funding.

Another concern is that an entity like ARPA-E would amount to the government's attempt to pick winning technologies instead of letting markets decide. Many find that the government has a poor record in that arena. Government, some believe, should focus on basic research rather than on developing commercial technology.

Others are more supportive of DOE research as it exists and are concerned that funding ARPA-E will take money away from traditional science programs funded by DOE's Office of Science in high-energy physics, fusion energy research, material sciences, and so forth that are of high quality and despite receiving limited funds produce Nobel-prize-quality fundamental research and commercial spinoffs. Some believe that DOE's model is more productive than DARPA's in terms of research quality per federal dollar invested.

Perhaps no experiment in the conduct of research and engineering has been more successful in recent decades than the Defense Advanced Research Projects Agency model. The new agency proposed herein is patterned after that model and would sponsor creative, out-of-the-box, transformational, generic energy research in those areas where industry by itself cannot or will not undertake such sponsorship, where risks and potential payoffs are high, and where success could provide dramatic benefits for the nation. ARPA-E would accelerate the process by which research is transformed to address economic, environmental, and security issues. It would be designed as a lean, effective, and agile—but largely independent—organization that can start and stop targeted programs based on performance and ultimate relevance. ARPA-E would focus on specific energy issues, but its work (like that of DARPA or NIH) would have significant spinoff benefits to national, state, and local government; to industry; and for the education of the next generation of researchers. The nature of energy research makes it particularly relevant to producing many spinoff benefits to the broad fields of engineering, the physical sciences, and mathematics, fields identified in this review as warranting special attention. Existing programs with similar goals should be examined to ensure that the nation is optimizing its investments in this area. Funding for ARPA-E would begin at $300 million for the initial year and increase to $1 billion over 5 years, at which point the program's effectiveness would be reevaluated. The committee picked this level of funding the basis of its review of the budget history of other new research activities and the importance of the task at hand.

The United States faces a variety of energy challenges that affect our economy, our security, and our environment (see Box 6-4). Fundamentally, those challenges involve science and technology. Today, scientists and engineers are already working on ideas that could make solar and wind power economical; develop more efficient fuel cells; exploit energy from tar sands, oil shale, and gas hydrates; minimize the environmental consequences of fossil-fuel use; find safe, affordable ways to dispose of nuclear waste; devise workable methods to generate power from fusion; improve our aging energy-distribution infrastructure; and devise safe methods for hydrogen storage.[43]

ARPA-E would provide an opportunity for creative "out-of-the box" transformational research that could lead to new ways of fueling the nation and its economy, as opposed to incremental research on ideas that have already been developed. One expert explains, "The supply [of fossil-fuel sources] is adequate now and this gives us time to develop alternatives, but

[43]M. S. Dresselhaus and I. L. Thomas. "Alternative Energy Technologies." *Nature* 414(2001):332-337.

BOX 6-4
Energy and the Economy

Capital, labor, and energy are three major factors that contribute to and influence economic growth in the United States. Capital is the equipment, machinery, manufacturing plants, and office buildings that are necessary to produce goods and services. Labor is the availability of the workforce to participate in the production of goods and services. Energy is the power necessary to produce goods and services and transport them to their destinations. These three components are used to compute a country's gross domestic product (GDP), the total of all output produced in the country. Without these three inputs, business and industry would not be able to transform raw materials into goods and services.

Energy is the power that drives the world's economy. In the industrialized nations, most of the equipment, machinery, manufacturing plants, and office buildings could not operate without an available supply of energy resources such as oil, natural gas, coal, or electricity. In fact, energy is such an important component of manufacturing and production that its availability can have a direct impact on GDP and the overall economic health of the United States.

Sometimes energy is not readily available because the supply of a particular resource is limited or because its price is too high. When this happens, companies often decrease their production of goods and services, at least temporarily. On the other hand, an increase in the availability of energy—or lower energy prices—can lead to increased economic output by business and industry.

Situations that cause energy prices to rise or fall rapidly and unexpectedly, as the world's oil prices have on several occasions in recent years, can have a significant impact on the economy. When these situations occur, the economy experiences what economists call a "price shock." Since 1970, the economy has experienced at least four such price shocks attributable to the supply of energy. Thus, the events of the last several decades demonstrate that the price and availability of a single important energy resource—such as oil—can significantly affect the world economy.

SOURCE: Adapted from Dallas Federal Reserve Bank at www.dallasfed.org/educate/everyday/ev2.html.

the scale of research in physics, chemistry, biology and engineering will need to be stepped up, because it will take sustained effort to solve the problem of long-term global energy security."[44]

[44]Ibid.

Although there are those who believe an organization like ARPA-E is not needed (Box 6-3), the committee concludes that it would play an important role in resolving the nation's energy challenges; in advancing research in engineering, the physical sciences, and mathematics; and in developing the next generation of researchers. A recent report of the Secretary of Energy Advisory Board's Task Force on the Future of Science Programs at the Department of Energy notes, "America can meet its energy needs only if we make a strong and sustained investment in research in physical science, engineering, and applicable areas of life science, and if we translate advancing scientific knowledge into practice. The current mix of energy sources is not sustainable in the long run."[45] Solutions will require coordinated efforts among industrial, academic, and government laboratories. Although industry owns most of the energy infrastructure and is actively developing new technologies in many fields, national economic and security concerns dictate that the government stimulate research to meet national needs (Box 6-4). These needs include neutralizing the provision of energy as a major driver of national security concerns. ARPA-E would invest in a broad portfolio of foundational research that is needed to invent transforming technologies that in the past were often supplied by our great industrial laboratories (see Box 6-5). Funding of research underpinning the provision of new energy sources is made particularly complex by the high-cost, high-risk, and long-term character of such work—all of which make it less suited to university or industry funding.

Among its many missions, DOE promotes the energy security of the United States, but some of the department's largest national laboratories were established in wartime and given clearly defense-oriented missions, primarily to develop nuclear weapons. Those weapons laboratories, and some of the government's other large science laboratories, represent significant national investments in personnel, shared facilities, and knowledge. At the end of the Cold War, the nation's defense needs shifted and urgent new agendas became clear—development of clean sources of energy, new forms of transportation, the provision of homeland security, technology to speed environmental remediation, and technology for commercial application. Numerous proposals over recent years have laid the foundation for more extensive redeployment of national laboratory talent toward basic and applied research in areas of national priority.[46]

---

[45]Secretary of Energy's Advisory Board, Task Force on the Future of Science Programs at the Department of Energy. *Critical Choices: Science, Energy and Security*. Final Report. Washington, DC: US Department of Energy, October 13, 2003. P. 5.

[46]Secretary of Energy Advisory Board. *Task Force on Alternative Futures for the Department of Energy National Laboratories* (the "Galvin Report"). Washington, DC: US Department of Energy, February 1995; President's Council of Advisors on Science and Technology.

BOX 6-5
The Invention of the Transistor

In the 1930s, the management of Bell Laboratories sought to develop a low-power, reliable, solid-state replacement for the vacuum tube used in telephone signal amplification and switching. Materials scientists had to invent methods to make highly pure germanium and silicon and to add controlled impurities with unprecedented precision. Theoretical and experimental physicists had to develop a fundamental understanding of the conduction properties of this new material and the physics of the interfaces and surfaces of different semiconductors. By investing in a large-scale assault on this problem, Bell announced the "invention" of the transistor in 1948, less than a decade after the discovery that a junction of positively and negatively doped silicon would allow electric current to flow in only one direction. Fundamental understanding was recognized to be essential, but the goal of producing an economically successful electronic-state switch was kept front-and-center. Despite this focused approach, fundamental science did not suffer: a Nobel Prize was awarded for the invention of the transistor. During this and the following effort, the foundations of much of semiconductor-device physics of the 20th century were laid.

Introducing a small, agile, DARPA-like organization could improve DOE's pursuit of R&D much as DARPA did for the Department of Defense. Initially, DARPA was viewed as "threatening" by much of the department's established research organization; however, over the years it has been widely accepted as successfully filling a very important role. ARPA-E would identify and support the science and technology critical to our nation's energy infrastructure. It also could offer several important national benefits:

- Promote research in the physical sciences, engineering, and mathematics.
- Create a stream of human capital to bring innovative approaches to areas of national strategic importance.

Federal Energy Research and Development for the Challenges of the Twenty-first Century. Report on the Energy Research and Development Panel, the President's Committee of Advisors on Science and Technology. Washington, DC, November 1997; Government Accounting Office. *Best Practices: Elements Critical to Successfully Reducing Unneeded RDT&E Infrastructure.* US GAO Report to Congressional Requesters. Washington, DC: US Government Accounting Office, January 8, 1998.

- Turn cutting-edge science and engineering into technology for energy and environmental applications.
- Accelerate innovation in both traditional and alternative energy sources and in energy-efficiency mechanisms.
- Foster consortia of companies, colleges and universities, and laboratories to work on critical research problems, such as the development of fuel cells.

The agency's basic administrative structure and goals would mirror those of DARPA, but there would be some important differences. DARPA exists mainly to provide a long-term "break-through" perspective for the armed forces. DOE already has some mechanisms for long-term research, but it sometimes lacks the mechanisms for transforming the results into technology that meets the government's needs. DARPA also helps develop technology for purchase by the government for military use. By contrast, most energy technology is acquired and deployed in the private sector, although DOE does have specific procurement needs. Like DARPA, ARPA-E would have a very small staff, would perform no R&D itself, would turn over its staff every 3 to 4 years, and would have the same personnel and contracting freedoms now granted to DARPA. Box 6-6 illustrates some energy technologies identified by the National Commission on Energy Policy as areas of research where federal research investment is warranted that is in research areas in which industry is unlikely to invest.

ACTION B-6: PRIZES AND AWARDS

The White House Office of Science and Technology Policy (OSTP) should institute a Presidential Innovation Award to stimulate scientific and engineering advances in the national interest. While existing Presidential awards address lifetime achievements or promising young scholars, the proposed awards would identify and recognize individuals who develop unique scientific and engineering innovations in the national interest at the time they occur.

A number of organizations currently offer prizes and awards to stimulate research, but an expanded system of recognition could push new scientific and engineering advances that are in the national interest. The current presidential honors for scientists and engineers are the National Medal of Science,[47] the National Medal of Technology, and the Presidential Early Career Awards for Scientists and Engineers. The National Medal of Science and the National Medal of Technology recognize career-long achievement.

The Presidential Early Career Awards for Scientists and Engineers pro-

[47]See http://www.nsf.gov/nsb/awards/nms/medal.htm.

BOX 6-6
Illustration of Energy Technologies

The National Commission on Energy Policy in its December 2004 report, *Ending the Energy Stalemate: A Bipartisan Strategy to Meet America's Energy Challenges,* recommended doubling the nation's annual direct federal expenditures on "energy research, development, and demonstration" (ERD&D) to identify better technologies for energy supply and efficient end use. Improved technologies, the commission indicates, will make it easier to

- Limit oil demand and reduce the fraction of it met from imports without incurring excessive economic or environmental costs.
- Improve urban air quality while meeting growing demand for automobiles.
- Use abundant US and world coal resources without intolerable impacts on regional air quality and acid rain.
- Expand the use of nuclear energy while reducing related risks of accidents, sabotage, and proliferation.
- Sustain and expand economic prosperity where it already exists—and achieve it elsewhere—without intolerable climatic disruption from greenhouse-gas emissions.

The commission identified what it believes to be the most promising technological options where private sector research activities alone are not likely to bring them to that potential at the pace that society's interests warrant. They fall into the following principal clusters:

- **Clean and efficient automobile and truck technologies,** including advanced diesels, conventional and plug-in hybrids, and fuel-cell vehicles
- **Integrated-gasification combined-cycle coal technologies** for polygeneration of electricity, steam, chemicals, and fluid fuels
- **Other technologies that achieve, facilitate, or complete carbon capture and sequestration,** including the technologies for carbon capture in hydrogen production from natural gas, for sequestering carbon in geologic formations, and for using the produced hydrogen efficiently
- **Technologies to efficiently** produce biofuels for the transport sector
- **Advanced nuclear technologies** to enable nuclear expansion by lowering cost and reducing risks from accidents, terrorist attacks, and proliferation
- **Technologies for increasing the efficiency of energy end use in buildings and industry.**

SOURCE: Chapter VI, Developing Better Energy Technologies for the Future. In National Commission on Energy Policy. 2004. *Ending the Energy Stalemate: A Bipartisan Strategy to Meet America's Energy Challenges.* Available at: http://www.energycommission.org.

gram, managed by the National Science and Technology Council, honors and supports the extraordinary achievements of young professionals for their independent research contributions.[48] The White House, following recommendations from participating agencies, confers the awards annually.

New awards could encourage risk taking; offer the potential for financial or non-remunerative payoffs, such as wider recognition for important work; and inspire and educate the public about current issues of national interest. The National Academy of Engineering has concluded that prizes encourage nontraditional participants, stimulate development of potentially useful but under funded technology, encourage new uses for existing technology, and foster the diffusion of technology.[49]

For those reasons, the committee proposes that the new Presidential Innovation Award be managed in a way similar to that of the Presidential Early Career Awards for Scientists and Engineers. OSTP already identifies the nation's science and technology priorities each year as part of the budget memorandum it develops jointly with the Office of Management and Budget. This year's topics are a good starting point for fields in which innovation awards (perhaps one award for each research topic) could be given:

- Homeland security R&D.
- High-end computing and networking R&D.
- National nanotechnology initiative.
- High-temperature and organic superconductors.
- Molecular electronics.
- Wide-band-gap and photonic materials.
- Thin magnetic films.
- Quantum condensates.
- Infrastructure (next-generation light sources and instruments with subnanometer resolution).
- Understanding complex biological systems (focused on collaborations with physical, computational, behavioral, social, and biological researchers and engineers).
- Energy and the environment (natural hazard assessment, disaster warnings, climate variability and change, oceans, global freshwater supplies, novel materials, and production mechanisms for hydrogen fuel).

---

[48]The participating agencies are the National Science Foundation, National Science and Technology Council, National Aeronautics and Space Administration, Environmental Protection Agency, Department of Agriculture, Department of Commerce, Department of Defense, Department of Energy, the Department of Health and Human Services' National Institutes of Health, Department of Transportation, and Department of Veterans Affairs.

[49]National Academy of Engineering. *Concerning Federally Sponsored Inducement Prizes in Engineering and Science*. Washington, DC: National Academy Press, 1999.

The proposed awards would be presented, shortly after the innovations occur, to scientists and engineers in industry, academe, and government who develop unique ideas in the national interest. They would illustrate the linkage between science and engineering and national needs and provide an example to students of the contributions they could make to society by entering the science and engineering profession.

Conclusion

Research sows the seeds of innovation. The influence of federally funded research in social advancement—in the creation of new industries and in the enhancement of old ones—is clearly established. But federal funding for research is out of balance: Strong support is concentrated in a few fields while other areas of equivalent potential languish. Instead, the United States needs to be among the world leaders in all important fields of science and engineering. But, new investigators find it increasingly difficult to secure funding to pursue innovative lines of research. An emphasis on short-term goals diverts attention from high-risk ideas with great potential that may take more time to realize. And the infrastructure essential for discovery and for the creation of new technologies is deteriorating because of failure to provide the funds needed to maintain and upgrade it.

7

What Actions Should America Take in Science and Engineering Higher Education to Remain Prosperous in the 21st Century?

BEST AND BRIGHTEST

Recommendation C: ***Make the United States the most attractive setting in which to study and perform research so that we can develop, recruit, and retain the best and brightest students, scientists, and engineers from within the United States and throughout the world.***

We live in a knowledge-intensive world. "The key strategic resource necessary for prosperity has become knowledge itself in the form of educated people and their ideas," as Jim Duderstadt and Farris Womack[1] put it. In this context, the focus of global competition is no longer only on manufacturing and trade but also on the production of knowledge and the development and recruitment of the "best and brightest" from around the world. Developed and developing nations alike are investing in higher education, often on the model of US colleges and universities. They are training undergraduate and graduate scientists and engineers[2] to provide the expertise they need to compete in creating jobs for their populations in the 21st-century economy. Numerous national public and private organizations[3]

[1]J. J. Duderstadt and F. W. Womack. *Beyond the Crossroads: The Future of the Public University in America.* Baltimore, MD: Johns Hopkins University Press, 2003.

[2]Natural sciences and engineering is defined by the National Science Foundation as natural (physical, biological, earth, atmospheric, and ocean sciences), agricultural, and computer sciences; mathematics; and engineering.

[3]Some examples are National Science Board. *The Science and Engineering Workforce: Realizing America's Potential.* NSB 03-69. Arlington, VA: National Science Foundation, 2003. Volume 1; Council on Competitiveness. *Innovate America.* Washington, DC: Council on Competitveness, 2004.

have recommended a national effort to increase the numbers of both domestic and international students pursuing science, technology, engineering, and mathematics degrees in the United States.[4]

There is concern that, in general, our undergraduates are not keeping up with those in other nations. The United States has increased the proportion of its college-age population earning first university degrees in the natural sciences and engineering over the last quarter-century, but it has still lost ground, now ranking 20th globally on this indicator.[5]

There are even more concerns about graduate education. In the 1990s, the enrollment of US citizens and permanent residents in graduate science and engineering programs declined substantially. Although enrollments began to rise again in 2001, by 2003 they had not yet returned to the peak numbers of the early 1990s.[6] Meanwhile, the United States faces new challenges in the recruitment of international graduate students and postdoctoral scholars. Over the past several decades, graduate students and postdoctoral scholars from throughout the world have come to the United States to take advantage of what has been the premier environment in which to learn and conduct research. As a result, international students now constitute more than a third of the students in US science and engineering graduate schools, up from less than one-fourth in 1982. More than half the international postdoctoral scholars are temporary residents, and half that group earned doctorates outside the United States.

Many of the international students educated in the United States choose to remain here after receiving their degrees, and they contribute much to our ability to create knowledge, produce technological innovations, and generate jobs throughout the economy. The proportion of international doctorate recipients remaining in the United States after receiving their degrees increased from 49% in the 1989 cohort to 71% in 2001.[7] But the consequences of the events of September 11, 2001, included drastic changes in visa processing, and the number of international students applying to and enrolling in US graduate programs declined substantially. More recently, there have been signs of recovery; however, we are still falling short of earlier trends in attracting and retaining such students. As other nations develop their own systems of graduate education to recruit and retain more highly skilled students and professionals, often modeled after the US sys-

---

[4]Another point of view presented in Box 7-1.

[5]National Science Board. *Science and Engineering Indicators 2004*. NSB 04-01. Arlington, VA: National Science Foundation, 2004.

[6]National Science Foundation. *Graduate Enrollment in Science and Engineering Programs Up in 2003, but Declines for First-Time Foreign Students: Info Brief*. NSF 05-317. Arlington, VA: National Science Foundation, 2005.

[7]The National Academies. *Policy Implications of International Graduate Students and Postdoctoral Scholars in the United States*. Washington, DC: The National Academies Press, 2005.

BOX 7-1
Another Point of View: Science and Engineering Human Resources

Some believe that calls for increased numbers of science and engineering students are based more on the fear of a looming crisis than on a reaction to reality. Indeed, skeptics argue that there is no current documented shortage in the labor markets for scientists and engineers. In fact, in some areas we have just the opposite.[a] For example, during the last decade, there have been surpluses of life scientists at the doctoral level, high unemployment of engineers, and layoffs in the information-technology sector in the aftermath of the "dot-bomb."

Although there have been concerns about declining enrollments of US citizens in undergraduate engineering programs and in science and engineering graduate education, and these concerns have been compounded by recent declines in enrollments of international graduate students, enrollments in undergraduate engineering and of US citizens in graduate science and engineering have recently risen.

All of this suggests that the recommendations for additional support for thousands of undergraduates and graduates could be setting those students up for jobs that might not exist. Moreover, there are those who argue that international students crowd out domestic students and that a decline in international enrollments could encourage more US citizens, including individuals from underrepresented groups, to pursue graduate education.

Over the last decade, there has been similar debate over the number of H-1B visas that should be issued, with fervent calls both for increasing and for decreasing the cap. A recent report of the National Academies argued that there was no scientific way to find the "right" number of H-1Bs and that determining the appropriate level is and must be a political process.[b]

---

[a]J. Mervis. "Down for the Count." *Science* 300(5622)(2003):1070-1074.

[b]National Research Council. *Building a Workforce for the Information Economy.* Washington, DC: National Academy Press, 2001.

tem, we face even further uncertainty about our ability to attract those students to our institutions and to encourage them to become US citizens.

We must also encourage and enable US students from all sectors of our own society to participate in science, mathematics, and engineering programs, at least at the level of those who would be our competitors. But given increased global competition and reduced access to the US higher education system, our nation's education and research enterprise must adjust so that it can continue to attract many of the best students from abroad.

The Committee on Prospering in the Global Economy of the 21st Century proposes four actions to improve the talent pool in postsecondary education in the sciences and engineering: stimulate the interest of US citizens in undergraduate study by providing a new program of 4-year undergraduate scholarships; facilitate graduate education by providing new, portable fellowships; provide tax credits to companies and other organizations that provide continuing education for their practicing scientists and engineers; and recruit and retain the best and brightest students, scientists, and engineers worldwide by making the United States the most attractive place to study, conduct research, and commercialize technological innovations.

ACTION C-1: UNDERGRADUATE EDUCATION

Increase the number and proportion of US citizens who earn bachelor's degrees in the physical sciences, the life sciences, engineering, and mathematics by providing 25,000 new 4-year competitive undergraduate scholarships each year to US citizens attending US institutions.

The Undergraduate Scholar Awards in Science, Technology, Engineering, and Mathematics (USA-STEM) program would help to increase the percentage of 24-year-olds with first degrees in the natural sciences or engineering from the current 6% to the 10% benchmark already met or substatially surpassed by Finland, France, Taiwan, South Korea, and the United Kingdom (see Figure 3-17).[8] To achieve this result, the committee recommends the following:

- The National Science Foundation should administer the program.
- The program should provide 25,000 new 4-year scholarships each year to US citizens attending domestic institutions to pursue bachelor's degrees in science, mathematics, engineering, or another field designated as a national need. (Eventually, there would be 100,000 active students in the program each year.)
- Eligibility for these awards and their allocation would be based on the results of a competitive national examination.
- The scholarships would be distributed to states based on the size of their congressional delegations and would be awarded by states.
- Recipients could use the scholarships at any accredited US institution.

[8]In 2000, there were 3,711,400 24-year-olds in the United States, of whom 5.67% held bachelor's degrees in the natural sciences and engineering.

- The scholarships would provide up to $20,000 per student to pay tuition and fees.
- The program would also grant the recipients' institutions $1,000 annually.
- The $1.1 billion program would phase in over 4 years beginning at $275 million per year.
- The federal government would grant funds to states to defray reasonable administrative expenses.
- Steps would be taken to ensure that the receipt of USA-STEM scholarships brought considerable prestige to the recipients and to the secondary institutions from which they are graduating.

The undergraduate years have a profound influence on career direction, and they can provide a springboard for students who choose to major and then pursue graduate work in science, mathematics, and engineering. However, many more undergraduates express an interest in science, mathematics, and engineering than eventually complete bachelor's degrees in those fields. A focused and sizeable national effort to stimulate undergraduate interest and commitment to these majors will increase the proportion of 24-year-olds achieving first degrees in the relevant disciplines.

The scholarship program's motivation is twofold. First, in the long run, the United States might not have enough scientists and engineers to meet its national goals if the number of domestic students from all demographic groups, including women and students from underrepresented groups, does not increase in proportion to our nation's need for them. It should be noted that there is always concern about the availability of jobs if the supply of scientists and engineers were to increase substantially. Although it is impossible to fine-tune the system such that supply and demand balance precisely in any given year, it is important to have sufficient numbers of graduates for the long-term outlook. Furthermore, it has been found that, for example, undergraduate training in engineering forms an excellent foundation for graduate work in such fields as business, law, and medicine. Finally, it is clear that an *inadequate* supply of scientists and engineers can be highly detrimental to the nation's well-being.

The second motivation for the program is to ensure that the fields of science, engineering, and mathematics recruit and develop a large share of the best and brightest US students. It should be considered a great achievement to participate in the USA-STEM program, and the honor of selection should be accompanied by significant recognition. To retain eligibility, recipients would be expected to maintain a specified standard of academic excellence in their college coursework.

Increasing participation of underrepresented minorities is critical to ensuring a high-quality supply of scientists and engineers in the United States over the long term. As minority groups increase as a percentage of the US population, increasing their participation rate in science and engineering is critical if we are just to maintain the overall participation rate in science among the US population.[9] Perhaps even more important, if some groups are underrepresented in science and engineering in our society, we are not attracting as many of the most talented people to an important segment of our knowledge economy.[10]

In postsecondary education, there are many principles that help minority-group students succeed, regardless of field. The Building Engineering and Science Talent[11] (BEST) committee outlined eight key principles to expand representation:

- Institutional leadership: Committing to inclusiveness across the campus community.
- Targeted recruitment: Investing in and supporting a K–12 feeder system.
- Engaged faculty: Rewarding faculty for the development of student talent.
- Personal attention: Addressing, through mentoring and tutoring, the learning needs of each student.
- Peer support: Giving students opportunities for interaction that builds support across cohorts and promotes allegiance to an institution, discipline, and profession.
- Enriched research experience: Offering beyond-the-classroom hands-on opportunities and summer internships that connect to the world of work.
- Bridge to the next level: Fostering institutional relationships to show students and faculty the pathways to career development.
- Continuous evaluation: Monitoring results and making appropriate program adjustments.

BEST goes on to note that even with all the design principles in place, comprehensive financial assistance for low-income students is critical be-

---

[9]National Science and Technology Council. *Ensuring a Strong US Scientific, Technical, and Engineering Workforce in the 21st Century*. Washington, DC: Executive Office of the President of the United States, 2000; Congressional Commission on the Advancement of Women and Minorities in Science, Engineering, and Technology Development. *Land of Plenty: Diversity as America's Competitive Edge in Science, Engineering, and Technology*. Arlington, VA: National Science Foundation, 2000.

[10]Fechter and Teitelbaum have argued that "underrepresentation is an indicator of talent that is not exploited to its fullest potential. Such underutilization, which can exist simultaneously with situations of abundance, represents a cost to society as well as to the individuals in these groups." A. Fechter and M. S. Teitelbaum. "A Fresh Approach to Immigration." *Issues in Science and Technology* 13(3)(1997):28-32.

[11]Building Engineering and Science Talent (BEST). 2004. *A Bridge for All: Higher Education Design Principles in Science, Technology, Engineering and Mathematics*. San Diego, CA: BEST. Available at: http://www.bestworkforce.com.

cause socioeconomic status also is an important determinant of success in higher education.

ACTION C-2: GRADUATE EDUCATION

The federal government should fund Graduate Scholar Awards in Science, Technology, Engineering, and Mathematics (GSA-STEM), a new scholarship program that would provide 5,000 new portable 3-year competitively awarded graduate fellowships each year for outstanding US citizens in science, mathematics, and engineering programs pursuing degrees at US universities. Portable fellowships would provide funds directly to students, who would choose where they wish to pursue graduate studies instead of having to follow faculty research grants.

Typically, college seniors and recent graduates consider several factors in deciding whether to pursue graduate study. An abiding interest in a field and the encouragement of a mentor often contribute to the positive side of the balance sheet. The availability of financial support, the relative lack of income while in school, and job prospects upon completing an advanced degree also weigh on students' minds, no matter how much society supports their choices. The National Defense Education Act was a tremendous stimulus to graduate study in the 1960s, 1970s, and early 1980s, but has been incrementally restricted to serve a broader set of goals (see Box 7-2). A similar effort is now called for to meet the nation's long-term need for scientists and engineers in universities, government, nonprofit organizations, the national laboratory system, and industry.

The committee makes the following recommendations:

- The National Science Foundation (NSF) should administer the program.
- Recipients could use the grants at any US institution to which they have been admitted.
- The program should be advised by a board of representatives from federal agencies who identify areas of national need.
- Tuition and fee reimbursement would be up to $20,000 annually, and each recipient would receive an annual stipend of $30,000. Those amounts would be adjusted over time for inflation.
- The program would be phased in over 3 years.
- The federal government would provide appropriate funding to academic institutions to defray reasonable administrative expenses.

There has been much debate in recent years about whether the United States is facing a looming shortage of scientists and engineers, including

BOX 7-2
National Defense Education Act

Adopted by Congress in response to the launch of Sputnik and the emerging threat to the United States posed by the Soviet Union in 1958, the original National Defense Education Act (NDEA) boosted education and training and was accompanied by simultaneous actions that created the National Aeronautics and Space Administration and the Advanced Research Project Agency (now the Defense Advanced Research Projects Agency) and substantially increased NSF funding. It was funded with federal funds of about $400-500 million (adjusted to US$ 2004 value). NDEA provided funding to enhance research facilities; fellowships to thousands of graduate students pursuing degrees in science, mathematics, engineering, and foreign languages; and low-interest loans for undergraduates in these fields.

By the 1970s the act had been largely superseded by other programs, but its legacy remains in the form of several federal student-loan programs.[a] The legislation ultimately benefited all higher education as the notion of defense was expanded to include most disciplines and fields of study.[b]

Today, however, there are concerns about the Department of Defense (DOD) workforce. This workforce has experienced a real attrition of more than 13,000 personnel over the last 10 years. At the same time, the DOD projects that its workforce demands will increase by more than 10% over the next 5 years (by 2010). Indeed, several major studies since 1999 argue that the number of US graduates in critical areas is not meeting national, homeland, and economic security needs.[c] Science, engineering, and language skills continue to have very high priority across governmental and industrial sectors.

Many positions in critical-skill areas require security clearances, meaning that only US citizens may apply. Over 95% of undergraduates are US citizens, but in many of the science and engineering fields fewer than 50% of those earning PhDs are US citizens. Retirements also loom on the horizon: over 60% of the federal science and engineering workforce is over 45 years old, and many of these people are employed by DOD. Department of Defense and other federal agencies face increased competition from domestic and global commercial interests for top-of-their-class, security-clearance-eligible scientists and engineers.

In response to those concerns, DOD has proposed in its budget submission a new NDEA. The new NDEA includes a number of new initiatives that some believe should be accomplished by 2008—the 50th anniversary of the original NDEA.[d]

---

[a]Association of American Universities. *A National Defense Education and Innovation Initiative: Meeting America's Economic and Security Challenges in the 21st Century.* Washington DC: AAU, 2006. Available at: http://www.aau.edu.

[b]M. Parsons. "Higher Education Is Just Another Special Interest." *The Chronicle of Higher Education* 51(22)(2005):B20. Available at: http://chronicle.com/prm/weekly/v51/i22/22b02001.htm.

[c]National Security Workforce. Challenges and Solutions Web page. Available at: http://www.defenselink.mil/ddre/doc/NDEA_BRIEFING.pdf.

[d]See http://www.defenselink.mil/ddre/nde2.htm and H.R. 1815, National Defense Authorization Act for Fiscal Year 2006, Sec. 1105. Science, Mathematics, and Research Transportation (SMART) Defense Education Program—National Defense Education Act (NDEA), Phase I.

those at the doctoral level. Although there is not a crisis at the moment and there are differences in labor markets by field that could lead to surpluses in some areas and shortages in others, the trends in enrollments and degrees are nonetheless cause for concern in a global environment wherein science and technology play an increasing role. The rationale for the fellowship is that the number of people with doctorates in the sciences, mathematics, and engineering awarded by US institutions each year has not kept pace with the increasing importance of science and technology to the nation's prosperity.

Currently, the federal government supports 7,000 full-time graduate fellows and trainees. Most of these grants are provided either to institutions or directly to students by the NSF's Graduate Research Fellowship program and Integrative Graduate Education and Research Traineeship Program (IGERT) or by the National Institutes of Health Ruth L. Kirschstein National Research Service Award program. The US Department of Education, through its Graduate Assistance in Areas of National Need program, also provides traineeships and has a mechanism for identifying areas for grantmaking to academic programs. Those are important sources of support, but they meet only a fraction of the need. The proposed 5,000 new fellowships each year eventually will increase to 22,000 the number of graduate students supported at any one time, thus helping to increase the number of US citizens and permanent residents earning doctorates in nationally important fields.

Portable graduate fellowships should attract high-quality students and offer them access to the best education possible. Students who have unencumbered financial support could select the US academic institutions that best meet their interests and that offer the best opportunities to broaden their experience before they begin focusing on specific research. The fellowships would offer substantial and steady financial support during the early years of graduate study, with the assumption that the recipients would find support from other means, such as research assistantships, once research subjects and mentors were identified.

An alternative point of view is that the support provided under this recommendation should be provided not—or not only—to *individuals* but also to *programs* that would use the funds both to develop a comprehensive approach to doctoral education and to support students through traineeships. Such institutional grants could be used by federal funders to directly require specific programmatic changes as well. They would also allow institutions to recruit promising students who might not apply for portable fellowships.

But, in the view of the committee, providing fellowships directly to students creates a greater stimulus to enroll and offers an additional positive effect: improvement of educational quality. The fellowships create competition among institutions that would lead to enhanced graduate programs (mentoring, course offerings, research opportunities, and facilities) and processes (time to degree, career guidance, placement assistance). To be sure, institutions can and should undertake many of those improvements in graduate programs even without this stimulus, and many have already implemented reforms to make graduate school more enticing. Institutional efforts to prepare graduate students for the jobs they will obtain in industry or academe and to improve the benefits and work conditions for postdoctoral scholars also could make career prospects more attractive.

The new program proposed here and led by NSF should draw advice from representatives of federal research agencies to determine its areas of focus. On the basis of that advice, NSF would make competitive awards either as part of its existing Graduate Research Fellowship program or through a separate program established specifically to administer the fellowships. The focus on areas of national need is important to ensure an adequate supply of suitably trained doctoral scientists, engineers, and mathematicians and appropriate employment opportunities for these students upon receipt of their degrees.

As discussed in Box 7-1, one question is whether these programs will simply produce science and engineering students who are unable to find jobs. There are also questions that the goal of increasing the number of domestic students is contrary to the committee's other concern about the potential for declining numbers of outstanding international students. As past National Academies reports have indicated, projecting supply and demand in science and engineering employment is prone to methodological difficulties. For example, the report *Forecasting Demand and Supply of Doctoral Scientists and Engineers: Report of a Workshop on Methodology* (2000) observed:

> The NSF should not produce or sponsor "official" forecasts of supply and demand of scientists and engineers, but should support scholarship to improve the quality of underlying data and methodology.

Those who have tried to forecast demand in the past have often failed abysmally. The same would probably be true today.

Other factors also influence the decisions of US students. As the recent COSEPUP study, *Policy Implications of International Graduate Students and Postdoctoral Scholars in the United States,* says:

> Recruiting domestic science and engineering (S&E) talent depends heavily on students' perception of the S&E careers that await them. Those perceptions can be solidified early in the educational process, before students graduate from high school. The desirability of a career in S&E is determined largely by the

> prospect of attractive employment opportunities in the field, and to a lesser extent by potential remuneration. Some aspects of the graduate education and training process can also influence students' decisions to enter S&E fields. The "pull factors" include time to degree, availability of fellowships, research assistantships, or teaching assistantships, and whether a long post doctoral appointment is required after completion of the PhD.

Taking those factors into account, the committee decided to focus its scholarships for domestic students on areas of national need as determined by federal agencies, with input from the corporate and business community.

In the end, the employment market will dictate the decisions students make. From a national perspective, global competition in higher education and research and in the recruitment of students and scholars means that the United States must invest in the development and recruitment of the best and brightest from here and abroad to ensure that we have the talent, expertise, and ideas that will continue to spur innovation and keep our nation at the leading edge of science and technology.

ACTION C-3: CONTINUING EDUCATION

To keep practicing scientists and engineers productive in an environment of rapidly changing science and technology, the federal government should provide tax credits to employers who help their eligible employees pursue continuing education.

The committee's recommendations are as follows:

- The federal government should authorize a tax credit of up to $500 million each year to encourage companies to sustain the knowledge and skills of their scientific and engineering workforce by offering opportunities for professional development.
- The courses to be pursued would allow employees to maintain and upgrade knowledge in the specific fields of science and engineering.
- The courses would be required to meet reasonable standards and could be offered internally or by colleges and universities.

Too often, business does not invest adequately in continuing education and training for employees, partly from the belief that investments could be lost if the training makes employees more marketable, and partly from the belief that maintaining skills is the personal responsibility of a professional. Tax credits would allow businesses to encourage continuing professional development—a benefit to employees, companies, and the economy.

Tax credits can also help industries adapt to technological change. The information-technology industry, for example, has continuing difficulty in matching worker skills and employer demand. The consequence is that employers cite worker shortages even when there is relatively high unemployment. That mismatch can be remedied by encouraging companies to invest in retraining capable employees whose skills have become obsolete as the technology landscape changes.

ACTION C-4: IMPROVE VISA PROCESSING

The federal government should continue to improve visa processing for international students and scholars to provide less complex procedures, and continue to make improvements on such issues as visa categories and duration, travel for scientific meetings, the technology alert list, reciprocity agreements, and changes in status.

Since 9/11, the nation has struggled to improve security by more closely screening international visitors, students, and workers. The federal government is now also considering tightening controls on the access that international students and researchers have to technical information and equipment. One consequence is that fewer of the best international scientists and engineers are able to come to the United States, and if they do enter the United States, their intellectual and geographic mobility is curtailed.

The post-9/11 approach fosters an image of the United States as a less than welcoming place for foreign scholars. At the same time, the home nations of many potential immigrants—such as China, India, Taiwan, and South Korea—are strengthening their own technology industries and universities and offering jobs and incentives to lure scientists and engineers to return to their nations of birth. Other countries have taken advantage of our tightened restrictions to open their doors more widely, and they recruit many who might otherwise have come to the United States to study or conduct research.

A growing challenge for policy-makers is to reconcile security needs with the flow of people and information from abroad. Restrictions on access to information and technology—much of it already freely available—could undermine the fundamental research that benefits so greatly from international participation. One must be particularly vigilant to ensure that thoughtful, high-level directives concerning homeland security are not unnecessarily amplified by administrators who focus on short-term safety while unintentionally weakening long-term overall national security. Any marginal benefits in the security arena have to be weighed against the ability of national research facilities to carry out unclassified, basic research and the ability of private companies with federal contracts to remain internationally competitive. An unbalanced increase in security will erode the nation's scientific and engineering productivity and economic strength and will destroy the welcoming atmosphere of our scientific and engineering institutions. Such restrictions would also add to the incentives for US companies to move operations overseas.

Many recent changes in visa processing and in the duration of Visas Mantis clearances have already made immigration easier. Visas Mantis is a program intended to provide additional security checks for visitors who may pose a security risk. The process, established in 1998 and applicable to all nonimmigrant visa categories, is triggered when a student or exchange-visitor applicant intends to study a subject on the technology alert list.

The committee endorses the recommendations made by the National Academies in *Policy Implications of International Graduate Students and Postdoctoral Scholars in the United States,*[12] particularly Recommendation 4-2, which states the following:

> If the United States is to maintain leadership in S&E, visa and immigration policies should provide clear procedures that do not unnecessarily hinder the inflow of international graduate students and postdoctoral scholars. New regulations should be carefully considered in light of national-security considerations and potential unintended consequences.
>
> a. Visa Duration: Implementation of the Student and Exchange Visitor Information System (SEVIS), by which consular officials can verify student and postdoctoral status, and of the United States Visitor and Immigrant Status Indicator Technology (US-VISIT), by which student and scholar status can be monitored at the point of entry to the United States, should make it possible for graduate students' and postdoctoral scholars' visas to be more commensurate with their programs, with a duration of 4-5 years.
>
> b. Travel for Scientific Meetings: Means should be found to allow international graduate students and postdoctoral scholars who are attending or appointed at US institutions to attend scientific meetings that are outside the United States without being seriously delayed in re-entering the United States to complete their studies and training.
>
> c. Technology Alert List: This list, which is used to manage the Visas Mantis program, should be reviewed regularly by scientists and engineers. Scientifically trained personnel should be involved in the security-review process.
>
> d. Visa Categories: New nonimmigrant-visa categories should be created for doctoral-level graduate students and postdoctoral scholars. The categories should be exempted from the 214b (see Box 7-3) provision whereby applicants must show that they have a residence in a foreign country that they have no intention of abandoning. In addition to providing a better mechanism for em-

---

[12]The National Academies. *Policy Implications of International Graduate Students and Postdoctoral Scholars in the United States.* Washington, DC: The National Academies Press, 2005.

BOX 7-3
The 214b Provision of the Immigration and Nationality Act: Establishing the Intent to Return Home

The Immigration and Nationality Act (INA) has served as the primary body of law governing immigration and visa operations since 1952. A potential barrier to visits by foreign graduate students is Section 214(b) of the INA, in accordance with which an applicant for student of exchange visa must provide convincing evidence that he or she plans to return to the home country, including proof of a permanent domicile in the home country. Legitimate applicants may find it hard to prove that they have no intention to immigrate, especially if they have relatives in the United States. In addition, both students and immigration officials are well aware that an F or J visa often provides entrée to permanent-resident status. It is not surprising that application and enforcement of the standard can depend on pending immigration legislation or economic conditions.[a]

---

[a]G. Chelleraj, K. E. Maskus, and A. Mattoo. *The Contributions of Skilled Immigration and International Graduate Students to US Innovation.* Working Paper N04-10. Boulder, CO: Center for Economic Analysis, University of Colorado at Boulder, September 2004. P. 18 and Table 1.

bassy and consular officials to track student and scholar visa applicants, these categories would provide a means for collecting clear data on numbers and trends of graduate-student and postdoctoral-scholar visa applications.

e. Reciprocity Agreements: Multiple-entry and multiple-year student visas should have high priority in reciprocity negotiations.

f. Change of Status: If the United States wants to keep the best students once they graduate, procedures for change of status should be clarified and streamlined.

ACTION C-5: EXTEND VISAS AND EXPEDITE RESIDENCE STATUS OF SCIENCE AND ENGINEERING PHDS

The federal government should provide a 1-year automatic visa extension to international students who receive doctorates or the equivalent in science, technology, engineering, mathematics, or other fields of national need at qualified US institutions to remain in the United States to seek employment. If these students are offered jobs by US-based employers and pass a security screening test, they should be provided automatic work permits and expedited residence status. If students are unable to obtain employment within 1 year, their visas would expire.

To create the most attractive setting for study, research and commercialization—and to attract international students, scholars, scientists, engineers, and mathematicians—the United States government needs to take steps to encourage international students and scholars to remain in the United States. These steps should be taken because of the contributions these people make to the United States and their home country.

As discussed in COSEPUP's international students report, a knowledge-driven economy is more productive if it has access to the best talent regardless of national origin. International graduate students and postdoctoral scholars are integral to the quality and effectiveness of the US science and engineering (S&E) enterprise. If the flow of these students and scholars were sharply reduced, research and academic work would suffer until an alternative source of talent were found. There would be a fairly immediate effect in university graduate departments and laboratories and a later cumulative effect on hiring in universities, industry, and government. There is no evidence that modest, gradual changes in the flow like those experienced in the recent past would have an adverse effect.

High-end innovation is a crucial factor for the success of the US economy. To maintain excellence in S&E research, which fuels high-end innovation, the United States must be able to recruit talented people. A substantial proportion of those talented people—students, postdoctoral scholars, and researchers—currently come from other countries.

The shift to staffing research and teaching positions at universities with nontenured staff, which depends in large part on a supply of international graduate students and postdoctoral scholars, should be the subject of a major study.

Multinational corporations (MNCs) hire international PhDs in similar proportion to the output of university graduate and postdoctoral programs. The proportion of international researchers in several large MNCs is around 30-50%. MNCs appreciate international diversity in their research staff. They pay foreign-born and domestic researchers the same salaries, which are based on degree, school, and benchmarks in the industry.

It is neither possible nor desirable to restrict US S&E positions to US citizens; this could reduce industries' and universities' access to much of the world's talent and remove a substantial element of diversity from our society.

One study of Silicon Valley illustrates the importance of international scientists and engineers to the US economy. It found that

> By the end of the 1990s, Chinese and Indian engineers were running 29 percent of Silicon Valley's technology businesses. By 2000, these companies collectively accounted for more than $19.5 billion in sales and 72,839 jobs. And the pace of immigrant entrepreneurship has accelerated dramatically in the last decade. . . .

> Far beyond their role in Silicon Valley, the professional and social networks that link new immigrant entrepreneurs with each other have become global institutions that connect new immigrants with their counterparts at home. These new transnational communities provide the shared information, contacts, and trust that allow local producers to participate in an increasingly global economy.
>
> Silicon Valley's Taiwanese engineers, for example, have built a vibrant two-way bridge connecting them with Taiwan's technology community. Their Indian counterparts have become key middlemen linking U.S. businesses to low-cost software expertise in India. These cross-Pacific networks give skilled immigrants a big edge over mainstream competitors who often lack the language skills, cultural know-how, and contacts to build business relationships in Asia. The long-distance networks are accelerating the globalization of labor markets and enhancing opportunities for entrepreneurship, investment, and trade both in the United States and in newly emerging regions in Asia.[13]

In response to those findings, the committee, in this proposed action, is endorsing a recommendation made by the Council on Competitiveness in its report *Innovate America*[14] to extend a 1-year automatic visa extension to international students who receive doctorates or the equivalent in science, technology, engineering, mathematics, or other fields of national need at qualified US institutions to remain in the United States to seek employment. If these students are offered jobs by US-based employers and pass a security screening test, they should be provided automatic work permits and expedited residence status. If students are unable to obtain employment within 1 year, their visas would expire.

ACTION C-6: SKILLS-BASED IMMIGRATION

The federal government should institute a new skills-based, preferential immigration option. Doctoral-level education and science and engineering skills would substantially raise an applicant's chances and priority in obtaining US citizenship. In the interim, the number of H-1B visas should be increased by 10,000, and the additional visas should be available for industry to hire science and engineering applicants with doctorates from US universities.[15]

---

[13]A. Saxenian. "Brain Circulation: How High-Skill Immigration Makes Everyone Better Off." *The Brookings Review* 20(1)(Winter 2002). Washington, DC: The Brookings Institute, 2002.

[14]Council on Competitiveness. *Innovate America*. Washington, DC: Council on Competitiveness, 2004.

[15]Since the report was released, the committee has learned that the Consolidated Appropriations Act of 2005, signed into law on December 8, 2004, exempts individuals that have received a master's or higher education degree from a US university from the statutory cap (up to

As discussed in the previous section, highly skilled immigrants make a major contribution to US education, research, entrepreneurship, and society. Therefore, it is important to encourage not only students and scholars to stay, but also other people with science, engineering, and mathematics PhDs regardless of where they receive their PhDs.

For the United States to remain competitive with Europe, Canada, and Australia in attracting these international highly skilled workers, the United States should implement a points-based immigration system. As discussed in a recent Organisation for Economic Co-operation and Development report,[16] skill-based immigration points systems, although not widespread, are starting to develop. Canada, Australia, New Zealand, and the UK use such systems to recruit highly skilled workers. The Czech Republic set up a pilot project that started in 2004.

In 2004, the European Union Justice and International Affairs council adopted a recommendation to facilitate researchers from non-EU countries, which asks member states to waive requirements for residence permits or to issue them automatically or through a fast-track procedure and to set no quotas that would restrict their admission. Permits should be renewable and family reunification facilitated. The European Commission has adopted a directive for a special admissions procedure for third-world nationals coming to the EU to perform research. This procedure will be in force in 2006.

• **Canada** has put into place a points-based program aimed at fulfilling its policy objectives for migration, particularly in relation to the labor-market situation. The admission of skilled workers depends more on human capital (language skills and diplomas, professional skills, and adaptability) than on specific abilities.[17] Canada has also instituted a business-immigrant selection program to attract investors, entrepreneurs, and self-employed workers.

20,000). The bill also raised the H-1B fee and allocated funds to train American workers. The committee believes that this provision is sufficient to respond to its recommendation—even though the 10,000 additional visas recommended is specifically for science and engineering doctoral candidates from US universities, which is a narrower subgroup.

[16]Unless otherwise noted, policies listed are from an overview presented in Organisation for Economic Co-operation and Development. *Trends in International Migration: 2004 Annual Report*. Paris: OECD, 2005. OECD members countries include Australia, Austria, Belgium, Canada, the Czech Republic, Denmark, Finland, France, Germany, Greece, Hungary, Iceland, Ireland, Italy, Japan, Korea, Luxembourg, Mexico, The Netherlands, New Zealand, Norway, Poland, Portugal, the Slovak Republic, Spain, Sweden, Switzerland, Turkey, the United Kingdom, and the United States.

[17]Applicants can check online their chances to qualify for migration to Canada as skilled workers. A points score is automatically calculated to determine entry to Canada under the Skilled Worker category. See Canadian Immigration Points Calculator Web site at: http://www.workpermit.com/canada/points_calculator.htm.

• **Germany** instituted a new immigration law on July 9, 2004. Among its provisions, in the realm of migration for employment, it encourages settlement by high-skilled workers, who are eligible immediately for permanent residence permits. Family members who accompany them or subsequently join them have access to the labor market. Like Canada, Germany encourages the immigration of self-employed persons, who are granted temporary residence permits if they invest a minimum of 1 million euros and create at least 10 jobs. Issuance of work permits and residence permits has been consolidated. The Office for Foreigners will issue both permits concurrently, and the Labor Administration subsequently approves the work permit.

• **UK**[18] The UK Highly Skilled Migrant Programme (HSMP) is an immigration category for entry to the UK for successful people with sought-after skills. It is in some ways similar to the skilled migration programs for entry to Australia and Canada. The UK has added an MBA provision to the HSMP. Eligibility for HSMP visas is assessed on a points system with more points awarded in the following situations:

- Preference for applicants under 28 years old.
- Skilled migrants with tertiary qualifications.
- High-level work experience.
- Past earnings.
- In a few rare cases, HSMP points are also awarded if one has an achievement in one's chosen field.
- One may also score bonus points if one is a skilled migrant seeking to bring a spouse or partner who also has high-level skills and work experience.

• **Australia** encourages immigration of skilled migrants, who are assessed on a points system with points awarded for work experience, qualifications, and language proficiency.[19] Applicants must demonstrate skills in specific job categories.

[18]The UK Highly Skilled Migrant Programme Web page also has a points calculator. Available at: http://www.workpermit.com/uk/highly_skilled_migrant_program.htm.

[19]See points calculator at: http://www.workpermit.com/australia/point_calculator.htm.

ACTION C-7: REFORM THE CURRENT SYSTEM OF "DEEMED EXPORTS"[20]

The current system of "deemed export" should be reformed. The new system should provide international students and researchers engaged in fundamental research in the United States with access to information and research equipment in US industrial, academic, and national laboratories comparable with the access provided to US citizens and permanent residents in a similar status. It would, of course, exclude information and facilities restricted under national security regulations. In addition, the effect of deemed-exports regulations on the education and fundamental research work of international students and scholars should be limited by removing from the deemed-exports technology list all technology items (information and equipment) that are available for purchase on the overseas open market from foreign or US companies or that have manuals that are available in the public domain, in libraries, over the Internet, or from manufacturers.

The controls governed by the Export Administration Act and its implementing regulations extend to the transfer of "technology." *Technology* is considered "*specific information* necessary for the 'development,' 'production,' or 'use' of a product," and providing such information to a foreign national within the United States may be considered a "deemed export" whose transfer requires an export license[21] (italics added). The primary responsibility for administering deemed exports lies with the Department of Commerce (DOC), but other agencies may have regulations to address the issue. Deemed exports are currently the subject of significant controversy.

---

[20]The controls governed by the Export Administration Act and its implementing regulations extend to the transfer of technology. Technology includes "*specific information* necessary for the 'development,' 'production,' or 'use' of a product" [emphasis added]. Providing information that is subject to export controls—for example, about some kinds of computer hardware—to a foreign national within the United States may be "deemed" an export, and that transfer requires an export license. The primary responsibility for administering controls on deemed exports lies with the Department of Commerce, but other agencies have regulatory authority as well.

[21]"Generally, technologies subject to the Export Administration Regulations (EAR) are those which are in the United States or of US origin, in whole or in part. Most are proprietary. Technologies which tend to require licensing for transfer to foreign nationals are also dual-use (i.e., have both civil and military applications) and are subject to one or more control regimes, such as National Security, Nuclear Proliferation, Missile Technology, or Chemical and Biological Warfare." (*"Deemed Exports" Questions and Answers*, Bureau of Industry and Security, Department of Commerce.) The International Traffic in Arms Regulations (ITAR), administered by the Department of State, control the export of technology, including technical information, related to items on the US Munitions List. Unlike the EAR, however, "publicly available scientific and technical information and academic exchanges and information presented at scientific meetings are not treated as controlled technical data."

In 2000, Congress mandated annual reports by agency offices of inspector general (IG) on the transfer of militarily sensitive technology to countries and entities of concern; the 2004 reports focused on deemed exports. The individual agency IG reports and a joint interagency report concluded that enforcement of deemed-export regulations had been ineffective; most of the agency reports recommended particular regulatory remedies.[22]

DOC sought comments from the public about the recommendations from its IG before proposing any changes. The department earned praise for this effort to reach out to potentially affected groups and is currently reviewing the 300 plus comments it received, including those from the leaders of the National Academies.[23]

On July 12, 2005, the Department of Defense (DOD) issued a notice in the *Federal Register* seeking comments on a proposal to amend the Defense Federal Acquisition Regulation Supplement (DFARS) to address requirements for preventing unauthorized disclosure of export-controlled information and technology under DOD contracts that follow the recommendations in its IG report. The proposed regulation includes a requirement for access-control plans covering unique badging requirements for foreign workers and segregated work areas for export-controlled information and technology, and it does not mention the fundamental-research exemption.[24] Comments were due by September 12, 2005.

Many of the comments in response to DOC expressed concern that the proposed changes were not based on systematic data or analysis and could have a significant negative effect on the conduct of research in both universities and the private sector, especially in companies with a substantial number of employees who are not US citizens.

CONCLUSION

The knowledge-driven global economy compels America to develop and recruit the finest experts available. Our students and our society prospered under a system of higher education and research that was the global leader in the second half of the 20th century. For a half-century at least, the United States has attracted graduate students and scholars from around the world. The system worked to our benefit, and it cannot now be taken for granted.

---

[22]Reports were produced by DOC, DOD, the Department of Energy (DOE), the Department of State, the Department of Homeland Security, and the Central Intelligence Agency. Only the interagency report and the reports from DOC, DOD, and DOE are publicly available.

[23]The letter from the presidents of the National Academies may be found at: http://www7.nationalacademies.org/rscans/Academy_Presidents_Comments_to_DOC.PDF.

[24]*Federal Register* 70(132)(July 2005):39976-39978. Available at: http://a257.g.akamaitech.net/7/257/2422/01jan20051800/edocket.access.gpo.gov/2005/05-13305.htm.

8

What Actions Should America Take in Economic and Technology Policy to Remain Prosperous in the 21st Century?

INCENTIVES FOR INNOVATION

Recommendation D: ***Ensure that the United States is the premier place in the world to innovate; invest in downstream activities such as manufacturing and marketing; and create high-paying jobs based on innovation by such actions as modernizing the patent system, realigning tax policies to encourage innovation, and ensuring affordable broadband access.***

As Wm. A. Wulf, President of the National Academy of Engineering, points out, "There is no simple formula for innovation. There is, instead, a multi-component 'environment' that collectively encourages, or discourages, innovation."[1] That environment encompasses such factors as research funding, an educated workforce, a culture that encourages risk taking, a financial system that provides patient capital for entrepreneurial activity, and intellectual property protection.[2] For more than a century, the United States has been a world leader in the development of new technology and the creation of new products. Its international competitive advantage rests in large part on a favorable environment for discovery and application of knowledge—its intellectual property.

---

[1]Wm. A. Wulf. "Review and Renewal of the Environment for Innovation." Unpublished paper, 2005.

[2]An alternative point of view is presented in Box 8-1.

Setting a policy framework that supports innovation is critical for at least two reasons. First, it enhances the competitiveness of US-based industries and supports domestic economic growth. Second, the nation stands to benefit from well-paying jobs if multinational corporations see the United States as the best place to perform research and development (R&D) and other activities related to innovation and ultimately to build factories and offices here.[3]

Our own history and contemporary international examples show that leadership in research is not a sufficient condition for gaining the lion's share of benefits from innovation. Recent developments in Japan illustrate what can happen to a science- and technology-based economy that does not adapt its innovation environment to changing conditions. Japan's growth trajectory in various science and engineering inputs and outputs (R&D investment, science and engineering workforce, patents) since the early 1990s has been similar to what it was before that time.[4] Yet its ability to profit from innovation in the form of higher productivity and income has recently fallen. Part of the explanation for the change is in the dual nature of the Japanese economy: World-class manufacturing that serves a global market exists side-by-side with inefficient industries, such as construction.[5] Economic mismanagement and a lack of flexibility in labor and capital markets also are to blame.

In contrast, in the middle 1990s the United States saw a jump in productivity growth from that which had prevailed since the first oil shock of the early 1970s.[6] In addition to continuous gains in manufacturing productivity and productivity growth generated by the use of information technology, the creation of new business methods that took advantage of information technology were widespread here.

Science and technology and the innovation process are not zero-sum games in the international context.[7] The United States has proved adept in

---

[3]National Research Council. *A Patent System for the 21st Century.* Washington, DC: The National Academies Press, 2004. P. 18.

[4]B. Steil, D. G. V. Nelson, and R. R. Nelson. *Technological Innovation and Economic Performance.* Princeton, NJ: Princeton University Press, 2002.

[5]D. W. Jorgenson and M. Kuroda. Technology, Productivity, and the Competitiveness of US and Japanese Industries. In T. Arrison, C. F. Bergsten, E. M. Graham, and M. C. Harris, eds. *Japan's Growing Technological Capability: Implications for the US Economy.* Washington, DC: National Academy Press, 1992. Pp. 83-97.

[6]W. Nordhaus. *The Sources of the Productivity Rebound and the Manufacturing Employment Puzzle.* Working Paper 11354. Cambridge, MA: National Bureau of Economic Research, 2005.

[7]Wm. A. Wulf. Observations on Science and Technology Trends: Their Potential Impacts on Our Future. In A. G. K. Solomon, ed. *Technology Futures and Global Wealth, Power and Conflict.* Washington, DC: Center for Strategic and International Studies, 2005. Pp. 9-16.

BOX 8-1
Another Point of View: Innovation Incentives

Some critics say the argument that the US economy is lagging in innovation compared with other nations, or even compared with its own historical performance, is not supported by the evidence. Indeed, comparing the current situation with that of 1989 is instructive and striking in this regard.

In 1989, the US economy had been suffering from extremely poor overall productivity growth for almost two decades.[a] By 2005, the United States had experienced almost a decade of accelerated productivity growth, briefly interrupted by the 2001 recession.[b]

In 1989, a panel of experts documented a long-term decline in US industrial performance in several critical sectors.[c] A decade later, a similar assessment showed US industry to be resurgent across a variety of sectors, including several that had been troubled in 1989.[d] In 2005, US-based companies—Google, Apple, Boeing, Genentech—remain at the global forefront in commercializing new technology and creating new markets based on innovation.

In contrast, the economies of most other developed nations have suffered from slower growth in gross domestic product (GDP), productivity, and income—and from higher unemployment and inflation.[e]

What accounts for this "American economic miracle," and will it continue? Various studies have identified key factors, although there is some disagreement over sustainability. In the area of innovation, structural US advantages include our system of research universities with both govern-

the past at taking advantage of breakthroughs and inventions from abroad.[8] But as other nations increase their innovation capacity, the United States must reassess its own environment for innovation and make adjustments to maintain leadership and to maximize the benefits of science and engineering for the public at large.

The innovation environment encompasses a broad range of policy areas. The Committee on Prospering in the Global Economy of the 21st century focused on intellectual property protection, the R&D tax credit, other tax incentives for innovation, and the availability of high-speed Internet access. Although some other important components of the innovation environment were not examined in detail, such as the corporate tax rate and tax-forgiveness policies in various nations, the committee believes the spe-

[8]NAS/NAE/IOM. *Capitalizing on Investments in Science and Technology*. Washington, DC: National Academy Press, 1999.

ment and private funding, the diverse portfolio of government-funded research awarded through peer review, strong intellectual property and securities regulation, and the financing of innovation "led by a uniquely dynamic venture capital industry."[f]

It is generally considered important for the United States to continue to reassess the environment for innovation and to address shortcomings wherever possible; some believe current incentives for companies to innovate and commercialize are strong and not in need of a significant overhaul.

---

[a]P. W. Bauer. "Are WE in a Productivity Boom? Evidence from Multifactor Productivity Growth." Cleveland, OH: Federal Reserve Bank of Cleveland, October 15, 1999. Table 1. Available at: http://www.clevelandfed.org/research/Com99/1015.pdf.

[b]D. W. Jorgenson, M. S. Ho, and K. J. Stiroh. "Projecting Productivity Growth: Lessons from the US Growth Resurgence." Discussion Paper 02-42. Washington, DC: Resources for the Future, July 2002. Available at: http://www.Rff.org/Documents/RFF-DP-02-42.pdf#search='U.S.%20productivity%20growth'; Bureau of Labor Statistics. "Productivity and Costs, 2nd Quarter 2005, Revised." News Release, September 7, 2005. Available at: http://www.bls.gov/news.release/prod2.nr0.htm.

[c]M. Dertouzos, R. Lester, and R. Solow. *Made in America: Regaining the Productive Edge.* Cambridge, MA: MIT Press, 1989.

[d]National Research Council. *US Industry in 2000: Studies in Competitive Renewal.* Washington, DC: National Academy Press, 1999.

[e]R. J. Gordon. *Why Was Europe Left at the Station When America's Productivity Locomotive Departed?* Working Paper 10661. Cambridge, MA: National Bureau of Economic Research, August 2004. Available at: http://www.nber.org/papers/w10661/.

[f]R. J. Gordon. The United States. In B. Steil, D. G. Victor, and R. R. Nelson, eds. *Technological Innovation and Economic Performance.* Princeton, NJ: Princeton University Press, 2002. Pp. 49-73.

cific changes recommended here create significant opportunities. It should be noted that several focus-group members and reviewers raised product liability and tort reform as areas for potential improvement. However, the committee determined that the Class Action Fairness Act of 2005, which represents a major policy change, is a step forward in the national approach to issues of product liability.[9]

ACTION D-1: ENHANCE THE PATENT SYSTEM

Enhance intellectual-property protection for the 21st century global economy to ensure that systems for protecting patents and other forms of intellectual property underlie the emerging knowledge economy but allow

---

[9]Statement on S.5, the Class-Action Fairness Act of 2005. White House press statement. February 18, 2005.

research to enhance innovation. The patent system requires reform of four specific kinds:

- **Provide the US Patent and Trademark Office with sufficient resources to make intellectual-property protection more timely, predictable, and effective.**
- **Reconfigure the US patent system by switching to a "first-inventor-to-file" system and by instituting administrative review *after* a patent is granted. Those reforms would bring the US system into alignment with patent systems in Europe and Japan.**
- **Shield research uses of patented inventions from infringement liability. One recent court decision could jeopardize the long-assumed ability of academic researchers to use patented inventions for research.**
- **Change intellectual-property laws that act as barriers to innovation in specific industries, such as those related to data exclusivity (in pharmaceuticals) and those which increase the volume and unpredictability of litigation (especially in information-technology industries).**

The US patent system is the nation's oldest intellectual-property policy.[10,11] A sound system for patents enhances social welfare by encouraging invention and the dissemination of useful technical information.[12] It also provides incentives for investment in commercialization that promotes economic growth, creates jobs, and advances other social goals.[13]

Balance is a critical element of a sound patent system. Without adequate intellectual-property protection, incentives to create are compromised. On the other hand, too much protection slows the application of valuable ideas. Thus, it is imperative that the US Patent and Trademark

[10]The US Patent and Trademark Office (USPTO), mandated by the US Constitution, awarded its first patent on July 31, 1790, to Samuel Hopkins for an improvement in "making Pot ash and Pearl ash by a new Apparatus and Process."

[11]Article I, section 8 of the Constitution reads, "Congress shall have power . . . to promote the progress of science and useful arts, by securing for limited times to authors and inventors the exclusive right to their respective writings and discoveries." Available at: http://www.uspto.gov/web/offices/pac/doc/general/#ptsc/.

[12]The USPTO offers this simplified definition: "A patent is an exclusive right granted for an invention, which is a product or a process that provides, in general, a new way of doing something, or offers a new technical solution to a problem. . . ." In addition, a patent item must be sufficiently different from what has been used or described before that it may be said to be non-obvious to a person having ordinary skill in the area of technology related to the invention. For example, the substitution of one color for another or changes in size, are ordinarily not patentable. Available at: http://www.uspto.gov/web/offices/pac/doc/general/#ptsc/.

[13]M. Myers, quoted in *Changes Needed to Improve Operation of US Patent System*. National Research Council News Release. Washington, DC: The National Academies, April 19, 2004.

Office (USPTO) and the courts scrupulously protect patent rights and rigorously enforce patent law.[14]

Concerns over questions of patent policy have previously led the National Academies to conduct an extensive study of the field, emphasizing questions related to innovation and technology.[15] That study explored stresses in the system and suggested remedies to promote vitality and improve the functioning of the patent system. This committee believes that several of those recommendations are particularly important, and they are reflected in the first three patent system action items contained herein.

The first priority with regard to patent reform is for Congress and the administration to increase the resources available to the USPTO. Patents are now acquired more frequently and asserted and enforced more vigorously than at any time in the past. That surge in activity is indicative that business, universities, and public entities attach great importance to patents and are willing to incur considerable expense to acquire, exercise, and defend them. There is evidence that the increased workload at the USPTO, with no significant concomitant increase in examiner staffing or other resources, has resulted in a decline in the quality of patent examinations and increased litigation costs after patents are granted.[16] Earlier reports by the National Academies and the Council on Competitiveness identify increasing USPTO capabilities having high priority.[17]

The National Academies report outlines how additional resources should be used. This includes having the USPTO hire and train additional examiners and implementing more capable electronic processing. It also notes that the USPTO should create a strong multidisciplinary analytical capability to assess management practices and proposed changes; provide an early warning of new technologies proposed for patenting; and conduct reliable, consistent reviews of reputable quality that address officewide performance and the performance of individual examiners.[18]

The second important action is to harmonize the US patent system with systems in other major economies by instituting postgrant review and moving from a first-to-invent to a first-inventor-to-file system. In addition to bringing the United States more in line with the patent policies of the rest of

---

[14]See http://www.federalreserve.gov/boarddocs/speeches/2004/200402272/default/.

[15]National Research Council. *A Patent System for the 21st Century*. Washington, DC: The National Academies Press, 2004. P. 18.

[16]J. L. King. Patent Examination Procedures and Patent Quality. In W. M. Cohen and S. A. Merrill, eds. *Patents in the Knowledge-Based Economy*. Washington, DC: The National Academies Press, 2003. Pp. 54-73.

[17]See National Research Council. *A Patent System for the 21st Century*. Washington, DC: The National Academies Press, 2004, especially pp. 103-108; Council on Competitiveness. *Innovate America*. Washington, DC: Council on Competitiveness, 2004, especially p. 69.

[18]See National Research Council. *A Patent System for the 21st Century*. Washington, DC: The National Academies Press, 2004. Pp. 103-108.

the world, these changes would increase the efficiency and predictability of the US system. Increased harmonization would aid US inventors who seek global protection for their inventions.

The only way to challenge a patent under the current system is by litigation. This has led to abuses, such as laying broad claims—sometimes without reason or merit—to patents in hopes of receiving a generous settlement from a competitor who wishes to avoid long and expulsive litigation. Often, competitors or other interested parties are the best available source of information about the state of the art. Inviting their input in a process of administrative review—the so-called opposition system—would allow for "peer review" of recently granted patents to serve as a second check or quality assurance of the initial examination by the patent office. Such opposition is much less expensive than litigation, open to anyone, and much faster—decisions can sometimes be made in 1 day. The 2004 National Academies report explains, in considerable detail, how such a system, which it calls "Open Review," would work.[19]

The United States still uses a first-to-invent rather than a first-to-file patent system. This requires a complex, expensive, and time-consuming (5-10 years) process to sort out who has the patent rights. It also absorbs the time of some of the most experienced patent examiners. Ultimately, the amount of resources devoted to resolving the priority question (which is resolved in favor of the first filer over two-thirds of the time)[20] outweighs the benefits, and the time and personnel required could be put to better use improving the quality of basic examinations.

Some might argue that the proposed changes would put smaller inventors at a disadvantage. However, resolving disputes through an opposition process is far less expensive than is litigation, and that alone would constitute a significant benefit to small companies and individual inventors with worthy claims. Periodic surveys by the American Intellectual Property Law Association indicate that patent litigation costs—now millions of dollars for each party in a case where the stakes are substantial—are increasing at double-digit annual rates. The relatively low cost of filing provisional applications to establish priority under a first-to-file system would not constitute a significant burden on small inventors.

The third recommended action is to preserve some existing research exemptions from infringement liability.[21] Until recently, it was widely be-

[19]Ibid., pp. 95-103.

[20]See http://www.oblon.com/media/index.php?id=181.

[21]The committee recognizes the interest of some reviewers in re-examining aspects of the technology transfer process governed by the Bayh–Dole Act and related legislation, but issues related to Bayh–Dole are controversial and have been under discussion for years. The committee believes that establishing a research exemption for infringement liability is a higher priority.

lieved, especially in the academic research community, that uses of patented inventions purely for research were shielded from infringement liability by an experimental-use exception first articulated in 19th-century case law. But in *Madey v. Duke University*,[22] a suit brought by a former Duke University professor and laboratory director, the Federal Circuit Court upended that notion by holding that there is no protection for research conducted as part of the university's normal "business" of investigation and education, regardless of its commercial or noncommercial character.

By the time *Madey* arrived before the court, most universities had established intellectual-property offices, and there were clear difficulties in distinguishing commercially motivated research from "pure" academic research. The court, without addressing that issue directly, decided that for a major research university even noncommercial research projects "unmistakably further the institution's legitimate business objectives, including educating and enlightening students and faculty participating in these projects."[23] Activities that further "business objectives," including research projects that "increase the status of the institution and lure lucrative research grants, students and faculty," are ineligible for an experimental use defense.

Thus, the court regarded virtually all research as a means of advancing the "legitimate business objectives" of a university. The result, wrote one observer, "is a seemingly disingenuous opinion that neither conforms to the implications of precedent nor explains the reasons for steering the law in a different direction, but pretends that prior courts never meant to give research science special treatment."[24] Because the courts have not traced the experimental-use defense, case by case, as a tool for mediating between the private interests of patent owners and the public interest of open scientific progress, that issue awaits resolution.

The 2004 National Academies study offers two alternatives.[25] The preferred solution would be the passage of appropriately narrow legislation to shield some research uses of patented inventions from infringement liability. If progress on the legislative front is delayed, the Office of Management and Budget might consider extending to grantees the "authorization and consent" protection that is provided to contractors, provided that such protection is strictly limited to research and does not extend to resulting commercial products or services.

---

[22]*Madey v. Duke Univ.* 307 F.3d 1351. Available at: 2002 U.S. App. LEXIS 20823, 64 U.S.P.Q.2d. (BNA) 1737 (Fed. Cir. 2002).

[23]Ibid.

[24]R. Eisenberg. "Science and the Law: Patent Swords and Shields." *Science* 299(5609)(2003): 1018-1019.

[25]National Research Council. *A Patent System for the 21st Century*. Washington, DC: The National Academies Press, 2004. P. 82.

The final action proposed herein for modernizing the patent system—and the only one our committee did not derive from the 2004 National Academies report—is to change intellectual-property laws that constitute barriers to innovation in specific industries. The two main problem areas are in the pharmaceutical and information-technology industries. It is particularly expensive to create and market new drugs and medicines, and the costs are unlikely to be recovered unless there is predictable intellectual-property protection of appropriate duration. The interaction of the US Food and Drug Administration (FDA) approval process and the patent system poses unique challenges to the pharmaceutical industry. The inherent risk to drug developers is illustrated by the reality that more than 90% of pharmaceutical candidates fail in clinical testing.[26] Furthermore, only 1 in 1,000 new formulations tested reach clinical trials,[27] and a relatively small minority of those, perhaps one-third, pay back the cost of even their own research.[28] It is critical that a balance be struck in finding an appropriate period of exclusivity such that innovation is stimulated and sustained but patients have access to generic-drug-pricing structures.

Current intellectual-property protection for new medicines is governed under the Hatch–Waxman law, enacted in 1984, to give 14 years of patent protection after FDA approval of a new medicine. However, the law does not provide the same period for sustained marketing exclusivity. It curtails the ability to extend patents and provides opportunities for early patent challenges. The protection of data under the law is roughly one-half as long as the period afforded in Europe, creating a relative disadvantage for the United States in attracting pharmaceutical businesses[29] (see Box 8-2).

In the near term, the United States should adopt the European period of 10-11 years. However, research should be undertaken to determine whether this period is adequate, given the complexity and length of drug development today.

Patent issues are also particularly important to the information-technology industry, especially in software and Internet-related activities. The volume and unpredictability of litigation have recently attracted considerable attention and are currently being reviewed by Congress. An

---

[26]C. Austin, L. Brady, T. Insel, and F. Collins. "NIH Molecular Libraries Initiative." *Science* 306(2004):1138-1139.

[27]Tufts Center for the Study of Drug Development. "Backgrounder: How New Drugs Move Through the Development and Approval Process." November 1, 2001. Available at: http://csdd.tufts.edu/NewsEvents/RecentNews.asp?newsid=4.

[28]H. Grabowski, J. Vernon, and J. DiMasi. "Returns on Research and Development for 1990s New Drug Introductions." *Pharmacoeconomics* 20(Supplement 3)(2002):11-29.

[29]International Association of Pharmaceutical Manufacturers & Associations. "A Review of Existing Data Exclusivity Legislation in Selected Countries." January 2004. Available at: http://www.who.int/intellectualproperty/topics/ip/en/Data.exclusivity.review.doc.

BOX 8-2
A Data-Exclusivity Case Study

Incentives to innovate could be considerably improved by enhancing data-package exclusivity. In the case of incentives to develop new medicines, data-package exclusivity protects for a period of years an innovator's regulatory submission package to the Food and Drug Administration from being used as a source of information by a company that produces generic products. The period in Europe is 10 years plus an additional year if the innovator has gained approval for more than one indication. The United States grants data exclusivity for a new chemical entity for 5 years; a second indication is entitled to 3 years of exclusivity. Those periods are generally too short to stimulate investment. Thus, innovation incentives in the United States are almost entirely patent-driven.

The current system has been successful in stimulating the creation of new molecules, but the limitations of the patent system sometimes result in denying patients the best that the pharmaceutical industry could offer. The limitations are due largely to the time constraints under which the patent system operates. Patents generally must be filed as quickly as possible after an invention occurs, and the ticking clock creates a tension with other aspects of drug development.[a]

The demands for data on a molecule's safety and efficacy are increasing. The generation of the necessary data requires time and money. It is to patients' benefit for as much time as appropriate to be devoted to the development of the data, but spending the time lessens the return on the developer's investment because it encroaches on the patent term. Bringing a new medicine to patients requires a sequence of major breakthroughs, which in the current system must be accomplished well before the life of a patent runs out. Often, the clock does run out, and the innovator must start over with a new molecule simply to get time "back on the clock." As a result, there is an ever-growing "graveyard" currently comprising more than 10 million compounds. There is no incentive to exhume these compounds in the absence of substantial data-package exclusivity, because patents will be either unavailable or of such narrow coverage that they would be easy to avoid in developing a related drug.

In addition, there is little incentive to pursue new indications for old molecules without appropriate data-package protection. Indeed, when no compound patent covers the product, there is a *disincentive* to develop new indications. Generic medicines may be approved for a smaller number of indications than those associated with the innovator's drug. If there is no compound patent and one of the indications is unpatentable, the generic medicine may be approved only for the unpatented indication. The innovator's entire market could then be eroded because

continued

BOX 8-2 Continued

physicians have the latitude to prescribe the generic compound for any indications, including patented ones. Every reasonable effort should be made to encourage the development of new indications for known compounds because of the greater level of knowledge about safety for already-marketed compounds than for brand-new ones.

[a]The pressure to file for patents as quickly as possible after an invention occurs is inevitable in a global knowledge economy, whether or not the United States stays with the current first-to-invent system or moves to a first-inventor-to-file system. Most of the world follows the latter system. Innovators seeking patent protection in the three major patenting regions (the United States, Europe, and Japan) must therefore manage their patent filings consistent with the first-inventor-to-file system.

additional complexity of sector-specific issues is that intellectual-property laws vary among nations, affecting innovation differently in different industries. The committee concludes that those issues are opportunities for Congress and other relevant federal entities to take productive actions, including those outlined above.

ACTION D-2: STRENGTHEN THE RESEARCH AND EXPERIMENTATION TAX CREDIT

Enact a stronger research and development tax credit to encourage private investment in innovation. The current Research and Experimentation (R&E) Tax Credit goes to companies that *increase* their research and development spending above a base amount calculated from their spending in prior years. Congress and the Administration should make the credit permanent,[30] and it should be increased from 20 to 40% of the qualifying increase so that the US tax credit is competitive with that of other countries. The credit should be extended to companies that have consistently spent large amounts on research and development so that they will not be subject to the current de facto penalties for having previously invested in research and development.

Much of the benefit of industry R&D spending accrues to society in ways that cannot be captured by individual firms. The R&E Tax Credit and similar policies in other nations are designed to promote more R&D investment and to encourage the creation and retention of jobs in the country that provides the tax incentive.

[30]The current R&D tax credit expired in December 2005.

Econometric studies have estimated that the tax credit encourages at least as much R&D spending as the credit costs in forgone tax revenue—and perhaps as much as twice that amount—particularly over the long term.[31] Political and community leaders traditionally have viewed R&D incentives primarily as a tax issue, but their effect on jobs could be even more significant. R&D incentives directly create or sustain high-wage, high-skill jobs in places where the research is conducted. When long-term gains in productivity, income, and tax revenue are added to the immediate gain in R&D spending encouraged by the tax credit, it seems clear that the credit is a cost-effective mechanism for encouraging innovation and creating quality jobs.

The first change the committee recommends, namely making the credit permanent, is perhaps the most straightforward. Since the introduction of the tax credit in 1980, it has been extended repeatedly, allowed to lapse, and periodically modified, all without being formalized as a permanent, reliable element of policy.[32] Over the years, numerous committees and groups have recommended that the credit be made permanent so that companies can plan longer term investments in US-based R&D with the knowledge that the credit will be available.[33] The Council on Competitiveness recently echoed the call to make the tax credit permanent.[34]

The second change, increasing the credit from 20 to 40%, would be more controversial and, in the near term, more costly. The cost of the current tax credit is estimated at $5.1 billion for fiscal year (FY) 2005. The cost for FY 2006 is estimated at about $4.2 billion, assuming the current credit, due to expire December 31, 2005, is extended once again.[35] The committee therefore estimated that permanent extension of the credit would cost about $5 billion per year (roughly what the credit currently costs), and that the other recommended changes (doubling the rate and expanding eligibility) could potentially result in doubling the cost.

There are several reasons to increase the rate, not the least of which is that the *effective* current credit is 13%, rather than 20%, for companies that deduct R&D expenses.[36] A higher percentage would raise the incentive effect of the credit.

---

[31]B. H. Hall and J. van Reenen. *How Effective Are Fiscal Incentives for R&D? A Review of the Evidence.* Working Paper 7098. Cambridge, MA: National Bureau of Economic Research, 1999.

[32]As currently extended, the R&D tax credit will expire on December 31, 2005.

[33]National Research Council. *Harnessing Science and Technology for America's Economic Future.* Washington, DC: National Academy Press, 1999. P. 46.

[34]Council on Competitiveness. *Innovate America.* Washington, DC: Council on Competitiveness, 2004. P. 59.

[35]See *Budget of the United States Government, Fiscal Year 2006, Analytical Perspectives.* Washington, DC: US Government Printing Office, 2005. P. 65. Available at: http://a255.g.akamaitech.net/7/255/2422/07feb20051415/www.gpoaccess.gov/usbudget/fy06/pdf/spec.pdf.

[36]This is due to the Section 280C limitation in the Internal Revenue Code. See J. R. Oliver. "Accounting and Tax Treatment of R&D: An Update." *The CPA Journal* 73(7)(2003):46-49.

It also is important to consider in international context the issue of whether the United States is keeping pace with other economies as an attractive location for R&D (see Table 8-1). Federal R&D tax credits rarely determine the type of research performed, but they can influence where the work is conducted.[37,38] As of 2000, the most recent year for which data are available, foreign-based multinational corporations (MNCs) performed $26 billion in R&D in the United States. US-based MNCs performed $19.8 billion in R&D overseas.[39] There is an obvious advantage in having MNCs locate operations in the United States as not only does it maintain the employment of the scientists and engineers at corporate research laboratories, but research activities are often located near production facilities that affect the employment of all workers where we already benefit from their contributions to US corporate R&D.[40]

The Organisation for Economic Co-operation and Development (OECD) has noted a trend in member countries toward more generous tax incentives for R&D investments.[41] By moving to a higher, permanent tax credit, the United States will be better positioned to compete against credits already offered elsewhere.

Likewise, national policy must be conformed to ensure appropriate revisions of regulations interpreting and implementing the federal R&E tax credit. Practical and uniform guidelines for the conduct of tax audits related to the federal R&E tax credit must also be adopted. Federal research tax-credit regulations should be updated to reflect the changing impact of technology on the character of R&D, such as expanded use of databases provided by external parties and the greater conduct of R&D through joint ventures. Any national policy on tax credits and related incentives should recognize the importance of having states and localities also conform their laws to embrace a focus on research and innovation.

---

[37]Organisation for Economic Co-operation and Development. "Tax Incentive for Research and Development: Trends and Issues." Available at: http://www.oecd.org/dataoecd/12/27/2498389.pdf.

[38]J. M. Poterba. Introduction. In J. M. Poterba, ed. *Borderline Case: International Tax Policy, Corporate Research and Development, and Investment.* Washington, DC: National Academy Press, 1997. P. 3. This is not to say that there is evidence that companies locate R&D in the country that has the best R&D tax credit. In fact, the industry perspectives in the Poterba volume suggest otherwise. And the second OECD paper referenced above indicates that the differential between the overall corporate tax rate and the credit is the key factor. For example, Ireland has a low overall corporate tax rate, so its R&D tax credit was not as effective as it would have been had the overall corporate tax rate been higher.

[39]National Science Board. *Science and Engineering Indicators 2004.* NSB 04-01. Arlington, VA: National Science Foundation, 2004. Pp. 4-64–4-65.

[40]Ibid., Tables 4-50, 4-51, and 4-52.

[41]Organisation for Economic Co-operation and Development. *Science, Technology, and Industry Outlook.* Paris: OECD, 2004. P. 67.

TABLE 8-1 Overview of R&D Tax Incentives in Other Countries

County	R&D Tax Incentive	Comment
Australia	• Allows a 125% deduction for R&D expenditures. • *Plus* a 175% deduction for R&D expenditures exceeding a base amount of prior-year spending.	The 125% deduction is the equivalent of a flat 7.5% R&D tax credit. In discussing its R&D-friendly environment, the Australian government's website (http://investaustralia.com) concludes, **"It's little surprise then, that many companies from around the world are choosing to locate their R&D facilities in Australia."** The government also points out that "50% of the most innovative companies in Australia are foreign-based."
Canada	• Offers a permanent 20% flat (i.e., first-dollar) R&D tax credit. • *Also*, many provincial governments offer various incentives (e.g., refundable credits) for R&D activities conducted in their provinces.	In 2003, US subsidiaries spent **$2.5 billion on R&D in Canada**, which has mounted an aggressive marketing campaign, including television and print advertisements, to lure more US companies to locate R&D operations north of the border. An Ontario print ad discusses **"R&D tax credits, among the most generous in the industrialized world"** and "a cost structure which KPMG confirms as lower than the U.S. and Europe"; the ad concludes, "You'll see why R&D in Ontario is clearly worth investigating."
China	• Offers foreign investment enterprises a 150% deduction for R&D expenditures, provided that R&D spending has increased by 10% from the prior year.	The 10% incremental-increase threshold should not be difficult to meet for US-owned companies growing start-up operations in China. China's Ningbo Economic & Technical Development Zone ("NETD") invites **global companies to "enjoy a number of preferential taxation policies,"** as well as other benefits.
France	• Allows a 50% R&D credit, includes a 5% flat credit and a 45% credit for R&D expenditures in excess of average R&D spending over the two previous years.	As is the case with the China R&D deduction, the incremental threshold governing the French 50% credit should be easy to meet for "inbound" companies growing their operations in France. In 2003, US subsidiaries spent **$1.8 billion on R&D in France.** "This is the first time in our industry that Americans are coming to Europe to join the R&D of Europeans," says Pasquale Pastore, President and CEO of STMicroelectronics, in *The New France, Where the Smart Money Goes.*
India	• Companies carrying on scientific research and development are entitled to a 100% deduction of profits for 10 years.	"More than 100 global companies . . . have established R&D centers in India in the past 5 years, and more are coming. . . . As I see it from my perch in India's science and technology leadership, **if India plays its cards right, it can become by 2020 the world's number-one knowledge production center,"** Raghunath

continued

TABLE 8-1 continued

County	R&D Tax Incentive	Comment
	• Automobile industry also is entitled to a 150% deduction for expenditures on in-house R&D facilities.	Mashelkar, Director General, Council for Scientific & Industrial Research, India, in *Science Magazine*.
Ireland	• Offers a 20% R&D tax credit, plus a full deduction, as well as a low generally applicable 12.5% corporate income tax rate. • Capital expenditure may also quality for a separate flat credit.	According to IDA Ireland, the government agency with responsibility for the promotion of direct investment by foreign companies into Ireland, "**Many leading global companies have found Ireland to be an excellent location for knowledge-based activities. . . .**" Nearly half of all IDA supported companies now have some expenditure on R&D and 7,300 people are engaged in the activity.
Japan	• Offers a flat 10% R&D tax credit (a 15% flat credit is provided for small companies), in addition to other incentives.	In 2003, US subsidiaries spent **$1.7 billion on R&D in Japan**. Junichiro Mimaki, an official from Japan's Ministry of Economy, Trade, and Industry, said in an August 26 interview with the Bureau of National Affairs that **R&D and IT tax relief has created 400,000 jobs and boosted gross domestic product by 6.1 trillion yen ($55 billion) over 3 years.**
Korea	• Tax holidays, up to 7 years, are provided for high-technology business. • In addition, a variety of tax credits are provided for R&D-type expenditures.	**Korea is moving aggressively to attract foreign R&D investment,** promoting not only tax incentives but also other benefits for foreign companies locating R&D in the Incheon Free Economic Zone ("IFEZ").
Singapore	• "R&D and Intellectual Property Management Hub Scheme" offers US companies a 5-year tax holiday for foreign income earned with respect to Singapore-based R&D.	According to Singapore's Economic Development Board Web site: "**Singapore does not just welcome business ideas; it actually seeks and nurtures them**. We play host to any shape and size of enterprise and innovation—startups with little more than the germ of an idea, global corporations with large R&D teams and complex production operations."
United Kingdom	• Allows a 125% deduction for R&D expenses, plus a 175% deduction for R&D expenditures exceeding a base amount of prior-year R&D spending.	The UK leads the world in attracting R&D investment by US affiliates—US subsidiaries spent more than **$4 billion on UK-based R&D** in 2003. The 125% deduction alone is the equivalent of a flat 7.5% R&D tax credit.

SOURCE: R&D Credit Coalition. "International R&D Incentives." Fact Sheet. September 15, 2005. Available at: http://www.investinamericasfuture.org/factsheets.html. Accessed October 11, 2005.

Finally, the definition of applicable expenses used to calculate the tax credit should be expanded to allow companies that have consistently maintained high levels of R&D spending to claim the credit. As currently written, the credit rewards companies that have high R&D expenditures compared with a base period. Companies that consistently invest large amounts, but do not appreciably increase those amounts over time, can be entitled to little or no credit. The formula should be amended so as not to penalize consistent R&D investors but rather to allow companies with significant and consistent R&D investments to receive tax credits.

Credit should be allowed for all relevant research expenditures (in contrast with the current incremental approach) by, for example, broadening the definition of qualifying expenditures. Qualifying expenditures could be broadened to include some legitimate costs of conducting research, such as employee benefit costs (defined benefits, retirement plans, healthcare plans, and so on) related to qualifying wages, as well as 100% of contract research costs (as opposed to the current 65%). In a different method, qualifying expenditures could be redefined to include all Internal Revenue Code (IRC) Section 174 expenditures (a much broader definition of R&D expenditures). A portion of the IRC (the Section 280C limitation) that reduces the federal R&D credit by 35% might also be repealed (the limitation has the result that the 20% tax credit available in the United States today is really only a 13% credit).

ACTION D-3: PROVIDE INCENTIVES FOR US-BASED INNOVATION

Many policies and programs affect innovation and the nation's ability to profit from it. It was not possible for the committee to conduct an exhaustive examination, but alternatives to current economic policies should be examined and, if deemed beneficial to the United States, pursued. These alternatives could include changes in overall corporate tax rates and special tax provisions, providing incentives for the purchase of high-technology research and manufacturing equipment, treatment of capital gains, and incentives for long-term investments in innovation. The Council of Economic Advisers and the Congressional Budget Office should conduct a comprehensive analysis to examine how the United States compares with other nations as a location for innovation and related activities with a view to ensuring that the United States is one of the most attractive places in the world for long-term innovation-related investment and for the jobs resulting from that investment. From a tax standpoint, that is not now the case.

Countries around the world are working to bolster innovation, often by improving the tax environment for high-technology business activities (see Box 8-3, Finland; Box 8-4, South Korea; Box 8-5, Ireland; Box 8-6,

BOX 8-3
Finland

The rapid growth of Finland's high-technology economy is often seen as testament to long-term strategic planning, systematic investment, and the ability to adopt innovative policies more quickly than other nations. In the 1970s, Finland's political leaders, research community, and labor unions engaged in planning to focus R&D funding in electronics, biotechnology, and materials technology. Sustained government support paid off, as electronics-based exports grew from 4% of Finland's economy in 1980 to 33% of all exports in 2003.[a] Today, Finland's private and public sectors invest 3.5% of GDP into R&D programs (ranked second in the world), and the proportion of its population working as research scientists is the highest in the world.[b]

---

[a]Organisation for Economic Co-operation and Development. *Innovation Policy and Performance: A Cross-Country Comparison*. Paris: OECD, 2005.

[b]Organisation for Economic Co-operation and Development. *Main Science & Technology Indicators*. Paris: OECD, 2005.

BOX 8-4
South Korea

South Korea recently established an agency to coordinate innovation policies and R&D strategies within the Ministry of Science and Technology. Almost 40% of all postsecondary degrees awarded there are in science and engineering, compared with 15% in the United States.[a] The government is seeking to double its expenditures on R&D between 2002 and 2007.

---

[a]Organisation for Economic Co-operation and Development. *Education Database*. Paris: OECD, 2005.

Singapore; and Box 8-7, Canada). There are strengthening signs that changes in US tax policy are needed to encourage investment in America. The flexibility of US capital markets, particularly for financing small, high-technology enterprises through venture capital and public stock offerings, had been one of our major strengths, encouraging companies to focus their innovation in the United States. The rapid rise of venture capital in the late 1990s, however, was followed by the precipitous collapse of the technology

BOX 8-5
Ireland

The success of the "Celtic Tiger" in the 1990s was remarkable, especially in comparison with other member nations of the European Union. In 1987, Irish GDP per capita was 69% of the European Union average, but by 2003 it had reached 136%.[a] Ireland's unemployment fell from 17% to 4% over the same period. How did Ireland go from being one of Europe's poorest nations to one of the richest? First, Ireland aggressively courted multinational corporations and maintained a business-friendly 12.5% corporate tax rate.[b] Most of the world's top pharmaceutical, medical device, and software concerns now have operations in Ireland.[c] Second, the government placed a strong emphasis on secondary and higher education, and tuition has been free since 1996. Participation in Irish higher education surpasses the OECD average. Today, Ireland is focused on increasing its public R&D spending and production of scientists and engineers to complement strong growth in R&D performance by foreign multinational corporations. The goal is to increase total R&D intensity in the economy from 1.4% of GDP in 2002 to 2.5% by 2010.[d]

---

[a]"Tiger, Tiger, Burning Bright." *The Economist* 373(8397)(2004):4-6.

[b]Heritage Foundation. "Ireland. 2005 Index of Economic Freedom." 2005. Available at: http://www.heritage.org.

[c]T. Friedman. *The End of the Rainbow.* New York Times, June 29, 2005. P. A-23.

[d]Organisation for Economic Co-operation and Development. *Science, Technology, and Industry Outlook.* Paris: OECD, 2005. P. 56.

BOX 8-6
Singapore

Singapore is continuing its long history of active government involvement to promote innovation. This includes a major investment in Biopolis, opened in October 2002, which Singapore intends to be a world-class biomedical sciences R&D hub for Asia.[a] It is backed with a portfolio of scholarships, fellowships, and grants to attract students and researchers from around the world. Another initiative is the Standards, Productivity, and Innovation Board,[b] which combines incentives and other help to increase the number of Singapore's small and medium-size high-technology and e-commerce businesses, improve national productivity and entrepreneurship, and expand the nation's position in retail markets.

---

[a]See http://www.one-north.com/pages/lifeXchange/index.asp. Accessed September 15, 2005.

[b]See http://www.spring.gov.sg/portal/main.html. Accessed September 15, 2005.

BOX 8-7
Canada

Canada's two-part innovation strategy covers almost every aspect of that nation's economic and educational systems. The first part, called Achieving Excellence: Investing in People, Knowledge, and Opportunity, is a plan to expand the Canadian economy.[a] The second part is Knowledge Matters: Skills and Learning for Canadians, which outlines plans to improve Canadian education.[b] The overall goal of the programs is to strengthen Canada's economy by improving quality in, and access to, elementary, secondary, and higher education; by promoting R&D in the sciences and engineering; and by extending the new programs and reforms from the federal government to the smallest township.

---

[a]Government of Canada. *Achieving Excellence: Investing in People, Knowledge and Opportunity*. Executive Summary. Ottawa, ON: Government of Canada, 2002. Available at: http://www.innovationstrategy.gc.ca/gol/innovation/site.nsf/en/in02425.html.

[b]Government of Canada. *Knowledge Matters: Skills and Learning for Canadians*. Executive Summary. Ottawa, ON: Government of Canada, 2002. Available at: http://www11.sdc.gc.ca/sl-ca/doc/summary.shtml.

stock bubble in 2001. Venture-capital investments have been fairly flat since then, so the United States no longer has that advantage.[42] Perhaps equally important is the fact that investment capital tends to be highly mobile and to follow opportunity irrespective of national borders.

The committee believes that the United States can and should do more, particularly in tax policy, to encourage long-term investments in innovation, but it was not able to examine all options and their implications within the schedule mandated for our study. Several creative new approaches to capital-gains taxation were discussed, including the option of reducing rates for very-long-term investments or offering more liberal allowances for loss writeoffs. The overall corporate tax rate, which some industry groups see as high by international standards (although there is controversy about this), is important for determining where companies invest in R&D and downstream activities. Finally, incentives for the purchase of high-tech manufacturing and research equipment—through tax credits and accelerated depreciation—were considered.

Those new approaches would have widespread consequences for the economy as a whole and for our national fiscal position. It would be neces-

---

[42]See the National Venture Capital Association Web site at: http://www.nvca.org.

sary to structure any new incentives as a comprehensive, integrated package. It would also be useful to compare the effects of various options, especially with reference to what other nations are doing. Any such analysis should examine US and foreign tax systems with a view to developing a package of incentives to ensure that the United States remains a highly attractive place for long-term innovation-related investments and for location of the follow-on jobs they produce.

ACTION D-4: ENSURE UBIQUITOUS BROADBAND INTERNET ACCESS

Several nations are well ahead of the United States in providing broadband access for home, school, and business. That capability can be expected to do as much to drive innovation, the economy, and job creation in the 21st century as did access to the telephone, interstate highways, and air travel in the 20th century. Congress and the administration should take prompt action—mainly in the regulatory arena and in spectrum management—to ensure widespread affordable broadband access in the near future.

The production of information-technology equipment and the use of information technology have been important engines for US productivity growth in a range of industries and for the resulting low-inflation economic expansion (briefly, but significantly, interrupted in 2001) that the nation has experienced since the mid-1990s.[43] The OECD estimates that the percentage of total capital investment accounted for by spending on that equipment is significantly higher in the United States than it is in other OECD economies.[44] Industries as diverse as financial services, retail, entertainment, and logistics and transportation are being transformed by information technology.

Although some believe that broadband access is not critical to US competitiveness, the committee disagrees. The information technology revolution will continue to fuel economic growth, the creation of high-paying jobs, and US leadership in science and engineering well into the future. Accelerating progress toward making broadband connectivity available and affordable for all US citizens and businesses is critical. Although penetration of broadband service in the United States is increasing rapidly, broadband leaders such as South Korea and Japan are still far ahead.[45]

---

[43]R. J. Gordon. *Technology and Economic Performance in the American Economy*. Working Paper 8771. Cambridge, MA: National Bureau of Economic Research, 2002.

[44]Organisation for Economic Co-operation and Development. *The Economic Impact of ICT*. Paris: OEDC, 2004. P. 67.

[45]P. Gralla. "U.S. Lags in Broadband Adoption Despite VoIP Demand, Says Report." *EE Times Online*, December 16, 2004. Available at: http://www.eet.com/showArticle.jhtml?articleID=55800449.

President Bush has announced a national goal of ubiquitous broadband access in the United States.[46] The committee urges the Administration and Congress to take the necessary steps to meet that goal. Many of the barriers to more rapid broadband penetration lie in the area of telecommunications regulation and spectrum policy, where in some cases entrenched industry interests are clashing to preserve and extend the advantages offered under policies promulgated in the past.[47]

Telecommunication infrastructure will be crucial to the competitiveness of any country in the 21st century. It is the medium by which data are accessed, consultations take place, and decisions are transmitted. One has only to look at the vast amounts of information transmitted by the financial community, the use of information in the retail market (for example, Wal-Mart, the largest retailer in the world, owes much of its competitiveness to its information-technology infrastructure for tracking sales, inventory, and consumer purchasing trends in real time), and the growth of online sales in almost every business segment of the economy.

As the Internet becomes more dominant in communication, information access, commerce, education, and entertainment, the key infrastructural factor will be broadband access. The potential effects on society and individuals of distance learning, telemedicine, Internet entertainment, and delivery of government services demonstrates how great the impact of broadband on the competitiveness of any country could be. The United States was an early leader in Internet broadband penetration but recently has fallen out of the top 10 countries in per capita broadband access. In fact, vast rural regions of the United States are devoid of affordable bidirectional broadband capability. Just as the United States was a leader in providing ubiquitous telecommunication capability to its citizens in the 20th century and reaped the benefits of voice-connectivity technology, it should be a leader in facilitating broadband Internet connectivity to its citizens in the 21st century. That infrastructure not only will support existing commerce but will facilitate the growth of new industries.

Broadband access clearly is not a "big-company issue;" large companies can generally afford the technology, and many have already put it in place in order to compete. Broadband is an important issue for ordinary citizens (providing, for example, the ability to telecommute on a national and international scale) as well as small and medium-sized businesses. As many of us have found when calling a company to help fix our computer, making an airline reservation, or getting guidance on how to help a sick child in the middle of the night, the person we call may be virtually any-

---

[46]"Bush Pushes Ubiquitous Broadband by 2007." *Reuters*, March 26, 2004.

[47]R. Hundt. "Why Is Government Subsidizing the Old Networks When 'Big Broadband' Convergence Is Inevitable and Optimal?" *New America Foundation Issue Brief*. December 2003.

where, whether rural or urban, at home or in a call center, in the United States or overseas. If we expect all of our citizens and companies to be competitive, universal availability of affordable broadband should be a matter of national policy.

Some of the programs and policies already being pursued in the United States, such as federal R&D funding and accelerated tax depreciation on equipment purchases, do cost the federal government money in terms of outlays and forgone direct revenue. However, the committee believes that the most important needed changes are in the regulatory and spectrum management areas. Policy changes in both of these areas have a broad impact on the incentives of private companies to invest in infrastructure and to develop competitive services. Recent examples of regulatory changes include Federal Communications Commission decisions to free newly deployed broadband infrastructure from legacy regulation and to develop a framework for deployment of Broadband over Power Lines (BPL). These sorts of regulatory changes do not entail financial investments by the federal government. The future of spectrum management is another particularly critical area.[48] And, as is the case with regulatory policy, changes in spectrum policy would not necessarily entail costs to the federal government and might even result in additional revenue.

CONCLUSION

The United States, if it is to ensure the continued high standard of living and security of its citizens, must maintain its position as the world's premier place for innovation, for investment in downstream activities such as manufacturing and marketing, and for creation of high-paying jobs. We can do this if, while implementing the other recommendations made herein, we modernize the patent system, realign tax policies to encourage innovation, and ensure the nation meets the goal of affordable broadband Internet access for all. The committee could not examine every possibility, but appropriate policy changes should be pursued in each of these areas. A comprehensive comparative analysis of tax rules, conducted by the Council of Economic Advisers and the Congressional Budget Office, could elucidate how we stack up against other nations as a location for innovation and related follow-on activities. The object of that examination and the adoption of the recommendations in this chapter would be to ensure that the United States provides the innovation-friendly environment needed to remain a highly attractive place to invest in the future.

---

[48]See US Department of Commerce. *Spectrum Policy for the 21st Century: The President's Spectrum Policy Initiative*. Report 1. Washington, DC: US Department of Commerce, June 2004. Available at: http://www.ntia.doc.gov/reports/specpolini/presspecpolini_report1_06242004.htm.

9

What Might Life in the United States Be Like if It Is Not Competitive in Science and Technology?

Since World War II, the United States has led the world in science and technology, and our significant investment in research and education has translated into benefits from security to healthcare and from economic competitiveness to the creation of jobs. As we enter the 21st century, however, our leadership is being challenged. Several nations have faster growing economies, and they are investing an increasing percentage of their resources in science and technology. As they make innovation-based development a central economic strategy, we will face profoundly more formidable competitors as well as more opportunities for collaboration. Our nation's lead will continue to narrow, and in some areas other nations might overtake us. How we respond to the challenges will affect our prosperity and security in the coming decades.

To illustrate the stakes of this new game, it is useful to examine the changing nature of global competition and to sketch three scenarios for US competitiveness—a baseline scenario, a pessimistic case, and an optimistic case. The scenarios demonstrate the importance of maintaining the nation's lead in science and technology.

"THE AMERICAN CENTURY"

In the second half of the 20th century, the United States led the world in many areas. It was the world's superpower, it had the highest per capita income of any major economy, it was first among developed countries in economic growth, and it generated the largest share of world exports—with less than 5% of the world's population, it consumes 24% of what the world produces.[1] US-based multinational corporations dominated most industrial sectors. In the 1990s, the United States experienced the longest economic boom in its history, driven in large part by investments in information technology and by accelerating productivity.

Central to prosperity over the last 50 years has been our massive investment in science and technology. Government spending on research and development (R&D) soared after World War II, and government spending on R&D as a percentage of the gross domestic product (GDP) reached a peak of 1.9% in 1964 (it has since fallen to 0.8%[2]). By 1970, the United States enrolled 30% of all postsecondary students in the world, and more than half the world's science and engineering doctorates were awarded here.[3]

Today, with just 5% of the world's population, the United States employs nearly one-third of the world's scientific and engineering researchers, accounts for 40% of all R&D spending, publishes 35% of science and engineering articles, and obtains 44% of science and engineering citations.[4] The United States comes out at or near the top of global rankings for competitiveness. The International Institute for Management Development ranks the United States first in global competitiveness; the World Economic Forum puts us second (after Finland) in overall competitiveness and first in technology and innovation.[5]

Leadership in science and technology has translated into rising standards of living. Technology improvements have accounted for up to one-half of GDP growth and at least two-thirds of productivity growth since 1946.[6] *Business Week* chief economist Michael Mandel argues that, without innovation, the long-term growth rate of the US economy would have been closer to 2.5% annually rather than the 3.6% that has been the average since the end of World War II. If our economy had grown at that lower

---

[1]Center for Sustainable Energy Systems, University of Michigan, "US Energy System Factsheet." August 2005. Available at: http://css.snre.umich.edu/css_doc/CSS03-11.pdf.

[2]American Association for the Advancement of Science. "US R&D as Percent of Gross Domestic Product, 1953-2003." May 2004. Available at: www.aaas.org/spp/rd. Based on National Science Foundation data in National Science Board. *Science and Engineering Indicators 2004*. NSB 04-01. Arlington, VA: National Science Foundation, 2004. Figure 4-5.

[3]R. B. Freeman. *Does Globalization of the Scientific/Engineering Workforce Threaten US Economic Leadership?* Working Paper 11457. Cambridge, MA: National Bureau of Economic Research, June 2005. P. 3.

[4]Ibid., p. 1.

[5]IMD. *World Competitiveness Yearbook* (2005); World Economic Forum. *The Global Competitiveness Report, 2004-2005*. New York: Oxford University Press, 2004.

[6]G. Tassey. *R&D Trends in the US Economy: Strategies and Policy Implications*. NIST Planning Report 99-2. Gaithersburg, MD: National Institute of Standards and Technology, April 1999.

rate over the last 50 years, he says, it would be 40% smaller today, with corresponding implications for jobs, wages, and the standard of living.[7]

NEW GLOBAL INNOVATION ECONOMY

The dominant position of the United States depended substantially on our own strong commitment to science and technology and on the comparative weakness of much of the rest of the world. But the age of relatively unchallenged US leadership is ending. The importance of sustaining our investments is underscored by the challenges of the 21st century: the rise of emerging markets, innovation-based economic development, the global innovation enterprise, the new global labor market, and an aging population with expanding entitlements.

Emerging Markets

Over the last two decades, the global economy has been transformed. With the fall of the Berlin Wall in 1989, the collapse of the Soviet Union in 1991, China's entry into the World Trade Organization in 2001, and India's recent engagement with international markets, almost 3 billion people have joined the global trading system in little more than a decade.

In the coming years, developing markets will drive most economic growth. Goldman Sachs projects that within 40 years the economies of Brazil, Russia, India, and China (the so-called BRICs) together could be larger than those of the G6 nations together—the United States, Japan, the United Kingdom, Germany, France, and Italy (Figure 9-1). The BRICs currently are less than 15% the size of the G6.[8] But India's economy could be larger than Japan's by 2032, and China could surpass every nation other than the United States by 2016 and reach parity with the United States by 2041.

The enormous populations of the BRICs (China's population is now 4.4 times and India's is 3.6 times the size of the US population[9]) mean that even though per capita income in those nations will remain well below that in the developed world, the BRICs will have a growing middle class of consumers. Within a decade, nearly 80% of the world's middle-income consumers could live in nations outside the currently industrialized world.

---

[7]M. J. Mandel. *Rational Exuberance: Silencing the Enemies of Growth and Why the Future Is Better Than You Think*. New York: Harper Business, 2004. P. 27.

[8]Goldman Sachs. *Dreaming with the BRICs: The Path to 2050*. Global Economics. Paper No. 99. New York: Goldman Sachs, October 2003.

[9]US Census Bureau Data Base. "Total Mid-Year Population, 2004-2050." Available at: http://www.census.gov/ipc/www/idbsprd.html.

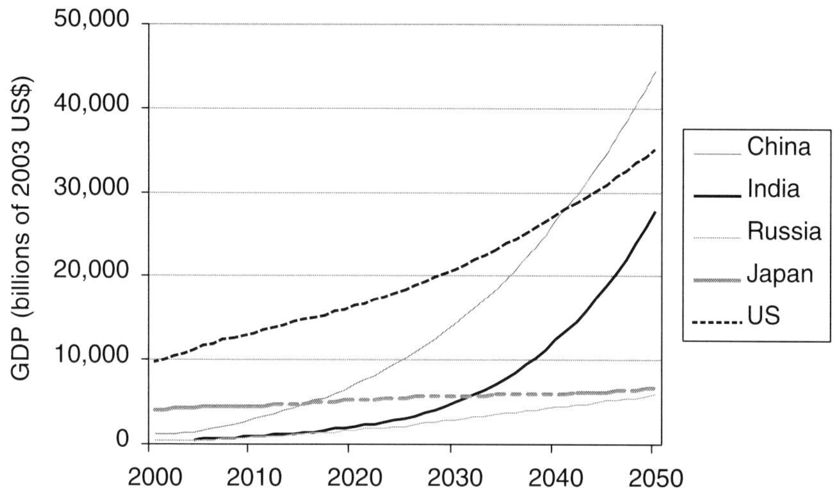

FIGURE 9-1 Projected growth of emerging markets for selected countries, in billions of constant 2003 US dollars, 2000-2050.
SOURCE: Goldman Sachs. *Dreaming with the BRICs: The Path to 2050.* Global Economics. Paper No. 99. New York: Goldman Sachs, October 2003.

China alone could have 595 million middle-income consumers and 82 million upper-middle-income consumers,[10] a combined number that is double the total projected population of the United States in that period. China's domestic market is already the largest in the world for more than 100 products. With 300 million subscribers and rising, China already is by far the biggest mobile-telephone market in the world. Only a small fraction of its population has Internet access, but China still has 100 million computer users, second only to the United States. China has become the second largest market for personal computers, and it will soon pass the United States.[11] Many US companies—including Google, Yahoo, eBay, and Cisco—expect China to be their largest market in the next 20 years.[12]

For decades, the United States has been the world's largest and most sophisticated market for an enormous range of goods and services. US consumers have stimulated productivity around the world with our apparently insatiable demand. Foreign multinational companies have invested in the

---

[10] P. A. Laudicina. *World Out of Balance: Navigating Global Risks to Seize Competitive Advantage.* New York: McGraw-Hill, 2005. P. 76.

[11] C. Prestowitz. *Three Billion New Capitalists: The Great Shift of Wealth and Power to the East.* New York: Basic Books, 2005. P. 74.

[12] D. Gillmor. *Now Is Time to Face Facts, Make Needed Investment.* San Jose Mercury News, March 14, 2004.

United States to gain access to our markets, giving this nation the largest stock of foreign direct investment in the world and employing 5.4 million Americans.[13] New products and services are designed, marketed, and launched here. Technical standards are set here. But as other markets overtake us, we could lose these advantages.

Innovation-Based Development

Driving the rapid growth in developed economies and in emerging markets is a new emphasis on science and technology. A report of the President's Council of Advisors on Science and Technology (PCAST) notes, "Other countries are striving to replicate the US innovation ecosystem model to compete directly against our own."[14] Through investments in R&D, infrastructure, and education and aided by foreign direct investment, many nations are rapidly retooling their economies to compete in technologically advanced products and services.

One sign of this new priority is increased R&D spending by many governments. The European Union (EU) has stated its desire to increase total R&D spending (government and industry) from less than 2% of GDP to 3% (the United States currently spends about 2.7%).[15] From 1992 to 2002, China more than doubled its R&D intensity (the ratio of total R&D spending to GDP), although the United States still spends significantly more than China does both in gross terms and as a percentage of GDP. Other nations also have increased their numbers of students, particularly in science and engineering. India and China are large enough that even if only relatively small portions of their populations become scientists and engineers, the size of their science and engineering workforce could still significantly exceed that of the United States. India already has nearly as many young professional engineers (university graduates with up to 7 years of experience) as the United States does, and China has more than twice as many.[16]

Multinational corporations are central to innovation-based development strategies, and nations around the world have introduced tax benefits, subsidies, science-based industrial parks, and worker-training programs to

---

[13]Organization for International Investment. "The Facts About Insourcing." Available at: http://www.ofii.org/insourcing/.

[14]President's Council of Advisors on Science and Technology. *Sustaining the Nation's Innovation Ecosystems, Information Technology Manufacturing and Competitiveness*. Washington, DC: White House Office of Science and Technology Policy, December 2004. P. 15.

[15]Organisation for Economic Co-operation and Development. *Science, Technology and Industry Outlook 2004*. Paris: OECD, 2004. P. 25. Available at: http://www.oecd.org/document/63/0,2340,en_2649_ 33703_33995839_1_1_1_1,00.html.

[16]McKinsey and Company. *The Emerging Global Labor Market: Part II—The Supply of Offshore Talent in Services*. New York: McKinsey and Company, June 2005.

lure the owners of high-technology manufacturing and R&D facilities. China uses those tools and its enormous potential market to encourage technology transfer to Chinese partner companies.[17] Most of the world's leading computer and telecommunications companies have R&D investments in China, and they are competing with local high-technology enterprises for market share. High-tech goods went from about 5% of China's exports in 1990 to 20% in 2000. Foreign enterprises accounted for 80% of China's exports in capital- and technology-intensive sectors in 1995, but they were only responsible for 50% by 2000. The United States now has a $30 billion advanced-technology trade deficit with China.

There was once a belief that developing nations would specialize in low-cost commodity products and developed economies would focus on high technology, allowing the latter to maintain a higher standard of living. Developing nations—South Korea, Taiwan, India, and China—have advanced so quickly that they can now produce many of the most advanced technologies at costs much lower than in wealthier nations. Most analysts believe that the United States, Europe, and Japan still maintain a lead in innovation—developing the new products and services that will appeal to consumers. But even here the lead is narrowing and temporary. And while the United States does currently maintain an advantage in terms of the availability of venture capital to underwrite innovation, venture capitalists are increasingly pursuing what may appear to be more promising opportunities around the world.

The Global Innovation Enterprise

Among the most powerful drivers of globalization has been the spread of multinational corporations. By the end of the 20th century, nearly 63,000 multinationals were operating worldwide.[18] Over the last few decades, corporations have used new information technologies and management practices to outsource production and business processes. Shifting from a vertically integrated structure to a network of partners allows companies to locate business activities in the most cost-efficient manner. The simultaneous opening of emerging markets and the rapid increase in workforce skill levels in those nations helped stimulate the offshore placement of key functions. First in manufacturing, then in technical support and back-office

---

[17]E. H. Preeg. *The Emerging Chinese Advanced Technology Superstate.* Arlington, VA: Manufacturers Alliance/MAPI and Hudson Institute, 2005; K. Walsh. *Foreign High-Tech R&D in China: Risks, Rewards, and Implications for US-China Relations.* Washington, DC: Henry L. Stimson Center, 2003.

[18]United Nations Conference on Trade and Development. *World Investment Report 2004: The Shift Towards Services.* New York and Geneva: United Nations, 2004.

operations, next in software design, increasingly sophisticated work is being performed in developing economies. Innovation itself is being both outsourced and sent offshore.[19] This is all part of the process that Thomas Friedman calls "the flattening of the world."[20]

Locations that combine strong R&D centers with manufacturing capabilities have a clear competitive advantage. Hence, in addition to the availability of scientists and engineers whose salaries are a fraction of the salaries of their US counterparts, India and China offer synergies between manufacturing and R&D. Top-level R&D and design are still conducted mostly in the United States, but global companies are becoming increasingly comfortable with offshore R&D, and other nations are rapidly increasing their capabilities.[21]

In 1997, China had fewer than 50 research centers that were managed by multinational corporations; by mid-2004, there were more than 600.[22] Much of the R&D currently performed in developing markets is designed to tailor products to local needs, but as local markets grow, the most advanced R&D could begin to migrate there. That said, it should be noted that the United States also benefits from offshore R&D—the amount of foreign-funded R&D conducted here has quadrupled since the mid-1980s. In fact, more corporate R&D investment now comes into the United States than is sent out of the country.[23]

The Emerging Global Labor Market

The three trends discussed already—the opening of emerging markets, innovation-based development, and the global innovation enterprise—have created a new global labor market, with far-reaching implications.

In the last few years, the phenomenon of sending service work overseas has garnered a great deal of attention in developed nations. The movement of US manufacturing jobs offshore through the 1980s and 1990s had major consequences for domestic employment in those sectors, although many argue that productivity increases were responsible for most of the reported

---

[19]Council on Competitiveness. *Going Global: The New Shape of American Innovation*. Washington, DC: Council on Competitiveness, 1998.

[20]T. L. Friedman. *The World Is Flat: A Brief History of the 21st Century*. New York: Farrar, Straus, and Giroux, 2005.

[21]President's Council of Advisors on Science and Technology. *Sustaining the Nation's Innovation Ecosystems, Information Technology Manufacturing and Competitiveness*. Washington, DC: White House Office of Science and Technology Policy, December 2004. P. 11.

[22]R. B. Freeman. *Does Globalization of the Scientific/Engineering Workforce Threaten US Economic Leadership?* Working Paper 11457. Cambridge, MA: National Bureau of Economic Research, June 2005. P. 9.

[23]K. Walsh. *Foreign High-Tech R&D in China: Risks, Rewards, and Implications for US-China Relations*. Washington, DC: Henry L. Stimson Center, 2003.

job losses.[24] Until recently, it seemed that jobs in the service sector were safe because most services are delivered face-to-face and only a small fraction is traded globally. But new technologies and business processes are opening an increasing number of services to global competition, from technical support to the reading of x-rays to stock research to the preparation of income taxes and even to the ordering of hamburgers at drive-through windows. There is a US company that uses a receptionist in Pakistan to welcome visitors to its office in Washington via flat-screen television.[25] The transformation of collaboration brought about by information and communications technologies means that the global workforce is now more easily tapped by global businesses. It is important to note, however, that a recent McKinsey Company report estimates that only 13% of the potential talent supply in low-wage nations is suited for work in multinational companies because the workers lack the necessary education or language skills.[26] But that is 13% of a very large number.

Forrester Research estimates that 3.4 million US jobs could be lost to offshoring by 2015.[27] Ashok Bardhan and Cynthia Kroll calculate that more than 14 million US jobs are at risk of being sent offshore.[28] The Information Technology Association of America (ITAA), Global Insight,[29] and McKinsey and Company[30] all argue that those losses will be offset by net gains in US employment—presuming that the United States takes the steps needed to maintain a vibrant economy. Many experts point out that the number of jobs lost to offshoring is small compared with the regular monthly churning of jobs in the US economy. McKinsey, for example, estimates that about 225,000 jobs are likely to be sent overseas each year, a small fraction of the total annual job churn. In 2004, the private sector created more than 30 million jobs and lost about 29 million; the net gain was 1.4 million jobs.[31]

---

[24]American Electronics Association. *Offshore Outsourcing in an Increasingly Competitive and Rapidly Changing World: A High-Tech Perspective.* Washington, DC: American Electronics Association, March 2004.

[25]S. M. Kalita. *Virtual Secretary Puts New Face on Pakistan.* Washington Post, May 10, 2005. P. A01.

[26]McKinsey and Company. *The Emerging Global Labor Market: Part II—The Supply of Offshore Talent in Services.* New York: McKinsey and Company, June 2005. P. 23.

[27]Forrester Research. *Near-Term Growth of Offshoring Accelerating.* Cambridge, MA: Forrester Research, May 14, 2004.

[28]A. Bardhan and C. Kroll. *The New Wave of Outsourcing.* Fisher Center Research Reports #1103. Berkeley, CA: University of California, Berkeley, Fisher Center for Real Estate and Urban Economics, November 2, 2003.

[29]Information Technology Association of America. *The Impact of Offshore IT Software and Services Outsourcing on the US Economy and the IT Industry.* Lexington, MA: Global Insight (USA), March 2004.

[30]McKinsey and Company. *Offshoring: Is It a Win-Win Game?* New York: McKinsey and Company, August 2003.

[31]US Bureau of Labor Statistics. "NEWS: Business Employment Dynamics: First Quarter 2005." November 18, 2005. Available at: http://www.bls.gov/rofod/3640.pdf.

Once again, this suggests that the US economy will continue to create new jobs at a constant rate, an assumption that in turn depends on our continued development of new technologies and training of workers for the jobs of the 21st century. Economists and others actively debate whether outsourcing or, more generally, free trade with low-wage countries with rapidly improving innovation capacities will help or hurt the US economy in the long term.[32] The optimists and the pessimists, however, agree on two fundamental points: in the short term, some US workers will lose their jobs and face difficult transitions to new, higher skilled careers; and in the long term, America's only hope for continuing to create new high-wage jobs is to maintain our lead in innovation.

Aging and Entitlements

The enormous and growing supply of labor in the developing world is but one side of a global demographic transformation. The other side is the aging populations of developed nations. The working-age population is already shrinking in Italy and Japan, and it will begin to decline in the United States, the United Kingdom, and Canada by the 2020s. More than 70 million US baby boomers will retire by 2020, but only 40 million new workers will enter the workforce.[33] Europe is expected to face the greatest period of depopulation since the Black Death, shrinking to 7% of world population by 2050 (from nearly 25% just after World War II).[34] East Asia (including China) is experiencing the most rapid aging in the world. At the same time, India's working-age population is projected to grow by 335 million people by 2030—almost equivalent to the entire workforce of Europe and the United States today.[35] Those extreme global imbalances suggest that immigration will continue to increase.

Population dynamics have major economic implications. The Organisation for Economic Co-operation and Development (OECD)

---

[32]W. C. Mann. *Globalization of IT Services and White Collar Jobs*. Washington, DC: Institute for International Economics, 2003; J. Bhagwati, A. Panagariya, and T. N. Srinivasan. "The Muddles Over Outsourcing." *Journal of Economic Perspectives* 18(Summer 2004):93-114 offer examples of the optimist view; R. Gomory and W. Baumol. *Global Trade and Conflicting National Interests*. Cambridge, MA: MIT Press, 2001; P. A. Samuelson. "Where Ricardo and Mill Rebut and Confirm Arguments of Mainstream Economists Supporting Globalization." *Journal of Economic Perspectives* 18(Summer 2004):135-146 offer a more pessimistic perspective.

[33]P. A. Laudicina. *World Out of Balance: Navigating Global Risks to Seize Competitive Advantage*. New York: McGraw-Hill, 2005. P. 49.

[34]United Nations, Department of Economic and Social Affairs, Population Division. "The World at Six Billion." October 12, 1999. Available at: http://www.un.org/esa/population/publications/sixbillion/sixbillion.htm.

[35]P. A. Laudicina. *World Out of Balance: Navigating Global Risks to Seize Competitive Advantage*. New York: McGraw-Hill, 2005. P. 62.

projects that the scarcity of working-age citizens will hamper economic growth rates between 2025 and 2050 for Europe, Japan, and the United States.[36] The Center for Strategic and International Studies (CSIS) estimates that the average cost of public pensions in the developed world will grow by 7% of GDP between now and the middle of the century; public health spending on the elderly will grow by about 6% of GDP.[37] There are now 3 pension-eligible elders in the developed world for every 10 working-age adults. Thirty-five years from now, the ratio will be 7 to 10. Here in the United States, the ratio of adults aged 60 and over to working-age adults aged 15-59 is expected to increase from .26 to .47 over the same period.[38]

Those trends have profound implications for US leadership in science and technology:

- The US science and engineering workforce is aging while the supply of new scientists and engineers who are US citizens is decreasing. Immigration will continue to be critical to filling our science and engineering needs.
- The rapidly increasing costs of caring for the aging population will further strain federal and state budgets and add to the expense columns of industries with large pension and healthcare obligations. It will thus become more difficult to allocate resources to R&D or education.
- Aging populations and rising healthcare costs will drive demand for innovative and cost-effective medical treatments.

Taken together, those trends indicate a significant shift in the global competitive environment. The importance of leadership in science and technology will intensify. As companies come to see innovation as the key to revenue growth and profitability, as nations come to see innovation as the key to economic growth and a rising standard of living, and as the planet faces new challenges that can be solved only through science and technology, the ability to innovate will be perhaps the most important factor in the success or failure of any organization or nation.

A recent report from the Council on Competitiveness argues that "innovation will be the single most important factor in determining America's success through the 21st century."[39] The United States cannot control such global forces as demographics, the strategies of multinational corporations,

---

[36]Central Intelligence Agency. *Long-Term Global Demographic Trends: Reshaping the Geopolitical Landscape.* Langley, VA: CIA, July 2001. P. 25.

[37]P. G. Peterson. "The Shape of Things to Come: Global Aging in the 21st Century." *Journal of International Affairs* 56(1)(Fall 2002). New York: Columbia University Press.

[38]R. Jackson and N. Howe. *The 2003 Aging Vulnerability Index.* Washington, DC: CSIS and Watson Wyatt Worldwide, 2003. P. 43.

[39]Council on Competitiveness. *Innovate America.* Washington, DC: Council on Competitiveness, December 2004.

and the policies of other nations, but we can determine how we want to engage with this new world, with all of its challenges and opportunities.

SCENARIOS FOR AMERICA'S FUTURE IN SCIENCE AND TECHNOLOGY

To highlight the choices we face, and their implications, it is useful to examine three scenarios that address the changing status of America's leadership in science and engineering.

Scenario 1: Baseline, America's Narrowing Lead

What is likely to happen if we do not change our current approach to science and technology? The US lead is so large that it is unlikely that any other nation would broadly overtake us in the next decade or so. The National Intelligence Council argues that the United States will remain the world's most powerful actor—economically, technologically, and militarily—at least through 2020.[40] But that does not mean the United States will not be challenged. The Center for Strategic and International Studies concludes, "Although US economic and technology leadership is reasonably assured out to 2020, disturbing trends now evident threaten the foundation of US technological strength."[41]

Over the last year or so, a virtual flood of books and articles has appeared expressing concern about the future of US competitiveness.[42] They identify trends and provide data to show that the relative position of the United States is declining in science and technology, in education, and in high-technology industry.[43] All of this leads to a few simple extrapolations

---

[40]National Intelligence Council. *Mapping the Global Future: Report of the National Intelligence Council's 2020 Project*. Pittsburgh, PA: Government Printing Office, December 2004.

[41]Center for Strategic and International Studies. *Technology Futures and Global Power, Wealth and Conflict*. Washington, DC: CSIS, May 2005. P. viii.

[42]Some of the most prominent publications include A. Segal. "Is America Losing Its Edge? Innovation in a Globalized World." *Foreign Affairs* (November/December 2004):2-8; G. Colvin. "America Isn't Ready." *Fortune*, July 25, 2005; K. H. Hughes. *Building the Next US Century: The Past and Future of US Economic Competitiveness*. Washington, DC: Woodrow Wilson Center Press, 2005; R. D. Atkinson. *The Past and Future of America's Economy: Long Waves of Innovation That Power Cycles of Growth*. Northampton, MA: E. Elgar, 2004; and R. Florida. *The Flight of the Creative Class: The New Global Competition for Talent*. New York: Harper Business, 2005.

[43]The Task Force on the Future of US Innovation. *The Knowledge Economy: Is the United States Losing Its Competitive Edge, Benchmarks for Our Innovation Future*. Washington, DC: The Task Force on the Future of US Innovation, February 2005.

for our global role over the next 30 years, assuming that we change nothing in our approach to science and education.

The US share of global R&D spending will continue to decline.

- US R&D spending will continue to lead the world in gross terms, but R&D intensity (spending as a percentage of GDP) will continue to fall behind that of other nations.
- US R&D will rely increasingly on corporate R&D spending.
- Industry spending now accounts for two-thirds of all US R&D.
- Total government spending on all physical sciences research is less than the $5 billion that a single company—IBM—spends annually on R&D, although an increasing amount of IBM's research, like that of most large corporations, is now performed abroad.
- Most corporate R&D is focused on short-term product development rather than on long-term fundamental research.
- US multinational corporations will conduct an increasing amount of their R&D overseas, potentially reducing their R&D spending in the United States, because other nations offer lower costs, more government incentives, less bureaucracy, high-quality educational systems, and in some cases superior infrastructure.

The US share of world scientific output will continue to decline.

- The share of US patents granted to US inventors is already declining, although the absolute number of patents to US inventors continues to increase.
- US researchers' scientific publishing will decline as authors from other nations increase their output.
    - The number of scientific papers published by US researchers reached a plateau in 1992.[44]
    - Europe surpassed the United States in the mid-1990s as the world's largest producer of scientific literature.
    - If current trends continue, publications from the Asia Pacific region could outstrip those from the United States within the next 6 or 7 years.[45]

---

[44]National Science Board. *Science and Engineering Indicators 2004*. NSB 04-01. Arlington, VA: National Science Foundation, 2004. Table 5-30.

[45]A. von Bubnoff. "Asia Squeezes Europe's Lead in Science." *Nature* 436(7049)(July 21, 2005):314.

The US share of scientists and engineers will continue to decline.

- Other nations will have larger numbers of students receiving undergraduate degrees in science and engineering. In 2000, more than 25 countries had a higher percentage of 24-year-olds with degrees in science and engineering than did the United States.[46]
- The number of graduate degrees awarded in science and engineering will decline.
  - The number of new doctorates in science and engineering peaked in the United States in 1998.
  - By 2010, China will produce more science and engineering doctorates than the United States does.[47]
  - The US share of world science and engineering doctorates granted will fall to about 15% by 2010, down from more than 50% in 1970[48] (Figure 9-2).
- International students and workers will make up an increasing share of those holding US science and engineering degrees and will fill more of our workforce.
  - In 2003, foreign students earned 38% of all US doctorates in science and engineering, and they earned 59% of US engineering doctorates.[49]
  - In 2000, foreign-born workers occupied 38% of all US doctoral-level science and engineering jobs, up from 24% just 10 years earlier.[50]

Our ability to attract the best international researchers will continue to decline.

- From 2002 to 2003, 1,300 international students enrolled in US science and engineering graduate programs. In each of the 3 years before that, the number had risen by more than 10,000.[51]

---

[46]National Science Foundation. *Science and Engineering Indicators 2004*. Arlington, VA: National Science Foundation, 2004. Appendix Table 2-33.

[47]R. B. Freeman. *Does Globalization of the Scientific/Engineering Workforce Threaten US Economic Leadership*? Working Paper 11457. Cambridge, MA: National Bureau of Economic Research, June 2005. P. 4.

[48]Ibid., p. 5.

[49]National Science Foundation. *Survey of Earned Doctorates, 2003*. Arlington, VA: National Science Foundation, 2005.

[50]R. B. Freeman. *Does Globalization of the Scientific/Engineering Workforce Threaten US Economic Leadership*? Working Paper 11457. Cambridge, MA: National Bureau of Economic Research, June 2005. P. 36.

[51]National Science Foundation. *Graduate Enrollment in Science and Engineering Programs Up in 2003, but Declines for First-Time Foreign Students*. NSF 05-317. Arlington, VA: National Science Foundation, 2005.

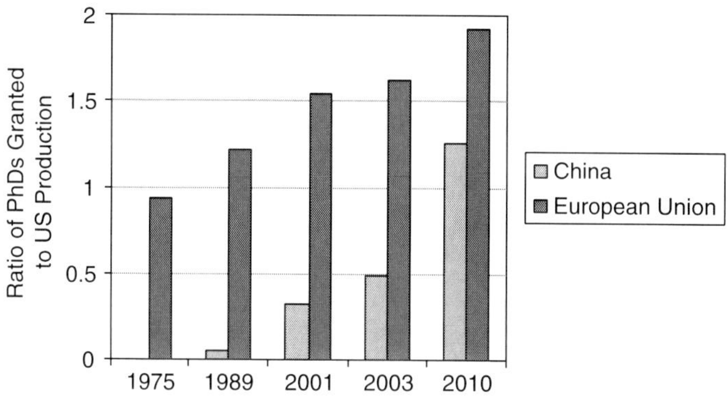

FIGURE 9-2 China and European Union production of science and engineering doctorates compared with US production, 1975-2010.
SOURCE: R. B. Freeman. *Does Globalization of the Scientific/Engineering Workforce Threaten US Economic Leadership*? Working Paper 11457. Cambridge, MA: National Bureau of Economic Research, June 2005.

• After a decline of 6% from 2001 to 2002, first-time, full-time enrollment of students with temporary visas fell 8% in 2003.[52]

• Snapshot surveys indicate international graduate student enrollments decreased again in 2004 by 6%[53] but increased by 1% in 2005.

• In the early 1990s, there were more science and engineering students from China, South Korea, and Taiwan studying at US universities than there were graduates in those disciplines at home. By the mid-1990s, the number attending US universities began to decline and the number studying in Asia increased significantly.[54]

PCAST observes, "While not in imminent jeopardy, a continuation of current trends could result in a breakdown in the web of 'innovation ecosystems' that drive the successful US innovation system."[55] Economist Ri-

[52]Ibid.

[53]H. Brown. *Council of Graduate Schools Finds Declines in New International Graduate Student Enrollment for Third Consecutive Year.* Washington, DC: Council of Graduate Schools, November 4, 2004; H. Brown. 2005. *Findings from 2005 CGS International Graduate Admissions Survey III: Admissions and Enrollment.* Washington, DC: Council of Graduate Schools. Available at: http://www.cgsnet.org/pdf/CGS2005IntlAdmitIII_Rep.pdf.

[54]The Task Force on the Future of US Innovation. *The Knowledge Economy: Is the United States Losing Its Competitive Edge, Benchmarks for Our Innovation Future.* Washington, DC: The Task Force on the Future of US Innovation, February 2005.

[55]President's Council of Advisors on Science and Technology. *Sustaining the Nation's Innovation Ecosystems, Information Technology Manufacturing and Competitiveness*, Washington, DC: White House Office of Science and Technology Policy, December 2004. P. 13.

chard Freeman says those trends foreshadow a US transition "from being a superpower in science and engineering to being one of many centers of excellence."[56] He adds that "the country faces a long transition to a less dominant position in science and engineering associated industries."[57]

The United States still leads the world in many areas of science and technology, and it continues to increase spending and output. But our share of world output is declining, largely because other nations are increasing production faster than we are, although they are starting from a much lower base. Moreover, the United States will continue to lead the world in other areas critical to innovation—capital markets, entrepreneurship, and workforce flexibility—although here as well our relative lead will shrink as other nations improve their own systems.

The biggest concern is that our competitive advantage, our success in global markets, our economic growth, and our standard of living all depend on maintaining a leading position in science, technology, and innovation. As that lead shrinks, we risk losing the advantages on which our economy depends. If these trends continue, there are several likely consequences:

- The United States will cease to be the largest market for many high-technology goods, and the US share of high-technology exports will continue to decline.
- Foreign direct investment will decrease.
- Multinational corporations (US-based and foreign) will increase their investment and hiring more rapidly overseas than they will here.
- The industries and jobs that depend on high-technology exports and foreign investment will suffer.
- The trade deficit will continue to increase, adding to the possibility of inflation and higher interest rates.
- Salaries for scientists, engineers, and technical workers will fall because of competition from lower-wage foreign workforces, and broader salary pressures could be exhibited across other occupations.
- Job creation will slow.
- GDP growth will slow.
- Growth in per capita income will slow despite our relatively high standard of living.
- Poverty rates and income inequality, already more pronounced here than in other industrialized nations, could increase.

---

[56]R. B. Freeman. *Does Globalization of the Scientific/Engineering Workforce Threaten US Economic Leadership*? Working Paper 11457. Cambridge, MA: National Bureau of Economic Research, June 2005. P. 2.

[57]Ibid., p. 3.

Today's leadership position is built on decisions that led to investments made over the past 50 years. The slow erosion of those investments might not have immediate consequences for economic growth and job creation, but the long-term effect is predictable and would be severe. Once lost, the lead could take years to recover, if indeed it could be recovered. Like a supertanker, the US economy does not turn on a dime, and if it goes off course it could be very difficult to head back in the right direction.

Given that they already have a commanding lead in many key sectors, it is likely that US multinational corporations will continue to succeed in the global marketplace. To do so, they will shift jobs, R&D funds, and resources to other places. Increasingly, it is no longer true that what is good for GM (or GE or IBM or Microsoft) is good for the United States. What it means to be a US company is likely to change as all multinationals continue to globalize their operations and ownership. As China and other developing nations become larger markets for many products and services, and as they maintain their cost advantages, US companies will increasingly invest there, hire there, design there, and produce there.

This nation's science and technology policy must account for the new reality and embrace strategies for success in a world where talent and capital can easily choose to go elsewhere.

Scenario 1 is the most likely case if current trends in government policies continue both here and in other nations and if corporate strategies remain as they are today. Two other scenarios represent departures from recent history. As such, they are more speculative and less detailed.

Scenario 2: Pessimistic Case, America Falls Decisively Behind

In Scenario 1, the United States continues to invest enough to maintain current trends in science and technology education and performance, leading to a slow decline in competitiveness. Scenario 2 considers what might happen if the commitment to science and technology were to lessen. Although that would run counter to our national history, several factors might lead to such an outcome:

- Rising spending on social security, Medicare, and Medicaid (now 42% of federal outlays compared with 25% in 1975) limit federal and state resources available for science and technology.[58] In 2005, Social Security, Medicare, and Medicaid accounted for 8.4% of GDP. If growth continues

[58] W. B. Bonvillian. "Meeting the New Challenge to US Economic Competitiveness." *Issues in Science and Technology* 21(1)(Fall 2004):75-82.

at the current rate, the federal government's total spending for Medicare and Medicaid alone would reach 22% of GDP by 2050.

- The war on terrorism refocuses government resources on short-term survival rather than long-term R&D.
- Increasingly attractive opportunities overseas draw industrial R&D funding and talented US scientists and engineers away from the United States.
- Higher US effective corporate tax rates discourage companies from investing in new facilities and research in the United States.
- Excessive regulation of research institutions reduces the amount of money available for actual research.

Those possibilities would exacerbate and accelerate the trends noted in Scenario 1:

- The availability of scientists and engineers could drop precipitously if foreign students and workers stop coming in large numbers, either because immigration restrictions make it more difficult or because better opportunities elsewhere reduce the incentives to work in the United States.
- US venture capitalists begin to place their funds abroad, searching for higher returns.
- Short-term cuts in funding for specific fields could lead to a rapid decline in the number of students in those disciplines, which could take decades to reverse.
- If they were faced with a lack of qualified workers, multinational corporations might accelerate their overseas hiring, building the capabilities of other nations while the US innovation system atrophies.
- Multinationals from China, India, and other developing nations, building on success in their domestic markets and on supplies of talented, low-cost scientists and engineers, could begin to dominate global markets, while US-based multinationals that still have a large percentage of their employees in the United States begin to fail, affecting jobs and the broader economy.
- Financing the US trade deficit, now more than $600 billion or about 6% of GDP, requires more than $2 billion a day of foreign investment. Many economists argue that such an imbalance is unsustainable in the long term.[59] A loss of competitiveness in key export industries could lead to a loss of confidence in the US ability to cover the debt, bringing on a crisis.

---

[59]C. Prestowitz. *Three Billion New Capitalists: The Great Shift of Wealth and Power to the East.* New York: Basic Books, 2005. P. xii.

• As innovation and investment move overseas, domestic job creation and wage growth could stall, lowering the overall standard of living in the United States.

The rapid pace of technological change and the increasing mobility of capital knowledge and talent mean that our current lead in science and technology could evaporate more quickly than is generally recognized if we fail to support it. The consequences would be enormous, and once lost, our lead would be difficult to regain.

Scenario 3: Optimistic Case, America Leads in Key Areas

The relative competitive lead enjoyed by the United States will almost certainly shrink as other nations rapidly improve their science and technology capacity. That means greater challenges for the United States, but it also presents an opportunity to raise living standards and improve quality of life around the world and to create a safer world. The United States might have a smaller share of the world's economy, but the economy itself will be larger. For that reason, the success of other nations need not imply the failure of the United States. But it does require that the United States maintain and extend its capacity to generate value as part of a global innovation system.

If we increase our commitment to leadership in science and technology, there are several likely results:

• Although the US share of total scientific output continues to decline, the United States maintains leadership across key areas.
• US researchers become leaders of global research networks.
• The US education system sets the standard for quality and innovation, giving graduates a competitive edge over the larger number of lower wage scientists and engineers trained in the developing world.
• Our universities and national laboratories act as centers for regional innovation, attracting and anchoring investment from around the world.
• Our economy generates sufficient growth to reduce our trade imbalances, reduce the federal budget deficit, and support an aging population.
• Investors continue to find it attractive to place their funds in US firms seeking to innovate and generate jobs in America.
• US leadership in science and technology supports our military leadership and addresses the major challenges of homeland security.

The rapid worldwide development that has resulted from advances in science and technology has raised global standards of living, but it also spawned a range of challenges that, paradoxically, will have to be solved through appropriate investments in research:

- To maintain its current rate of growth, by 2020 China will need to boost energy consumption by 150%, and India will need to do so by 100%.[60] It will be essential to develop clean, affordable, and reliable energy.
- The increased movement of people around the world will lead to more outbreaks of communicable diseases. Meanwhile, aging populations will require new treatments for chronic diseases.
- As the means to develop weapons of mass destruction become more widely available, security measures must advance.
- In an increasingly interconnected economy, even small disruptions to communications, trade, or financial flows can have major global consequences. Methods to manage complex systems and respond quickly to emergencies will be essential.

The strains of managing global growth will require global collaboration. Around the world, the growing scale and sophistication of science and technology mean that we are much more likely to be able to solve those and other problems that will confront us. Advances in information technology, biotechnology, and nanotechnology will improve life for billions of people. The leadership of the United States in science and technology will make a critical contribution to those efforts and will benefit the lives of Americans here at home. Each challenge offers an opportunity for the United States to position itself as the leader in the markets that will be created for solutions to global challenges in such fields as energy, healthcare, and security.

It is important to recognize that all nations in the global economy are now inextricably linked. Just as global health, environmental, and security issues affect everyone, so are we all dependent on the continued growth of other economies. It is clearly in America's interest for China, India, the EU, Japan, and other nations to succeed. Their failure would pose a far greater threat to US prosperity and security than would their success. In the global economy, no nation can prosper in isolation. However, it is the thesis of this report that it is important that such global prosperity be shared by the citizens of the United States.

---

[60]National Intelligence Council. *Mapping the Global Future: Report of the National Intelligence Council's 2020 Project*. Pittsburgh, PA: Government Printing Office, December 2004. P. 62.

CONCLUSION

It is easy to be complacent about US competitiveness and pre-eminence in science and technology. We have led the world for decades, and we continue to do so in many fields. But the world is changing rapidly, and our advantages are no longer unique. Without a renewed effort to bolster the foundations of our competitiveness, it is possible that we could lose our privileged position over the coming decades. For the first time in generations, our children could face poorer prospects for jobs, healthcare, security, and overall standard of living than have their parents and grandparents. We owe our current prosperity, security, and good health to the investments of past generations. We are obliged to renew those commitments to ensure that the US people will continue to benefit from the remarkable opportunities being opened by the rapid development of the global economy.

Appendix A

Committee and Professional Staff Biographic Information

NORMAN R. AUGUSTINE [NAE] (Chair) retired in 1997 as chair and chief executive officer of Lockheed Martin Corporation. Previously, he served as chair and chief executive officer of the Martin Marietta Corporation. On retiring, he joined the faculty of the Department of Mechanical and Aerospace Engineering at Princeton University. Earlier in his career, he had served as under secretary of the Army and as assistant director of defense research and engineering. Mr. Augustine has been chair of the National Academy of Engineering and served 9 years as chairman of the American Red Cross. He has also been president of the American Institute of Aeronautics and Astronautics and served as chairman of the Jackson Foundation for Military Medicine. He has been a trustee of the Massachusetts Institute of Technology and Princeton. He is a trustee emeritus of Johns Hopkins University and serves on the President's Council of Advisors on Science and Technology and on the Department of Homeland Security's Advisory Council. He is a former chairman of the Defense Science Board. He is on the boards of Black and Decker, Lockheed Martin, Procter and Gamble, and Phillips Petroleum, and he has served as chairman of the Business Roundtable Taskforce on Education. He has received the National Medal of Technology and the Department of Defense's highest civilian award, the Distinguished Service Medal, five times. Mr. Augustine holds a BSE and an MSE in aeronautical engineering, both from Princeton University, and has received 19 honorary degrees. He is the author or coauthor of four books.

CRAIG R. BARRETT [NAE] is chief executive officer of Intel Corporation. He received a BSc in 1961, an MS in 1963, and a PhD in 1964, all in materials science from Stanford University. After graduation, he joined the faculty of Stanford University in the Department of Materials Science and Engineering and remained through 1974, rising to the rank of associate professor. Dr. Barrett was a Fulbright Fellow at Danish Technical University in Denmark in 1972 and a North Atlantic Trade Organization Postdoctoral Fellow at the National Physical Laboratory in England from 1964 to 1965. He was elected to the National Academy of Engineering in 1994 and became NAE chair in July 2004. Dr. Barrett joined Intel in 1974 as a technology-development manager. He was named a vice president in 1984, and was promoted to senior vice president in 1987 and executive vice president in 1990. Dr. Barrett was elected to Intel's board of directors in 1992 and was named the company's chief operating officer in 1993. He became Intel's fourth president in May 1997 and chief executive officer in 1998. Dr. Barrett is a member of the boards of directors of Qwest Communications International Inc., the National Forest Foundation, Achieve, Inc., the Silicon Valley Manufacturing Group, and the Semiconductor Industry Association. In addition to serving as cochairman of the National Alliance of Business Coalition for Excellence in Education, Dr. Barrett served on the National Commission on Mathematics and Science Teaching for the 21st Century (also known as the Glenn Commission). Dr. Barrett is the author of over 40 technical papers dealing with the influence of microstructure on the properties of materials and of a textbook on materials science, *Principles of Engineering Materials*. He was the recipient of the American Institute of Mining, Metallurgical, and Petroleum Engineers Hardy Gold Medal in 1969.

GAIL CASSELL [IOM] is vice president of scientific affairs and Distinguished Lilly Research Scholar for Infectious Diseases of Eli Lilly and Company. She was previously the Charles H. McCauley Professor and chairman of the Department of Microbiology at the University of Alabama Schools of Medicine and Dentistry at Birmingham, a department that ranked first in research funding from the National Institutes of Health under her leadership. She is a current member of the Director's Advisory Committee of the National Centers for Disease Control and Prevention. She is a past president of the American Society for Microbiology (ASM), a former member of the National Institutes of Health (NIH) Director's Advisory Committee, and a former member of the Advisory Council of the National Institute of Allergy and Infectious Diseases of NIH. Dr. Cassell served 8 years on the Bacteriology-Mycology 2 Study Section and as chair for 3 years. She also was previously chair of the Board of Scientific Councilors of the Center for Infectious Diseases of the Centers for Disease Control and Prevention. Dr.

Cassell has been intimately involved in establishment of science policy and legislation related to biomedical research and public health. She is the chairman of the Public and Scientific Affairs Board of ASM, is a member of the Institute of Medicine, has served as an adviser on infectious diseases and indirect costs of research to the White House Office of Science and Technology Policy, and has been an invited participant in numerous congressional hearings and briefings related to infectious diseases, antimicrobial resistance, and biomedical research. She has served on several editorial boards of scientific journals and has written over 250 articles and book chapters. Dr. Cassell has received several national and international awards and an honorary degree for her research in infectious diseases.

STEVEN CHU [NAS] is the director of E.O. Lawrence Berkeley National Laboratory, and a professor of physics and cellular and molecular biology at the University of California, Berkeley. Previously, he held positions at Stanford University and AT&T Bell Laboratories. Dr. Chu's research in atomic physics, quantum electronics, polymer physics, and biophysics includes tests of fundamental theories in physics, the development of methods to laser-cool and trap atoms, atom interferometry, and the manipulation and study of polymers and biologic systems at the single-molecule level. While at Stanford, he helped to start Bio-X, a multidisciplinary initiative that brings together the physical and biologic sciences with engineering and medicine. Dr. Chu has received numerous awards and is a cowinner of the Nobel Prize in physics (1997). He is a member of the National Academy of Sciences, the American Philosophical Society, the American Academy of Arts and Sciences, and the Academica Sinica and is a foreign member of the Chinese Academy of Sciences and the Korean Academy of Science and Engineering. Dr. Chu also serves on the boards of the William and Flora Hewlett Foundation, the University of Rochester, NVIDIA, and the (planned) Okinawa Institute of Science and Technology. He has served on numerous advisory committees, including the Executive Committee of the National Academy of Sciences Board on Physics and Astronomy, the National Institutes of Health Advisory Committee to the Director, and the National Nuclear Security Administration Advisory Committee to the Director. Dr. Chu received his AB degrees in mathematics and physics from the University of Rochester, a PhD in physics from the University of California, Berkeley, and a number of honorary degrees.

ROBERT M. GATES has been the president of Texas A&M University, a land-grant, sea-grant, and space-grant university, since August 2002. Dr. Gates served as interim dean of the George Bush School of Government and Public Service at Texas A&M from 1999 to 2001. He served as director of central intelligence from November 1991 until January 1993. In that position, he headed all foreign-intelligence agencies of the United States and directed the Central Intelligence Agency (CIA). Dr. Gates is the only career officer in CIA's history to rise from entry-level employee to director. He served as deputy director of central intelligence from 1986 to 1989 and as assistant to the president and deputy national security adviser at the White House from January 1989 to November 1991. Dr. Gates joined the CIA in 1966 and spent nearly 27 years as an intelligence professional, serving six presidents. During that period, he spent nearly 9 years at the National Security Council, serving four presidents of both political parties. Dr. Gates has been awarded the National Security Medal and the Presidential Citizens Medal, has twice received the National Intelligence Distinguished Service Medal, and has three times received CIA's highest award, the Distinguished Intelligence Medal. He is the author of the memoir *From the Shadows: The Ultimate Insider's Story of Five Presidents and How They Won the Cold War*, published in 1996. He serves as a member of the board of trustees of the Fidelity Funds and on the board of directors of NACCO Industries, Inc., Brinker International, Inc., and Parker Drilling Company, Inc. Dr. Gates received his bachelor's degree from the College of William and Mary, his master's degree in history from Indiana University, and his doctorate in Russian and Soviet history from Georgetown University.

NANCY S. GRASMICK is Maryland's first female state superintendent of schools. She has served in that post since 1991. Dr. Grasmick's career in education began as a teacher of deaf children at the William S. Baer School in Baltimore City. She later served as a classroom and resource teacher, principal, supervisor, assistant superintendent, and associate superintendent in the Baltimore County Public Schools. In 1989, she was appointed special secretary for children, youth, and families, and in 1991, the state Board of Education appointed her state superintendent of schools. Dr. Grasmick holds a PhD from the Johns Hopkins University, an MS from Gallaudet University, and a BS from Towson University. She has been a teacher, an administrator, and a child advocate. Her numerous board and commission appointments include the President's Commission on Excellence in Special Education, the US Army War College Board of Visitors, the Towson University Board of Visitors, the state Planning Committee for Higher Education, and the Maryland Business Roundtable for Education. Dr. Grasmick has received numerous awards for leadership, including the Harold W. McGraw, Jr. Prize in Education.

CHARLES O. HOLLIDAY, JR. [NAE] is the chairman of the board and chief executive officer of DuPont. He became chief executive officer in 1998 and chairman in 1999. He started at DuPont in 1970 at DuPont's Old Hickory site after receiving a BS in industrial engineering from the University of Tennessee. He is a licensed professional engineer. In 2004, he was elected a member of the National Academy of Engineering and became chairman of the Business Roundtable's Task Force for Environment, Technology, and Economy the same year. Mr. Holliday is a past chairman of the World Business Council for Sustainable Development (WBCSD), the Business Council, and the Society of Chemical Industry–American Section. While chairman of WBCSD, Mr. Holliday was coauthor of *Walking the Talk*, which details the business case for sustainable development and corporate responsibility. Mr. Holliday also serves on the board of directors of HCA, Inc. and Catalyst and is a former director of Analog Devices.

SHIRLEY ANN JACKSON [NAE] is the 18th president of Rensselaer Polytechnic Institute, the oldest technologic research university in the United States, and has held senior leadership positions in government, industry, research, and academe. Dr. Jackson is immediate past president of the American Association for the Advancement of Science (AAAS) and chairman of the AAAS board of directors, a member of the National Academy of Engineering, and a fellow of the American Academy of Arts and Sciences and the American Physical Society, and she has advisory roles in other national organizations. She is a trustee of the Brookings Institution, a life member of the Massachusetts Institute of Technology Corporation, a member of the Council on Foreign Relations, and a member of the Executive Committee of the Council on Competitiveness. She serves on the boards of Georgetown University and Rockefeller University, on the board of directors of the New York Stock Exchange, and on the board of regents of the Smithsonian Institution, and she is a director of several major corporations. Dr. Jackson was chairman of the US Nuclear Regulatory Commission in 1995-1999; at the Commission, she reorganized the agency and revamped its regulatory approach by articulating and moving strongly to risk-informed, performance-based regulation. Before then, she was a theoretical physicist at the former AT&T Bell Laboratories and a professor of theoretical physics at Rutgers University. Dr. Jackson holds an SB in physics, a PhD in theoretical elementary-particle physics from the Massachusetts Institute of Technology, and 31 honorary doctoral degrees.

ANITA K. JONES [NAE] is Lawrence R. Quarles Professor of Engineering and Applied Science. She received her PhD in computer science from Carnegie Mellon University (CMU) in 1973. She left CMU as an associate professor when she cofounded Tartan Laboratories. She was vice-president of Tartan from 1981 to 1987. In 1988, she joined the University of Virginia as a professor and the chair of the Computer Science Department. From 1993 to 1997 she served at the US Department of Defense, where as director of defense research and engineering, she oversaw the department's science and technology program, research laboratories, and the Defense Advanced Research Projects Agency. She received the US Air Force Meritorious Civilian Service Award and a Distinguished Public Service Award. She served as vice chair of the National Science Board and cochair of the Virginia Research and Technology Advisory Commission. She is a member of the Defense Science Board, the Charles Stark Draper Laboratory Corporation, National Research Council Advisory Council for Policy and Global Affairs, and the Massachusetts Institute of Technology Corporation. She is a fellow of the Association for Computing Machinery, the Institute of Electrical and Electronics Engineers, and American Association for the Advancement Science, and she is the author of 45 papers and two books.

JOSHUA LEDERBERG [NAS/IOM] is Sackler Foundation Scholar at Rockefeller University in New York. He is a cowinner of the Nobel Prize in 1958 for his research in genetic structure and function in microorganisms. As a graduate student at Yale University, Dr. Lederberg and his mentor showed that the bacterium *Escherichia coli* could share genetic information through recombinant events. He went on to show in 1952 that bacteriophages could transfer genetic information between bacteria in *Salmonella*. In addition to his contributions to biology, Dr. Lederberg did extensive research in artificial intelligence, including work in the National Aeronautics and Space Administration experimental programs seeking life on Mars and the chemistry expert system DENDRAL. Dr. Lederberg is professor emeritus of molecular genetics and informatics. He received his PhD from Yale University in 1948.

RICHARD LEVIN is the president of Yale University and Frederick William Beinecke Professor of Economics. In his writings and public testimony, Dr. Levin has described the substantial benefits of government funding of basic scientific research conducted by universities. A specialist in the economics of technologic change, Dr. Levin has written extensively on such subjects as intellectual-property rights, the patent system, industrial research and development, and the effects of antitrust and public regulation on private industry. Before his appointment as president, he devoted himself for two decades to teaching, research, and administration. He chaired Yale's Economics Department and served as dean of the Graduate School of Arts and Sciences. Dr. Levin is a director of Lucent Technologies and a trustee of the William and Flora Hewlett Foundation, one of the largest philanthropic organizations in the United States. He served on a presidential commission reviewing the US Postal Service and as a member of the bipartisan commission reviewing US intelligence capabilities. As a member of the Board of Science, Technology, and Economic Policy at the National Academy of Sciences, Dr. Levin cochaired a committee that examined the effects of intellectual-property rights policies on scientific research and made recommendations for a patent system meeting the needs of the 21st century. He received his bachelor's degree in history from Stanford University in 1968 and studied politics and philosophy at Oxford University, where he earned a bachelor of letters. In 1974, he received his PhD in economics from Yale and was named to the Yale faculty. He holds honorary degrees awarded by Peking, Harvard, Princeton, and Oxford Universities. He is a fellow of the American Academy of Arts and Sciences.

C. D. (DAN) MOTE, JR. [NAE] began his tenure as president of the University of Maryland and as Glenn L. Martin Institute Professor of Engineering in 1998. Before assuming the presidency at Maryland, Dr. Mote served on the University of California, Berkeley (UCB) faculty for 31 years. From 1991 to 1998, he was vice chancellor at UCB, held an endowed chair in mechanical systems, and was president of the UC Berkeley Foundation. He earlier served as chair of UCB's Department of Mechanical Engineering. Dr. Mote's research is in dynamic systems and biomechanics. Internationally recognized for his research on the dynamics of gyroscopic systems and the biomechanics of snow skiing, he has produced more than 300 publications; holds patents in the United States, Norway, Finland, and Sweden; and has mentored 56 PhD students. He received his BS, MS, and PhD in mechanical engineering from UCB. Dr. Mote has received numerous awards and honors, including the Humboldt Prize awarded by the Federal Republic of Germany. He is a recipient of the Berkeley Citation, an award from the University of California similar to an honorary doctorate, and was named distinguished engineering alumnus. He has received three honorary degrees. He is a member of the National Academy of Engineering and serves on its Council. He was elected to honorary membership in the American Society of Mechanical Engineers International, its most distinguished recognition, and is a fellow of the American Academy of Arts and Sciences, the International Academy of Wood Science, the Acoustical Society of America, and the American Association for the Advancement of Science. He serves as director of the Technology Council of Maryland and the Greater Washington Board of Trade. In its latest survey, *Washington Business Forward* magazine named him one of the 20 most influential people in the metropolitan Washington area.

CHERRY MURRAY [NAS, NAE] is the deputy director for science and technology at Lawrence Livermore National Laboratory (LLNL), in which she is the senior executive responsible for overseeing the quality of science and technology in the laboratory's scientific and technical programs and disciplines. Dr. Murray came to LLNL from Bell Labs, Lucent Technologies, where she served as senior vice president for physical sciences and wireless research. She joined Bell Labs in 1978 as a member of the technical staff. She was promoted to a number of positions over the years, including department head for low-temperature physics, department head for condensed-matter physics and semiconductor physics, and director of the physical research laboratory. In 2000, Dr. Murray became vice president for physical sciences, and in 2001, senior vice president. Dr. Murray received her BS and PhD in physics from the Massachusetts Institute of Technology.

PETER O'DONNELL, JR. is president of the O'Donnell Foundation of Dallas, a private foundation that develops and funds model programs designed to strengthen engineering and science education and research. In higher education, the O'Donnell Foundation provided the challenge grant that led to the creation of 32 science and engineering chairs at the University of Texas (UT) at Austin. Also at UT-Austin, it developed the plan that created the Institute for Computational Engineering and Science, and it constructed the Applied Computational Engineering and Science Building to foster interdisciplinary research at the gradate level. In medicine, Mr. O'Donnell endowed the Scholars in Medical Research Program, designed to launch the most promising new assistant professors on their biomedical careers and thereby help to develop future leaders of medical science. In public education, Mr. O'Donnell created the Advanced Placement Incentive Program, which has increased the number of students, especially Hispanic and Black students, who pass college-level courses in mathematics, science, and English while still in high school. The incentive program is now in 43 school districts in Texas and served as the model for both the state of Texas and the federal Advanced Placement (AP) incentive programs. Mr. O'Donnell is chairman of Advanced Placement Strategies, Inc., a nonprofit organization he founded to manage and implement the AP incentive program in Texas schools. He served as a member of President Reagan's Foreign Intelligence Advisory Board, as commissioner of the Texas National Research Laboratory Commission, and on the State of Texas Select Committee on Higher Education. He is a trustee of the Cooper Institute, a member of the Presidents' Circle of the National Academy of Sciences, and a founding member of the National Innovation Initiative Council on Competitiveness. Mr. O'Donnell has pursued a career in investments and philanthropy. He received his BS in mathematics from the University of the South and an MBA from the Wharton School of the University of Pennsylvania.

LEE R. RAYMOND [NAE] is the chairman of the board and chief executive officer of Exxon Mobil Corporation. Dr. Raymond was chairman of the board and chief executive officer of Exxon Corporation from 1993 until its merger with Mobil Oil Corporation in 1999. He served as a director of

Exxon Corporation from 1984 until the merger. Since joining the organization in 1963, Dr. Raymond has held a variety of management positions in domestic and foreign operations, including Exxon Company, USA; Creole Petroleum Corporation; Exxon Company, International; Exxon Enterprises; and Esso Inter-America, Inc. He served as the president of Exxon Nuclear Company, Inc., in 1979 and moved to New York in 1981, when he was named executive vice president of Exxon Enterprises. In 1983, Dr. Raymond was named president and director of Esso Inter-America Inc. with responsibilities for Exxon's operations in the Caribbean and Central and South America. He served as the senior vice president of Exxon Corporation from 1984 to 1987 and as its president from 1987 to 1993 and in 1996. Dr. Raymond has been a director of J.P. Morgan Chase & Co. or a predecessor institution since 1987 and served as a member of the Committee on Director Nominations and Board Affairs and Chairman of the Committee on Management Development and Executive Compensation. He serves as a director of the United Negro College Fund, the chairman of the American Petroleum Institute, trustee and vice chairman of the American Enterprise Institute, and trustee of the Wisconsin Alumni Research Foundation. He is a member of the Business Council, the Business Roundtable, the Council on Foreign Relations, the National Academy of Engineering, the Emergency Committee for American Trade, and the National Petroleum Council. He is secretary of the Energy Advisory Board, the Singapore-US Business Council, the Trilateral Commission, and the University of Wisconsin Foundation. Dr. Raymond graduated in 1960 from the University of Wisconsin with a bachelor's degree in chemical engineering. In 1963, he received a PhD in chemical engineering from the University of Minnesota.

ROBERT C. RICHARDSON [NAS] is the F. R. Newman Professor of Physics and the vice provost for research at Cornell University. He received a BS and an MS in physics from Virginia Polytechnic Institute. After serving in the US Army, he obtained his PhD from Duke University in 1966. He is a member of the National Academy of Sciences. He is also member of the Governing Board at Duke University, the American Association for the Advancement of Science, and Brookhaven Science Associates. Dr. Richardson has served as chair of various committees of the American Physical Society (APS) and recently completed a term on the Governing Board of the National Science Board. Dr. Richardson was awarded the Nobel Prize for the discovery that liquid helium-3 undergoes a pairing transition similar to that of superconductors. He has also received a Guggenheim fellowship, the Eighth Simon Memorial Prize (of the British Physical Society), the Buckley Prize of the APS, and an honorary doctor of science degree from the Ohio State University. He has published more than 95 scientific articles in major research journals.

P. ROY VAGELOS [NAS, IOM] is retired chairman and chief executive officer of Merck & Co., Inc. He received an AB in 1950 from the University of Pennsylvania and an MD in 1954 from Columbia University. After a residency at the Massachusetts General Hospital in Boston, he joined the National Institutes of Health, where from 1956 to 1966 he served as senior surgeon and then section head of comparative biochemistry. In 1966, he became chairman of the Department of Biological Chemistry at Washington University School of Medicine in St. Louis; in 1973, he founded the university's Division of Biology and Biomedical Sciences. He joined Merck Research Laboratories in 1975, where he was president until 1985, when he became CEO and later chairman of the company. He retired in 1994. Dr. Vagelos is a member of the National Academy of Sciences, the American Academy of Arts and Sciences, and the American Philosophical Society. He has received many awards in science and business and 14 honorary doctorates. He has been chairman of the board of the University of Pennsylvania, a member of the Business Council and the Business Roundtable, and a member of the boards of TRW, McDonnell Douglas, Estee Lauder, and Prudential Finance. He also served as cochair of the New Jersey Performing Arts Center and president and CEO of the American School of Classical Studies in Athens. He is chairman of Regeneron Pharmaceuticals and Theravance, two biotechnology companies. He is also chairman of the Board of Visitors at Columbia University Medical Center, where he chairs the capital campaign. He serves on a number of public-policy and advisory boards, including the Donald Danforth Plant Science Center and Danforth Foundation.

CHARLES M. VEST [NAE] is president emeritus at the Massachusetts Institute of Technology (MIT) and is a life member of the MIT Corporation, the institute's board of trustees. He was president of MIT from 1990 to 2004. During his presidency, he emphasized enhancing undergraduate education, exploring new organizational forms to meet emerging directions in research and education, building a stronger international dimension in education and research programs, developing stronger relations with industry, and enhancing racial and cultural diversity at MIT. He also devoted considerable energy to bringing issues concerning education and research to broader public attention and to strengthening national policy on science, engineering, and education. With respect to the latter, Dr. Vest chaired the President's Advisory Committee on the Redesign of the Space Station and served as a member of the President's Committee of Advisors on Science and Technology, the Massachusetts Governor's Council on Economic Growth and Technology, and the National Research Council Board on Engineering Education. He chairs the US Department of Energy Task Force on the Future of Science Programs and is vice chair of the Council on

Competitiveness and immediate past chair of the Association of American Universities. He sits on the board of directors of IBM and E.I. du Pont de Nemours and Co. In 2004, he was asked by President Bush to serve as a member of the Commission on the Intelligence Capabilities of the United States Regarding Weapons of Mass Destruction. He earned his BS in mechanical engineering from West Virginia University in 1963 and his MS and PhD degrees from the University of Michigan in 1964 and 1967, respectively. His research interests are the thermal sciences and the engineering applications of lasers and coherent optics.

GEORGE M. WHITESIDES [NAS, NAE] is the Woodford L. and Ann A. Flowers University Professor of Chemistry at Harvard University, where his research interests include materials science, biophysics, complexity, surface science, microfluidics, self-assembly, microtechnology and nanotechnology, and cell-surface biochemistry. He received an AB from Harvard University in 1960 and a PhD from the California Institute of Technology in 1964. He was a member of the faculty of the Massachusetts Institute of Technology from 1963 to 1982. He joined the Department of Chemistry of Harvard University in 1982 and was department chairman in 1986-1989. He is a member of the American Academy of Arts and Sciences, the National Academy of Sciences, and the American Philosophical Society. He is also a fellow of the American Association for the Advancement of Science and the New York Academy of Science, a foreign fellow of the Indian National Science Academy, and an honorary fellow of the Chemical Research Society of India. He has served as an adviser to the National Research Council, the National Science Foundation, and the Defense Advanced Research Projects Agency at the Department of Defense.

RICHARD N. ZARE [NAS] is the Marguerite Blake Wilbur Professor in Natural Science at Stanford University. He is a graduate of Harvard University, where he received his BA in chemistry and physics in 1961 and his PhD in chemical physics in 1964. In 1965, he became an assistant professor at the Massachusetts Institute of Technology. He moved to the University of Colorado in 1966 and remained there until 1969 while holding joint appointments in the Departments of Chemistry and Physics and Astrophysics. In 1969, he was appointed to a full professorship in the Chemistry Department at Columbia University, becoming the Higgins Professor of Natural Science in 1975. In 1977, he moved to Stanford University. Dr. Zare is renowned for his research in laser chemistry, which resulted in a greater understanding of chemical reactions at the molecular level. He has received numerous honors and awards and is a member of the American Philosophical Society, the National Academy of Sciences, the American Academy of Arts and Sciences, and the American Chemical Society. He served as the chair of the President's Committee on the National Medal of Science in 1997-2000; chaired the National Research Council's Commission on Physical Sciences, Mathematics, and Applications in 1992-1995; and was chair of the National Science Board for the last 2 years of his 1992-1998 service. He is the chairman of the Board of Directors of Annual Reviews, Inc., and he will chair the Department of Chemistry at Stanford University in 2005-2008.

STAFF

DEBORAH D. STINE (Study Director) is associate director of the Committee on Science, Engineering, and Public Policy; director of the National Academies Christine Mirzayan Science and Technology Policy Fellowship Program; and director of the Office of Special Projects. Dr. Stine has received both group and individual achievement awards for her work on various projects throughout the National Academies since 1989. She has directed studies and other activities on science and security in an age of terrorism, human reproductive cloning, presidential and federal advisory committee science and technology appointments, facilitating interdisciplinary research, setting priorities for the National Science Foundation's large research facilities, advanced research instrumentation and facilities, evaluating federal research programs, international benchmarking of US research, and many other issues. Before coming to the National Academies, she was a mathematician for the Air Force, an air-pollution engineer for the state of Texas, and an air-issues manager for the Chemical Manufacturers Association. She holds a BS in mechanical and environmental engineering from the University of California, Irvine, an MBA from what is now Texas A&M at Corpus Christi, and a PhD in public administration with a focus on science and technology policy analysis from American University. She received the Mitchell Prize Young Scholar Award for her research on international environmental decision-making.

ALAN ANDERSON has worked as a consultant writer for the National Academies since 1994, contributing to reports on science policy, education and training, government-industry partnerships, scientific evidence, and other topics primarily for the Committee on Science, Engineering, and Public Policy and the Board on Science, Technology, and Economic Policy. He is also editorial director of the Millennium Science Initiative, an independent nongovernmental organization whose mission is to strengthen science and technology in developing countries. He has worked in science and medical journalism for over 25 years, serving as reporter, writer, and foreign correspondent for *Time* magazine, the *New York Times Magazine*,

Saturday Review, and other publications. He holds a BA in English from Yale University and an MS in journalism from Columbia University.

THOMAS ARRISON is director of the Forum on Information Technology and Research Universities at the National Academies. He holds MAs in public policy and Asian studies and a BA in political science from the University of Michigan. He studied in Japan for 2 years, completing business internships in the banking and semiconductor industries and intensive training in Japanese language. Before being named director of the new forum in 2002, he was associate director of the Government-University-Industry Research Roundtable. Mr. Arrison joined the National Academies in 1990 and has served as the study director for numerous activities and publications, including nine committee consensus reports.

DAVID ATTIS is director of policy studies at the Council on Competitiveness. He serves as the deputy director of the National Innovation Initiative, a multiyear effort to increase the United States's capacity for innovation across all sectors of the economy. Before joining the council, Dr. Attis was a consultant with A.T. Kearney, Inc., in its general consulting practice and its Global Business Policy Council. His work included business turnarounds, strategy consulting, information-systems implementation, global risk assessments, and policy analysis. He holds a PhD in the history of science from Princeton University, an MPhil in the history and philosophy of science from Cambridge University, and a BA in physics from the University of Chicago. His doctoral thesis explored the development of mathematics in Ireland from the surveyors of the 17th century through the Celtic Tiger economy of the 1990s.

RACHEL COURTLAND is a research associate for the National Academies Committee on Science, Engineering, and Public Policy. She earned her BA in physics from the University of Pennsylvania in May 2003 and her MS in physics from Emory University in 2004. In graduate school, she studied the local perturbation of supercooled colloidal suspensions using two-dimensional confocal microscopy and conducted preparatory work for a National and Aeronautics Space Administration payload project. As an undergraduate, she led Women Interested in the Study of Physics, an organization created to help to foster a more comfortable environment for women scientists at undergraduate and graduate levels and dedicated to raising awareness of issues facing women in academe.

LAUREL L. HAAK is a program officer for the National Academies Committee on Science, Engineering, and Public Policy. She received a BS and an MS in biology from Stanford University. She was the recipient of a pre-doctoral NIH National Research Service Award and received a PhD in neuroscience in 1997 from Stanford University Medical School, where her research focused on calcium signaling and circadian rhythms. She was awarded a National Research Council research associateship to work at NIH on intracellular calcium dynamics in oligodendrocytes. From 2002 to 2003, she was editor of *Science's* Next Wave Postdoc Network at the American Association for the Advancement of Science. While a postdoctoral scholar, she was editor of the *Women in Neuroscience* newsletter and served as president of the organization from 2003 to 2004. She is an ex officio member of the Society for Neuroscience Committee on Women in Neuroscience, has served on the Biophysics Society Early Careers Committee, and was an adviser for the National Postdoctoral Association.

PETER HENDERSON is director of the National Academies Board on Higher Education and Workforce (BHEW). His specializations include postsecondary education, the labor market for scientists and engineers, and federal science and technology research funding. He oversees BHEW's Evaluation of the Lucille P. Markey Trust Programs in Biomedical Science and Assessment of NIH Minority Research Training Programs and supervises BHEW staff working on studies that examine the community-college pathway to engineering careers. He has contributed as a study director or staff member to *Building a Workforce for the Information Economy*, *Measuring the Science and Engineering Enterprise: Priorities for the Division of Science Resource Studies*, *Attracting Science and Mathematics PhDs to Secondary School Education*, *Monitoring International Labor Standards*, *Trends in Federal Support of Research and Graduate Education*, and *Observations on the President's Federal Science and Technology Budget*. Dr. Henderson holds a master's degree in public policy (1984) from Harvard University's John F. Kennedy School of Government and a PhD in American political history from the Johns Hopkins University (1994). He joined the National Academies staff in 1996 and was a recipient of the National Academies Distinguished Service Award in 2003.

JO L. HUSBANDS is a senior project director with Development, Security, and Cooperation of the Policy and Global Affairs division. In that capacity, she is working on a project to engage the international scientific community in addressing the possibility that the results of biotechnology research will be misused to support terrorism or biologic weapons. She is also developing new projects related to defense economics and the proliferation of conventional weapons and technologies. From 1991 through 2004, she was director of the National Academies Committee on International Security and Arms Control and its Working Group on Biological Weapons Control. Dr. Husbands is an adjunct professor in the security studies program at Georgetown University, where she teaches a course on "The International Arms Trade." She holds a PhD in political science from the University of Minnesota and a master's degree in international public policy (international economics) from the Johns Hopkins University School of Advanced International Studies. She is a member of the Advisory Board of Women in International Security and a fellow of the International Union of Pure and Applied Chemistry.

BENJAMIN A. NOVAK (Policy Fellow) is pursuing his MS in public policy and management at Carnegie Mellon University. He received his BA in political science and his BS in biomedical engineering from the University of Pittsburgh, where he was a member of the University Honors College. As an undergraduate student, Mr. Novak had the unusual experience of completing internships in both technical and policy fields working in a variety of places, including the US Congress House of Representatives Committee on Science, the Vascular Research Center of David Vorp, and the Artificial Liver Laboratory of Jack Patzer.

STEVE OLSON is the author of *Mapping Human History: Genes, Race, and Our Common Origins* (Houghton Mifflin), which was one of five finalists for the 2002 nonfiction National Book Award and received the Science-in-Society Award from the National Association of Science Writers. His most recent book, *Count Down: Six Kids Vie for Glory at the World's Toughest Math Competition* (Houghton Mifflin), was named a best science book of 2004 by *Discover* magazine. He has written several other books, including *Evolution in Hawaii* and *On Being a Scientist.* He has been a consultant writer for the National Academy of Sciences and National Research Council, the Howard Hughes Medical Institute, the National Institutes of Health, the Institute for Genomic Research, and many other organizations. He is the author of articles in *The Atlantic Monthly, Science, The Washington Post, Scientific American, Washingtonian, Slate, Teacher, Astronomy, Science 82-86*, and other magazines. He also is coauthor of an article published in *Nature* in September 2004 that presented a fundamentally new perspective on human ancestry. From 1989 through 1992, he served as special assistant for communications in the White House Office of Science and Technology Policy. He earned a bachelor's degree in physics from Yale University in 1978.

JOHN B. SLANINA (Policy Fellow) is a graduate student at the Georgia Institute of Technology (Georgia Tech) and a Christine Mirzayan Science and Technology Policy Fellow at the National Academies. He is pursuing an MS in public policy, and his research encompasses the incorporation of innovative practices in the manufacturing sector and regional economic development. He previously received an MS in mechanical engineering at Georgia Tech in 2002, where he performed research in sensor design for bioengineering applications. During the 2000-2001 school year, he studied engineering at the École Nationale Supérieure d'Arts et Métiers in Metz, France. He earned his undergraduate degrees in mechanical engineering and mathematics from Youngstown State University in 2000.

Appendix B

Statement of Task and Congressional Correspondence

STATEMENT OF TASK

This congressionally-requested study will address the following questions:

What are the top 10 actions, in priority order, that federal policy makers could take to enhance the science and technology enterprise so the United States can successfully compete, prosper, and be secure in the global community of the 21st Century?

What implementation strategy, with several concrete steps, could be used to implement each of those actions?

United States Senate

WASHINGTON, DC 20510

May 27, 2005

Dr. Bruce Alberts
President
National Academy of Sciences
2101 Constitution Avenue
Washington, DC 20418

Dear Dr. Alberts:

The Energy Subcommittee of the Senate Energy and Natural Resources Committee has been given the latitude by Chairman Pete Domenici to hold a series of hearings to identify specific steps our government should take to ensure the preeminence of America's scientific and technological enterprise.

The National Academies could provide critical assistance in this effort by assembling some of the best minds in the scientific and technical community to identify the most urgent challenges the United States faces in maintaining leadership in key areas of science and technology. Specifically, we would appreciate a report from the National Academies by September 2005 that addresses the following:

- Is it essential for the United States to be at the forefront of research in broad areas of science and engineering? How does this leadership translate into concrete benefits as evidenced by the competitiveness of American businesses and an ability to meet key goals such as strengthening national security and homeland security, improving health, protecting the environment, and reducing dependence on imported oil?

  What specific steps are needed to ensure that the United States maintains its leadership in science and engineering to enable us to successfully compete, prosper, and be secure in the global community of the 21st century? How can we determine whether total federal research investment is adequate, whether it is properly balanced among research disciplines (considering both traditional research areas and new multidisciplinary fields such as nanotechnology), and between basic and applied research?

- How do we ensure that the United States remains at the epicenter of the ongoing revolution in research and innovation that is driving 21st century economies? How can we assure investors that America is the preferred site for investments in new or expanded businesses that create the best jobs and provide the best services?

- How can we ensure that critical discoveries across all the scientific disciplines are predominantly American and exploited first by firms producing and hiring in America? How can we best encourage domestic firms to invest in invention and innovation to meet new global competition and how can public research investments best supplement these private sector investments?

- What specific steps are needed to develop a well-educated workforce able to successfully embrace the rapid pace of technological change?

Your answers to these questions will help Congress design effective programs to ensure that America remains at the forefront of scientific capability, thereby enhancing our ability to shape and improve our nation's future.

We look forward to reviewing the results of your efforts.

Sincerely,

Lamar Alexander

Lamar Alexander
Chairman
Energy Subcommittee

Jeff Bingaman
Ranking Member
Committee on Energy and
Natural Resources

U.S. HOUSE OF REPRESENTATIVES

COMMITTEE ON SCIENCE

SUITE 2320 RAYBURN HOUSE OFFICE BUILDING
WASHINGTON, DC 20515-6301
(202) 225-6371
TTY: (202) 226-4410
http://www.house.gov/science/welcome.htm

June 30, 2005

Dr. Bruce Alberts
President
National Academy of Sciences
2101 Constitution Avenue
Washington, DC 20418

Dear Dr. Alberts:

We understand that the National Academies, in response to a request from Senators Alexander and Bingaman, are in the early stages of developing a study related to the urgent challenges facing the United States in maintaining leadership in key areas of science and technology. Because the Science Committee considers ensuring the strength and vitality of the Nation's scientific and technology enterprise an important part of its broad oversight responsibility, we are writing to endorse the request for this study and to encourage the National Academies to carry it forward expeditiously.

In addition, we would like to suggest some specific questions we hope to see addressed by the study:

- What skills will be required by the future U.S. science and engineering workforce in order for it to command a salary premium over foreign scientists and engineers? Are alternative degree programs needed, such as professional science masters degrees, to meet the needs of industry and to lead to attractive career paths for students?

- Are changes needed in the current graduate education system, such as: a different mix in graduate support among fellowships, traineeships and research assistantships; and more research faculty positions and fewer postdocs and graduate students in traditional graduate programs?

- Should a greater proportion of federal research funding be allocated to high-risk, exploratory research and should funding priorities among broad fields of science and engineering be readjusted?

- What policies and programs will help ensure the rapid flow of research results into the marketplace and promote the commercialization of research in a way that leads to the creation of good jobs for Americans?

The Committee looks forward to reviewing the results of this effort, and hopes that a draft response would be available by September 30, 2005. We hope that the new and innovative ideas you produce as the result of this effort will be able to translate into policies that will enhance U.S. prosperity in the 21st century. If you have any questions, please contact Dan Byers of the Majority Staff or Jim Wilson of the Minority Staff.

Sincerely,

SHERWOOD BOEHLERT
Chairman

BART GORDON
Ranking Member

Appendix C

Focus-Group Sessions

The Committee on Prospering in the Global Economy of the 21st Century convened focus groups on Saturday, August 6, 2005, from 9 am to 4 pm. The purpose of the focus groups was to gather experts in five broad subjects—K–12 education, higher education, science and engineering research, innovation and workforce, and national and homeland security—to provide input to the committee on how the United States can successfully compete, prosper, and be secure in the global community.

Each focus-group participant was provided background on the committee members and on other focus-group members, 13 issue papers (see Appendix D) that summarized past reports on the various topics that were discussed, and a list of recommendations gleaned from past reports and interviews with committee and focus-group members.

The charge to focus-group participants is listed in full on page 252. Essentially, each group was asked to define and set priorities for the top three actions for its subject that federal policy-makers could take to ramp up the innovative capacity of the United States. Each focus group was chaired by a member of the committee, who presented the group's priorities to the full committee during an open discussion session. The content of those presentations is listed starting on page 254. Focus-group biographies are listed starting on page 264.

Prospering in the Global Economy of the 21st Century: An Agenda for American Science and Technology

Agenda

Focus-Group Meeting

August 6, 2005

Keck Center of the National Academies
500 5th Street, NW
Washington, DC

9:00	Continental Breakfast Available (Room 100)
9:30	Study Overview and Charge to Focus Groups *Norman Augustine*, Chair, Committee on Prospering in the Global Economy of the 21st Century
10:00	Focus Groups Meet

K–12 Education	Room 110	*Roy Vagelos*, Chair
Higher Education	Room 101	*Chuck Vest*, Chair
Research	Room 201	*Dan Mote*, Chair
Innovation	Room 204	*Gail Cassell*, Chair
Security	Room 105	*Anita Jones*, Chair

12:00	Lunch (Available in meeting rooms)
2:45	Break (Move to Room 100)
3:00	Focus Groups Report on Results of their Deliberations (Room 100)
4:00	Adjourn

Focus Group Charge

The Committee on Prospering in the Global Economy of the 21st Century would like to thank you for helping it in its important task to address the following questions:

What are the top 10 actions, in priority order, that federal policy-makers could take to enhance the science and technology enterprise so that the United States can successfully compete, prosper, and be secure in the global community of the 21st century? What implementation strategy, with several concrete steps, could be used to implement each of those actions?

Your role, as a focus-group participant, is to help the committee, in your area of expertise:

- Identify existing ideas the federal government (President, Congress, or federal agencies) could take. The ideas should not be too general—they need to be sufficiently actionable that they could be turned into congressional language.
- Brainstorm new ideas.
- Evaluate all ideas.
- Prioritize all ideas to propose to the committee the top 3 actions the federal government could take so that the United States can successfully compete, prosper, and be secure in the global community of the 21st century.

Since there are five focus groups, we expect a total of 15 prioritized recommendations to result from the focus-group session, which will be presented and discussed at a plenary session at the end of the day. These 15 recommendations would then be used by the committee as input to its decision-making process as it comes up with a "top 10" list on Sunday.

Each focus group is chaired by a committee member and has a staff member with expertise in the issue and a science and technology (S&T) policy fellow (graduate student) to assist them. The staff is available to put together any action list that is produced (no summary of the discussion is planned).

In evaluating each proposal, here are some evaluation criteria to keep in mind:

Minimum Selection Criteria

- **Can the actions be taken by those who requested the study?** The President, Congress, or the federal agencies?

Evaluation Criteria

- **Cost**—What is a rough estimate of how much the action will cost? Is the cost reasonable relative to the financial resources likely to be available? Can resources for this action be diverted from an existing activity as opposed to "new money"?
- **Impact**—Which degree of impact is the action likely to have on the problem of concern?
- **Cost-effectiveness**—Which actions provide the most "bang for the buck"?
- **Timeframe**—What is the desired timeframe for the action to have an impact? Is the action likely to have impact in the short- or long-term or both?
- **Distributional Effects**—Who are the winners and the losers? Is this the best action for the nation as a whole?
- **Ease of Implementation**—To what degree is the challenge easy, medium, or hard to implement?
- **History**—Has the action been suggested by another committee or policy-maker before? If so, why has it not been implemented? Can the challenges be overcome this time?
- **Is the Moment Right for This Action?** Are they likely to be viable in the near-term political and policy context?

K–12 Education Focus Group Top Recommendation Summary
Roy Vagelos, *Chair*

National Objectives

- Lay a foundation for a workforce that is capable in science, technology, engineering, and mathematics (STEM)—including those who can create, support, and sustain innovation.
- Develop a society that embraces STEM literacy.
- Develop and sustain K–12 teacher corps capable of and motivated to teach science and mathematics.
- Establish meaningful measures.

Top Recommendations

1. **The federal government should provide peer-reviewed long-term support for programs to develop and support a K–12 teacher core that is well-prepared to teach STEM subjects.**
   a. Programs for in-service teacher development that provide in-depth content and pedagogical knowledge; some examples include summer programs, master's programs, and mentor teachers.
   b. Provide scholarship funds to in-service teachers to participate in summer institutes and content-intensive degree programs.
   c. Provide seed grants to universities and colleges to provide summer institute and content-intensive degree programs for in-service teachers.
2. **Establish a program to encourage undergraduate students to major in STEM and teach in K–12 for at least 5 years. The program should include support mechanisms and incentives to enable teacher retention.**
   a. Provide a scholarship for joint STEM bachelor's degree and teacher certification program. Mandate a service requirement and pay a federal signing bonus.
   b. Encourage collaboration between STEM departments and education departments to train STEM K–12 teachers.
3. **Provide incentives to encourage students, especially minorities and women, to complete STEM K–12 coursework, including**
   a. Monetary incentives to complete advanced coursework.
   b. Tutoring and after-school programs.
   c. Summer engineering and science academies, internships, and research opportunities.
   d. Support school and curriculum organization models (statewide specialty schools, magnet schools, dual-enrollment models, and the like).

4. Support the design of state public school assessments that measure necessary workplace skills to meet innovation goals and ensure No Child Left Behind assessments include these goals.
5. Provide support to research, develop, and implement a new generation of instructional materials (including textbooks, modules, computer programs) based on research evidence on student learning outcomes, with vertical alignment and coherence across assessments and frameworks. Link teacher development and curricular development.

K–12 Focus Group Participants

Roy Vagelos, *Chair*
Carolyn R. Bacon, Executive Director, O'Donnell Foundation
Susan Berardi, Consultant
Rolf K. Blank, Director of Education Indicators, Council of Chief State School Officers
Rodger W. Bybee, Executive Director, Biological Sciences Curriculum Study
Hai-Lung Dai, Hirschmann-Makineni Chair Professor of Chemistry, University of Pennsylvania
Joan Ferrini-Mundy, Associate Dean for Science and Mathematics Education and Outreach, College of Natural Science, Michigan State University
Bruce Fuchs, Director, Office of Science Education, National Institutes of Health
Ronald Marx, Professor of Educational Psychology and Dean of Education, University of Arizona
David H. Monk, Professor of Educational Administration and Dean of College of Education, Pennsylvania State University
Carlo Parravano, Executive Director, Merck Institute for Science Education
Anne C. Petersen, Senior Vice President for Programs, W. K. Kellogg Foundation
Helen R. Quinn, Physicist, Stanford Linear Accelerator Center, Stanford University
Deborah M. Roudebush, Physics Teacher, Fairfax County Public Schools
Daniel K. Rubenstein, Mathematics Teacher, New York City Collegiate School
J. Stephen Simon, Senior Vice President, Exxon Mobil Corporation

Higher Education Focus Group Top Recommendation Summary
Charles Vest, *Chair*

National Objective

The United States should lead in the discovery of new scientific and technological knowledge and its efficient translation into new products and services in order to sustain its preeminence in technology-based industry and job creation.

Our higher education system has a critical role in meeting this objective.

Recommendation

We recommend that Congress enact the Innovation Development Education and Acceleration Act (IDEA Act). Its purpose is to increase the number of US students, consistent with our demography, who will become innovation leaders; professional scientists and engineers; and science, mathematics, and engineering educators at all levels.

1. **Undergraduate Education: Increase the number and proportion of citizens who hold STEM degrees to meet international benchmarks, i.e., migrate, over 5 years, from 5 to 10% of earned first (bachelor's-level) degrees.**
   a. Provide competitive multiagency (nonthematic) scholarships for undergraduates in science, engineering, mathematics, technology, and other critical areas. The scholarships would carry with them supplemental support for pedagogical innovation for the departments, programs, or institutions in which the students study. This program should support students at 2-year and 4-year colleges and research universities.
2. **Graduate Education: Increase the number of US graduate students in science, engineering, and mathematics programs in areas of strategic national needs.**
   a. Create a new multiagency support program for graduate students in STEM areas related to strategic national needs. This support should include an appropriate mix of competitive portable fellowships and competitive training grants.
3. **Faculty Preparation and Support: Support the propagation of effective and creative programs that develop scientific and technological leaders who understand the innovation process.**
   a. Support workshops, preparation of educational materials, and experience-based programs.

4. Create global scientific and technological leaders.

a. Provide a globally-oriented education and opportunity for US students, and maintain the US as the most desirable place to pursue graduate education and/or scientific and technological careers.

b. Define the policies that will maintain our long-term security and vitality through the openness of American education and research and the free flow of talent and ideas.

Higher Education Focus Group

Chuck Vest, *Chair*
M. R. C. Greenwood, Provost and Senior Vice President for Academic Affairs, University of California
Daniel Hastings, Professor of Aeronautics and Astronautics and Engineering Systems, Massachusetts Institute of Technology
Randy H. Katz, United Microelectronics Corporation Distinguished Professor in Electrical Engineering and Computer Science, University of California, Berkeley
George M. Langford, E. E. Just Professor of Natural Sciences and Professor of Biological Sciences, Dartmouth College
Joan F. Lorden, Provost and Vice Chancellor for Academic Affairs, University of North Carolina-Charlotte
Claudia Mitchell-Kernan, Vice Chancellor for Graduate Studies and Dean of Graduate Division, University of California, Los Angeles
Stephanie Pfirman, Chair, Department of Environmental Science, Barnard College
Paul Romer, STANCO 25 Professor of Economics, Graduate School of Business, Stanford University
James M. Rosser, President and Professor of Health Care Management, California State University, Los Angeles
Tim Stearns, Associate Professor of Biological Sciences and Genetics, Stanford University
Debra Stewart, President, Council of Graduate Schools
Orlando L. Taylor, Vice Provost for Research, Dean of Graduate School, and Professor of Communications, Howard University
Isiah M. Warner, Vice Chancellor for Strategic Initiatives, Louisiana State University
Dean Zollman, University Distinguished Professor, Distinguished University Teaching Scholar, and Head of Department of Physics, Kansas State University

Research Focus Group Top Recommendation Summary
Dan Mote, *Chair*

National Objective

America's leadership in S&T has created our prosperity, security, and health. That leadership is now threatened. Our leadership resulted from a long-term investment in basic research. In order to keep our leadership position we must revitalize our investments, particularly in the physical and mathematical sciences and engineering.

Recommendations

1. **Set the federal research budget to 1% of gross domestic product (GDP) within the next 5 years to sustain US leadership in innovation for prosperity, security, and quality of life.**
   a. Address 21st-century global economy grand challenges in energy, security, health, and environment through interagency initiatives.
   b. Bring physical sciences, engineering, mathematics, and information science up to the levels of health sciences.
   c. All agencies would expand their basic research programs.
   d. Replace decaying infrastructure in universities, national labs, and other research organizations.
   e. Longer-term, stable funding.
2. **To foster breakthroughs in science and technology, allocate at least 5% of federal agency research portfolios to high-risk basic research.**
   a. Allow for discretionary distribution for basic research with program oversight.
   b. Provide at least 5 years of adequate support for early-career researchers.
   c. Provide technical program managers in federal agencies with discretionary funding.
3. **Make S&T an attractive career to the best and the brightest.**
   a. Create an undergraduate loan forgiveness program for students who complete a PhD in S&T and work as STEM researchers (e.g., $25,000 per year).
   b. Create training grants for graduate and postgraduate education across federal research budgets.
   c. Provide 5 years of transition funding for early career research.
   d. Cultivate K–12 students to careers in science and technology.
   e. Actively recruit and support the world's best students and researchers and make it attractive for them to stay: address problems with visas, deemed exports, and other barriers.

Research Focus Group

Dan Mote, *Chair*
Paul Avery, Professor of Physics, University of Florida
Gary Bachula, Vice President for External Relations, Internet2
Angela Belcher, John Chipman Associate Professor of Materials Science and Engineering and Biological Engineering, Massachusetts Institute of Technology
Elsa M. Garmire, Sydney E. Jenkins Professor of Engineering, Dartmouth College
Heidi E. Hamm, Earl W. Sutherland, Jr., Professor and Chair of Pharmacology, Vanderbilt University
Mark S. Humayun, Professor of Ophthalmology, Biomedical Engineering, and Cell and Neurobiology, University of Southern California
Madeleine Jacobs, Executive Director and Chief Executive Officer, American Chemical Society
Cato T. Laurencin, Lillian T. Pratt Distinguished Professor and Chair of Department of Orthopaedic Surgery, University of Virginia
David LaVan, Assistant Professor of Mechanical Engineering, Yale University
Philip LeDuc, Assistant Professor of Mechanical Engineering, Carnegie Mellon University
Deirdre R. Meldrum, Professor and Director of Genomation Laboratory, Department of Electrical Engineering, University of Washington

Innovation and Workforce Focus Group Top Recommendation Summary
Gail Cassells, *Chair*

National Objective

Accelerate the process of innovation to:

- Solve national problems
- Create and retain well-paying jobs
- Ensure prosperity

Recommendations

1. **Tax Policy: Make the R&D tax credit permanent, and extend coverage to research conducted in university-industry consortia.**
2. **National Energy Initiative.**
   a. Sharp increase in agency R&D related to energy prosperity.
   b. National Energy Prosperity fellowships.
   c. Cabinet-level National Council on Energy Prosperity.
3. **National Agency for Innovation.**
   a. New independent, project-based agency, reports to president.
   b. University–industry projects on specific goals.
   c. Broad, nonmilitary, national interest.
   d. \$3-\$5 billion per year.
   e. Outputs: functional prototypes and processes, training, monitoring of US innovation and competitiveness.
   f. Issues to resolve: metrics, intellectual property (IP), governance.
4. **Stimulate interest of young people in S&T.**
   a. National scholarships program for first-generation college students who major in S&E.
   b. Scholarship recipients available for national S&E role models program to explain to elementary and secondary students what they do and how success in school prepared them.

Innovation and Workforce Focus Group

Gail Cassell, *Chair*
Miller Adams, Vice President, Boeing Technology Ventures
Robert J. Aiken, Director of Engineering, International Academic Research and Technology Initiatives, Cisco Systems, Inc.
Ron Blackwell, Chief Economist, American Federation of Labor and Congress of Industrial Unions (AFL-CIO)
Craig Blue, Distinguished Research Engineer and Group Leader, Materials Processing Group, Metals and Ceramics Division, Oak Ridge National Laboratory
Susan Butts, Director, External Technology, Dow Chemical Company
Paul Citron, Vice President (retired), Technology Policy and Academic Relations, Medtronic, Inc.
Chad Evans, Vice President, National Innovation Initiative, Council on Competitiveness
Kent H. Hughes, Director, Program on Science, Technology, America and the Global Economy, Woodrow Wilson International Center for Scholars
Marvin Kosters, Resident Scholar, American Enterprise Institute
Mark B. Myers, Visiting Executive Professor of Management, Wharton School of the University of Pennsylvania
Juliana C. Shei, Global Technology Manager, General Electric
Nancy Vorona, Vice President, Research Investment, Virginia's Center for Innovative Technology

National and Homeland Security Focus Group
Top Recommendation Summary
Anita Jones, *Chair*

Globalization Is a Fact of Life

- S&T provides our qualitative national security advantage.
- S&T enables our prosperity, which in turn finances strong security.
- S&T increasingly originates abroad.
- Isolation damages our security and our economy.
- Need to engage with and ensure access to innovators and innovation abroad.

National Objectives

- Stimulate innovation and its adoption to serve security.
- Rebalance security S&T research funding invested in basic research.
- Accelerate creation of knowledge in the United States and acquisition of knowledge from abroad.
- Attract and retain global best and brightest.

Only the federal government can provide the framework/strategy for balancing contending national interests.

Recommendations

1. **To stimulate innovation and its adoption to serve security, create new mechanisms to discover, develop, and exploit new ideas.**
   a. Legal reform—extend liability protection for homeland security providers.
   b. Create new prototypes for university-industry-national lab partnerships.
      i. Experiment with mix of funding mechanisms, e.g., SEMATECH, InQTel, for security.
      ii. Streamlined, standardized IP provisions based on best practices for universities and national labs.
2. **To rebalance security S&T research funding invested in basic research, dedicate 3% of national defense/homeland security budget to S&T and 20% of S&T budget to long-term research.**
   a. Cost: Δ of $ in research spending.
   b. Caveats/concerns: Need institutional champion in each agency?

3. **Create a single national strategy to attract and retain the global best and brightest to US S&T enterprise.**
   a. Increase support for the National Defense Education Act (NDEA-21).
      i. Double the number of US students going into science and engineering and related security fields.
      ii. Provide a national service educational benefit incentive.
   b. Redesign visa, deemed-export, and immigration policies to attract and retain foreign talent.

National and Homeland Security Focus Group

Anita Jones, *Chair*
Ronald M. Atlas, Graduate Dean, Professor of Biology, and Codirector, Center for the Deterrence of Biowarfare and Bioterrorism, University of Louisville
Pierre Chao, Senior Fellow and Director of Defense Industrial Initiatives, Center for Strategic and International Studies
Richard T. Cupitt, Senior Consultant, MKT, and Scholar-In-Residence, School of International Service, American University
Kenneth Flamm, Dean Rusk Professor of International Affairs, Lyndon B. Johnson School, University of Texas at Austin
Alice P. Gast, Robert T. Haslam Professor, Department of Chemical Engineering, and Vice President for Research and Associate Provost, Massachusetts Institute of Technology
William Happer, Professor, Department of Physics, Princeton University
Robert Hermann, Senior Partner, Global Technology Partners, LLC *(via videoconference)*
Richard Johnson, Senior Partner, Arnold and Porter, LLP
James A. Lewis, Senior Fellow and Director of Technology Public Policy, Center for Strategic and International Studies
Daniel B. Poneman, Principal, The Scowcroft Group
Sheila R. Ronis, President, The University Group, Inc.
General Larry Welch (retired), Senior Associate, Institute for Defense Analyses *(via videoconference)*
Rear Admiral Robert H. Wertheim (retired), Consultant

Focus Group Participant Biographies

MILLER ADAMS is vice president of Boeing Technology Ventures, a unit of Boeing Phantom Works, the research and development organization of the Boeing Company. He leads a team responsible for the overall Enterprise Technology Planning Process for Boeing. He also is responsible for some aspects of external-technology acquisition strategies for Boeing, including the Evaluation of External Technology Solutions, International Industrial Technology Programs, Strategic Technology Alliances, Global University Research Collaborations, and Boeing's overall Global R&D Strategy. Mr. Adams is responsible for Boeing's internal incubator program known as the Chairman's Innovation Initiative and for value-creating strategies around spin-in business opportunities built on Boeing technologies. He received a BA from Seattle University and a law degree from the University of Puget Sound (now Seattle University School of Law). At Boeing, he serves as the executive focal between Boeing and Tuskegee University. In 2003, Mr. Adams received the Chairman's Award at the annual Black Engineer of the Year Awards Conference. He is involved in a broad array of professional and community organizations.

ROBERT J. AIKEN is the director of engineering for Cisco's International Academic Research and Technology Initiatives (ARTI). He manages a team of Internet and network technology experts who help to identify, define, and develop Cisco's next-generation Internet strategy and technologies via Cisco's university research and advanced network research infrastructure programs. He helped to design and deploy the Department of Energy's (DOE) international multi-protocol Energy Sciences Network and was the National Science Foundation's (NSF) manager for and coauthor of NSF's very high performance Backbone Network Service and Network Access Points architecture, which commercialized the Internet in the early 1990s. He was a major contributor at both DOE and NSF to the development and implementation of the federal government's High Performance Computing and Communications Council and Next Generation Internet programs, specifically with respect to network research and distributed systems. With Javad Boroumand, he is responsible for Cisco's leadership role in the National Lambda Rail. He has also been an assistant professor of computer science and a college information technology director, and he serves on the National Research Council's Transportation Research Board Subcommittee on Telecommuting and Internet2's Industry Advisory Council.

RONALD M. ATLAS is the graduate dean, professor of biology, and codirector of the Center for the Deterrence of Biowarfare and Bioterrorism at the University of Louisville. He has his BS from the State University of

New York at Stony Brook and his MS and PhD from Rutgers University. He was a postdoctoral fellow at the Jet Propulsion Laboratory, where he worked on Mars life detection. He is a member of the Department of Homeland Security Science and Technology Advisory Committee, the National Aeronautics and Space Administration's Planetary Protection Board, and the Federal Bureau of Investigation's Scientific Working Group on Microbial Genetics and Forensics. He previously served as president of the American Society for Microbiology (ASM), cochaired the ASM Task Force on Biological Weapons, and was a member of the National Institutes of Health Recombinant DNA Advisory committee. His early research focused on oil spills, and he discovered bioremediation as part of his doctoral studies. Later, he turned to the molecular detection of pathogens in the environment, which forms the basis for biosensors to detect biothreat agents. He is the author of nearly 300 manuscripts and 20 books. He is a fellow of the American Academy of Microbiology and has received the ASM Award for Applied and Environmental Microbiology, the ASM Founders Award, and the Edmund Youde Lectureship Award in Hong Kong. He regularly advises the US government on policy issues related to the deterrence of bioterrorism.

PAUL AVERY is professor of physics at the University of Florida. He received his PhD in high-energy physics from the University of Illinois in 1980. His research is in experimental high-energy physics, and he participates in the CLEO experiment at Cornell University and the Compact Muon Solenoid experiment at CERN, Geneva. Avery is the director of two NSF-funded Grid projects, Grid Physics Networks and the International Virtual Data Grid Laboratory. Both are collaborations of computer scientists, physicists, and astronomers conducting grid research applied to several frontier experiments in physics and astronomy with massive computational and data needs. He is co-principal investigator of the NSF-funded projects, Center for High Energy Physics Research and Education Outreach and UltraLight, and is one of the principals seeking to establish the Open Science Grid.

GARY BACHULA is the vice president for external relations for Internet2. He has substantial government and not-for-profit experience and an extensive history of leadership in technology development. Most recently, Dr. Bachula served as acting under secretary of commerce for technology at the US Department of Commerce, where he led the formation of government–industry partnerships around such programs as GPS and the Partnership for a New Generation of Vehicles. As vice president for the Consortium for International Earth Science Information Network (CIESIN) from 1991 to 1993, he managed strategic planning and program development for the organization designated to build a distributed information network as part of the National Aeronautics and Space Administration's (NASA's) Mission to Planet Earth. From 1986 to 1990, he chaired the Michigan governor's Cabinet Council, and from 1974 to 1986, he served as chief of staff to US Representative Bob Traxler of Michigan and advised on appropriations for NASA, the Environmental Protection Agency, the National Science Foundation, and other federal R&D agencies. Dr. Bachula holds undergraduate and law (JD) degrees from Harvard University. He served at the Pentagon in the US Army during the Vietnam War.

CAROLYN R. BACON is executive director of the O'Donnell Foundation in Dallas. The purpose of the foundation is to support quality education, especially in science and engineering. She previously served as administrative assistant to former Senator John Tower of Texas. In 1989, she was appointed to the White House Education Policy and Advisory Council. President George H. W. Bush also appointed her to the Board of the Corporation for Public Broadcasting, where she served as chairman of the Education Committee. Texas Governor Clements appointed her to a 6-year term on the Texas Higher Education Coordinating Board and former Governor George W. Bush named her the first chairman of the Telecommunications Infrastructure Fund Board of Texas. In 2003-2004 she served as the governor's public member on the Texas Joint Select Committee on Public School Finance. Her board memberships include the National Center for Educational Accountability, the College of Computing at the Georgia Institute of Technology, Advanced Placement Strategies, Inc., of Dallas, and the Foundation for the Education of Young Women. She is a member of the Junior League of Dallas and Charter 100 of Dallas. She holds a BA in political science from the College of William and Mary.

ANGELA BELCHER is the John Chipman Associate Professor of Materials Science and Engineering and Biological Engineering at the Massachusetts Institute of Technology. She is a materials chemist with expertise in biomaterials, biomolecular materials, organic-inorganic interfaces, and solid-state chemistry. She received her BS in creative studies with an emphasis in biochemistry and molecular biology and a PhD in inorganic chemistry from the University of California, Santa Barbara (UCSB). After a year of postdoctoral research in electrical engineering at UCSB, Dr. Belcher joined the faculty at the University of Texas at Austin in the Department of Chemistry and Biochemistry in 1999. Her interest focuses on interfaces, including the interfaces of scientific disciplines and the interfaces of materials. Dr. Belcher and her students have pioneered a novel, noncovalent self-organizational approach that uses evolutionarily selected and engineered peptides to recognize and bind electronic and magnetic building blocks. She was recently awarded an annual MacArthur Foundation Fellowship. Her recent awards include the 2004 Four Star General Recognition Award (US Army), 2003 Top 10 Innovators Under 40 (*Fortune* magazine), the 2002 World Technology Award (*Materials* magazine), 2002 *Popular Science* Brilliant Ten, and 2002 *Technology Review* Top 100 Inventors. In 2002, she was named as 1 of 12 women expected to make the biggest impact in chemistry in the next century by *Chemical and Engineering News* and was runner-up for Innovator of the Year and runner-up for Researcher of the Year by *Small Times Magazine*, and finalist for Scientist of the Year by *Wired* magazine. She is a 2001 Packard Fellow, 2001 Alfred P. Sloan Research Fellow, and has received the 2000 Presidential Early Career Award for Science and Engineering, 2000 Beckman Young Investigator Award, 1999 DuPont Young Investigator Award, and a 1999 Army Research Office Young Investigators Award.

SUSAN BERARDI worked in management and employee development for nearly 10 years before leaving corporate America to become a full-time mother of three young boys. At such companies as FMC Defense Systems, Motorola, and IDX Systems Corporation, she worked with managers and technical teams to improve the intangible assets that drove performance and bottom-line results. In addition to one-on-one executive coaching, she facilitated and trained numerous technical teams to resolve customer-service and team-performance issues that were hindering company profitability. She also designed selection and retention programs to attract and keep best-in-class technical and managerial talent. As an independent consultant, Ms. Berardi provided leadership training and facilitation for several start-up technology companies in Massachusetts and California. She has been a guest speaker for the Society of Concurrent Engineering and the International Council on Systems Engineering. Most recently, Ms. Berardi has been working pro bono for the Reading and North Andover School Districts in Massachusetts, facilitating administrative retreats and bringing teachers and parents together to improve student reading, mathematics, and arts capabilities. She worked with school administrators to create a tool to measure and improve the return on investment of a school district. She has also written several articles on behalf of these schools in an effort to educate taxpayers on budget and curriculum issues, special-education costs and legal requirements, and the importance of foreign languages and the arts in early education. Ms. Berardi has an MA degree in labor relations and a BA from the University of Illinois.

RON BLACKWELL is chief economist of the American Federation of Labor and Congress of Industrial Unions (AFL-CIO), where he coordinates the economic agenda of the federation and represents AFL-CIO on corporate and economic issues affecting American workers and union strategies. From 1996 to 2004, he was the director of the AFL-CIO Corporate Affairs Department. Before coming to the AFL-CIO, Mr. Blackwell was assistant to the president of the Amalgamated Clothing and Textile Workers Union and chief economist of UNITE. Before joining the labor movement, he was an academic dean in the Seminar College of the New School for Social Research in New York, where he taught economics, politics, and philosophy. Mr. Blackwell represents the American labor movement on the Economic Policy Working Group of the Trade Union Advisory Committee to the Organisation for Economic Co-operation and Development (OECD) and participated in formulation of the *OECD Principles of Corporate Governance* and the recent review of the *OECD Guidelines for Multinational Enterprises*. He serves on the Board of Directors of the Industrial Relations Research Association; the Research Advisory Council of the Economic Policy Institute; the Board on Manufacturing and Engineering Design of the National Academies; the advisory boards of the Jackson Hole Center for Global Affairs and the International Center for Corporate Governance and Accountability at the George Washington University Law School; and the editorial boards of *Perspectives on Work* and the *New Labor Forum*. He recently received the Nat Weinberg Award from the Walter P. Reuther Library for service to the labor movement and social justice. He is author of "Corporate Accountability or Business as Usual," in *New Labor Forum* (summer 2003) and "Globalization and the American Labor Movement" in the book edited by Steve Fraser and Joshua Freeman, *Audacious Democracy: Labor, Intellectuals and the Social Reconstruction of America*. He is also coeditor of *Worldly Philosophy: Essays in Political and Historical Economics*, a festschrift for Robert Heilbroner.

ROLF K. BLANK is director of education indicators at the Council of Chief State School Officers where he has been a senior staff member for 17 years. He is responsible for developing, managing, and reporting a system of state-by-state and national indicators of the condition and quality of education in public schools. Dr. Blank is directing the council's work with the US Department of Education on state education indicators and accountability systems, which provides annual trends for each state on student outcomes, school programs, and staff and school demographics. In addition, he is directing a 3-year experimental design study on improving effectiveness of instruction in mathematics and science with data on enacted curriculum, supported by the National Science Foundation. He coordinates two state collaborative projects—one on accountability systems and one on surveys of enacted curriculum—that provide technical assistance and professional development to state education leaders and staff. In his council leadership role, Blank collaborates with state education leaders, researchers, and professional organizations in directing program-evaluation studies and technical-assistance projects aimed at improving the quality of K–12 public education. He holds a PhD from Florida State University and an MA from the University of Wisconsin-Madison.

CRAIG BLUE [NAE] is a Distinguished Research Engineer and the group leader of the Materials Processing Group of the Metals and Ceramics Division at Oak Ridge National Laboratory (ORNL). He received his PhD in materials science from the University of Cincinnati and finished his studies while under a NASA Fellowship at NASA Lewis Research Center. He came to ORNL in March 1995, where he initiated and developed the Infrared Processing Center in the Materials Processing Group. The center has projects with the Defense Advanced Research Projects Agency, the US Army, the Department of Energy, NASA, and industry. The center has two of the most powerful plasma arc lamps in the world and has enabling technology of functionalization of nanomaterials with collaborations across the laboratory and across the United States. Dr. Blue has been instrumental in the revitalization and evolution of the Materials Processing Group, became group leader in January 2004, and is developing a new Advanced Materials Processing Laboratory and associated programs. He has over 60 open-literature publications, 5 patents, and 60 technical presentations. He has received numerous honors, including an R&D 100 Award on the development of advanced infrared heating, and UT/Battelle Distinguished Engineer of the Year. He was selected to attend the National Academy of Engineering's Ninth Annual Symposium on Frontiers of Engineering in 2003, and the International Symposium on Frontiers of Engineering in Japan in 2004. He serves on the steering committee for the National Space and Missile Materials Symposium and on a technical board for the Next Generation Manufacturing Initiative. He is working with colleagues in the evolution of an enabling pulse thermal processing technique for flexible electronics, titanium processing, and bulk amorphous materials.

SUSAN BUTTS is the director of external technology at the Dow Chemical Company. She is responsible for Dow's sponsored research programs at over 150 universities, institutes, and national laboratories worldwide and for Dow's contract research activities with US and European government agencies. She also holds the position of global staffing leader for R&D with responsibility for recruiting and hiring programs. Before joining the external-technology group, Dr. Butts held several other positions at Dow, including senior resource leader for atomic spectroscopy and inorganic analysis in the Analytical Sciences Laboratory, manager of PhD hiring and placement, safety and regulatory affairs manager for Central Research, and principal investigator on various catalysis research projects in Central Research.

RODGER W. BYBEE is executive director of the Biological Sciences Curriculum Study (BSCS), a nonprofit organization that develops curriculum materials, provides professional development, and conducts research and evaluation for the science education community. Before joining BSCS, he was executive director of the National Research Council's Center for Science, Mathematics, and Engineering Education. Between 1986 and 1995, he was associate director of BSCS. Dr. Bybee participated in the development of the National Science Education Standards, and in 1993-1995 he chaired its content working group. At BSCS, he was principal investigator for four new NSF programs: the elementary school program, Science for Life and Living: Integrating Science, Technology, and Health; the middle school program, Middle School Science and Technology; the high school biology program, Biological Science: A Human Approach; and the college program, Biological Perspectives. His work at BSCS also included serving as principal investigator for programs to develop curriculum frameworks for teaching about the history and nature of science and technology in high schools, community colleges, and 4-year colleges and curriculum reform based on national standards. From 1990 to 1992, Dr. Bybee chaired the curriculum and instruction study panel for the National Center for Improving Science Education (NCISE). From 1972 to 1985, he was professor of education at Carleton College in Northfield, Minnesota. He has taught science in the elementary school, junior and senior high school, and college. Dr. Bybee has written widely in education and psychology. He is coauthor of the leading textbook, *Teaching Secondary School Science: Strategies for Developing Scientific Literacy*. His most recent book is *Achieving Scientific Literacy: From Purposes to Practices*, published in 1997. He has received several awards, including Leader of American Education and Outstanding Educator in America; in 1979 he was Outstanding Science Educator of the Year, and in 1998 the National Science Teachers Association presented him its Distinguished Service to Science Education Award.

PIERRE CHAO is a senior fellow and director of defense industrial initiatives at the Center for Strategic and International Studies (CSIS). Before joining CSIS, Mr. Chao was a managing director and senior aerospace-defense analyst at Credit Suisse First Boston (CSFB) in 1999-2003, where he was responsible for following the US and global aerospace-defense industry. He remains a CSFB senior adviser. Before joining CFSB, he was the senior aerospace-defense analyst at Morgan Stanley Dean Witter in 1995-1999. He served as the senior industry analyst at Smith Barney during 1994 and as a director at JSA International, a Boston and Paris-based management-consulting firm that focused on the aerospace-defense industry (1986-1988 and 1990-1993). Mr. Chao was also a cofounder of JSA Research, an equity research boutique specializing in the aerospace-defense industry. Before signing on with JSA, he worked in the New York and London offices of Prudential-Bache Capital Funding as a mergers and acquisitions banker focusing on aerospace and defense (1988-1990). Mr. Chao garnered numerous awards while working on Wall Street. *Institutional Investor* ranked his team the number 1 global aerospace-defense group in 2000-2002, and he was on the *Institutional Investor* All-America Research Team every year he was eligible in 1996-2002. He was ranked the number 1 aerospace-defense analyst by corporations in the 1998-2000 Reuters Polls and the number 1 aerospace-defense analyst in the 1995-1999 Greenwich Associates polls, and appeared on the *Wall Street Journal* All-Star list in 4 of 7 eligible years. In 2000, Mr. Chao was appointed to the Presidential Commission on Offsets in International Trade. He is also a guest lecturer at the National Defense University and the Defense Acquisition University. He has been sought out as an expert analyst of the defense and aerospace industry by the Senate Committee on Armed Services, the House Committee on Science, the Office of the Secretary of Defense, Department of Defense (DOD) Defense Science Board, the Army Science Board, the National Aeronautics and Space Administration, the French General Delegation for Armament, North Atlantic Treaty Organization, and the Aerospace Industries Association Board of Governors. Mr. Chao earned dual BS degrees in political science and management science from the Massachusetts Institute of Technology.

PAUL CITRON [NAE] retired as vice president of Technology Policy and Academic Relations at Medtronic, Inc., in 2003 after 32 years with the company. His previous position was vice president of science and technology; he had responsibility for corporationwide assessment and coordination of technology initiatives and for priority-setting in corporate research. Citron was awarded a BS in electrical engineering from Drexel University in 1969 and an MS in electrical engineering from the University of Minnesota in 1972. He was elected to the National Academy of Engineering in 2003 for "innovations in technologies for monitoring cardiac rhythm and for patient-initiated cardiac pacing, and for outstanding contributions to industry-academia interactions." Mr. Citron was elected founding fellow of the American Institute of Medical and Biological Engineering in January 1993, has twice won the American College of Cardiology Governor's Award for Excellence, and in 1980 was inducted as a fellow of the Medtronic Bakken Society, the company's highest technical recognition. He has written numerous publications and holds eight US medical-device patents. In 1980, he was given Medtronic's Invention of Distinction award for his role as coinventor of the tined pacing lead. He has been a visiting professor at Georgia Institute of Technology and the University of California, San Diego where he taught corporate entrepreneurship.

RICHARD T. CUPITT is a senior consultant to MKT and a scholar-in-residence in the School of International Service of American University. He served as the special adviser to the under secretary of commerce for industry and security. Before joining the Department of Commerce in January 2002, Dr. Cupitt worked as the associate director and Washington liaison for the Center for International Trade and Security of the University of Georgia, and as a visiting scholar at the Center for Strategic and International Studies in Washington, DC. Dr. Cupitt received his PhD from the University of Georgia in 1985 and taught at Emory University and the University of North Texas before returning to the University of Georgia. In addition to his most recent book, *Reluctant Champions: U.S. Presidential Policy and Strategic Export Controls—Truman, Eisenhower, Bush and Clinton* (Routledge, 2000), Cupitt has coedited two books on export controls and is a coauthor of a forthcoming book. His articles on export controls have appeared in many scholarly journals. He has contributed to the work of several national study commissions, served on US delegations to international export control conferences, and regularly testified before Congress on export controls. Dr. Cupitt has conducted fieldwork on export controls in more than a dozen countries and has served as a consultant to Lawrence Livermore National Laboratory, Argonne National Laboratory, and the Organisation for Economic Co-operation and Development. Dr. Cupitt is a former governor's fellow with the Georgia World Congress Institute and a National Merit Scholar.

HAI-LUNG DAI is the Hirschmann-Makineni Chair Professor of Chemistry at the University of Pennsylvania. He came to the University of California, Berkeley, for graduate study in 1976 after graduating from the National Taiwan University and military service. Dai did postdoctoral research at the Massachusetts Institute of Technology. He joined the University of Pennsylvania faculty as assistant professor in 1984, and was promoted to full professor in 1992. He served as chairman of the Chemistry Department from 1996-2002. In addition to his academic appointment, Dr. Dai currently holds a gubernatorial appointment in the Pennsylvania State Board on Drugs, Devices and Cosmetics. He is a fellow of the American Physical Society and is chair-elect of its Chemical Physics Division. Dr. Dai has published more than 140 papers in molecular and surface sciences. His major research accomplishments include the discovery of the dominating contribution of long-range interactions in collision energy transfer, the development of Fourier transform spectroscopy with fast time resolution and multiple-resonance spectroscopy for detecting unstable molecules and transient radicals, and the development of nonlinear optical techniques for probing molecule-surface interactions. He has received many honors, including the Coblentz Prize in Molecular Spectroscopy, the Morino Lectureship of Japan, the American Chemical Society Philadelphia Section Award, and a Guggenheim Fellowship. In 2000, Dr. Dai established a pioneering master's degree program at the University of Pennsylvania for inservice high school chemistry teachers to receive content-intensive training. In 2004, the program became the Penn Science Teacher Institute with Dr. Dai as director, and the Institute enlarged to include middle school teachers.

CHAD EVANS is vice president of the Council on Competitiveness National Innovation Initiative (NII), a private-sector effort aimed at developing and implementing a national innovation agenda for the United States. Cochaired by IBM Chairman and Chief Executive Officer Samuel J. Palmisano and Georgia Institute of Technology President G. Wayne Clough, the NII involves the active participation of nearly 400 innovation thought-leaders and stakeholders across the country. Mr. Evans also spearheads the council's benchmarking efforts, including its flagship publication, *The Competitiveness Index,* chaired by Michael Porter, of the Harvard Business School. Mr. Evans' work at the council has focused on understanding the globalization of R&D investments, assessing the strengths and weaknesses of the US innovation platform, and benchmarking national innovative capacities in developed and emerging economies. He was a senior associate with the Council during the 1990s and returned to the Council and Washington, DC, after a stint in Deloitte & Touche's National Research and Analysis Office, where he provided the firm's senior leadership with daily competitive-intelligence briefings. He holds a MS in foreign service from the Georgetown University School of Foreign Service, with an honors concentration in international business diplomacy from Georgetown's Landegger Program, and a BA from Emory University.

JOAN FERRINI-MUNDY is associate dean for science and mathematics education and outreach in the College of Natural Science at Michigan State University (MSU). Her faculty appointments are in mathematics and teacher education. She holds a PhD in mathematics education from the University of New Hampshire and was a faculty member in mathematics there in 1983-1995. Dr. Ferrini-Mundy taught mathematics at Mount Holyoke College from 1982-1983, where she cofounded the Summer Math for Teachers program. She served as a visiting scientist at the National Science Foundation in 1989-1991. She has chaired the National Council of Teachers of Mathematics (NCTM) Research Advisory Committee and the American Educational Research Association in Special Interest Group for Research in Mathematics Education, and she was a member of the NCTM Board of Directors. Dr. Ferrini-Mundy came to MSU in 1999 from the National Research Council's Center for Science, Mathematics, and Engineering Education, where she served as director of the Mathematical Sciences Education Board. Her research interests are in calculus learning and K–14 mathematics education reform. She chairs the writing group for Standards 2000, the revision of the NCTM standards.

KENNETH FLAMM is the Dean Rusk Professor of International Affairs at the Lyndon B. Johnson School of Public Affairs at the University of Texas at Austin. Earlier, he worked at the Brookings Institution in Washington, DC, where he served for 11 years as a senior fellow in the Foreign Policy Studies Program. He is a 1973 honors graduate of Stanford University and received a PhD in economics from the Massachusetts Institute of Technology in 1979. From 1993 to 1995, Dr. Flamm served as principal deputy assistant secretary of defense for economic security and special assistant to the deputy secretary of defense for dual use technology policy. He was awarded the department's Distinguished Public Service Medal by Defense Secretary William J. Perry in 1995. Dr. Flamm has been a professor of economics at the Instituto Tecnológico de México in Mexico City, the University of Massachusetts, and George Washington University. He has also been an adviser to the director general of income policy in the Mexican Ministry of Finance and a consultant to the Organisation for Economic Cooperation and Development, the World Bank, the National Academy of Sciences, the Latin American Economic System, the US Department of Defense, the US Department of Justice, the US Agency for International Development, and the Office of Technology Assessment of the US Congress. He has played an active role in the National Research Council's committee on Government-Industry Partnerships and played a key role in that committee's review of the Small Business Innovation Research Program at the Department of Defense. Dr. Flamm has made major contributions to our understanding of the growth of the electronics industry, with a particular focus on the development of the computer and the US semiconductor industry. He is working on an analytic study of the post-Cold War defense industrial base and has expert knowledge of international trade and high-technology industry issues.

BRUCE FUCHS, an immunologist who did research on the interaction between the brain and the immune system, is the director of the National Institutes of Health (NIH) Office of Science Education. Dr. Fuchs directs the creation of a series of K–12 science education curriculum supplements that highlight the medical research findings of NIH. The supplements are designed to meet teacher educational goals as outlined in the *National*

Science Education Standards and are available free to teachers across the nation. The office is also creating innovative science and career education Web resources that will be accessible to teachers and students with a variety of disabilities. Before coming to NIH, Dr. Fuchs was a researcher and teacher at the Medical College of Virginia with grant support from the National Institute of Mental Health and the National Institute on Drug Abuse. He has a BS in biology from the University of Illinois and a PhD in immunology from Indiana State University. Dr. Fuchs has organized and participated in numerous science education outreach efforts directed at students, teachers, and the public. Dr. Fuchs has organized more than a dozen "Mini-Med School" and "Science in the Cinema" programs for the public and Congress since his arrival at NIH.

ELSA M. GARMIRE [NAE] is Sydney E. Jenkins Professor of Engineering at Dartmouth College. She received her AB at Harvard and her PhD at the Massachusetts Institute of Technology, both in physics. After postdoctoral work at the California Institute of Technology, she spent 20 years at the University of Southern California, where she was eventually named William Hogue Professor of Electrical Engineering and director of the Center for Laser Studies. She came to Dartmouth in 1995 and served 2 years as dean of Thayer School. Author of over 250 journal papers and holder of 9 patents, she has been on the editorial boards of five technical journals. Dr. Garmire is a member of the National Academy of Engineering and the American Academy of Arts and Sciences and a fellow of the Institute of Electrical and Electronic Engineers, the American Physical Society, and the Optical Society of America, of which she was president; she has served on the boards of three other professional societies. In 1994, she received the Society of Women Engineers Achievement Award. She has been a Fulbright senior lecturer and a visiting faculty member in Japan, Australia, Germany, and China. She has been chair of the NSF Advisory Committee on Engineering Technology and served on the NSF Advisory Committee on Engineering and the Air Force Science Advisory Board.

ALICE P. GAST is the Robert T. Haslam Professor in the Department of Chemical Engineering and the vice president for research and associate provost of the Massachusetts Institute of Technology. Until 2001, she was a professor of chemical engineering at Stanford University, and professor of the Stanford Synchrotron Radiation Laboratory and professor, by courtesy, of chemistry at Stanford. Dr. Gast earned her BS in chemical engineering at the University of Southern California in 1980 and her PhD in chemical engineering from Princeton University in 1984. She spent a postdoctoral year on a North Atlantic Treaty Organization fellowship at the École Supérieure de Physique et de Chimie Industrielles in Paris. She was on the faculty at Stanford from 1985 to 2001 and returned to Paris for a sabbatical as a John Simon Guggenheim Memorial Foundation Fellow in 1991 and to Munich, Germany, as a Humboldt Fellow in 1999. In Dr. Gast's research, the aim is to understand the behavior of complex fluids through a combination of colloid science, polymer physics, and statistical mechanics. In 1992, she received the National Academy of Sciences Award for Initiative in Research and the Colburn Award of the American Institute of Chemical Engineers. She was the 1995 Langmuir Lecturer for the American Chemical Society. Dr. Gast is a member of the American Academy of Arts and Sciences. She served as a member and then cochair of the National Research Council's Board on Chemical Sciences and Technology and now serves on the Division on Earth and Life Studies Committee. She also serves on the Homeland Security Science and Technology Advisory Committee.

M. R. C. GREENWOOD [IOM] is provost and senior vice president for academic affairs for the 10-campus University of California (UC) system. She previously served as chancellor of UC, Santa Cruz (UCSC), a position she held from July 1996 to March 2004. In addition to her administrative responsibilities, Dr. Greenwood holds a UCSC appointment as professor of biology. Before her UCSC appointments, Dr. Greenwood served as dean of graduate studies, vice provost for academic outreach, and professor of biology and internal medicine at UC, Davis. Previously, she taught at Vassar College, where she was the John Guy Vassar Professor of Natural Sciences and chair of the Biology Department. Dr. Greenwood is a member of the Institute of Medicine, a fellow of the California Academy of Sciences, and a member of the board of directors of the California Healthcare Institute. She is a fellow and past president of the American Association for the Advancement of Science and a member of the Board of Directors of the National Association of State Universities and Land-Grant Colleges. Among her numerous distinctions, she was a member of the National Oceanic and Atmospheric Administration Science Advisory Board and of the Task Force on the Future of Science Programs at the US Department of Energy. She is a former member of the National Science Board and the Laboratory Operations Board of the US Department of Energy. She was chairman of the National Research Council's Office of Science and Engineering Policy Advisory Board and now serves as chair of its Policy and Global Affairs Division. She is a member of the National Commission on Writing for America's Families, Schools, and Colleges, appointed by the College Board. From November 1993 to May 1995, Dr. Greenwood was associate director for science at the Office of Science and Technology Policy. In that position, she supervised the Science Division, directing budget development for the multibillion dollar fundamental-science national effort and development of science-policy documents, including *Science in the National Interest*. She was also responsible for interagency coordination, cochaired two National Science and Technology Council committees, and provided advice on a $17 billion budget for fundamental science. Dr. Greenwood graduated summa cum laude from Vassar College and received her PhD from the Rockefeller University. Her research interests are in developmental cell biology, genetics, physiology, nutrition, and science and higher education policy.

HEIDI E. HAMM is the Earl W. Sutherland, Jr. Professor and chair of pharmacology at Vanderbilt University. Hamm obtained her PhD in zoology in 1980 from the University of Texas at Austin and performed her postdoctoral training at the University of Wisconsin-Madison from 1980 to 1983. Her initial research centered around circadian clocks and melatonin synthesis in the avian retina; her postdoctoral work investigated the role of transducin in visual transduction using blocking monoclonal antibodies. She held faculty appointments at the University of Illinois at Chicago School of Medicine and Northwestern University before moving to Vanderbilt in 2000 to chair the Department of Pharmacology. Hamm studies a specific mechanism of neuronal communication known as G-protein signaling. G-protein-mediated signaling is a critical part of biologic function in the brain and other body systems. Because many pharmaceuticals are targeted to G-protein signaling cascades, gaining a better understanding of their function is crucial to developing more efficient treatments and designing better drugs. Her research focuses on the structure and function of guanine triphosphate binding proteins and the molecular mechanisms of signal transduction. Dr. Hamm has received numerous awards, including the Glaxo Cardiovascular Discovery Award, two Distinguished Investigator Awards from the National Alliance for Research in Schizophrenia and Depression, the Faculty of the Year award from the University of Illinois College of Medicine, and the Stanley Cohen Award "For Research Bringing Diverse Disciplines, such as Chemistry or Physics, to Solving Biology's Most Important Fundamental Problems" from Vanderbilt University in 2003. She gave the Fritz Lipmann Lecture at the American Society for Biochemistry and Molecular Biology (ASBMB) in 2001. She is president-elect of the ASBMB; she previously served as the organization's secretary (1995-1998) and program chair (1998). She has served on the editorial boards of the *Journal of Biological Chemistry*, *Biochemistry*, and *Investigative Ophthalmology and Visual Science*. She is a member of the editorial boards of *Molecular Pharmacology* and the *American Journal of Physiology—Lung Cellular and Molecular Physiology*. She was a member of the scientific advisory board of Medichem Life Sciences in 2000-2002. She is a founder and member of the scientific advisory board of Cue BIOtech.

WILLIAM HAPPER [NAS] is a professor in the Department of Physics at Princeton University. He is a specialist in modern optics, optical and radio-frequency spectroscopy of atoms and molecules, and spin-polarized atoms and nuclei. He received a BS in physics from the University of North Carolina in 1960 and a PhD in physics from Princeton University in 1964. Dr. Happer began his academic career in 1964 at Columbia University as a member of the research and teaching staff of the Physics Department. While serving as a professor of physics, he also served as codirector of the Columbia Radiation Laboratory from 1971 to 1976 and director from 1976 to 1979. In 1980, he joined the faculty at Princeton University. He was named the Class of 1909 Professor of Physics in 1988. In 1991, he was appointed director of energy research in DOE by President Bush. While serving in that capacity under Secretary of Energy James Watkins, he oversaw a basic research budget of some $3 billion, which included much of the federal funding for high-energy and nuclear physics, materials science, magnetic confinement fusion, environmental science, biology, the Human Genome Project, and other work. He remained at DOE until 1993 to help during the transition to the Clinton administration. He was reappointed professor of physics at Princeton University in 1993 and named Eugene Higgens Professor of Physics and chair of the University Research Board in 1995. Dr. Happer has maintained an interest in applied, as well as basic, science and has served as a consultant to numerous firms, charitable foundations, and government agencies. From 1987 to 1990, he served as chairman of the Steering Committee of JASON, a group of scientists and engineers who advise agencies of the federal government on defense, intelligence, energy policy, and other technical matters. He is a trustee of the MITRE Corporation and the Richard Lounsbery Foundation and a cofounder in 1994 of Magnetic Imaging Technologies Incorporated (MITI), a small company specializing in the use of laser polarized noble gases for magnetic resonance imaging. MITI was purchased by Nycomed Amersham in 1999. Dr. Happer is a fellow of the American Physical Society and the American Association for the Advancement of Science, and a member of the American Academy of Arts and Sciences, the National Academy of Sciences, and the American Philosophical Society. He was awarded an Alfred P. Sloan Fellowship in 1966, an Alexander von Humboldt Award in 1976, the 1997 Broida Prize, the 1999 Davisson-Germer Prize of the American Physical Society, and the Thomas Alva Edison Patent Award in 2000.

DANIEL HASTINGS is professor of aeronautics and astronautics and engineering systems at the Massachusetts Institute of Technology (MIT). He joined the MIT faculty as an assistant professor in 1985, advancing to associate professor in 1988 and full professor in 1993. He earned a PhD and an SM from MIT in aeronautics and astronautics in 1980 and 1978, respectively, and received a BA in mathematics from Oxford University,

England, in 1976. Dr. Hastings served as chief scientist to the US Air Force from 1997 to 1999. In that role, he served as chief scientific adviser to the chief of staff and the secretary and provided assessments on a wide array of scientific and technical issues affecting the Air Force mission. He led several influential studies on where the Air Force should invest in space, global energy projection, and options for a science and technology workforce for the 21st century. Dr. Hastings' recent research has concentrated on space systems and space policy and on issues related to spacecraft-environment interactions, space propulsion, space-systems engineering, and space policy; and he has published many papers and a book on those subjects. He has led several national studies on government investment in space technology. Dr. Hastings is a fellow of the American Institute of Aeronautics and Astronautics and a member of the International Academy of Astronautics. He is a member of the National Science Board and of the Applied Physics Laboratory Science and Technology Advisory Panel, and the chair of Air Force Scientific Advisory Board. He is a member of the MIT Lincoln Laboratory Advisory Committee and is on the Board of Trustees of the Aerospace Corporation. He has served on several national committees on issues in national security space.

ROBERT HERMANN is a senior partner of Global Technology Partners, LLC, which specializes in investments in technology, defense, aerospace, and related businesses worldwide. In 1998, Hermann retired from United Technologies Corporation (UTC), where he held the position of senior vice president for science and technology. In that role, he was responsible for ensuring the development of the company's technical resources and the full exploitation of science and technology by the corporation. He was also responsible for the United Technologies Research Center. Hermann joined the company in 1982 as vice president for systems technology in the electronics sector and later served in a series of assignments in the defense and space systems groups before being named vice president for science and technology. Before joining UTC, he served for 20 years with the National Security Agency with assignments in research and development, operations, and the North Atlantic Treaty Organization. In 1977, he was appointed principal deputy assistant secretary of defense for communications, command, control, and intelligence. In 1979, he was named assistant secretary of the Air Force for research, development, and logistics and in parallel was director of the National Reconnaissance Office. He received his BS, MS, and PhD in electrical engineering from Iowa State University.

KENT H. HUGHES is the director of the Woodrow Wilson International Center for Scholar's Program on Science, Technology, America, and the Global Economy. He served as US associate deputy secretary of commerce from 1993 to 1999. He was also president of the Council on Competitiveness, senior economist of the Congressional Joint Economic Committee, and chief economist to Senate Majority Leader Robert C. Byrd. He is the author of *Building the Next American Century: The Past and Future of American Economic Competitiveness*. He holds a PhD in economics from Washington University in St. Louis, an LLB from Harvard Law School, and a BA from Yale University.

MARK S. HUMAYUN is professor of ophthalmology, biomedical engineering, and cell and neurobiology at the University of Southern California (USC). He received his BS from Georgetown University in 1984, his MD from Duke University in 1989, and his PhD from the University of North Carolina at Chapel Hill in 1994. He finished his training by completing an ophthalmology residency at Duke and a fellowship in vitreoretinal diseases at Johns Hopkins Hospital. He stayed on as a faculty member at Johns Hopkins and rose to the rank of associate professor before moving to USC in 2001. Humayun is the director of USC's National Science Foundation Biomimetic MicroElectronics Systems Engineering Research Center. He is also the codeveloper of a retinal implant that has received wide attention for its potential to restore sight and is the director of the DOE Artificial Retina Project that is a consortium of five DOE laboratories, four universities, and industry. Dr. Humayun's research projects focus on the most challenging eye diseases: retinal degeneration, including macular degeneration, and retinitis pigmentosa. He is a member of 11 academic organizations, including Institute of Electrical and Electronics Engineers-Engineering in Medicine and Biology Society, the Biomedical Engineering Society, the Association for Research in Vision and Ophthalmology, the American Society of Retinal Specialists, the Retina Society, the American Ophthalmological Society, and the American Academy of Ophthalmology. In the last 5 years, as a principal investigator, he has held multiple research grants from the National Science Foundation, DOE, and Second Sight, and oversight on three grants totalling $20 million in funding. He also holds three patents in the retinal prosthesis artificial-vision field. Humayun has written more than 70 peer-reviewed papers and more than 19 chapters. He has been a guest speaker in 90 lectures around the world.

MADELEINE JACOBS has been executive director and chief executive officer of the American Chemical Society (ACS) since January 2004. Before then, she served for 8½ years as editor-in-chief of *Chemical & Engineering News* magazine, the weekly news magazine of the chemical world published by ACS, and 2 years as managing editor. She has held other senior management positions in a wide variety of scientific and educational organizations, including the National Institutes of Health, the National Institute of Standards and Technology, and the Smithsonian Institution, where she served as the director of public affairs. Her professional interests include trends in the chemical industry, the public image of chemistry, employment trends, minority-group representation, and equality of the sexes in science.

RICHARD JOHNSON is a senior partner in the Washington, DC, office of Arnold & Porter, LLP. He specializes in legal, regulatory, and public-policy issues related to fundamental research, technology, innovation, and innovative strategic relationships, especially with respect to biotechnology and life sciences, nanotechnology, and other emerging technologies; intellectual property, trade, and innovation matters; and research-university and independent-research institute legal and policy issues. He formerly served as general counsel for international trade at the US Department of Commerce, where he was responsible for both trade-policy and international-technology issues. Dr. Johnson has served as a US delegate to numerous international trade, health-innovation, and international-technology meetings, and he has testified before the US Congress and international organizations. In addition to receiving his JD from the Yale Law School, where he was editor of the *Yale Law Journal*, he received his MS from MIT where he was a National Science Foundation national fellow. He is a member of the MIT Corporation's Visiting Committee and several other university and think-tank advisory boards. Dr. Johnson serves as chairman of the Organisation for Economic Co-operation Development/Business and Industry Advisory Committee Biotechnology Committee, vice chairman of the OECD Technology and Innovation Committee, and cochair of its health innovation and nanotechnology task forces, and he participates on a wide range of advisory committees and task forces related to health innovation, intellectual-property and innovation policy, science and security, and the globalization of research.

RANDY H. KATZ [NAE] is the United Microelectronics Corporation Distinguished Professor in Electrical Engineering and Computer Science at the University of California, Berkeley (UCB). He received his undergraduate degree from Cornell University and his MS and PhD from UCB. He joined the faculty at UCB in 1983. He is a fellow of the Association for Computing Machinery (ACM) and the Institute of Electrical and Electronics Engineers (IEEE), and a member of the National Academy of Engineering and the American Academy of Arts and Sciences. He has published over 230 refereed technical papers, book chapters, and books. His hardware-design textbook, *Contemporary Logic Design*, has sold over 85,000 copies worldwide and has been in use at over 200 colleges and universities. A second edition, cowritten with Gaetano Borriello, published in 2005. He has supervised 41 MS theses and 27 PhD dissertations, and he leads a research team of over a dozen graduate students, technical staff, and industrial and academic visitors. He has won numerous awards, including 12 best paper awards, one "test of time" paper award, one paper selected for a 50-year retrospective on IEEE communications publications, three best-presentation awards, the Outstanding Alumni Award of the Berkeley Computer Science Division, the Computing Research Association Outstanding Service Award, the Berkeley Distinguished Teaching Award, the Air Force Exceptional Civilian Service Decoration, the IEEE Reynolds Johnson Information Storage Award, the American Society for Engineering Education Frederic E. Terman Award, and the ACM Karl V. Karlstrom Outstanding Educator Award. With colleagues at Berkeley, he developed Redundant Arrays of Inexpensive Disks (RAID), which is now a $25-billion-per-year industry sector. While on leave for government service in 1993-1994, he established whitehouse.gov and connected the White House to the Internet. His current research interests are in reliable, adaptive distributed systems supported by new services deployed on network appliances (also known as programmable network elements). Prior research interests have included database management, VLSI Computer Aided Design, high-performance multiprocessor and storage architectures, transport and mobility protocols spanning heterogeneous wireless networks, and Internet service architectures for converged data and telephony.

MARVIN KOSTERS is a resident scholar at the American Enterprise Institute (AEI) and editor of the AEI *Evaluative Studies* series. He served as a senior economist on the President's Council of Economic Advisers and at the White House Office of the Assistant to the President for Economic Affairs. Mr. Kosters held a senior policy position at the US Cost of Living Council and a research position at the RAND Corporation. He is the author of *Wage Levels and Inequality* (1998). He edited *The Effects of the Minimum Wage on Employment* (1996), *Personal Saving, Consumption, and Tax Policy* (1992), and *Workers and Their Wages* (1991). He was also the coeditor of *Trade and Wages: Leveling Wages Down?* (1994) and of *Reforming Regulation* (1980). Mr. Kosters has contributed to the *American Economic Review* and *Public Interest.* He is coauthor of *Closing the Education Achievement Gap: Is Title I Working?*, published by AEI Press (2003).

GEORGE M. LANGFORD is the E. E. Just Professor of Natural Sciences and professor of biological sciences at Dartmouth College. He is also an adjunct professor of physiology at the Dartmouth Medical School. Dr. Langford received his PhD from the Illinois Institute of Technology in Chicago and completed postdoctoral training at the University of Pennsylvania. He was professor of physiology in the School of Medicine of the University of North Carolina at Chapel Hill before joining the faculty at Dartmouth College. Dr. Langford is a cell biologist and neuroscientist who studies cellular mechanisms of learning and memory. His research program will help to understand how the brain remembers and what makes it forget when neurodegenerative diseases, such as Alzheimer's, take hold. He served on the National Science Board (NSB), the governing board of the National Science Foundation from 1998 to 2004, was chair of the NSB Education and Human Resources Committee from 2002 to 2004, and was vice-chair of the NSB National Workforce Taskforce Subcommittee from 1999 to 2004. He serves on the National Nanotechnology Infrastructure Network, the Burroughs Wellcome Fund Career Awards in the Biomedical Sciences Advisory Committee, the National Institutes of Health Synapses, Cytoskeleton and Trafficking Study Section, the National Research Council Associateships Program Committee, and the Sherman Fairchild Foundation Scientific Advisory Board.

CATO T. LAURENCIN [IOM] is the Lillian T. Pratt Distinguished Professor and chair of the Department of Orthopaedic Surgery at the University of Virginia. He is also a university professor at the University of Virginia, and holds professorships in biomedical engineering and chemical engineering. Dr. Laurencin earned his BSE in chemical engineering from Princeton University and his MD from Harvard Medical School, where he earned the Robinson Award for Excellence in Surgery. Simultaneously, he earned a PhD in biochemical engineering/biotechnology from MIT, where he was a Hugh Hampton Young Scholar. After completing his doctoral programs, Dr. Laurencin continued clinical training at the Harvard University Orthopaedic Surgery Program and ultimately became chief resident in orthopaedic surgery at the Beth Israel Hospital, Harvard Medical School. Simultaneously, he was an instructor in the Harvard–MIT Division of Health Sciences and Technology, where he directed a biomaterials laboratory at MIT. Dr. Laurencin later completed a clinical fellowship in sports medicine and shoulder surgery at the Hospital for Special Surgery in New York, working with the team physicians for the New York Mets, and at St. John's University in New York. Board-certified in orthopaedic surgery, Laurencin is a fellow of the American College of Surgeons, a fellow of the American Academy of Orthopaedic Surgeons, fellow of the American Institute for Medical and Biological Engineering, and an International Fellow in Biomaterials Science and Engineering. Dr. Laurencin's research interests are in biomaterials, tissue engineering, drug delivery, and nanotechnology. He received the Presidential Faculty Fellowship Award from President Clinton in recognition of his research involving biodegradable polymers. He most recently received the William Grimes Award for Excellence in Chemical Engineering from the American Institute of Chemical Engineers and the Leadership in Technology Award from the New Millennium Foundation. He is a member of the Institute of Medicine.

DAVID LaVAN is assistant professor of mechanical engineering at Yale University, where he teaches machine design at the freshman and senior levels. His approach is derived from a background in materials science and mechanical engineering and experience as a consulting engineer. He incorporates failure analysis, product liability, codes and standards, and forensic engineering in his design classes. He also introduces students to the latest generation of analysis and simulation software. His research focuses on materials and devices at the nano, micro, and macro scales. Of particular interest is the development of biologic applications of microsystems. His laboratory is working on the development of in vivo sensors and novel materials and devices for microelectromechnical systems. Some projects are long-term implantable sensors for cancer detection and monitoring, injectable sensors, and the micromachining of biopolymers for applications in tissue engineering and neuroscience. In addition to new devices, his laboratory is developing novel methods to characterize materials and devices at the microscale.

PHILIP LeDUC is a McGowan faculty member and an assistant professor in mechanical engineering at Carnegie Mellon University. Dr. LeDuc earned his BS from Vanderbilt University in 1993 and his MS from North Carolina State in 1995. He obtained his PhD at Johns Hopkins University and was a postdoctoral fellow at Children's Hospital/Harvard Medical School in 1999. Using computational biology through collaboration with colleagues at the University of Pittsburgh Medical Center, Dr. LeDuc anticipates "developing a computational framework to look at how cells and molecules interact, for the purpose of improving drugs for disease treatment." His research focuses on linking mechanics to biochemistry by exploring the science of molecular to cellular biomechanics through nanotechnology and microtechnology, control theory, and computational biology. The link between mechanics and biochemistry has been implicated in myriad scientific and medical problems, from orthopaedics and cardiovascular medicine to cell motility and division to signal transduction and gene expression. Most of the studies have focused on organ-level issues, but cellular and molecular research has become essential over the last decade in this field because of the revolutionary developments in genetics, molecular biology, microelectronics, and biotechnology.

JAMES A. LEWIS is a senior fellow and director of the Center for Strategic and International Studies (CSIS) Technology and Public Policy Program. Before joining CSIS, he was a career diplomat who worked on a variety of national security issues during his federal service. Dr. Lewis's extensive diplomatic and regulatory experience includes negotiations on military basing in Southeast Asia, the Cambodia peace process, the five power talks on arms transfer restraint, the Wassenaar Arrangement, and several bilateral agreements on security and technology. Dr. Lewis was the head of the delegation of the Wassenaar Experts Group for advanced civil and military technologies and a political adviser to the US Southern Command (for Just Cause), to US Central Command (for Desert Shield), and to the US Central America Task Force. He was responsible for the 1993 redrafting of the International Traffic in Arms Regulations, the 1997 regulations implementing the Wassenaar Agreement, numerous regulations on high-performance computing and satellites, and the 1999 and 2000 regulations liberalizing US controls on encryption products. Since going to CSIS, he has written numerous publications, including *China as a Military Space Competitor* (2004), *Globalization and National Security* (2004), *Spectrum Management for the 21st Century* (2003), *Perils and Prospects for Internet Self-Regulation* (2002), *Assessing the Risk of Cyber Terrorism, Cyber War, and Other Cyber Threats* (2002), *Strengthening Law Enforcement Capabilities for Counterterrorism* (2001), and *Preserving America's Strength in Satellite Technology* (2001). His current research involves digital identity, innovation, military space, and China's information-technology industry. In 2004, Dr. Lewis was elected the first chairman of the Electronic Authentication Partnership, an association of companies, nonprofits, and government organizations that develops rules for federated authentication. He received his PhD from the University of Chicago in 1984.

JOAN F. LORDEN joined the University of North Carolina (UNC)-Charlotte as provost and vice chancellor for academic affairs in August 2003. She received a BA and a PhD in psychology from Yale University. Before coming to UNC-Charlotte, she served as associate provost for research and dean of the Graduate School at the University of Alabama at Birmingham (UAB), where she was professor of psychology. She has published extensively on brain-behavior relationships and specialized in the study of animal models of human neurologic disease. In 1991, she was awarded the Ireland Prize for Scholarly Distinction. She has served on peer-review panels and scientific advisory boards at NIH, NSF, and private agencies. At UAB, she organized the doctoral program in behavioral neuroscience and directed the universitywide interdisciplinary Graduate Training Program in Neuroscience. In addition to her work in research and graduate education at UAB, Dr. Lorden founded an Office of Postdoctoral Education, programs for professional development of graduate students, an undergraduate honors program, and several programs designed to improve the recruitment of women and minority-group members into doctoral programs in science and engineering. Dr. Lorden was elected chair of the Board of Directors of the Council of Graduate Schools (2003) and during 2002-2003 was the dean in residence in the Division of Graduate Education at

NSF. She has chaired the Board of Directors of Oak Ridge Associated Universities, was a trustee of the Southeastern Universities Research Association, and chaired the executive committee of the National Association of State Universities and Land-Grant Colleges Council on Research Policy and Graduate Education. Dr. Lorden is a member of the National Research Council's Committee on the Methodology for the Study of the Research Doctorate. She is a member of the Society for Neuroscience, the American Psychological Association, and the American Psychological Society.

RONALD MARX is professor of educational psychology and dean of education at the University of Arizona. His previous appointments were at Simon Fraser University and the University of Michigan, where he served as the chair of the Educational Studies Program and later as the codirector of the Center for Highly Interactive Computing in Education and the Center for Learning Technologies in Urban Schools. His research focuses on how classrooms can be sites for learning that is highly motivated and cognitively engaging. Since 1994, Dr. Marx has been engaged in large-scale urban school reform in Detroit and Chicago. With his appointment as dean in 2003, he has been working to link the college's research, teaching, and outreach activities closely to K–12 schools and school districts. Dr. Marx received his PhD from Stanford University.

DEIRDRE R. MELDRUM is professor and director of the Genomation Laboratory in the Department of Electrical Engineering and adjunct professor of bioengineering and mechanical engineering at the University of Washington. She received a BS in civil engineering from the University of Washington in 1983, an MS in electrical engineering from Rensselaer Polytechnic Institute in 1985, and a PhD in electrical engineering from Stanford University in 1993. As an engineering cooperative student at the National Aeronautics and Space Administration Johnson Space Center in 1980 and 1981, she was an instructor for the astronauts on the shuttle-mission simulator. From 1985 to 1987, she was a member of the technical staff at the Jet Propulsion Laboratory and performed theoretical and experimental work in identification and control of large flexible space structures and robotics. Her research interests include genome automation, microscale systems for biologic applications, robotics, and control systems. Dr. Meldrum is a member of the American Association for the Advancement of Science (AAAS), the American Chemical Society, the Association for Women in Science, the Human Genome Organization, Sigma Xi, and the Society of Women Engineers. She was awarded an NIH Special Emphasis Research Career Award in 1993 to train in biology and genetics, bring her engineering expertise to the genome project, and develop automated laboratory instrumentation. In December 1996, she was the recipient of a Presidential Early Career Award for Scientists and Engineers for recognition of innovative research using a broad set of interdisciplinary approaches to advance DNA-sequencing technology. Since August 2001, she has directed an NIH center of excellence in genomic sciences, the Microscale Life Sciences Center (MLSC). The MLSC includes 10 investigators from the University of Washington and one from the Fred Hutchinson Cancer Research Center. In 2003, Meldrum became a fellow of the AAAS; and in 2004, a fellow of the Institute of Electrical and Electronic Engineers.

CLAUDIA MITCHELL-KERNAN has been vice chancellor for graduate studies and dean of the Graduate Division at the University of California, Los Angeles (UCLA), since 1989. As chief academic and administrative officer of the Graduate Division, she has responsibility for graduate admissions, campuswide student support and fellowship programs, and graduate academic affairs and works to ensure that standards of excellence, fairness, and equity are maintained across all graduate programs. She is concurrently a professor in the Departments of Anthropology and Psychiatry and Biobehavioral Sciences. She received her PhD from the University of California, Berkeley, and her BA and MA from Indiana University and was a member of the faculty at Harvard University before coming to UCLA in 1973. Much of Dr. Mitchell-Kernan's early work was in linguistic anthropology, and her classic sociolinguistic studies of black communities continue to be widely cited. Her most recent book, *The Decline in Marriage Among African Americans*, coedited with M. Belinda Tucker, was published in 1995 by Russell Sage. Other books on children's discourse, television and the socialization of ethnic-minority children, and linguistic patterns of black children reflect the breadth of her scholarly interests. She conducts research on marriage and family-formation patterns in the United States among Americans and West Indian immigrants. Throughout her career, she has maintained an active record of service to federal agencies that sponsor research. President Clinton appointed her to the NSB for a 6-year term in 1994. At the national level, she is serving as the dean in residence for the Council of Graduate Schools (CGS), is on the Board of Higher Education and Workforce of the National Research Council, and is on the board of directors of the Consortium of Social Science Associations. She has recently served on the board of directors of the CGS and chaired its Advisory Committee on Minorities in Graduate Education, as chair of the board of directors of the Graduate Record Examination, on the advisory board of the National Security Education Program, and on the Board of Deans of the African American Institute. She has been a member of the Board of Directors of the Los Angeles-based Golden State Minority Foundation and the board of directors of the Venice Family Clinic.

DAVID H. MONK is professor of educational administration and dean of the College of Education at the Pennsylvania State University (PSU). He earned his AB in 1972 at Dartmouth College and his PhD in 1979 at the University of Chicago, and he was a member of the Cornell University faculty for 20 years before becoming dean at PSU in 1999. He has also been a third-grade teacher and has taught in a visiting capacity at the University of Rochester and the University of Burgundy in Dijon, France. Dr. Monk is the author of *Educational Finance: An Economic Approach* (1990), *Raising Money for Education: A Guide to the Property Tax* (1997) (with Brian O. Brent), and *Cost Adjustments in Education* (2001) (with William J. Fowler, Jr.), in addition to numerous articles in scholarly journals. He is a coeditor of *Education Finance and Policy,* the journal of the American Education Finance Association, and *Leadership and Policy in Schools.* He also serves on the editorial boards of *Economics of Education Review*, the *Journal of Education Finance*, *Educational Policy*, and the *Journal of Research in Rural Education.* He consults widely on matters related to educational productivity and the organizational structuring of schools and school districts and is a past president of the American Education Finance Association.

MARK B. MYERS is visiting executive professor in the Management Department at the Wharton School of the University of Pennsylvania. His research interests include identifying emerging markets and technologies to enable growth in new and existing companies with emphases on technology identification and selection, product development and technology competences. Dr. Myers serves on the Science, Technology and Economic Policy Board of the National Research Council and cochairs, with Yale President Richard Levin, the National Research Council's study of Intellectual Property in the Knowledge-Based Economy. Dr. Myers retired from the Xerox Corporation at the beginning of 2000, after a 36-year career in its R&D organizations. He was the senior vice president in charge of corporate research, advanced development, systems architecture, and corporate engineering from 1992 to 2000. During this period he was a member of the senior management committee in charge of the strategic direction setting of the company. His responsibilities included the corporate research centers: PARC in Palo Alto, California; the Webster Center for Research and Technology near Rochester, New York; the Xerox Research Centre of Canada, Mississauga, Ontario; and the Xerox Research Centre of Europe in Cambridge, England, and Grenoble, France. Dr. Myers is chairman of the Board of Trustees of Earlham College and has held visiting faculty positions at the University of Rochester and at Stanford University. He holds a bachelor's degree from Earlham College and a doctorate from Pennsylvania State University.

CARLO PARRAVANO has served as executive director of the Merck Institute for Science Education since 1992. He is responsible for the planning, development, and implementation of numerous initiatives to improve science education. Before assuming that position, Dr. Parravano was professor of chemistry and chair of the Division of Natural Sciences at the State University of New York (SUNY) at Purchase. While at SUNY/Purchase, he taught courses in general, physical, analytic, and environmental chemistry. In addition to his academic and administrative appointments, he served as director of the Center for Mathematics and Science Education of the SUNY/Purchase-Westchester School Partnership. Dr. Parravano is a recipient of the SUNY Chancellor's Award for Excellence in Teaching. In 1999, he was elected an AAAS fellow; and in 2003, he received the National Science Teachers Association's (NSTA's) Distinguished Service to Science Education Award. In 2004, he was designated a national associate of the National Academy of Sciences and appointed to the Steering Committee for the 2009 National Assessment of Educational Progress in Science. Dr. Parravano earned a BA in chemistry at Oberlin College and a PhD in physical chemistry in 1974 at the University of California, Santa Cruz. His research has been in molecular-beam studies of excited atoms and molecules and the application of physical-chemical techniques to the solution of biochemical and environmental problems. Dr. Parravano is a member of a number of professional organizations, including AAAS (chair, Education Section, 2003), the American Chemical Society, and NSTA. He served as founding vice chair of the New Jersey Professional Teaching Standards Board (1999-2003) and as cochair of the New Jersey Science Curriculum Standards Group. He is a member of the National Research Council's Board on Science Education (Executive Committee) and is on the advisory boards of the National Science Resources Center, Biological Sciences Curriculum Study (chair), and the New Jersey Business Coalition for Educational Excellence. In 2005, Dr. Parravano was appointed to the New Jersey Mathematics Task Force and to the Quality Teaching and Learning Task Force. He also serves as principal investigator for an NSF-funded mathematics-science partnership award.

ANNE C. PETERSEN [IOM] is the senior vice president for programs at the W. K. Kellogg Foundation of Battle Creek, Michigan. As a senior member of the executive staff since 1996, she provides leadership for all programming, including the development of effective programming strategies, teamwork, policies, philosophies, and organizationwide systems to accomplish the programmatic mission of the foundation. Previously, Dr. Petersen was deputy director and chief operating officer of NSF, then a $3.6 billion federal research agency with 1,300 employees. Before joining NSF, she served as vice president for research and dean of the Graduate School at the University of Minnesota where she was professor of adolescent development and pediatrics. Before that, she was the first dean of the College of Health and Human Development at Pennsylvania State University. She has written more than a dozen books and 200 articles on adolescent and sex issues, including evaluation, health, adolescent development, and higher education. Her honors include election to the Institute of Medicine. She is a founding member of the Society for Research on Adolescence and was president and council member. She was president of developmental psychology in the American Psychological Association and is a fellow of the American Association for the Advancement of Science, the American Psychological Association, and the American Psychological Society. She is president-elect of the International Society for the Study of Behavioral Development. Dr. Petersen holds a BS in mathematics, an MS in statistics, and a PhD in measurement, evaluation, and statistical analysis from the University of Chicago.

STEPHANIE PFIRMAN chairs the Department of Environmental Science at Barnard College. Her current research interests include environmental aspects of sea ice in the Arctic, interdisciplinary research and education, and advancing women scientists. As the first chair of NSF's Advisory Committee for Environmental Research and Education, Dr. Pfirman oversaw analysis of a 10-year outlook for environmental research and education at NSF. She is also a co-principal investigator of NSF's ADVANCE grant (to advance women scientists) to Columbia's Earth Institute. Before joining Barnard, Dr. Pfirman was a senior scientist at Environmental Defense and codeveloper of the award-winning traveling exhibition, "Global Warming: Understanding the Forecast," developed jointly with the American Museum of Natural History. She was research scientist and coordinator of Arctic programs for the University of Kiel and GEOMAR, Research Center for Marine Geoscience in Germany; staff scientist for the US House of Representatives Committee on Science Subcommittee on Environment; and oceanographer with the US Geological Survey in Woods Hole, Massachusetts. Dr. Pfirman received her PhD from the Massachusetts Institute of Technology/Woods Hole Oceanographic Institution Joint Program in Oceanography and Oceanographic Engineering, Department of Marine Geology and Geophysics, and a BA from Colgate University's Geology Department.

DANIEL B. PONEMAN is a principal of The Scowcroft Group, which provides strategic advice to the group clients in the energy, aerospace, information-technology, and manufacturing industries, and others. For 9 years, he practiced law in Washington, DC, assisting clients in a wide variety of regulatory and policy matters, including export controls, trade policy, and sanctions issues. From 1993 through 1996, Dr. Poneman served as special assistant to the president and senior director for nonproliferation and export controls at the National Security Council (NSC), with responsibilities for the development and implementation of US policy in such fields as peaceful nuclear cooperation, missile-technology and space-launch activities, sanctions determinations, chemical and biologic arms-control efforts, and conventional-arms transfer policy. During that period, he participated in negotiations and consultations with governments in Africa, Asia, Europe, Latin America, and the former Soviet Union. Dr. Poneman joined the NSC staff in 1990 as director of defense policy and arms control after service in the Department of Energy. He has served as a member of the Commission to Assess the Organization of the Federal Government to Combat the Proliferation of Weapons of Mass Destruction and other federal advisory panels. He received AB and JD degrees from Harvard University and an MLitt degree in politics from Oxford University. Dr. Poneman is the author of books on nuclear-energy policy, Korea, and Argentina and is a member of the Council of Foreign Relations.

HELEN R. QUINN started her college career at the University of Melbourne, Australia. Two years into her degree, she moved to the United States and joined the physics department of Stanford University, where she completed both her BSc and her PhD in physics. After a postdoctoral fellowship at Deutsche Elektronen-Synchrotron in Hamburg, Germany, she briefly taught high school physics and then joined the staff and then the faculty of Harvard University. A few years later, she returned to Stanford to join the Stanford Linear Accelerator Center, and she has been there since 1977. Her research concentrates on theoretical particle physics with a focus on phenomenology of the weak interactions; she is involved in outreach activities to encourage interest in physics. Her work with Robert Peccei resulted in what is now known as the Peccei-Quinn symmetry. Dr. Quinn was president of the American Physical Society for 2003. She was named a fellow of the American Academy of Arts and Sciences in 1996 and was elected to the National Academy of Sciences in 2003. She was awarded the Dirac Medal of the International Centre for Theoretical Physics in 2000 for her work with Peccei and in the Georgi-Quinn-Weinberg computation of how different types of interactions may be unified. In addition to her research Dr. Quinn has maintained a steady involvement in precollege education, working chiefly with local efforts to improve science teaching. She was a coauthor of the Investigation and Experimentation strand of the California science standards.

PAUL ROMER is the STANCO 25 Professor of Economics in the Graduate School of Business at Stanford University and a senior fellow of the Hoover Institution. Dr. Romer was the lead developer of "new growth theory." This body of work, which grew out of his 1983 PhD dissertation, provides a better foundation for business and government thinking about the dynamics of wealth creation. It addresses one of the oldest questions in economics: What sustains economic growth in a physical world characterized by diminishing returns and scarcity? It also sheds new light on current economic issues. Among these, Dr. Romer is studying how government policy affects innovation and how faster technologic change might influence asset prices. Dr. Romer was named one of America's 25 most influential people by *Time* magazine in 1997. He was elected a fellow of the American Academy of Arts and Sciences in 2000. He is also a fellow of the Econometric Society and a research associate with the National Bureau of Economic Research (NBER). He was a member of the National Research Council Panel on Criteria for Federal Support of Research and Development (1995), a member of the Executive Council of the American Economics Association, and a fellow of the Center for Advanced Study in the Behavioral Sciences. Before coming to Stanford, Dr. Romer was a professor of economics at the University of California, Berkeley, and the University of Chicago. Dr. Romer holds a PhD in economics from the University of Chicago.

SHEILA R. RONIS is president of The University Group, Inc., a management consulting firm and think tank specializing in strategic management, visioning, national security, and public policy. She is also an adjunct professor at the University of Detroit Mercy and at Oakland University, where she teaches "Strategic Management and Business Policy," "Managing the Global Firm," and "Issues of Globalization" in the MBA programs. She often lectures at the Industrial College of the Armed Forces (ICAF) at the National Defense University in Washington, DC, and participates in its annual National Security Strategy Exercise. In June 2005, she chaired at ICAF the Army's Eisenhower National Security Series event "The State of the U.S. Industrial Base: National Security Implications in a World of Globalization." Her BS is in physics and mathematics and her MA and PhD from Ohio State University are in organizational behavior and general social systems theory.

JAMES M. ROSSER has served as president and professor of healthcare management at California State University, Los Angeles, since 1979 and as professor of microbiology since 2004. He has served in many civic and community organizations, including the Los Angeles Area Council of the Boy Scouts of America, the Los Angeles County Alliance for College Ready

Public Schools, the California Chamber of Commerce, Americans for the Arts, Community Television of Southern California (KCET), Los Angeles After-School Education and Child Care Program—LA's BEST, the Music Center Performing Arts Council/Education Council, and the California Community Foundation. His professional affiliations have included the American Association of State Colleges and Universities, the American Council on Education, the Western Association of Schools and Colleges, the Woodrow Wilson National Fellowship Foundation, the California Council on Science and Technology, Edison International, the United California Bank, the FEDCO, Inc. Foundation, and numerous committees and commissions of the California State University system. He is a past chair of the Education and Human Resources Advisory Committee of the National Science Foundation. He was chair of the National Academy of Engineering Forum on Diversity in the Engineering Workforce in 2000-2002.

DEBORAH M. ROUDEBUSH has been a physics teacher for 21 years. She holds national board certification in adolescent and young adult science. She was a 2001 Presidential Awardee for Excellence in Science Teaching. She has been a physics-teacher resource agent through the American Association of Physics Teachers since 1992 and is the associate member for Virginia to the National Academy of Sciences Teacher Advisory Council. She has been a reader for advanced placement for computer science and physics since 1996. She has a keen interest in physics education research and the implications for improving physics teaching at all levels. She is an advocate for the importance of physics and science education for all students to enable data-driven decision-making at all levels of government.

DANIEL K. RUBENSTEIN is currently the head of the Mathematics Department at Collegiate School in New York City. He has worked in secondary education for 13 years. His first faculty position was teaching mathematics at Sidwell Friends School in Washington, DC. In addition, he spent a semester as assistant director and mathematics teacher at School Year Abroad Beijing. After 8 years of independent-school teaching, a Sidwell alumnus recruited Mr. Rubenstein to help build the mathematics program of the fledgling SEED Foundation Public Charter School in southeast Washington, DC, where he remained for 2 years. He is a nationally board-certified mathematics teacher and an associate member of the National Academy of Sciences Teacher Advisory Council. In 2002, he received the Presidential Award for Excellence in Mathematics Teaching. He holds a bachelor's degree in mathematics from Hamilton College and a master's degree from St. Johns College in Santa Fe, New Mexico, and he is enrolled in a doctoral program at Columbia University in education leadership.

JULIANA C. SHEI joined the General Electric Global Research Center in 1991. In 1995, she was appointed global technology manager and is responsible for the management of the R&D Center's Global Technology Acquisition Programs. In that role, she has established research collaborations with organizations around the world. Ms. Shei was the project manager to establish a GE Research Center in Shanghai, China, in June 2000 and now leads Japan Technology Initiative in Japan. Ms. Shei is a member of the American Chemical Society and cochair of the Industrial Research Institute External Technology Directors' Network. She is a board member for the United States Industry Coalition. She was a member of the Gore-Chernomyrdin Science & Technology delegation in 1997 and served as an industry representative for the President's Council of Advisers on Science and Technology in 2002. Shei is very active in community service. She was a founder and the president of the Network, a professional women's organization affiliated with the National Association for Female Executives, served as the board chair for the Chinese Community Center of the Capital District of New York, and is a board member of the Japanese Cultural Association of the Capital District. A native of Tokyo, Japan, Ms. Shei obtained her undergraduate degree from National Cheng Kung University in Taiwan, her MS from the University of Massachusetts, and her MBA from Rensselaer Polytechnic Institute. Before joining General Electric, she worked at Ames Laboratory, the Research Center at the US Steel Corporation, and the Sterling Winthrop Research Institute (Eastman Kodak's Pharmaceutical Division).

J. STEPHEN SIMON is a senior vice president of Exxon Mobil Corporation. Mr. Simon holds a BS degree in civil engineering from Duke University and an MBA from Northwestern University. He joined Exxon Company, USA in July 1967 and shortly thereafter began a 2-year assignment in the US Army. He returned to Exxon USA in July 1969 as a business analyst in the Baton Rouge refinery. After holding a variety of supervisory and managerial positions throughout the Baton Rouge and Baytown refineries and in Exxon USA's refining and controller's departments, Mr. Simon became executive assistant to Exxon USA's executive vice president in Houston. In 1980, he returned to the Baton Rouge refinery as Operations Division manager and then became refinery manager. In 1983, Mr. Simon moved to New York, where he was executive assistant to the president of Exxon corporation. In 1984, he moved to London, England, as supply manager in the Petroleum Products Department of Esso Europe Inc. and then supply and transportation manager. Mr. Simon returned to Houston in 1986 as general manager of Exxon USA's Supply Department. In 1988, he became chief executive and general manager, Esso Caribbean and Central America, in Coral Gables, Florida. Simon moved to Italy in 1992 to become executive vice president and then president of Esso Italiana. He returned to the

United States in 1997 and was named an executive vice president of Exxon Company, International, headquartered in Florham Park, New Jersey. In December 1999, he was appointed president of Exxon Mobil Refining & Supply Company and vice president of Exxon Mobil Corporation. In December 2004, he assumed his current position as senior vice president of the Corporation. Mr. Simon has served on the local boards of many voluntary organizations—including United Way, Boy Scouts, and the Salvation Army—and is a member of the Governance Committee of the National Action Council for Minorities in Engineering. He has also served on the boards of the American Petroleum Institute and the National Association of Manufacturers. He is a member of the board of visitors for Duke University's School of Engineering and a member of the president's council. In addition, he is on the Kellogg Advisory Board of Northwestern University.

TIM STEARNS is an associate professor in the Department of Biological Sciences and the Department of Genetics at Stanford University. He is also a member of the Committee on Cancer Biology, the steering group for the cancer-biology graduate training program, and he is chair of the Committee on Graduate Admissions and Policy, which oversees all graduate programs in the biosciences at Stanford. Dr. Stearns is the recipient of a Howard Hughes Medical Institute Professor Award, which he has used to develop a program for research-oriented undergraduates. The laboratory course for this program, Biosci 54/55, draws sophomore-level students from diverse intellectual backgrounds and has them use interdisciplinary approaches to solve problems in cell biology. Dr. Stearns recently cofounded the Advanced Imaging Lab in Biophysics course, and he has taught advanced summer laboratory courses at Cold Spring Harbor Laboratory at Woods Hole, and in Chile and South Africa. His research involves using a combination of imaging, genetics, biochemistry, and structural biology to understand the cytoskeleton. His laboratory was one of the first to use green fluorescent protein to visualize cytoskeletal dynamics and is a leader in understanding microtubule organization and its relationship to the cell cycle.

DEBRA STEWART became the fifth president of the Council of Graduate Schools (CGS) in July 2000. Before coming to the CGS, Dr. Stewart was vice chancellor and dean of the Graduate School at North Carolina State University. She also served as interim chancellor at the University of North Carolina at Greensboro (1997) and as graduate dean and then vice provost (1988-1998) at North Carolina State. Among its 11 international members, CGS includes 9 major Canadian universities. Dr. Stewart received her PhD in political science from the University of North Carolina at Chapel Hill, her master's degree in government from the University of Maryland, and her BA from Marquette University. She is the author or coauthor of numerous scholarly articles on administrative theory and public policy. Her disciplinary research focuses on ethics and managerial decision-making. With sustained support from the National Science Foundation, Dr. Stewart has conducted research on political attitudes and moral reasoning among public officials in Poland and Russia.

ORLANDO L. TAYLOR is vice provost for research, dean of the graduate school, and professor of communications at Howard University. Before joining the Howard faculty in 1973, Dr. Taylor was a faculty member at Indiana University. He has also served as a visiting professor at Stanford University. Dr. Taylor has served on the board of directors of the Council of Graduate Schools and was board chair in 2001. He is a past president of the Northeastern Association of Graduate Schools and the National Communication Association. He is the immediate past president of the Consortium of Social Science Associations and chairman of the board of the Jacob Javits Fellowship Program in the Humanities for the US Department of Education. He also serves as a member of the board of trustees of the University Corporation for Atmospheric Research. Dr. Taylor has served in many capacities at Howard University: he has served as executive assistant to the president, interim vice president for academic affairs, dean of the School of Communications, and chair of the Department of Communication Arts and Sciences. Dr. Taylor's pioneering work in communication disorders, sociolinguistics, educational linguistics, and intercultural communication has led to the development of new theories and applications. In most of his scholarly work, he has focused on the rich cultural and linguistic diversity of the American people. He is the author of numerous articles, chapters, and books. The American Speech-Language-Hearing Association awarded him its highest award, Honors of the Association, and the Alumni Association of the University of Michigan awarded him its Distinguished Service Alumni Award. The University of Massachusetts, Amherst, has awarded him the Chancellor's Medal, and Yale University its Bouchet Medal for Leadership in Minority Graduate Education. Dr. Taylor received his bachelor's degree from Hampton University, his master's degree from Indiana University, and his PhD degree from the University of Michigan.

NANCY VORONA is vice president of research investment at the Center for Innovative Technology (CIT). Her responsibilities include strategy and program development for CIT's initiatives in nanotechnology and life sciences. Before her current appointment, she was CIT's senior industry director for advanced materials and electronics. Ms. Vorona joined CIT in 1998. Ms. Vorona's professional experience in electronics includes several years in marketing and sales management with International Rectifier Corporation, a US manufacturer of power semiconductors based in California. She was also responsible for international marketing and sales for Integrated Display Technology Ltd., a Hong Kong manufacturer of consumer electronic products. In 1993, she joined the Virginia Economic Development Partnership to establish and increase the international business of Virginia's information-technology and telecommunications companies. Ms. Vorona received a BA from the University of North Carolina at Chapel Hill and a master's degree in international management from Thunderbird, the American Graduate School of International Management in Glendale, Arizona.

ISIAH M. WARNER is Boyd Professor and vice chancellor for strategic initiatives of the Louisiana State System (LSU). He graduated cum laude from Southern University with a BS in chemistry in 1968. After working for Battelle Northwest in Richland, Washington, for 5 years, Dr. Warner attended graduate school in chemistry at the University of Washington, receiving his PhD in chemistry (analytical) in June 1977. He was assistant professor of chemistry at Texas A&M University from 1977 to 1982 and was awarded tenure and promotion to associate professor effective September 1982. However, he elected to join the faculty of Emory University as associate professor and was promoted to full professor in 1986. Dr. Warner was named to an endowed chair at Emory University in September 1987 and was the Samuel Candler Dobbs Professor of Chemistry until he left in August 1992. During the 1988-1989 academic year, he was on leave to the National Science Foundation as program officer for analytical and surface chemistry. In August 1992, Dr. Warner joined LSU as Philip W. West Professor of Analytical and Environmental Chemistry. He was chair of the Chemistry Department from 1994 to 1997 and was appointed Boyd Professor of the LSU System in July 2000, and Vice Chancellor for Strategic Initiatives in 2001. The primary research emphasis of Warner's research group is the development and application of improved methodologies (chemical, mathematical, and instrumental) for the study of complex chemical systems. His research interests include fluorescence spectroscopy, guest-host interactions, studies in organized media, spectroscopic applications of multi-channel detectors, chromatography, environmental analyses, and mathematical analyses and interpretation of chemical data using chemometrics.

GENERAL LARRY WELCH (retired) was the 12th chief of staff of the US Air Force. As chief, he served as the senior uniformed Air Force officer responsible for the organization, training, and equipage of a combined active-duty, Guard, reserve, and civilian force serving at locations in the United States and overseas. Formerly president of the Institute for Defense Analyses, General Welch now serves as a senior associate. In addition, he provides expertise to a number of organizations, including the Council on Foreign Relations, the Defense Science Board, the Joint Committee on

Nuclear Weapons Surety, the National Missile Defense Independent Review Team, the US Space Command Independent Strategic Advisory Group, and the US Strategic Command Strategic Advisory Group. General Welch received a BS in business administration from the University of Maryland and an MS in international relations from George Washington University.

REAR ADMIRAL ROBERT H. WERTHEIM (retired) [NAE] is a consultant on national security and related issues. During his 38 years in the Navy, he was director of strategic systems programs, responsible for the research, development, production, and operational support of the Navy's submarine-launched ballistic-missile program. After retirement from the Navy, he served for 7 years as Lockheed Corporation senior vice president for science and engineering; for the last 17 years, he has been a private consultant. He is a member of advisory groups serving the US Strategic Command, the Los Alamos and Livermore National Laboratories, and Draper Laboratory. Other current service includes membership on the joint Department of Defense and Department of Energy (DOE) Advisory Committee on Nuclear Weapons Surety and on the University of California President's Council on the National Laboratories. He is a former member of the National Academy of Sciences Committee on International Security and Arms Control, the DOE Laboratory Operations Board, and the Defense Science Board. Admiral Wertheim graduated with honors from New Mexico Military Institute in 1942. He graduated with distinction from the Naval Academy in 1945 and received an MS in physics from the Massachusetts Institute of Technology in 1954. He has been elected a member of the National Academy of Engineering and of the scientific and engineering societies, Sigma Xi and Tau Beta Pi, an honorary member of the American Society of Naval Engineers; and a fellow of the American Institute of Aeronautics and Astronautics and the California Council on Science and Technology. Admiral Wertheim has been honored with the Navy Distinguished Service Medal (twice), the Legion of Merit, the Gold Medal of the American Society of Naval Engineers, the Rear Admiral William S. Parsons Award of the Navy League, the Chairman of the Joint Chiefs of Staff Distinguished Public Service Medal, and the Secretary of Defense Medal for Outstanding Public Service. He was inducted into the New Mexico Military Institute Hall of Fame in 1987 and has been honored by the US Naval Academy with its 2005 Distinguished Graduate Award for his lifetime of service to the Navy and the nation.

DEAN ZOLLMAN is University Distinguished Professor, Distinguished University Teaching Scholar, and head of the Department of Physics at Kansas State University (KSU). He has focused his scholarly activities on research and development in physics education since 1972. He has received the NSF Director's Award for Distinguished Teacher Scholars (2004), the Carnegie Foundation for the Advancement of Teaching Doctoral University Professor of the Year (1996), and American Association of Physics Teachers' Robert A. Millikan Medal (1995). His research concentrates on investigating the mental models and operations that students develop as they learn physics and how students transfer knowledge in the learning process. He also applies cutting-edge technology to the teaching of physics and to providing instructional and pedagogic materials to physics teachers, particularly teachers whose background does not include a substantial amount of physics. He has twice been a Fulbright Fellow in Germany. In 1989, he worked at Ludwig-Maximilians University in Munich on development of measurement techniques for digital video. In 1998, he visited the Institute for Science Education at the University in Kiel, where he investigated student understanding of quantum physics. Dr. Zollman is coauthor of six videodisks for physics teaching, the Physics InfoMall database, and a textbook. He leads the Visual Quantum Mechanics project, which develops materials for teaching quantum physics to three groups of students: nonscience students, science and engineering students, and students interested in biology and medicine. His present instructional and research projects include Modern Miracle Medical Machines, Physics Pathway, and research on student learning.

Appendix D

Issue Briefs

The issue briefs presented in this appendix summarize findings and recommendations from a variety of recently published reports and papers as input to the deliberations of the Committee on Prospering in the Global Economy of the 21st Century. The papers were provided as background information to the study committee and focus group participants.

The 13 papers, written by members of the committee's staff, are included here only as a historical record and a useful summary of relevant reports, scientific literature, and data analysis. Statements in this brief should not be seen as the conclusions of the National Academies or the committee.

Each issue brief provides an overview of the findings and recommendations of previously released studies from the National Academies and other groups. The issue briefs cover topics relevant to the committee's charge, including K–12 education, higher education, research policy, and national and homeland security policy.

Specifically, the topics addressed are:

- K–12 Science, Mathematics, and Technology Education
- Attracting the Most Able US Students to Science and Engineering
- Undergraduate, Graduate, and Postgraduate Education in Science, Engineering, and Mathematics
- Implications of Changes in the Financing of Public Higher Education
- International Students and Researchers in the United States
- Achieving Balance and Adequacy in Federal Science and Technology Funding
- The Productivity of Scientific and Technological Research
- Investing in High-Risk and Breakthrough Research
- Ensuring That the United States Is at the Forefront in Critical Fields of Science and Technology
- Understanding Trends in Science and Technology Critical to US Prosperity
- Ensuring That the United States Has the Best Environment for Innovation
- Scientific Communication and Security
- Science and Technology Issues in National and Homeland Security

K–12 Science, Mathematics, and Technology Education

SUMMARY

US education in science, technology, engineering, and mathematics is undergoing great scrutiny. Just as the launch of Sputnik 1 in 1957 led the United States to undertake the most dramatic educational reforms of the 20th century, the rise of new international competitors in science and technology is forcing the United States to ask whether its educational system is suited to the demands of the 21st century.

These concerns are particularly acute in K–12 education. In comparison with their peers in other countries, US students on average do worse on measures of mathematics and science performance the longer they are in school. On comparisons of problem-solving skills, US students perform more poorly overall than do the students in most of the countries that have participated in international assessments. Some believe the United States has failed to achieve the objective established in the Goals 2000: Educate America Act—for US students to be first in the world in mathematics and science achievement in the year 2000.

National commissions, industrial groups, and leaders in the public and private sectors are in broad agreement with policy initiatives that the federal government could undertake to improve K–12 science, mathematics, and technology education. Some of these are listed below:

This issue paper summarizes findings and recommendations from a variety of recently published reports and papers as input to the deliberations of the Committee on Prospering in the Global Economy of the 21st Century. Statements in this paper should not be seen as the conclusions of the National Academies or the committee.

Increasing the Number of Excellent Teachers

- Allocate federal professional-development funds to summer institutes that address the most pressing professional-development needs of mathematics and science teachers.
- Keep summer-institute facilitators—teachers current with the most effective teaching methods in their disciplines and who have shown demonstrable results of higher student achievement in mathematics and science—abreast of new insights and research in science and mathematics teaching by providing funding for training them.
- Encourage higher education institutions to establish mathematics and science teaching academies that include faculty from science, mathematics, and education departments through a competitive grant process.
- Support promising students to study science, mathematics, and engineering teaching—particularly those obtaining degrees in science, mathematics, or engineering who plan to teach at the K–12 level following graduation through scholarships and loan programs for students as well as institutional funding. Qualified college students and midcareer professionals need to be attracted into teaching and given the preparation they require to succeed. Experts in mathematics, science, and technology should be able to become teachers by completing programs to acquire and demonstrate fundamental teaching skills. Recruitment, preparation, and retention of minority-group teachers are particularly important as groups underrepresented in science, mathematics, and engineering become a larger percentage of the student population.
- Conduct an aggressive, national-outreach media campaign to attract young people to teaching careers in mathematics and science.
- Work for broad improvements in the professional status of science, mathematics, and technology teachers. Structured induction programs for new teachers, district–business partnerships, award programs, and other incentives can inspire teachers and encourage them to remain in the field. Most important, salaries for science, mathematics, and technology teachers need to reflect what they could receive in the private sector and be in accord with their contributions to society, and teachers need to be treated as professionals and as important members of the science and engineering communities.

Enhancing the Quality and Cohesion of Educational Standards

- Help colleges, businesses, and schools work together to link K–12 standards to college admissions criteria and workforce needs to create a seamless K–16 educational system.
- Provide incentives for states and coalitions of states to conduct benchmarking studies between their standards and the best standards available.

• Foster the development of high-quality curricula and assessments that are closely aligned with world-class standards.

• Establish ambitious but realistic goals for student performance—for example, that 30% of high school seniors should be proficient in science by 2010 as measured by the National Assessment of Educational Progress (NAEP).

Changing the Institutional Structure of Schools

• Provide seed money or incentives for new kinds of schools and new forms of schooling. Promising ideas include small high schools, dual-enrollment programs in high schools and colleges, colocation of schools with institutions of higher education, and wider use of Advanced Placement and International Baccalaureate courses.

• Help districts institute reorganization of the school schedule to support teaching and learning. Possibilities include devoting more time to study of academic subjects, keeping schools open longer in the day and during parts of the summer, and providing teachers with additional time for development and collaboration.

• Provide scholarships for low-income students who demonstrate that they have taken a core curriculum in high school that prepares them to study science, mathematics, or engineering in college.

The challenge for policy-makers is to find ways of generating meaningful change in an educational system that is large, complex, and pluralistic. Sustained programs of research, coordination, and oversight can channel concerns over K–12 science, mathematics, and technology education in productive directions.

THE CHALLENGE OF K–12 SCIENCE, MATHEMATICS, AND TECHNOLOGY EDUCATION

The state of US K–12 education in science, mathematics, and technology has become a focus of intense concern. With the economies and broader cultures of the United States and other countries becoming increasingly dependent on science and technology, US schools do not seem capable of producing enough students with the knowledge and skills needed to prosper.

On the 1996 NAEP, fewer than one-third of students performed at or above the proficiency level in mathematics and science—with "proficiency" denoting competence in challenging subject matter.[1] Alarmingly, more than

[1]National Center for Education Statistics. *NAEP 1999 Trends in Academic Progress: Three Decades of Academic Performance*. NCES 2000-469. Washington, DC: US Department of Education, 2000.

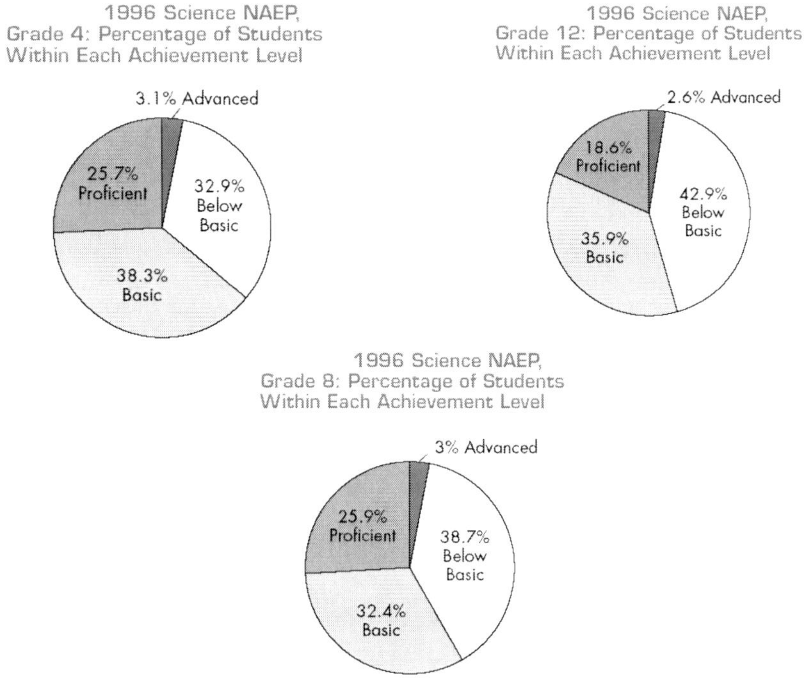

FIGURE K–12-1A NAEP 1996 science results, grades 4, 8, and 12. Studies suggest that a large portion of US students are lacking in science skills. In 1996, at least one-third of students in 4th, 8th, and 12th grade performed below basic in national tests. SOURCE: S. C. Loomis and M. L. Bourque, eds. *National Assessment of Educational Progress Achievement Levels, 1992-1998 for Science*. Washington, DC: National Assessment Governing Board, July 2001. Available at: http://www.nagb.org/pubs/sciencebook.pdf.

one-third of students scored below the basic level in these subjects, meaning they lack the fundamental knowledge and skills they will need to get good jobs and participate fully in our technologically sophisticated society (see Figures K–12-1A and K–12-1B).

International comparisons document a gradual decline in performance and interest in mathematics and science as US students get older. Though 4th graders in the United States perform well in math and science compared with their peers in other countries (see Tables K–12-1 and K–12-2), 12th graders in 1999 were almost last in performance among the countries that participated in the Third International Mathematics and Science Study

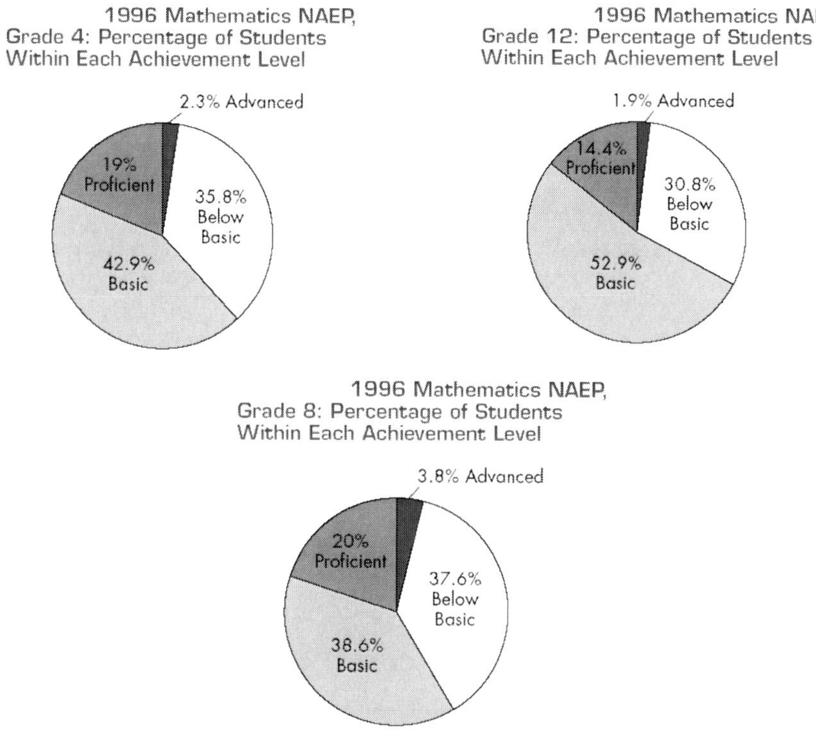

FIGURE K–12-1B NAEP 1996 mathematics results, grades 4, 8, and 12. The results are similar for mathematics: 30% of students scored below basic.
SOURCE: S. C. Loomis and M. L. Bourque, eds. *National Assessment of Educational Progress Achievement Levels, 1992-1998 for Science.* Washington, DC: National Assessment Governing Board, July 2001. Available at: http://www.nagb.org/pubs/sciencebook.pdf.

(TIMSS).[2] Among the 20 countries assessed in advanced mathematics and physics, none scored significantly lower than the United States in mathematics, and only one scored significantly lower in physics.

There has been some good news about student achievement.[3] US 8th graders did better on an international assessment of mathematics and science in 2003 than they did in 1995 (see Tables K–12-3 and K–12-4). The

[2]National Center for Education Statistics. *Pursuing Excellence: A Study of Twelfth-Grade Mathematics and Science Achievement in International Context.* NCES 98-049. Washington, DC: US Government Printing Office, 1998.

[3]R. W. Bybee and E. Stage. "No Country Left Behind." *Issues in Science and Technology* (Winter 2005):69-75.

TABLE K–12-1 Average TIMSS Mathematics Scale Scores of 4th-Grade Students, by Country: 1995 and 2003

Country	1995	Country	2003
Singapore	590	Singapore	594
Japan	567	Hong Kong SAR[1,2]	575
Hong Kong SAR[1,2]	557	Japan	565
(Netherlands)	549	Netherlands[1]	540
(Hungary)	521	Latvia-LSS	533
United States	**518**	England	531
(Latvia-LSS)[3]	499	Hungary	529
(Australia)	495	**United States[1]**	**518**
Scotland	493	Cyprus	510
England	484	Australia[1]	499
Norway	476	New Zealand	496
Cyprus	475	Scotland	490
New Zealand[4]	469	Slovenia	479
(Slovenia)	462	Norway	451
Iran, Islamic Republic of	387	Iran, Islamic Republic of	389

■ Average is higher than the U.S. average
□ Average is not measurably different from the U.S. average
□ Average is lower than the U.S. average

[1]Met international guidelines for participation rates in 2003 only after replacement schools were included.
[2]Hong Kong is a Special Administrative Region (SAR) of the People's Republic of China.
[3]Designated LSS because only Latvian-speaking schools were included in 1995. For this analysis, only Latvian-speaking schools are included in the 2003 average.
[4]In 1995, Maori-speaking students did not participate. Estimates in this table are computed for students taught in English only, which represents between 98-99 percent of the student population in both years.
NOTE: Countries are ordered based on the average score. Parentheses indicate countries that did not meet international sampling or other guidelines in 1995. All countries met international sampling and other guidelines in 2003, except as noted. See NCES (1997) for details regarding 1995 data. The tests for significance take into account the standard error for the reported difference. Thus, a small difference between the United States and one country may be significant while a large difference between the United States and another country may not be significant. Countries were required to sample students in the upper of the two grades that contained the most number of 9-year-olds. In the United States and most countries, this corresponds to grade 4. See table A1 in appendix A for details.
SOURCE: International Association for the Evaluation of Educational Achievement (IEA), Trends in International Mathematics and Science Study (TIMSS), 1995 and 2003.

SOURCE: National Center for Education Statistics. *Highlights from the Trends in International Mathematics and Science Study: TIMSS 2003*. Washington, DC: United States Department of Education, December 2004. P. 8. Available at: http://nces.ed.gov/pubs2005/2005005.pdf.

TABLE K–12-2 Differences in Average TIMSS Science Scale Scores of 4th-Grade Students, by Country: 1995 and 2003

Country	1995	2003	Difference[1]
Singapore	523	565	42 ▲
Japan	553	543	-10 ▼
Hong Kong SAR[2,3]	508	542	35 ▲
England[3]	528	540	13 ▲
United States[3]	**542**	**536**	**-6**
(Hungary)	508	530	22 ▲
(Latvia-LSS)[4]	486	530	43 ▲
(Netherlands)[3]	530	525	-5
New Zealand[5]	505	523	18 ▲
(Australia)[3]	521	521	-1
Scotland[2]	514	502	-12 ▼
(Slovenia)	464	490	26 ▲
Cyprus	450	480	30 ▲
Norway	504	466	-38 ▼
Iran, Islamic Republic of	380	414	34 ▲

▲ p<.05, denotes a significant increase.
▼ p<.05, denotes a significant decrease.
[1]Difference calculated by subtracting 1995 from 2003 estimate using unrounded numbers.
[2]Hong Kong is a Special Administrative Region (SAR) of the People's Republic of China.
[3]Met international guidelines for participation rates only after replacement schools were included.
[4]Designated LSS because only Latvian-speaking schools were included in 1995. For this analysis, only Latvian-speaking schools are included in the 2003 average.
[5]In 1995, Maori-speaking students did not participate. Estimates in this table are computed for students taught in English only, which represents between 98-99 percent of the student population in both years.
NOTE: Countries are ordered based on the 2003 average scores. Parentheses indicate countries that did not meet international sampling or other guidelines in 1995. All countries met international sampling and other guidelines in 2003, except as noted. See NCES (1997) for details regarding 1995 data. The tests for significance take into account the standard error for the reported difference. Thus, a small difference between averages for one country may be significant while a large difference for another country may not be significant. Countries were required to sample students in the upper of the two grades that contained the largest number of 9-year-olds. In the United States and most countries, this corresponds to grade 4. See table A1 in appendix A for details. Detail may not sum to totals because of rounding.
SOURCE: International Association for the Evaluation of Educational Achievement (IEA), Trends in International Mathematics and Science Study (TIMSS), 1995 and 2003.

SOURCE: National Center for Education Statistics. *Highlights from the Trends in International Mathematics and Science Study: TIMSS 2003*. Washington, DC: United States Department of Education, December 2004. P. 16. Available at: http://nces.ed.gov/pubs2005/2005005.pdf.

TABLE K–12-3 Average TIMSS Mathematics Scale Scores of 8th-Grade Students, by Country: 1995 and 2003

Country	1995	Country	2003
Singapore	609	Singapore	605
Japan	581	Korea, Republic of	589
Korea, Republic of	581	Hong Kong SAR[2]	586
Hong Kong SAR[1]	569	Japan	570
Belgium-Flemish	550	Belgium-Flemish	537
Sweden	540	Netherlands[2]	536
Slovak Republic	534	Hungary	529
(Netherlands)	529	Russian Federation	508
Hungary	527	Slovak Republic	508
(Bulgaria)	527	Latvia-LSS[3]	505
Russian Federation	524	Australia	505
(Australia)	509	**(United States)**	**504**
New Zealand	501	Lithuania[4]	502
Norway	498	Sweden	499
(Slovenia)	494	Scotland[2]	498
(Scotland)	493	New Zealand	494
United States	**492**	Slovenia	493
(Latvia-LSS)[3]	488	Bulgaria	476
(Romania)	474	Romania	475
Lithuania[4]	472	Norway	46
Cyprus	468	Cyprus	45
Iran, Islamic Republic of	418	Iran, Islamic Republic of	411

■ Average is higher than the U.S. average
□ Average is not measurably different from the U.S. average
□ Average is lower than the U.S. average

[1]Hong Kong is a Special Administrative Region (SAR) of the People's Republic of China.
[2]Met international guidelines for participation rates in 2003 only after replacement schools were included.
[3]Designated LSS because only Latvian-speaking schools were included in 1995. For this analysis, only Latvian-speaking schools are included in the 2003 average.
[4]National desired population does not cover all of the international desired population.
NOTE: Countries are ordered by average score. Parentheses indicate countries that did not meet international sampling or other guidelines in 1995 or 2003. See appendix A for details regarding 2003 data. See NCES (1997) for details regarding 1995 data. The tests for significance take into account the standard error for the reported difference. Thus, a small difference between the United States and one country may be significant while a large difference between the United States and another country may not be significant. Countries were required to sample students in the upper of the two grades that contained the largest number of 13-year-olds. In the United States and most countries, this corresponds to grade 8. See table A1 in appendix A for details.
SOURCE: International Association for the Evaluation of Educational Achievement (IEA), Trends in International Mathematics and Science Study (TIMSS), 1995 and 2003.

SOURCE: National Center for Education Statistics. *Highlights from the Trends in International Mathematics and Science Study: TIMSS 2003*. Washington, DC: United States Department of Education, December 2004. P. 19. Available at: http://nces.ed.gov/pubs2005/2005005.pdf.

TABLE K–12-4 Difference in Average TIMSS Science Scale Scores of 8th-Grade Students, by Country: 1995, 1999, and 2003

Country	1995	1999	2003	Difference[1] (2003-1995)	Difference[1] (2003-1999)
Singapore	580	568	578	-3	10
Chinese Taipei	—	569	571	†	2
Korea, Republic of	546	549	558	13 ▲	10 ▲
Hong Kong SAR[2,3]	510	530	556	46 ▲	27 ▲
Japan	554	550	552	-2	3
Hungary	537	552	543	6	-10 ▼
(Netherlands)[2]	541	545	536	-6	-9
(United States)	**513**	**515**	**527**	**15 ▲**	**12 ▲**
(Australia)[4]	514	—	527	13 ▲	†
Sweden	553	—	524	-28 ▼	†
(Slovenia)[4]	514	—	520	7 ▲	†
New Zealand	511	510	520	9	10
(Lithuania)[5]	464	488	519	56 ▲	31 ▲
Slovak Republic	532	535	517	-15 ▼	-18 ▼
Belgium-Flemish	533	535	516	-17 ▼	-19 ▼
Russian Federation	523	529	514	-9	-16 ▼
(Latvia-LSS)[6]	476	503	513	37 ▲	11
(Scotland)[2]	501	—	512	10	†
Malaysia	—	492	510	†	18 ▲
Norway	514	—	494	-21 ▼	†
Italy[7]	—	493	491	†	-2
(Israel)[7]	—	468	488	†	20 ▲
(Bulgaria)	545	518	479	-66 ▼	-39 ▼
Jordan	—	450	475	†	25 ▲
Moldova, Republic of	—	459	472	†	13 ▲
(Romania)	471	472	470	-1	-2
Iran, Islamic Republic of	463	448	453	-9 ▼	5
(Macedonia, Republic of)	—	458	449	†	-9
Cyprus	452	460	441	-11 ▼	-19 ▼
Indonesia[5]	—	435	420	†	-15 ▼
Chile	—	420	413	†	-8
Tunisia	—	430	404	†	-26 ▼
Philippines	—	345	377	†	32 ▲
South Africa[8]	—	243	244	†	1

—Not available.
†Not applicable.
▲ p<.05, denotes a significant increase.
▼ p<.05, denotes a significant decrease.
[1]Difference calculated by subtracting 1995 or 1999 from 2003 estimate using unrounded numbers.
[2]Met international guidelines for participation rates in 2003 only after replacement schools were included.
[3]Hong Kong is a Special Administrative Region (SAR) of the People's Republic of China.
[4]Because of national-level changes in the starting age/date for school, 1999 data for Australia and Slovenia cannot be compared to 2003.
[5]National desired population does not cover all of the international desired population in all years for Lithuania, and in 2003 for Indonesia.
[6]Designated LSS because only Latvian-speaking schools were included in 1995 and 1999. For this analysis, only Latvian-speaking schools are included in the 2003 average.
[7]Because of changes in the population tested, 1995 data for Israel and Italy are not shown.
[8]Because within classroom sampling was not accounted for, 1995 data are not shown for South Africa.
NOTE: Countries are sorted by 2003 average scores. The tests for significance take into account the standard error for the reported difference. Thus, a small difference between averages for one country may be significant while a large difference for another country may not be significant. Parentheses indicate countries that did not meet international sampling and/or other guidelines in 1995, 1999, and/or 2003. See appendix A for details regarding 2003 data. See Gonzales et al. (2000) for details regarding 1995 and 1999 data. Countries were required to sample students in the upper of the two grades that contained the largest number of 13-year-olds. In the United States and most countries, this corresponds to grade 8. See table A1 in appendix A for details. Detail may not sum to totals because of rounding.
SOURCE: International Association for the Evaluation of Educational Achievement (IEA), Trends in International Mathematics and Science Study (TIMSS), 1995, 1999, and 2003.

SOURCE: National Center for Education Statistics. *Highlights from the Trends in International Mathematics and Science Study: TIMSS 2003*. Washington, DC: United States Department of Education, December 2004. P. 17. Available at: http://nces.ed.gov/pubs2005/2005005.pdf.

TABLE K–12-5 Trends in Average NAEP Mathematics Scale Scores for Students Ages 9, 13, and 17: 1973-2004

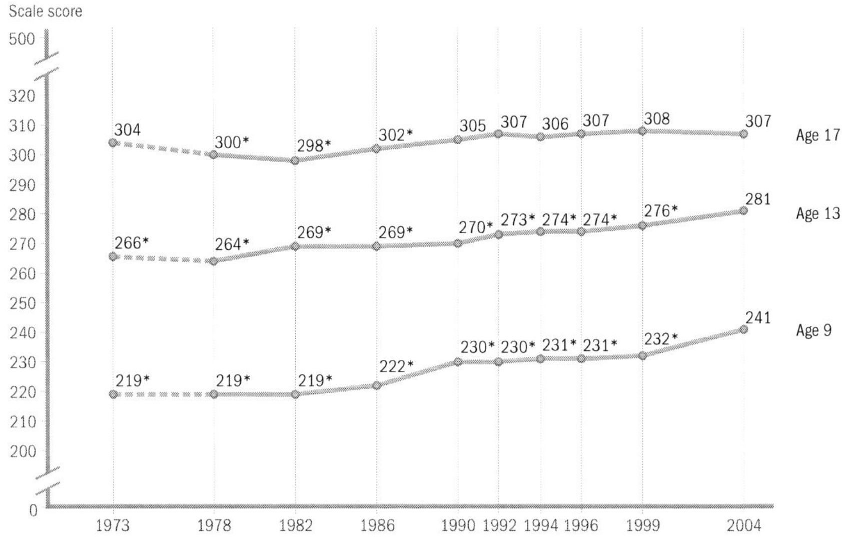

NOTE: *Significantly different from 2004.
SOURCE: National Assessment Governing Board. *National Assessment of Educational Progress 2004: Trends in Academic Progress Three Decades of Student Performance in Reading and Mathematics*. Washington, DC: United States Department of Education, July 14, 2005.

achievement gap separating Black and Latino students from European-American students narrowed during that period (see Figure K–12-2). However, a recent assessment by the Program for International Student Assessment found that US 15-year-olds are near the bottom of all countries in their ability to solve practical problems requiring mathematical understanding. Additionally, testing for the last 30 years has shown that although scores among 9- and 13-year-olds have increased, scores for 17-year-olds have remained stagnant (see Table K–12-5) and there is a gender gap (see K–12-6).

Perhaps the hardest trend to document is a sense of disillusionment with careers based on science and technology.[4] Fewer children respond posi-

[4]Committee for Economic Development, Research and Policy Committee. *Learning for the Future: Changing the Culture of Math and Science Education to Ensure a Competitive Workforce*. New York: Committee for Economic Development, 2003.

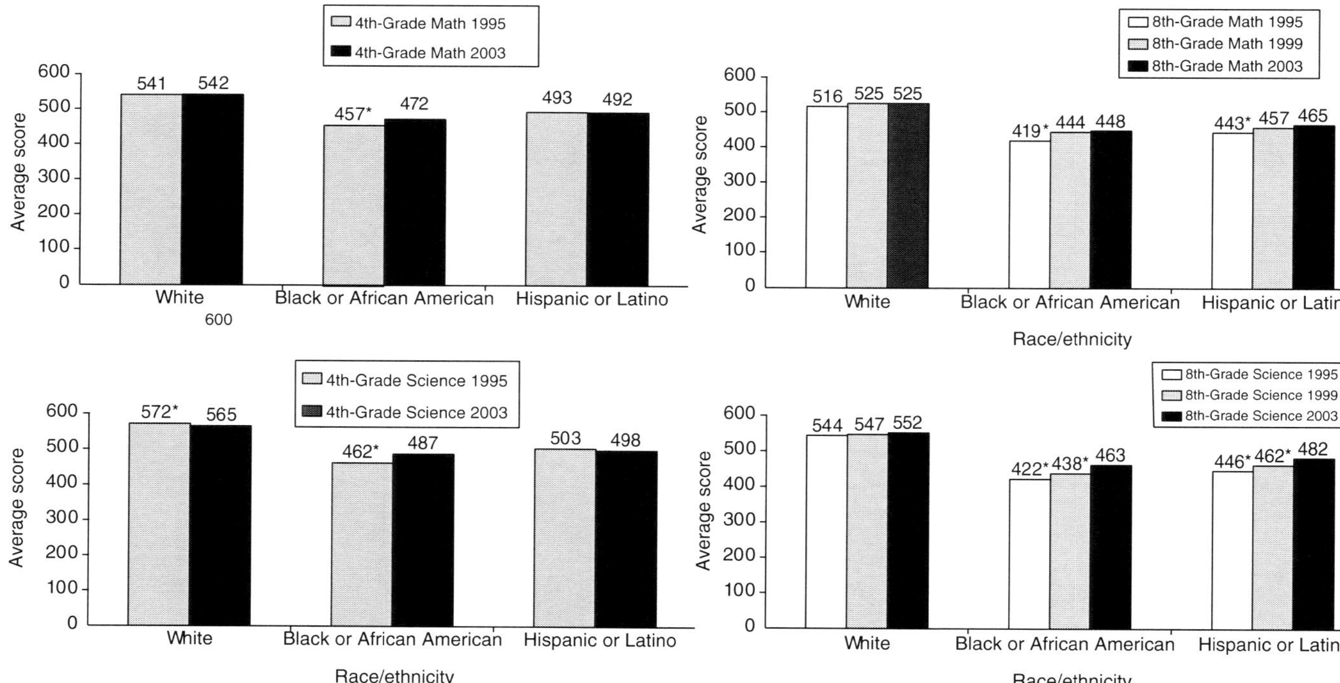

FIGURE K–12-2 TIMSS math and science scores, 4th (1995 and 2003) and 8th (1995, 1999, and 2003) graders.
SOURCE: National Center for Education Statistics. *Highlights from the Trends in International Mathematics and Science Study: TIMSS 2003*. Washington, DC: US Department of Education, December 2004. Figure 1-4. Available at: http://nces.ed.gov/pubs2005/2005005.pdf.

TABLE K–12-6 Students at or Above Basic and Proficient Levels as Measured in NAEP Mathematics and Science Tests, Grades 4, 8, and 12, by Sex: 1996 and 2000

	1996			2000		
Variable	Grade 4	Grade 8	Grade 12	Grade 4	Grade 8	Grade 12
Mathematics						
At or above basic						
Male	65*	62*	70*	70	67	66
Female	63*	63	69*	68	65	64
At or above pro cient						
Male	24*	25*	18	28	29	20
Female	19*	23	14	24	25	14
Science						
At or above basic						
Male	68	62	60*	69	64	54
Female	67	61	55	64	57	51
At or above pro cient						
Male	31	31*	25	33	36	21
Female	27	27	17	26	27	16

NOTE: *Significantly different from 2000.
SOURCE: National Science Board. *Science and Engineering Indicators 2004*. NSB 04-01. Arlington, VA: National Science Foundation, 2004. Appendix Table 1-4. This table was based on US Department of Education, National Center for Education Statistics (NCES). *The Nation's Report Card: Mathematics 2000*. NCES 2001-517. Washington, DC: US Department of Education, 2001; National Center for Education Statistics (NCES). *The Nation's Report Card: Science 2000*. NCES 2003-453. Washington, DC: US Department of Education, 2003.

tively when surveyed to statements such as "I like math" than has been the case in the past. The number of schools offering advanced courses, such as Advanced Placement and International Baccalaureate has increased dramatically, but the vast majority of students in high school will never take an advanced science or mathematics course (see Tables K–12-7 and K–12-8; see Figure K–12-3). And a lack of interest in science, mathematics, and technology is particularly pronounced among disadvantaged groups that have been underrepresented in those fields.

In general, many Americans do not know enough about science, technology, and mathematics to contribute to or benefit from the knowledge-based society that is taking shape around us. At the same time, other countries have learned from our example that preeminence in science and engineering pays immense economic and social dividends, and they are boosting their investments in these critical fields.

The traditions of autonomy and pluralism in American education limit

TABLE K–12-7 High-School Graduates Completing Advanced Mathematics Courses (1990, 1994, and 1998), by Students and School Characteristics in 1998

Year and characteristic	Any trigonometry/ algebra III	Any precalculus/ analysis	Any statistics/ probability	Calculus	
				Any	AP/IB
1990	20.7	13.6	1.0	7.2	NA
1994	24.0	17.4	2.1	10.2	NA
1998	20.8	23.1	3.7	11.9	6.3
Male	19.4	23.1	3.4	12.0	6.8
Female	22.5	22.9	4.0	11.6	6.0
White	23.6	25.1	4.3	13.1	7.0
Asian/Pacific Islander	18.0	41.4	3.8	20.1	13.1
Black	15.5	14.0	2.1	7.2	3.3
Hispanic	10.9	15.4	1.7	7.1	3.7
School urbanicity					
Urban	19.0	28.5	3.6	13.2	7.7
Suburban	20.9	26.7	4.0	12.1	7.5
Rural	22.6	13.4	3.4	10.4	3.5
School size[a]					
Small	22.2	21.9	3.6	10.8	3.4
Medium	21.9	22.8	3.8	12.9	6.9
Large	16.7	25.1	3.4	10.3	7.7
School poverty[b]					
Very low	26.3	35.4	6.5	15.6	8.8
Low	18.1	23.6	4.3	12.0	6.7
Medium	22.4	14.9	1.7	9.2	3.9
High	23.6	9.8	0.8	6.9	4.9

[a]Small = fewer than 600 students enrolled, medium = 600-1,800, and large = more than 1,800.
[b]Measured by percentage of students eligible for free or reduced-price lunches: very low = 5 percent or less, low = 6-25 percent, medium = 26-50 percent, and high = 51-100 percent.
NOTE: AP = Advanced Placement, IB = International Baccalaureate, NA = not available. AP and IB courses were coded separately in 1998 and 2000 but not in prior years. AP/IB calculus courses are counted both in their specific column and in the "any calculus" column. Before 1998, AP and IB courses were coded with the general set of courses.
SOURCE: National Science Board. *Science and Engineering Indicators 2004*. NSB 04-01. Arlington, VA: National Science Foundation, 2004. Appendix Table 1-10. This table was based on US Department of Education, Center for National Education Statistics, High School Transcript Studies, various years.

the influence that the federal government can exert on state educational systems, school districts, and individual schools. Nevertheless, the federal government can enable change by leveraging its investments in K–12 education, by providing information and other resources to organizations, and by helping to coordinate the many groups and individuals with a stake in science, mathematics, and technology education. Three policy arenas seem particularly promising: teacher preparation, educational standards, and institutional change.

TABLE K–12-8 High-School Graduates Completing Advanced Courses (1990, 1994, and 1998), by Students and School Characteristics in 1998

Year and characteristic	Advanced biology		Chemistry		Physics		Advanced biology, chemistry, and physics
	Any	AP/IB	Any	AP/IB	Any	AP/IB	
1990	27.5	NA	45.0	NA	21.5	NA	7.4
1994	34.8	NA	50.4	NA	24.5	NA	9.9
1998	37.4	4.9	56.4	2.9	28.6	1.7	12.1
Male	33.8	4.0	53.3	3.3	31.0	2.3	11.8
Female	40.8	5.8	59.2	2.6	26.6	1.2	12.3
White	38.5	5.0	58.8	2.9	31.1	1.8	13.4
Asian/Pacific Islander	43.0	14.0	63.7	9.5	37.4	4.8	15.7
Black	35.8	3.4	51.1	1.2	20.3	0.8	7.6
Hispanic	31.2	3.1	45.5	2.9	19.4	1.3	8.2
School urbanicity							
Urban	43.0	5.9	62.4	3.9	30.8	2.7	14.0
Suburban	39.4	5.9	56.1	3.2	31.2	2.0	14.6
Rural	29.3	2.6	50.9	1.6	23.1	0.4	7.3
School size[a]							
Small	36.4	2.9	57.7	0.9	25.7	0.3	11.7
Medium	36.8	4.9	56.6	2.9	31.0	1.9	13.4
Large	40.1	6.6	55.0	4.8	24.8	2.6	9.2
School poverty[b]							
Very low	37.9	6.4	71.2	4.8	43.0	3.6	17.8
Low	39.4	4.6	54.2	1.9	26.9	0.9	11.7
Medium	34.1	3.4	52.4	2.2	23.6	1.3	10.2
High	37.7	5.3	50.7	2.1	17.4	1.0	7.5

[a]Small = fewer than 600 students enrolled, medium = 600-1,800, and large = more than 1,800.
[b]Measured by percentage of students eligible for free or reduced-price lunches: very low = 5 percent or less, low = 6-25 percent, medium = 26-50 percent, and high = 51-100 percent.
NOTE: AP = Advanced Placement, IB = International Baccalaureate, NA = not available. AP and IB courses were coded separately in 1998 and 2000 but not in prior years. AP/IB courses are counted both in their specific columns and in columns that correspond to the general course category. For example, AP chemistry is included in the "any chemistry" column in addition to being listed in its own column. Before 1998, AP and IB courses were coded with the general set of courses.
SOURCE: National Science Board. *Science and Engineering Indicators 2004*. NSB 04-01. Arlington, VA: National Science Foundation, 2004. Appendix Table 1-11. This table was based on US Department of Education, Center for National Education Statistics, High School Transcript Studies, various years.

IMPROVING THE QUALITY OF MATHEMATICS, SCIENCE, AND TECHNOLOGY TEACHING

Students learn about science, mathematics, and technology first and foremost through interactions with teachers. Changing the nature of those

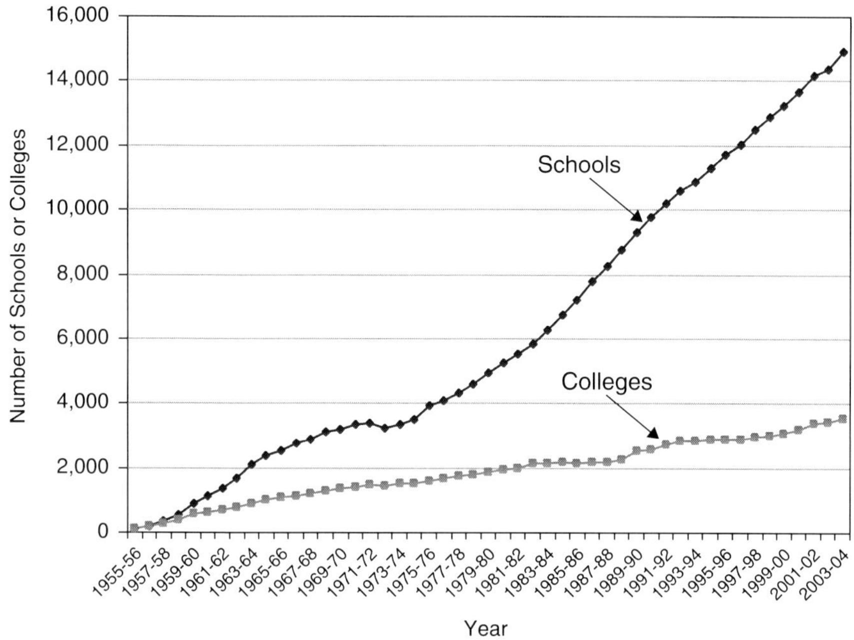

FIGURE K–12-3 Number of schools and colleges participating in AP programs.
SOURCE: National Research Council. *Learning and Understanding: Improving Advanced Study of Mathematics and Science in US High Schools*. Washington, DC: National Academy Press, 2002. Data courtesy of Jay Labov, Center for Education, National Academies.

interactions is the surest way to improve education in these subjects in the United States.

Many mathematics and science teachers in US schools do not have backgrounds needed to teach these subjects well (see Figure K–12-4).[5] Many of these teachers at the high school level—and even more at the middle school level—do not have a college degree in the subject they are teaching (see Tables K–12-9 and K–12-10). Many lack certification to teach mathematics and science, and a subset of teachers start in the classroom without any formal training. The lack of adequate training and background is especially severe at schools serving large numbers of disadvantaged students, creating a vicious circle in which a substandard education and low achievement are intertwined (see Table K–12-11). The stresses on teachers are equally se-

[5]US Department of Education, The National Commission on Mathematics and Science Teaching for the 21st Century. *Before It's Too Late*. Washington, DC: US Department of Education, 2000.

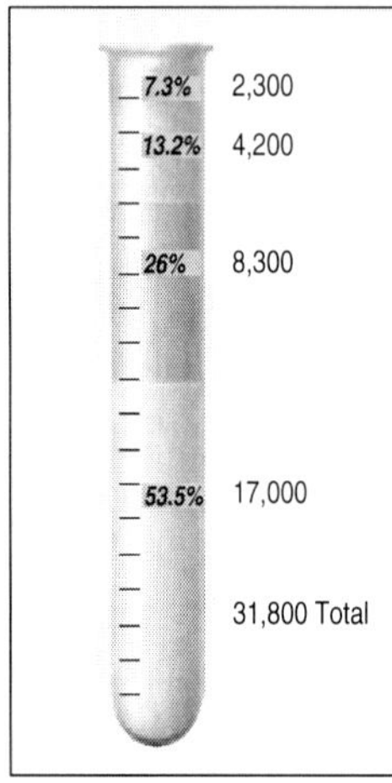

FIGURE K–12-4 Middle and high school mathematics and science positions filled during the 1993-1994 school year by certified and noncertified teachers.
SOURCE: National Center for Education Statistics. *Schools and Staffing Survey (1993-1994)*. Washington, DC: United States Department of Education, 2006.

vere: Of new mathematics and science teachers, about one-third leave teaching within the first 3 years.

The best predictors of higher student achievement in mathematics and science are (1) full certification of the teacher and (2) a college major in the field being taught.[6] Teachers need a high-quality education and continued development as professionals throughout their careers. Federal policy initiatives that could help meet these objectives include the following:

• Allocate federal professional-development funds to summer institutes that address the most pressing professional-development needs of mathematics and science teachers.[7]

• Keep summer-institute facilitators—teachers current with the most effective teaching methods in their disciplines and who have shown demonstrable results of higher student achievement in mathematics and science—abreast of new insights and research in science and mathematics teaching by providing funding for training them.[8]

---

[6]Ibid.
[7]Ibid.
[8]Ibid.

TABLE K–12-9 Public High School Students Whose Mathematics and Science Teachers Majored or Minored in Various Subject Fields, by Poverty Level and Minority Enrollment in School: 1999-2000

Subject and school characteristics	Mathematics/ statistics major	Mathematics/ statistics minor	Mathematics education major	Science, computer science, or engineering major	Other major
Mathematics					
Students in poverty (percent)					
0–10	45.1	3.7	31.3	4.0	15.9
More than 10 to 50	37.6	5.2	34.4	4.0	18.8
More than 50	43.4	5.8	23.6	10.3	17.0
Minority enrollment (percent)					
0–5	42.5	3.7	35.3	2.4	16.2
More than 5 to 45	39.4	4.4	35.7	4.1	16.3
More than 45	41.6	6.6	24.5	7.3	19.9
Biology/life sciences	Biology/life science major	Biology/life science minor	Other science major or minor	Science education major	Other major
Students in poverty (percent)					
0–10	62.6	5.7	7.0	7.8	16.9
More than 10 to 50	61.2	7.1	8.0	11.6	12.0
More than 50	62.5	6.4	2.7	7.0	21.4
Minority enrollment (percent)					
0–5	59.8	7.9	5.4	13.5	13.4
More than 5 to 45	64.2	4.6	6.0	8.4	16.7
More than 45	64.4	7.8	7.0	6.5	14.3
Physical sciences	Physical science major	Physical science minor	Biology/life science major or minor	Science education major	Other major
Students in poverty (percent)					
0–10	41.8	10.7	14.4	15.5	17.6
More than 10 to 50	40.9	14.2	13.1	15.2	16.6
More than 50	30.8	15.7	26.1	6.0	21.5
Minority enrollment (percent)					
0–5	41.4	11.2	14.4	19.3	13.6
More than 5 to 45	41.7	14.3	15.0	13.4	15.7
More than 45	40.7	14.3	18.2	7.4	19.4

NOTE: Students in poverty are those who are approved to receive free or reduced-price lunches. Percents may not sum to 100 because of rounding. Physical sciences include chemistry, geology/earth sciences, other natural sciences (except biology/life sciences), and engineering.
SOURCE: National Science Board. *Science and Engineering Indicators 2004*. NSB 04-01. Arlington, VA: National Science Foundation, 2004. Appendix Table 1-13. This table was based on US Department of Education, National Center for Education Statistics, Schools and Staffing Survey, 1999-2000.

• Encourage higher-education institutions to establish mathematics and science teaching academies that include faculty from science, mathematics, and education departments through a competitive grant process.[9]

• Support promising students to study science, mathematics, and engineering teaching—particularly those obtaining degrees in science, mathematics, or engineering who plan to teach at the K–12 level following graduation through scholarships and loan programs for students as well as institutional funding.[10] Qualified college students and midcareer profession-

[9]Ibid.
[10]Ibid.

TABLE K–12-10 Public Middle and High School Mathematics and Science Teachers Who Entered Profession Between 1995-1996 and 1999-2000 and Reported Feeling Well Prepared in Various Aspects of Teaching in First Year, by Participating in Induction and Mentoring Activities: 1999-2000

Subject and activity	Handle classroom management and discipline	Use variety of instructional methods	Teach subject matter	Use computers in classroom instruction	Plan lessons effectively	Assess students	Select/adapt curriculum and instructional materials
All mathematics teachers	50.5	65.1	90.1	41.5	77.5	69.7	53.9
Induction program							
Yes	50.8	67.2	89.4	45.1	78.2	70.9	55.8
No	50.0	61.8	91.1	35.7	76.4	67.9	51.0
Mentor							
Yes	53.5	68.7	89.6	41.8	79.3	72.6	57.1
No	44.6	58.2	91.0	40.8	74.0	64.1	47.8
All science teachers	50.8	66.0	82.9	48.0	69.4	68.8	58.6
Induction program							
Yes	51.7	70.1	83.8	51.3	74.7	72.5	63.6
No	49.1	58.0	81.0	41.5	59.0	61.6	48.6
Mentor							
Yes	56.6	73.6	84.9	51.5	75.3	74.3	64.6
No	42.0	54.4	79.7	42.6	60.3	60.4	49.2

SOURCE: National Science Board. *Science and Engineering Indicators 2004*. NSB 04-01. Arlington, VA: National Science Foundation, 2004. Appendix Table 1-15. This table was based on US Department of Education, National Center for Education Statistics, Schools and Staffing Survey, 1999-2000.

TABLE K–12-11 Public School Students, Teachers, and Cost Data

Fall 2003 enrollment K–12[a]	48,132,518
High school graduates—2003-2004[a]	2,771,781
Male graduates going to college—2001[b]	60%
Female graduates going to college—2001[b]	64%
Total number of school teachers—2003-2004[a]	3,044,012
Total number of math and science teachers (K–12)[c]	1,700,000
Total number of math teachers (6–12) 1999-2000[d]	191,214
Total number of science teachers (6–12) 1999-2000[d]	159,488
Average public school teacher salary—2003-2004[a]	$46,752
Average spent per student[a]	$8,248
Operating school districts in the United States[a]	15,397

SOURCES:
[a]National Education Association. *Rankings & Estimates: Rankings of the States 2004 and Estimates of School Statistics 2005*. Atlanta, GA: NEA Research, June 2005. Available at: http://www.nea.org/edstats/images/05rankings.pdf.
[b]National Science Board. *Science and Engineering Indicators 2004*. NSB 04-01. Arlington, VA: National Science Foundation, 2004. Appendix Table 1-19.
[c]National Commission on Mathematics and Science Teaching for the 21st Century. *Before It's Too Late: A Report to the Nation*. Washington, DC: National Assessment of Education Progress, September 27, 2000. Available at: http://www.ed.gov/inits/Math/glenn/toc.html.
[d]National Center for Education Statistics. *Digest of Education Statistics 2003*. Washington, DC: US Department of Education, 2004. Table 67.

als need to be attracted into teaching and given the preparation they require to succeed. Experts in mathematics, science, and technology should be able to become teachers by completing programs to acquire and demonstrate fundamental teaching skills. Recruitment, preparation, and retention of minority-group teachers are particularly important as groups underrepresented in science, mathematics, and engineering become a larger percentage of the student population.[11]

• Conduct an aggressive national-outreach media campaign to attract young people to teaching careers in mathematics and science.[12]

• Work for broad improvements in the professional status of science, mathematics, and technology teachers.[13] Structured induction programs for new teachers, district–business partnerships, award programs, and other incentives can inspire teachers and encourage them to remain in the field. Most

[11]National Research Council, Committee on Science and Mathematics Teacher Preparation. *Educating Teachers of Science, Mathematics, and Technology: New Practices for the New Millennium*. Washington, DC: National Academy Press, 2000.

[12]Ibid.

[13]National Science Foundation, National Science Board. *The Science and Engineering Workforce: Realizing America's Potential*. Arlington, VA: National Science Foundation, 2003.

important, salaries for science, mathematics, and technology teachers need to reflect what they could receive in the private sector and be in accord with their contributions to society, and teachers need to be treated as professionals and as important members of the science and engineering communities.

ENHANCING THE QUALITY AND COHESION OF EDUCATIONAL STANDARDS

Since the early 1990s, states have been developing academic standards in mathematics, science, and technology education based in part on national standards developed by the National Council of Teachers of Mathematics, the National Research Council, the American Association for the Advancement of Science, and other organizations. The use of these standards in curriculum development, teaching, and assessment has had a positive effect on student performance and probably contributed to the recent increased performance of 8th-grade students in international comparisons.[14]

But standards still vary greatly from state to state and across districts and often are not well aligned with the tests used to measure student performance. In addition, many sets of standards remain focused on lower-level skills that may be easier to measure but are not necessarily linked to the knowledge and skills that students will need to do well in college and in the modern workforce. A common flaw in mathematics and science curricula and textbooks is the attempt to cover too much material, which leads to superficial treatments of subjects and to needless repetition when hastily taught material is not learned the first time. Standards need to identify the most important "big ideas" in mathematics, science, and technology, and teachers need to ensure that those subjects are mastered.

The No Child Left Behind legislation requires testing of students' knowledge of science beginning in 2006-2007, and the science portion of the NAEP is being redesigned. Development of such assessments raises profound methodologic issues, such as how to assess inquiry and problem-solving skills using traditional large-scale testing formats.

Several federal initiatives can serve the national interest in establishing and maintaining high educational standards while respecting local responsibility for teaching and learning:

- Help colleges, businesses, and schools work together to link K–12 standards to college admissions criteria and workforce needs to create a seamless K–16 educational system.[15]

---

[14]Bybee and Stage, 2005.

[15]National Science Foundation, National Science Board. *Preparing Our Children: Math and Science Education in the National Interest.* Arlington, VA: National Science Foundation, 1999.

- Provide incentives for states and coalitions of states to conduct benchmarking studies between their standards and the best standards available.
- Foster the development of high-quality curricula and assessments that are closely aligned with world-class standards.
- Establish ambitious but realistic goals for student performance—for example, that 30% of high school seniors should be proficient in science by 2010 as measured by the NAEP.

CHANGING THE INSTITUTIONAL STRUCTURE OF SCHOOLS

Students and teachers remain constrained by several of the key organizational features of schools.[16] The structure of the curriculum, of individual classes, of schools, and of the school day keeps many students from taking advantage of opportunities that could build their interest in science and technology.

Possible federal initiatives include these:

- Provide seed money or incentives for new kinds of schools and new forms of schooling. Promising ideas include small high schools, dual-enrollment programs in high schools and colleges, colocation of schools with institutions of higher education, and wider use of Advanced Placement and International Baccalaureate courses.
- Help districts institute reorganization of the school schedule to support teaching and learning.[17] Possibilities include devoting more time to study of academic subjects, keeping schools open longer in the day and during parts of the summer, and providing teachers with additional time for development and collaboration.
- Provide scholarships for low-income students who demonstrate that they have taken a core curriculum in high school that prepares them to study science, mathematics, or engineering in college.

CATALYZING CHANGE

The federal government has an important role to play in catalyzing the efforts of states, school districts, and schools to improve science, mathematics, and technology education. Promising actions include the following:

---

[16]US Department of Education, National Education Commission on Time and Learning. *Prisoners of Time*. Washington, DC: US Department of Education, 1994.

[17]Ibid.

• Launch a large-scale program of research, demonstration, and evaluation in K–12 science, mathematics, and technology education.[18] Such a program should include distinguished researchers working in partnership with practitioners and policy-makers and supported by a national coalition of public and private funding organizations and other stakeholders.
• Help create a nongovernment Coordinating Council for Mathematics and Science Teaching that would bring together groups with a stake in mathematics and science teaching and monitor progress on teacher recruitment, preparation, retention, and rewards.[19]
• Support the creation of state councils of business leaders, higher-education representatives, and K–12 educators to achieve comprehensive, coordinated, system-level improvement in science, mathematics, and technology education from prekindergarten through college.[20]

The United States brings unique strengths to the challenge of reforming K–12 science, mathematics, and technology education, including the flexibility of its workforce and its unparalleled legacy of achievement in science and technology. The challenge facing policy-makers is to find ways of generating meaningful change in an educational system that is large, complex, and pluralistic.

[18]National Research Council, Committee on a Feasibility Study for a Strategic Education Research Program. *Improving Student Learning: A Strategic Plan for Education Research and Its Utilization*. Washington, DC: National Academy Press, 1999.

[19]US Department of Education, The National Commission on Mathematics and Science Teaching for the 21st Century. *Before It's Too Late*. Washington, DC: US Department of Education, 2000.

[20]Business-Higher Education Forum. *A Commitment to America's Future: Responding to the Crisis in Mathematics and Science Education*. Washington, DC: American Council on Education, 2005.

Attracting the Most Able US Students to Science and Engineering

SUMMARY

The world economy is growing rapidly in fields that require science, engineering, and technologic skills. The United States can remain a leader in science and engineering (S&E) only with a well-educated and effectively trained population. The most innovative S&E work is done by a relatively small number of especially talented, knowledgeable, and accomplished individuals. Because of the importance of S&E to our nation, attracting and retaining individuals capable of such achievements ought to be a goal of federal policy.

It follows that a key component of national and economic security policy must be US S&E students. The United States has relied on drawing the best and brightest from an international talent pool. However, recent events have led some to be concerned that the United States cannot rely on a steady flow of international students. Furthermore, as other developed countries encourage international students to come to their countries and developing countries enhance their postsecondary educational capacity, there is increased competition for the best students, which could further reduce the flow of international students to the United States. Therefore, any policies aimed at encouraging student interest in S&E must have a significant component that focuses on domestic talent.

This paper summarizes findings and recommendations from a variety of recently published reports and papers as input to the deliberations of the Committee on Prospering in the Global Economy of the 21st Century. Statements in this paper should not be seen as the conclusions of the National Academies or the committee.

Fundamentally, policy levers designed to influence the number of US S&E workers fall into two categories: supply-side and demand-side. Among supply-side issues are K–12 science, mathematics and technology teaching, undergraduate S&E educational experience, graduate training experience, opportunity costs compared with those of other fields and professions, and length of postdoctoral training period. On the demand side are funding for research and availability of research jobs, both of which are powerfully influenced by public policies and by public and private expenditures on research and development.

Past reports have identified a number of options the federal government could take to influence the education and career decisions of top US students, including the following:

- Double the number of magnet high schools specializing in science, technology, engineering, and mathematics from approximately 100 to 200 over the next 10 years.
- Support competitive undergraduate scholarships for students interested in science, mathematics, and engineering.
- Provide scholarships to all qualified students majoring in science or mathematics at a 4-year college who have an economic need and who maintain high levels of academic achievement.
- Provide at least 5,000 portable graduate fellowships, each with a duration of up to 5 years, for training in emerging fields, to encourage US students to pursue S&E graduate studies.
- Provide graduate student stipends competitive with opportunities in other venues.
- Support a significant number of selective research assistant professorships in the natural sciences and engineering open to postdoctoral scholars who are US citizens or permanent residents.
- Partner with industry to sponsor a series of public-service announcements exalting science and technology careers.

GETTING AN EARLY START: K–12 S&E PROGRAMS

One proven way of fostering students' interest in science and technology is through magnet high schools that emphasize those subjects. There are approximately 100 such schools in the United States, and studies have shown that graduates from these schools are more likely to study science, mathematics, or engineering in college and enter those fields during their careers.[1] It is not known, however, whether these students would have had similar career trajectories even if they had not attended magnet schools.

[1]K. Powell. "HothoUSe High," *Nature* 435(2005):874-875.

During the undergraduate years, involvement in research projects and the guidance of experienced mentors are powerful means of retaining students in S&E.[2] Mentors can provide advice, encouragement, and information about people and issues in a particular field. An early exposure to research can demonstrate to students the kinds of opportunities they will encounter if they pursue research careers.

TRENDS IN UNDERGRADUATE AND GRADUATE STUDENT INTEREST IN S&E

When one examines the issue, it becomes clear that there is a great deal of domestic student interest in undergraduate S&E programs. About 30% of students entering college in the United States (of whom over 95% are US citizens or permanent residents) intend to major in S&E fields. This proportion has remained fairly constant over the last 20 years. However, a considerable gap exists between freshman intentions and successful degree completion. Undergraduate S&E programs report the lowest retention rate among all academic disciplines. A National Center for Educational Statistics (NCES) longitudinal study of first-year S&E students in 1990 found that fewer than 50% of undergraduate students entering college declaring a S&E major had completed S&E degrees within 5 years.[3] Indeed, approximately 50% of such undergraduate students changed their major field within the first 2 years.[4] Undergraduates who opt out of S&E programs are among the most highly qualified college entrants.[5] They are also disproportionately women and nonwhite students, indicating that many potential entrants are discouraged before they can join the S&E workforce.[6]

---

[2]R. F. Subotnik, K. M. Stone, and C. Steiner. "Lost Generation of Elite Talent in Science." *Journal of Secondary Gifted Education* 13(2001):33-43.

[3]L. K. Berkner, S. Cuccaro-Alamin, and A. C. McCormick. *Descriptive Summary of 1989-1990 Beginning Postsecondary Students: 5 Years Later with an Essay on Postsecondary Persistence and Attainment*. NCES 96155. Washington, DC: National Center for Education Statistics, 1996.

[4]T. Smith. *The Retention and Graduation Rates of 1993-1999 Entering Science, Mathematics, Engineering, and Technology Majors in 175 Colleges and Universities*. Norman, OK: Center for Institutional Data Exchange and Analysis (C-IDEA), University of Oklahamo, 2001.

[5]S. Tobias. *They're Not Dumb, They're Different. Stalking the Second Tier*. Tucson, AZ: Research Corporation, 1990; E. Seymour and N. Hewitt. *Talking About Leaving: Why Undergraduates Leave the Sciences*. Boulder, CO: Westview Press, 1997; M. W. Ohland, G. Zhang, B. Thorndyke, and T. J. Anderson. "Grade-Point Average, Changes of Major, and Majors Selected by Students Leaving Engineering." *34th ASEE/IEEE Frontiers in Education Conference*, 2004. Session T1G:12-17.

[6]M. F. Fox and P. Stephan. "Careers of Young Scientists: Preferences, Prospects, and Reality by Gender and Field." *Social Studies of Science* 31(2001):109-122; D. L. Tan. *Majors in Science, Technology, Engineering, and Mathematics: Gender and Ethnic Differences in Persis-*

Graduate enrollment in S&E programs has been a relatively level 22-26% of total enrollments since 1993 (see Figures TS-1A, B, C, and D and TS-2). Growth in the number of S&E doctorates awarded is due primarily to the increased numbers of international students but also to the increasing participation of women and underrepresented minority groups.[7] If the primary objective of the US S&E enterprise is to maintain excellence, a major challenge is to determine how to continue to attract the best international students and at the same time encourage the best domestic students to enter S&E undergraduate and graduate programs.

DECISION POINTS AND DISINCENTIVES

There are inherent disincentives that push students away from S&E programs and careers. These disincentives fall into three broad categories: curriculum, economics, and environment. Undergraduate attrition may be due partly to a disconnect between the culture and curricula in high schools compared with those at colleges and universities.[8] For example, poor mathematics preparation in high school may underlie attrition in undergraduate physics programs. Underrepresented groups such as Blacks and American Indians, who are educated disproportionately in underserved communities, are on the whole less well prepared for college.[9] These types of problems suggest transitional programs to bridge the gap between high school and college, but the value of such strategies has not been compared with those at other levels in the educational system.

Higher education is costly, and employment opportunities fluctuate. Whether a student perceives that a degree will lead to a viable career is a major factor determining choice of field.[10] This is illustrated particularly

tence and Graduation. Norman, OK: University of Oklahoma, 2002. Available at: http://www.ou.edu/education/csar/literature/tan_paper3.pdf; Building Engineering and Science Talent (BEST). *The Talent Imperative: Diversifying America's S&E Workforce*. San Diego: BEST, 2004; G. D. Heyman, B. Martyna, and S. Bhatia. "Gender and Achievement-Related Beliefs Among Engineering Students." *Journal of Women and Minorities in S&E* 8(2002):33-45.

7National Science Foundation. *Graduate Enrollment Increases in S&E Fields, Especially in Engineering and Computer Sciences*. NSF 03-315. Arlington, VA: National Science Foundation, 2003.

8A. Venezia, M. W. Kirst, and A. L. Antonio. *Betraying the College Dream: How Disconnected K–12 and Postsecondary Education Systems Undermine Student Aspirations*. Stanford, CA: The Bridge Project, Stanford University, 2003. Available at: http://www.stanford.edu/group/bridgeproject/betrayingthecollegedream.pdf.

9E. Babco. *Trends in African American and Native American Participants in STEM Higher Education*. Washington, DC: Commission on Professionals in Science and Technology, 2002.

10C. T. Clotfeltner, R. G. Ehrenberg, M. Getz, and J. J. Siegfried. *Economic Challenges in Higher Education*. Chicago, IL: University of Chicago Press, 1991; M. S. Teitelbaum. "Do We Need More Scientists?" *The Public Interest* 153(2003):40-53.

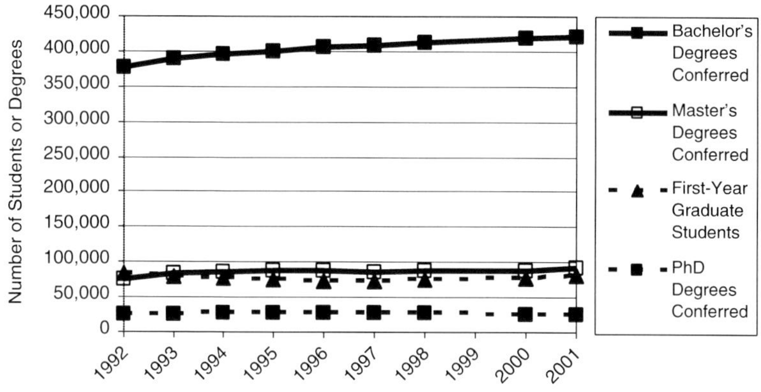

FIGURE TS-1A Number of first-year graduate students and number of S&E degrees conferred, by degree type, 1992-2001.
SOURCE: Data on first-year graduates are from National Science Foundation. *Survey of Graduate Student and Postdoctorates in Science and Engineering*. NSF 03-320. Arlington, VA: National Science Foundation, 2003. Degree data from National Science Foundation. *Science and Engineering Degrees: 1966-2001*. NSF 04-311. Arlington, VA: National Science Foundation, 2003.

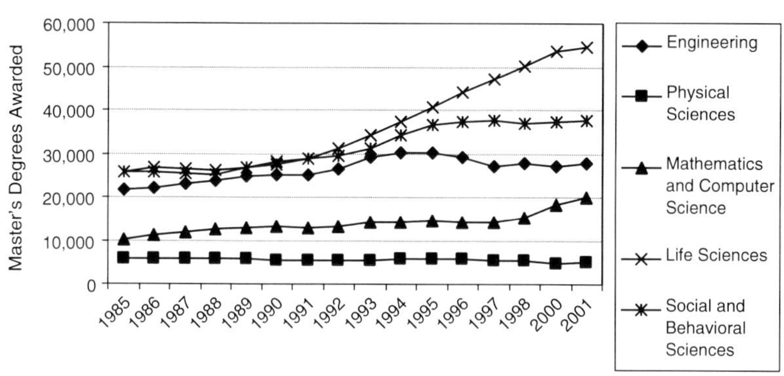

FIGURE TS-1B Number of S&E master's degrees awarded, by field, 1985-2001.
SOURCE: National Science Foundation. *Science and Engineering Degrees: 1966-2001*. NSF 04-311. Arlington, VA: National Science Foundation, 2003.

well in engineering: undergraduate student decisions to major in particular fields vary depending on business cycles.

Research indicates that large schools, which often foster a competitive "weeding out" environment, have a much higher attrition rate than smaller schools. This environment can be compounded by the culture of specific

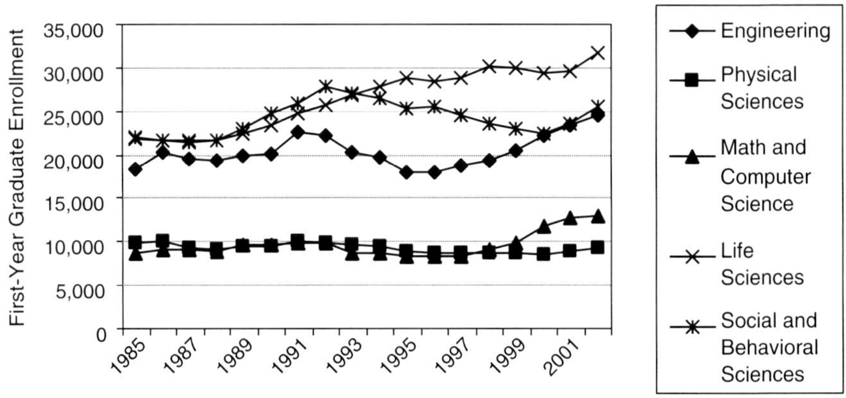

FIGURE TS-1C Number of first-year S&E graduate enrollments, by field, 1985-2001.
SOURCE: Data on first-year graduates are from National Science Foundation. *Survey of Graduate Student and Postdoctorates in Science and Engineering*. NSF 03-320. Arlington, VA: National Science Foundation, 2003.

fields. Some researchers argue that a key factor in stemming attrition is feeling connected to the intellectual and social life of the college.[11] Another researcher writes of three types of university cultures—the *elite* (scientific excellence), the *pluralist* (research, teaching, and service), and the *communitarian* (citizenship)—each carrying its own set of values and signals, some of which are competing.[12] Departments, colleges and universities, and professional societies each have a role in providing a high-quality, engaging learning environment.

After a student's determination of an undergraduate major or concentration, another key transition point is a decision to enter and complete graduate training.[13] Major factors to consider include time to degree and economics.[14] Unclear job prospects and lost earning potential are major

[11]V. Tinto. *Leaving College: Rethinking the CaUSes and Curses of Student Attrition*. Chicago, IL: University of Chicago Press, 1993; J. M. Braxton. *Reworking the Student Departure Puzzle*. Nashville, TN: Vanderbilt University Press, 2000.

[12]M. F. Fox and P. Stephan. "Careers of Young Scientists: Preferences, Prospects, and Reality by Gender and Field." *Social Studies of Science* 31(2001):109-122.

[13]A. Lu. *The Decision Cycle for People Going to Graduate School*. Stamford, CT: Peterson's Thomson Learning, 2002.

[14]NAS/NAE/IOM. *Reshaping the Graduate Education of Scientists and Engineers*. Washington, DC: National Academy Press, 1995.

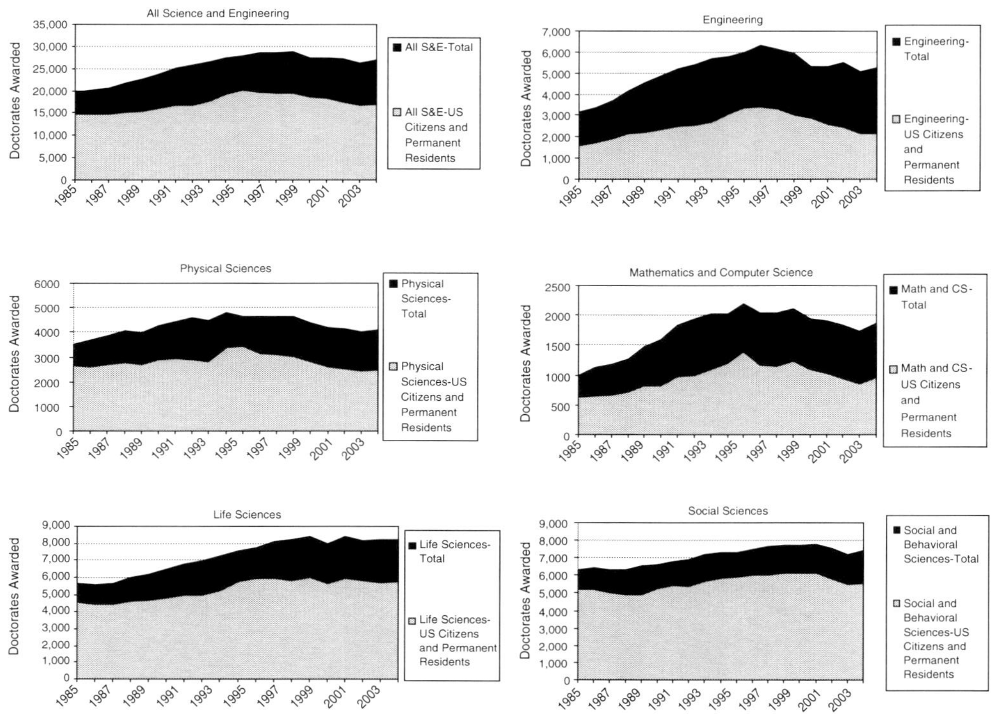

FIGURE TS-1D Number of doctorates awarded, by field and citizenship, 1985-2003. US citizens and permanent residents earn on average about 60-70% of S&E doctoral degrees; about 80% in life sciences and social sciences, 60% in physical sciences, and 50% in engineering and mathematics and computer sciences.
SOURCE: National Science Foundation. *Survey of Earned Doctorates.* Arlington, VA: National Science Foundation, 2005.

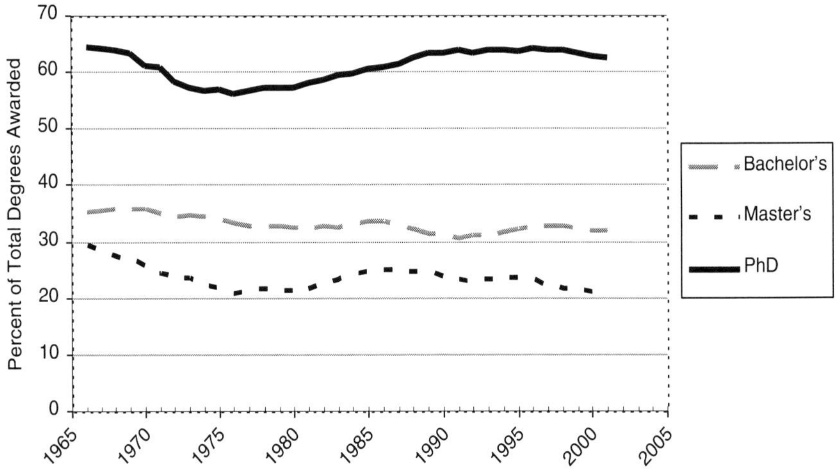

FIGURE TS-2 Percent of total degrees awarded which are S&E degrees, by degree type: 1966 to 2001. Most US doctorate degrees are awarded in S&E fields.
SOURCE: Based on National Science Foundation. *Science and Engineering Degrees: 1966-2001*. NSF 04-311. Arlington, VA: National Science Foundation, 2003. Table 1. Data from National Center for Education Statistics, Integrated Postsecondary Education Data System. Completions Survey and National Science Foundation/ Division of Science Resources Statistics Survey of Earned Doctorates.

disincentives for many considering an advanced S&E degree.[15] An issue raised in several studies on doctoral education is that prospective students are underinformed. A large, cross-disciplinary, multi-institutional survey on the experiences of doctoral students indicated that students entering doctoral programs entered their programs "without having a good idea of the time, money, clarity of purpose, and perseverance that doctoral education entails."[16] The burden of being informed does not rest solely on the prospective student. While professional schools make a point to inform prospective students of the salary and employment levels of graduates, it

[15]R. Freeman, E. Weinstein, E. Marincola, J. Rosenbaum, and F. Solomon. "CAREERS: Competition and Careers in Biosciences." *Science* 294(5550)(2001):2293-2294; W. Butz, G. A. Bloom, M. E. Gross, T. K. Kelly, A. Kofner, and H. E. Rippen. *Is There a Shortage of Scientists and Engineers?: How Would We Know?* IP-241-OSTP. Santa Monica, CA: RAND Corporation, 2003. Available at: http://www.rand.org/publications/IP/IP241/IP241.pdf; M. S. Teitelbaum. "Do We Need More Scientists?" *The Public Interest* 153(2003):40-53.

[16]C. M. Golde and T. M. Dore. *At Cross Purposes: What the Experiences of Doctoral Students Reveal About Doctoral Education*. Philadelphia, PA: A Report Prepared for The Pew Charitable Trusts, 2001.

appears that S&E graduate programs rarely make such information available.[17]

Career Prospects in S&E

Students considering research careers can face daunting prospects. Graduate and postdoctoral training may take over a decade, usually with low pay and few benefits. Most researchers do not become full-fledged members of the profession until their mid-30s or later—an especially onerous burden for those who are trying to balance the demands of work and family.

Even at the end of this long training period, many do not find the jobs for which they have been trained. The stagnation of funding for the physical sciences, mathematics, engineering, and the social sciences over the last decade has led to fewer academic faculty positions in these fields. Even in expanding fields, such as the biosciences, the number of permanent academic research and teaching positions has not kept up with the growing number of students who are entering these fields. As a result, more and more researchers languish in temporary positions.[18] The fastest-growing employment category since the early 1980s has been "other academic appointments," which is currently increasing at about 4.9% annually.[19] These jobs are essentially holding positions filled by young researchers coming from postdoctoral positions who would like to join an academic faculty on a tenure track and are willing to wait. It is an increasingly long wait as institutions are decreasing the number of faculty appointments to decrease the long-term commitments that they entail. From 1993 to 2001, the number of biomedical tenure-track appointments increased by 13.8%, while those for nontenure-track faculty increased by 45.1% and other appointments by 38.9% (see Figures TS-3A and B).

In fields outside the life sciences, most doctorates go on to careers in industry or government (see Figures TS-4A and B). Increasingly, these sectors are providing research opportunities for the best students. At the same time that biotechnology firms are gearing up their R&D operations, top industrial research laboratories, such as Bell Labs and Xerox PARC are

---

[17]P. Romer. *Should the Government Subsidize Supply or Demand in the Market for Scientists and Engineers?* Working Paper 7723. Cambridge, MA: National Bureau for Economic Research, 2000. Available at: http://www.nber.org/papers/w7723/; National Research Council. *Trends in the Early Careers of Life Scientists*. Washington, DC: National Academy Press, 1998.

[18]National Research Council. *Trends in the Early Careers of Life Scientists*. Washington, DC: National Academy Press, 1998.

[19]National Research Council. *Advancing the Nation's Health Needs*. Washington, DC: The National Academies Press, 2005.

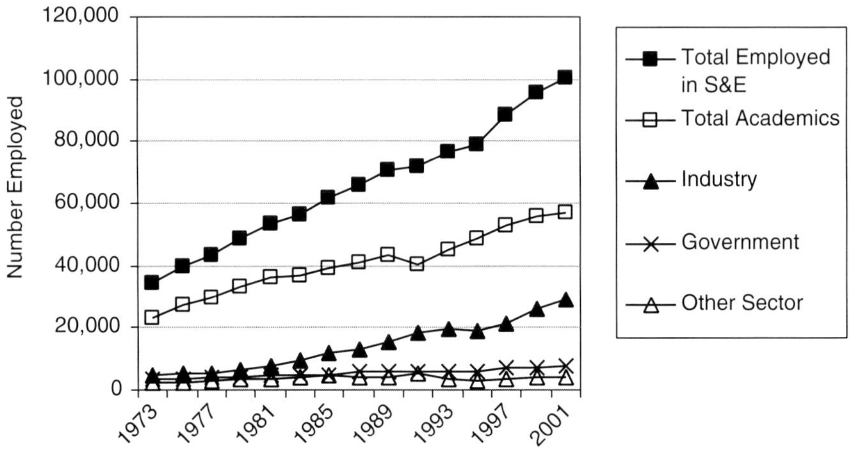

FIGURE TS-3A Number of biomedical jobs, by sector, 1973-2001.
SOURCE: National Research Council. *Advancing the Nation's Health Needs.* Washington, DC: The National Academies Press, 2005. Appendix E.

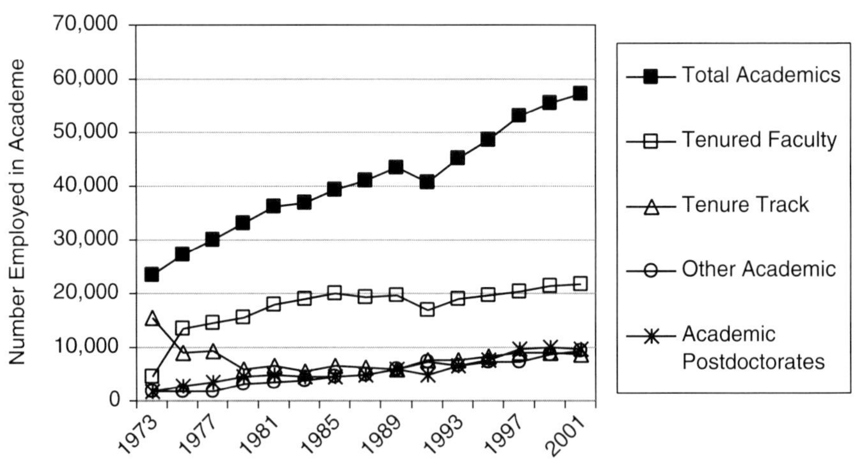

FIGURE TS-3B Number of biomedical academic jobs, by tenure-track status, 1973-2001.
SOURCE: National Research Council. *Advancing the Nation's Health Needs.* Washington, DC: The National Academies Press, 2005. Appendix E.

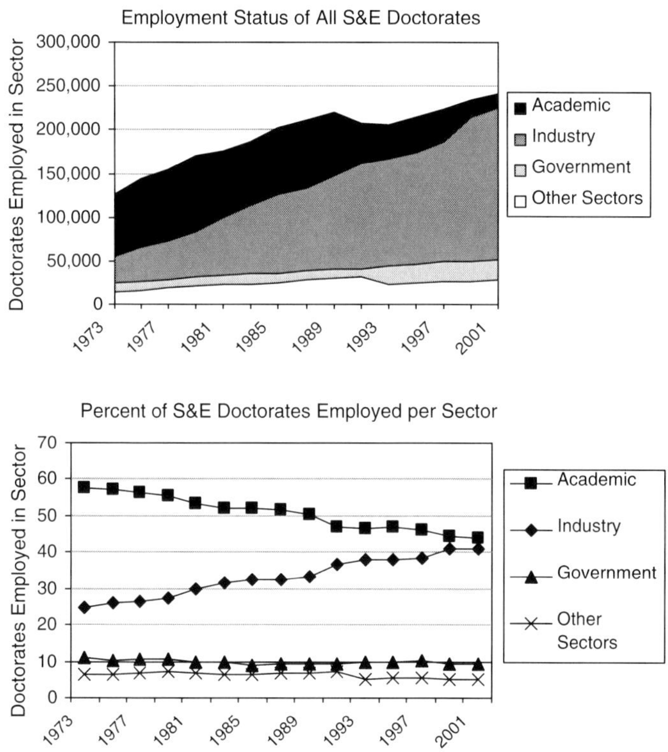

FIGURE TS-4A Number and percentage of S&E doctorates employed, by sector, 1973-2001.
SOURCE: National Science Foundation. *Survey of Doctoral Recipients 2003*. Arlington, VA: National Science Foundation, 2005.

closing down, leaving physical-science graduates with few options. Increasingly, mathematics and computer-science graduates are turning to finance and Wall Street. Given these shifts in workforce opportunities, top US students may consider options other than S&E very attractive. Careers in such professions as law, medicine, business, and health services require less training, offer more secure job prospects, and have much higher lifetime earning potential (see Tables TS-1A and B).

INTEREST IN RESEARCH CAREERS BY TOP STUDENTS TRACKS JOB MARKET

The current contrast between these options and research is influencing career decisions. According to available sources of data, accomplished US

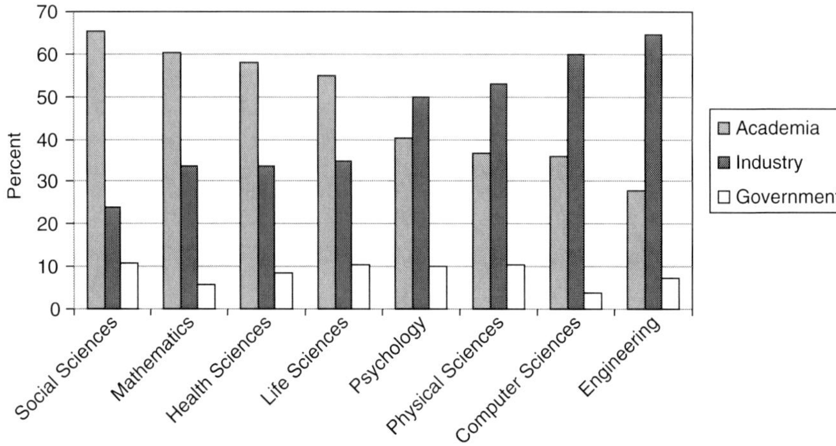

FIGURE TS-4B Work sector of PhDs by field, 2001.
SOURCE: National Science Foundation. *Survey of Doctoral Recipients 2003*. Arlington, VA: National Science Foundation, 2005.

students are increasingly turning away from S&E, especially during their undergraduate years.[20] In the 1990s, surveys of science majors from top universities showed a striking decline of interest in S&E careers. Between 1984 and 1998, the percentage of college seniors planning to go to graduate school in the next fall in S&E fields dropped from 17 to 12%. Among those students with A or A- grade-point averages, the declines were comparably steep—from 25 to 18%.[21]

Between 1992 and 2000, the number of college seniors who scored highly on the Graduate Record Examination (GRE) and indicated that they intended to study S&E in graduate school fell by 8%. The number of these top students planning to go to graduate school in fields other than S&E grew by 7% (Figure TS-5). The greatest declines were in engineering (25%) and mathematics (19%). Among top GRE scorers, however, enrollment in biological sciences programs showed a 59% gain. When it came to careers outside S&E, the researchers found that the fields attracting the largest growth in top GRE scorers were short training programs in health profes-

[20]W. Zumeta and J. S. Raveling. "Attracting the Best and the Brightest." *Issues in Science and Technology* (Winter 2002):36-40.

[21]E. I. Holmstrom, C. D. Gaddy, V. V. Van Horne, and C. M. Zimmerman. *Best and Brightest: Education and Career Paths of Top S&E Students*. Washington, DC: Commission on Professionals in Science and Technology, 1997.

sions, such as physical therapy, speech and language pathology, and public health—drawing 88% more top scorers in 2000 than in 1992.

Where are top students going if not into S&E? The top US students do not appear to be headed in large numbers into law school or medical school, where enrollments have been flat or declining. But more do seem to be attracted to graduate business schools, where the number of MBAs awarded annually grew by nearly one-third during the 1990s. During this period, many S&E undergraduate students also may have entered directly into the workforce after graduating, attracted in part by the booming economy. As the economy slowed in the early part of this decade, some of these students may have returned to graduate school, and more undergraduates may have opted to continue their studies.[22]

Indeed, 1999 appears to have been the nadir for student interest in S&E graduate study. The economy's recent slump has prompted growing numbers of top US college graduates to attend graduate school, new data show, sharply reversing course from the late 1990s, when more of the brightest young Americans headed for quicker-payoff careers in business and health. By 2001, with fewer high-technology jobs beckoning, the share of top US citizen scorers (above 750) on the GRE quantitative scale heading to graduate school in the natural sciences and engineering increased by about 31% compared with 1998, after having declined by 21% in the previous 6 years.[23] This recent increase is comparable with the 29% gain in the number of all score levels of examinees who intended to enroll in graduate school in S&E. And the total number of GRE examinees increased by 9% between 1998 and 2001, suggesting that more students in a variety of fields were preparing for graduate school.

Enrollments of International Students[24]

As the number of US students studying S&E in graduate schools has dropped, these schools and employers of scientists and engineers have compensated by enrolling and employing more students and trained personnel from other countries. In 2003, foreign students earned 38% of doctorates

---

[22]W. Zumeta and J. S. Raveling. The Best and the Brightest for Science: Is There a Problem Here? In M. P. Feldman and A. N. Link, eds. *Innovation Policy in the Knowledge-Based Economy*. Boston: Kluwer Academic Publishers, 2001. Pp. 121-161.

[23]W. Zumeta and J. S. Raveling. "The Market for Ph.D. Scientists: Discouraging the Best and Brightest? Discouraging All?" AAAS Symposium, February 16, 2004. Press release available at: http://www.eurekalert.org/pub_releases/2004-02/uow-rsl021304.php.

[24]See also the International Students Issue Brief elsewhere in this report.

TABLE TS-1A Median PhD Salaries of Engineering and Science Graduates, by Occupation and Field of Doctorate in 1997

	Occupation	
	All Sectors	University
Economics	$75,000	55,000
Computer Science	75,000	56,000
Engineering	73,000	65,000
Physical Science	65,000	52,000
Biological Sciences	56,000	40,000
S&E PhDs in Management, Median Net Income. MDs	92,000	85,000
	Field	
	All Sectors	University
Economics	$69,000	62,000
Computer Science	72,000	57,000
Engineering	75,000	68,000
Physical Science	70,000	54,300
Biological Sciences	60,000	53,000

SOURCE: R. B. Freeman, E. Weinstein, E. Marincola, J. Rosenbaum, and F. Solomon. *Careers and Rewards in Bio Sciences: The Disconnect Between Scientific Progress and Career Progression*. Bethesda, MD: American Society for Cell Biology, 2001. Available at: http://www.ascb.org/publications/competition.html.

in S&E, including 59% of engineering doctorates.[25] In 2000, foreign-born professionals occupied 22% of all US S&E jobs, up from 14% just 10 years before.

But relying on foreign sources of students and research professionals is risky. As systems of higher education and research continue to develop in other countries, it is likely that fewer scientists and engineers will want to come to the United States to study or work. Security concerns also have led to a drop in applications to US graduate programs from international students. Over time, multinational firms may decide simply to locate their R&D facilities overseas, closer to their sources of scientists and engineers.

[25]The National Academies. *Policy Implications of International Graduate Students and Postdoctoral Scholars in the United States*. Washington, DC: The National Academies Press, 2005.

TABLE TS-1B Bioscience Salary Case Study on Lifetime Income Disadvantage

Lifetime earnings for most doctorates are lower than in other high-level careers, particularly for bioscientists, who are paid less than other highly educated workers at any given level of job experience and who take longer to obtain full-time jobs. The two factors cumulate to a huge lifetime economic disadvantage—on the order of $400,000 in earnings compared with high-paying PhD fields, such as engineering, which also require many years of preparation but in which graduates do not in general delay entry into the job market to take postdoctoral postions. This is equivalent to a salary disadvantage of ~$25,000 per year for every year of working life. Medicine, which has a similar career as the biosciences because of residency in hospitals after completion of training, has about twice the lifetime income.

The economic disadvantage is greater when we compare bioscience with professions that require less preparatory training. Consider, for example, a person who has just graduated from a 2-year MBA program in 2000, earning $77,000 in base salary and $12,560 in signing bonus (without stock options). A bioscience PhD who completed postdoctoratal training might earn $50,000 as a starting assistant professor. But the MBA graduate would have spent 2 years in school compared with the 10-12 years that students spend as graduate students and postdoctoral fellows. The salary differential cumulates to a lifetime difference in earnings, exclusive of stock options, conservatively estimated at $1 million discounted at 3%—comparable with $62,000 per year of working life. Add in the options and bonuses that managers get, and this differential could easily double.

SOURCE: Based on Freeman et al., 2001.

Finally, an overreliance on foreign-born scientists and engineers may have the subtle effect of discouraging US students from entering these fields, both because of cultural differences they might encounter during their education (about 20% of the faculty members in S&E were not born in the United States[26]) and because of a downward pressure on wages caused by an abundance of international scientists and engineers eager to work in this country.

Possible federal actions include the following:

- Double the number of magnet high schools specializing in science, technology, engineering, and mathematics from approximately 100 to 200 over the next 10 years. Federal support for these schools would send a powerful message to the entire K–12 system about the importance of science and technology.
- Sponsor regional, national, and international meetings and competitions for high-school students and undergraduates interested in science,

[26]National Science Board. *Science and Engineering Indicators 2004*. NSB 04-01. Arlington, VA: National Science Foundation, 2004.

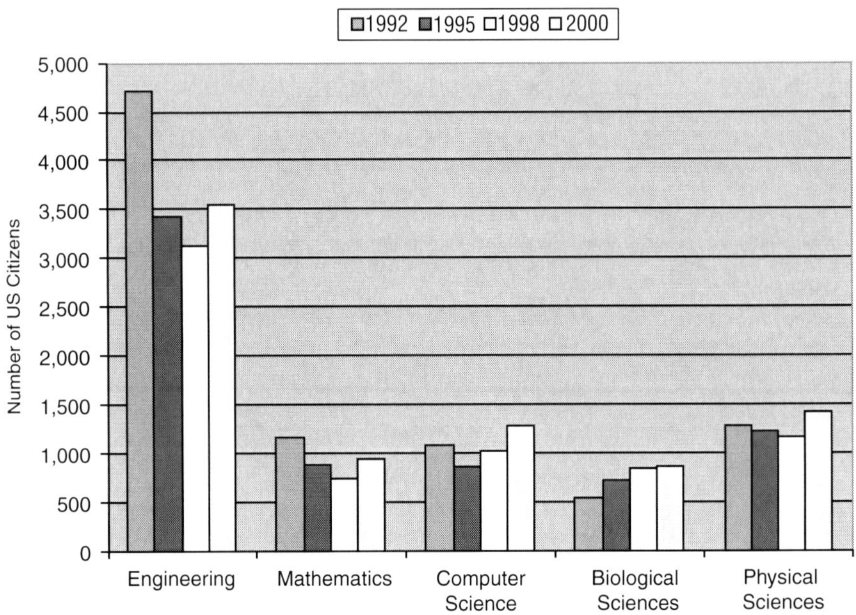

FIGURE TS-5 Number of US citizen GRE examinees scoring over 750 on the quantitative scale by intended S&E field, 1992, 1995, 1998, and 2000.
SOURCE: W. Zumeta and J. Raveling. "The Best and Brightest: Is There a Problem Here?" 2002. Available at: http://www.cpst.org/BBIssues.pdf.

mathematics, and engineering. Extracurricular activities and interactions with established scientists, mathematicians, and engineers can be powerful motivating forces for students interested in these subjects.

- Partner with industry to sponsor a series of public-service announcements exalting S&E careers.[27]
- Provide scholarships to all qualified students majoring in science or mathematics at 4-year colleges who have an economic need and who maintain high levels of academic achievement.[28] Financial assistance also should be provided to 2-year colleges and to students at those institutions to pre-

[27]American Electronics Association. *Losing the Competitive Challenge?* Washington, DC, 2005.

[28]Council on Competitiveness. *Innovate America.* Washington, DC: Council on Competitiveness, 2004.

pare for careers in S&E and to transfer to 4-year programs. Tax credits could be provided to companies or individuals who contribute to scholarship funds for S&E students.

• Provide at least 5,000 portable graduate fellowships, each with a duration of up to 5 years, for training in emerging fields.[29]

• Support prestigious fellowships for graduate study in S&E at US universities that would inspire the best US students in these fields. Though these grants should be linked to the student and therefore portable, an institutional component of each grant would spur competition for these students among institutions.

• Provide graduate-student stipends competitive with opportunities in other venues.[30]

• Substantially increase the number of undergraduate and graduate S&E students drawn from the "underrepresented majority."[31] Today, women, Blacks, Hispanics, American Indians, and persons with disabilities make up two-thirds of the US workforce but only 25% of the technical workforce.

• Support a significant number of selective research assistant professorships in the natural sciences and engineering at universities.[32] These would be highly competitive positions open to postdoctoral scholars who are US citizens or permanent residents. They would provide young and creative scholars with opportunities to pursue research of their own choosing even if they cannot secure positions at research institutions. This would expand the pool of good jobs in S&E in a way that would be expected to affect young people who are trying to decide whether to go to graduate school.

• Develop prizes for research goals of particular national interest, such as curing AIDS or going into space cheaply. Such prizes can provide flexibility for the researchers striving to achieve them and inspire and educate the public in current research interests.[33]

---

[29]Ibid.

[30]National Science Board, 2003.

[31]Building Engineering & Science Talent. *The Talent Imperative*, San Diego: BEST, 2004.

[32]W. Zumeta and J. S. Raveling. "Attracting the Best and the Brightest." *Issues in Science and Technology* (Winter 2002):36-40.

[33]National Academy of Engineering. *Concerning Federally Sponsored Inducement Prizes in Engineering and Science*. Washington, DC: National Academy Press, 1999.

Undergraduate, Graduate, and Postgraduate Education in Science, Engineering, and Mathematics

SUMMARY

As educators of the nation's future scientists, engineers, mathematicians, and K–12 teachers, US 2-year and 4-year colleges and universities are the central institutions in building the human resources needed for scientific and technological leadership.

However, these institutions face a number of challenges in producing knowledgeable graduates and trained professionals. Today, the United States ranks 17th globally in the proportion of its college-age population that earns science and engineering (S&E) degrees, down from third several decades ago.[1] Many other nations now have a higher fraction of 24-year-olds with S&E degrees (see Figure HE-1). And even though the proportion of its population who attends graduate school is small, because of its large population China graduates three times as many engineers from its colleges as does the United States.

In the past, the United States has relied on international students and scientific and engineering professionals to maintain its base of human resources in these fields. But global competition for S&E talent is intensifying, and enrolling higher percentages of US students in these programs would have many benefits.

This paper summarizes findings and recommendations from a variety of recently published reports and papers as input to the deliberations of the Committee on Prospering in the Global Economy of the 21st Century. Statements in this paper should not be seen as the conclusions of the National Academies or the committee.

[1]Council on Competitiveness. *Innovate America*. Washington, DC: Council on Competitiveness, 2004.

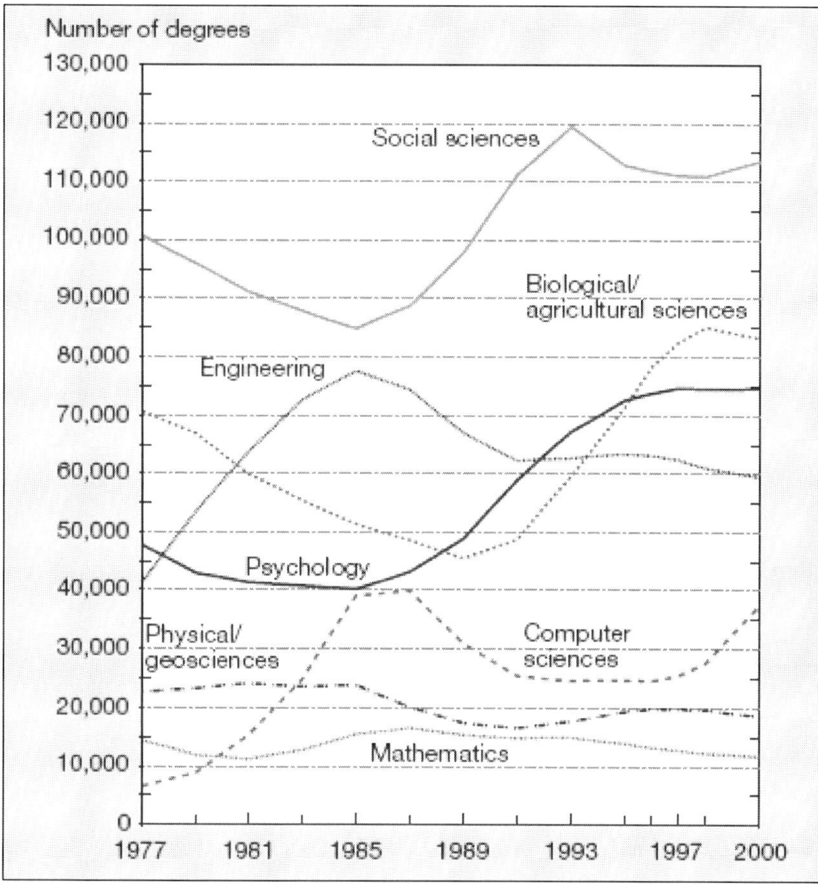

FIGURE HE-1 S&E bachelor's degrees, by field: selected years, 1977-2000.
NOTE: Geosciences include earth, atmospheric, and ocean sciences.
SOURCE: National Science Board. *Science and Engineering Indicators 2004*. NSB 04-01. Arlington, VA: National Science Foundation, 2004. Figure 2-11.

To meet this goal, many believe that the United States will need to attract S&E students from all demographic groups. Today, Blacks, Hispanics, and other underrepresented minority groups are about a quarter of the US population but make up only 17.9% of the undergraduate population, 2.5% of the these majors, and 6% of the S&E workforce (see Table HE-1 and Figure HE-2). Only a quarter of this workforce consists of women, though women are almost half the total US workforce. By 2020, more than

TABLE HE-1 Ratio of Bachelor's Degree to the 24-Year-Old Population, by Selected Fields, Sex, and Race/Ethnicity: 1990 and 2000

	Degree					Degree		
Sex and race/ethnicity	All bachelor's degrees	All S&E	NS&E	Social/ behavioral sciences	24-year-old population	Bachelor's	NS&E[a]	Social/ behavioral science
	Number					Ratio to 24-year-old population[b]		
1990 total	1,062,160	345,794	169,938	175,856	3,722,737	28.5	4.6	4.7
Male	495,876	199,917	117,249	82,668	1,855,513	26.7	6.3	4.5
Female	566,284	145,877	52,689	93,188	1,867,224	30.3	2.8	5.0
White	856,686	270,225	127,704	142,521	2,628,439	32.6	4.9	5.4
Asian/Pacific Islander	38,027	19,437	13,338	6,099	120,797	31.5	11.0	5.0
Underrepresented minority	107,377	33,419	15,259	18,160	973,500	11.0	1.6	1.9
Black	59,301	18,230	7,854	10,376	484,754	12.2	1.6	2.1
Hispanic	43,864	13,918	6,868	7,050	459,073	9.6	1.5	1.5
American Indian/Alaskan Native	4,212	1,271	537	734	29,674	14.2	1.8	2.5
2000 total	1,253,121	398,622	210,434	188,188	3,703,200	33.8	5.7	5.1
Male	536,158	197,669	128,111	69,558	1,886,400	28.4	6.8	3.7
Female	716,963	200,953	82,323	118,630	1,816,800	39.5	4.5	6.5
White	895,129	270,416	142,400	128,016	2,433,400	36.8	5.9	5.3
Asian/Pacific Islander	75,265	12,368	23,185	12,368	148,800	50.6	15.6	8.3
Underrepresented minority	200,967	63,519	27,939	35,559	1,121,000	17.9	2.5	3.2
Black	104,212	32,924	13,795	19,129	527,600	19.8	2.6	3.6
Hispanic	88,324	27,984	12,919	15,065	560,200	15.8	2.3	2.7
American Indian/Alaskan Native	8,431	2,611	1,246	1,365	33,200	25.4	3.8	4.1

[a]NS&E degrees include natural (physical, biological, earth, atmospheric, and ocean sciences), agricultural, and computer sciences; mathematics, and engineering.

[b]Number of degrees per 100 24-year-olds.

NOTE: NS&E = natural sciences and engineering.

SOURCE: National Science Board. *Science and Engineering Indicators 2004*. NSB 04-01. Arlington, VA: National Science Foundation, 2004. Table 2-8. This table was based on US Department of Education, Completions Survey; National Science Foundation, Division of Science Resources Statistics, WebCASPAR database system, available at: http://caspar.nsf.gov/; and US Bureau of the Census, Population Division.

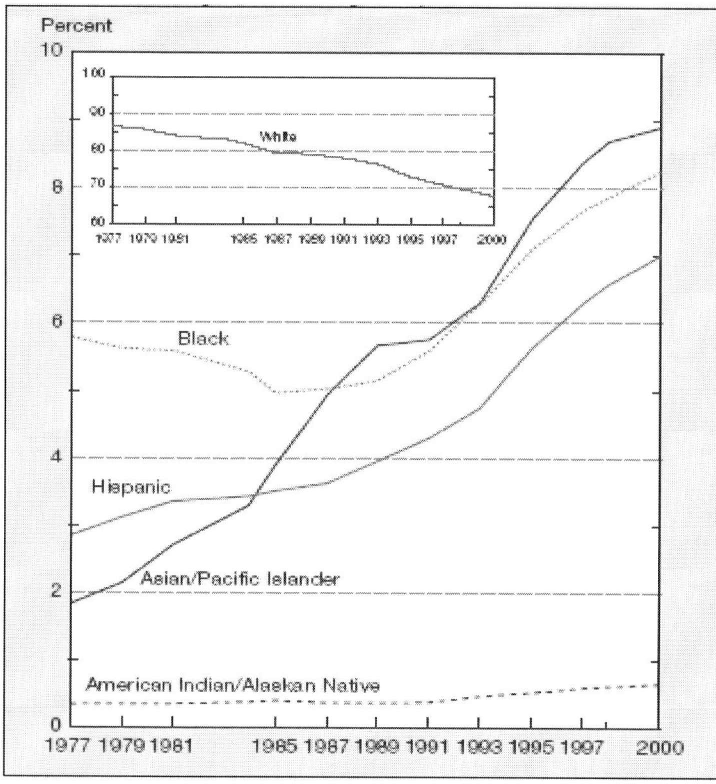

FIGURE HE-2 Minority share of S&E bachelor's degrees, by race/ethnicity: selected years, 1977-2000.
SOURCE: National Science Board. *Science and Engineering Indicators 2004*. NSB 04-01. Arlington, VA: National Science Foundation, 2004. Figure 2-13.

40% of the US college-age population will be members of currently underrepresented minorities.

The federal government has a key role in establishing workforce policies that address national needs and opportunities. Given how many years of education and training are required for someone to become a scientist, engineer, or mathematician, policies may need to focus on long-term opportunities that may help to smooth short-term labor-market dynamics. Among the federal actions that organizations have recommended are the following:

Undergraduate Education

- Provide incentives for all institutions of higher education to provide diverse internship opportunities for all undergraduates to study science, mathematics, engineering, and technology as early in their academic careers as possible.
- Expand funding for programs at 2-year and 4-year colleges that succeed in attracting and retaining women and members of minority groups underrepresented in science, mathematics, and engineering.

Graduate Education

- Establish education and traineeship grants to institutions focused on frontier research areas and multidisciplinary or innovation-oriented studies.
- Require institutions applying for federal grants to report on the size, scope, and performance (student completion rates and career outcomes) of their graduate programs to determine whether these programs are meeting the interests of students in preparing them for diverse careers in academe, industry, government, and the nonprofit sector.

Postdoctoral Training

- Develop federal policies and standards for postdoctoral fellows supported on federal research grants, including letters of appointment, performance evaluations, benefits and leave, and stipend support.
- Help develop creative solutions to the problems faced by dual-career couples so that more US students opt to pursue research careers.
- Create standards for and require the submission of demographic information on postdoctoral scholars supported on federal research grants by investigators awarded such grants. Collect data on postdoctoral working conditions, prospects, and careers.

The following discusses these issues in greater depth.

UNDERGRADUATE EDUCATION

The undergraduate years have a profound influence both on future professionals in science and mathematics and on broader public support of those fields. Undergraduate education acts as a springboard for students who choose to major in and then pursue graduate work in science and mathematics. Undergraduate institutions and community colleges train the technical support personnel who will keep our technological society functioning smoothly in the years ahead. And colleges and universities prepare the elementary and secondary teachers who impart lifelong knowledge and attitudes about science and mathematics to their students. For many, the undergraduate years are the last opportunity for rigorous academic study of these subjects.

Precollege education needs to include quality instruction in standards-based classrooms and a clear awareness that achievement in science and mathematics will be expected for admission to college. In addition, faculty in these disciplines should assume greater responsibility for the pre-service and in-service education of K–12 teachers.

Many introductory undergraduate courses in science and mathematics fields have been taught to select out the best, most committed students and discard the rest. This strategy is being questioned: Are introductory courses the appropriate place and time for such filtering? Are the students being turned away any less good than those who stay? Evidence indicates that undergraduates who opt out of S&E programs are among the most highly qualified college entrants.[2] Can the United States afford to turn away talented students interested in these fields?

Some argue more broadly that all college students should gain an awareness, understanding, and appreciation of the natural and human-constructed worlds and have at least one laboratory experience. Therefore, introductory science and mathematics courses must find ways to provide students both with a broad education in these fields and with the specific skills they need to continue studying these subjects, as is the case with most other introductory courses in colleges. Students who decide to pursue non-S&E majors would then have the background and education to make informed decisions about S&E in their personal lives and professional careers.

To serve these multiple objectives, many introductory and lower-level courses and programs would need to be designed to encourage students to continue, rather than end, their study of S&E subjects. Institutions should continually and systematically evaluate the efficacy of courses in these subjects for promoting student learning.

Many of these issues are also highly relevant to students who enter 2-year colleges after graduating from high school. For example, about a quarter of the students who earn bachelor's degrees in engineering have taken a substantial number of their lower-level courses at a community college, and nearly half have taken at least one community college course.

---

[2]S. Tobias. *They're Not Dumb, They're Different. Stalking the Second Tier*. Tucson, AZ: Research Corporation, 1990; E. Seymour and N. Hewitt. *Talking About Leaving: Why Undergraduates Leave the Sciences*. Boulder, CO: Westview Press, 1997; M. W. Ohland, G. Zhang, B. Thorndyke, and T. J. Anderson. "Grade-Point Average, Changes of Major, and Majors Selected by Students Leaving Engineering." *34th ASEE/IEEE Frontiers in Education Conference*, 2004. Session T1G:12-17.

As more students make community colleges their point of entry to postsecondary education, the quality of the S&E education they receive in 2-year institutions becomes increasingly important. Community college students need access to the kinds of lower-division courses that can prepare them for upper-division coursework in science, mathematics, and engineering, either at their own institutions or through partnerships between institutions, distance learning, or other means. Two-year colleges need to provide students with access to the kinds of equipment, laboratories, and other infrastructure they need to succeed.

The federal government can help promote these institutional changes through the following actions:

- Provide incentives for all institutions of higher education to provide diverse internship opportunities for all undergraduates to study science, mathematics, engineering, and technology as early in their academic careers as possible.[3] Introductory courses should be integral parts of the standard curriculum, and all colleges should routinely evaluate the success of these courses.
- Encourage science, mathematics, and engineering departments to work with education departments and surrounding school districts to improve the preparation of K–12 students.
- Expand funding for science, mathematics, and engineering programs at 2-year and 4-year colleges that succeed in attracting and retaining women and members of minority groups underrepresented in these programs.[4]

MASTER'S AND PROFESSIONAL EDUCATION

The baccalaureate has been the entry-level degree for many professional positions over the last century, but many employers in our increasingly complex economy now recognize the value of employees who have advanced training (see Figure HE-3). Master's degree programs provide students with S&E knowledge that is more in-depth than that provided in baccalaureate programs and supplements this knowledge with skills that have application in business, government, and nonprofit settings. Master's degree programs also can provide the interdisciplinary training necessary for real-world jobs and can be structured to provide job-relevant skills in teamwork, project management, business administration, communication, statistics, and informatics. Moreover, master's programs have the potential

---

[3]National Research Council. *Transforming Undergraduate Education in Science, Mathematics, Engineering, and Technology*. Washington, DC: National Academy Press, 1999.

[4]National Science Foundation, National Science Board. *The Science and Engineering Workforce: Realizing America's Potential*. Arlington, VA: National Science Foundation, 2003.

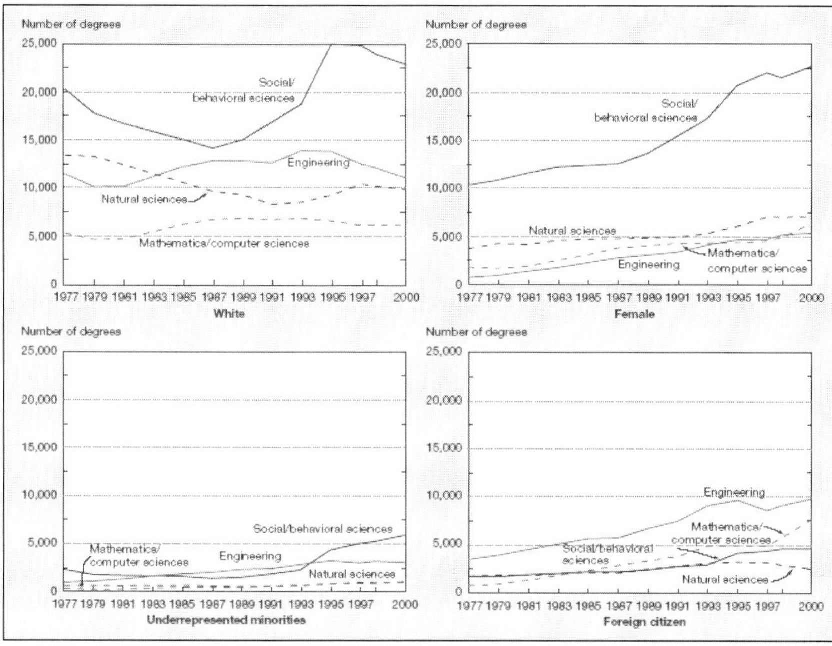

FIGURE HE-3 Master's degree in S&E fields earned by selected groups, 1977-2000. SOURCE: National Science Board. *Science and Engineering Indicators 2004*. NSB 04-01. Arlington, VA: National Science Foundation, 2004. Figure 2-17.

to attract greater numbers of women and minority-group members than do doctoral programs.

A number of reports since the mid-1990s have argued that master's degree programs for students in S&E with appropriate career aspirations can develop a cadre of professionals who meet employer needs. These reports have called for changes in master's education to make these programs more appropriate, cost effective, and attractive to students. In engineering, for example, the emphasis on increased skill in communications, business, the social sciences, cross-cultural studies, and important technologies has meant that the first professional degree should not be at the baccalaureate but at the master's level, as is the case in business, law, and medicine.

Options for the federal government include the following:

- Direct the National Science Foundation to fund professional science master's programs at institutions that demonstrate innovative approaches to orienting master's-level degree programs toward scientific or technical skills needed in the US workforce.

GRADUATE EDUCATION

Graduate education in the United States is widely seen as the best in the world. America's universities produce most of the scientists, engineers, and mathematicians who will maintain our preeminence in science and technology (see Figure HE-4). They educate the college faculty and K–12 teachers who will critically influence public support for scientific and technological endeavors And the intensive research experiences that are at the heart of graduate education at the doctoral level produce much of the new knowledge that drives scientific and technological progress.

Students from many nations travel to the United States to enroll in science, engineering, and mathematics graduate programs and to serve as postdoctoral fellows. For example, international students account for nearly half of all graduate enrollments in engineering and computer science. The presence of large numbers of international students in US graduate schools has both positive and negative consequences.[5] These students enhance the intellectual and cultural environments of the programs in which they are enrolled. Many remain in the United States after their training is finished and contribute substantially to our scientific and technological enterprise. However, the large numbers of foreign students in US graduate schools may have the effect of discouraging US students from pursuing this educational pathway because the rapidly increasing number of students has diminished the relative rewards of becoming a scientist or engineer.[6] US colleges and universities have an important role to play in encouraging more US students to pursue graduate education in science, engineering, and mathematics.

The federal government helps support graduate education through research assistantships funded through federal research project grants, fellowship and traineeship programs, and student loans (see Figure HE-5). The availability, level, and timing of this funding have implications for determining who can pursue a graduate education and how long it will take to complete that education. Also, the type of support—whether a research assistantship, teaching assistantship, traineeship, or fellowship—affects the content of graduate education and the kinds of skills one learns during graduate school.

---

[5]NAS/NAE/IOM. *Reshaping the Graduate Education of Scientists and Engineers.* Washington, DC: National Academy Press, 1995.

[6]R. E. Gomory and H. T. Shapiro. "Globalization: Causes and Effects." *Issues in Science and Technology* (Summer 2003):18-20.

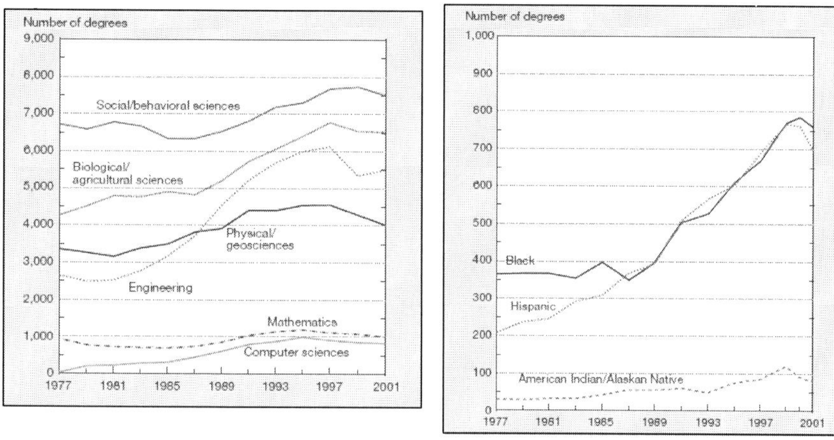

FIGURE HE-4 S&E doctoral degrees earned in US universities, by field and underrepresented minority S&E doctoral degrees, by race/ethnicity: selected years, 1977-2001.
SOURCE: National Science Board. *Science and Engineering Indicators 2004*. NSB 04-01. Arlington, VA: National Science Foundation, 2004. Figures 2-19 and 2-21.

In the 1990s, several events led to a national discussion of the content and process of doctoral education that continues today. In the late 1980s, labor-market forces pointed toward an impending shortage of PhDs in the arts and sciences in the early to mid-1990s. When the end of the Cold War, a national recession, state budget cuts, and the end of mandatory retirement for college faculty led instead to disappointing job prospects for new PhDs in the early 1990s, a national debate on the doctorate and the job prospects of PhD recipients ensued.

Also, in the 1990s, for the first time, more than half of PhDs in science and engineering reported that they held positions outside academe (see Figure HE-6). This trend has generated interest in providing graduate students with more information about their career options, including whether they should pursue a master's or doctoral degree and whether they should seek opportunities in government, industry, or nonprofit organizations as well as academe. In turn, this trend has focused attention on the need for training that provides the practical career skills needed in the workplace: pedagogic skills, technological proficiency, the ability to communicate well in writing or oral presentations, experience working in teams, and facility in grant writing and project management.

One great problem in discussions of workforce issues is the paucity of reliable, representative, and timely data. Often policy-makers are making decisions about the future based on data that are 2-3 years old.

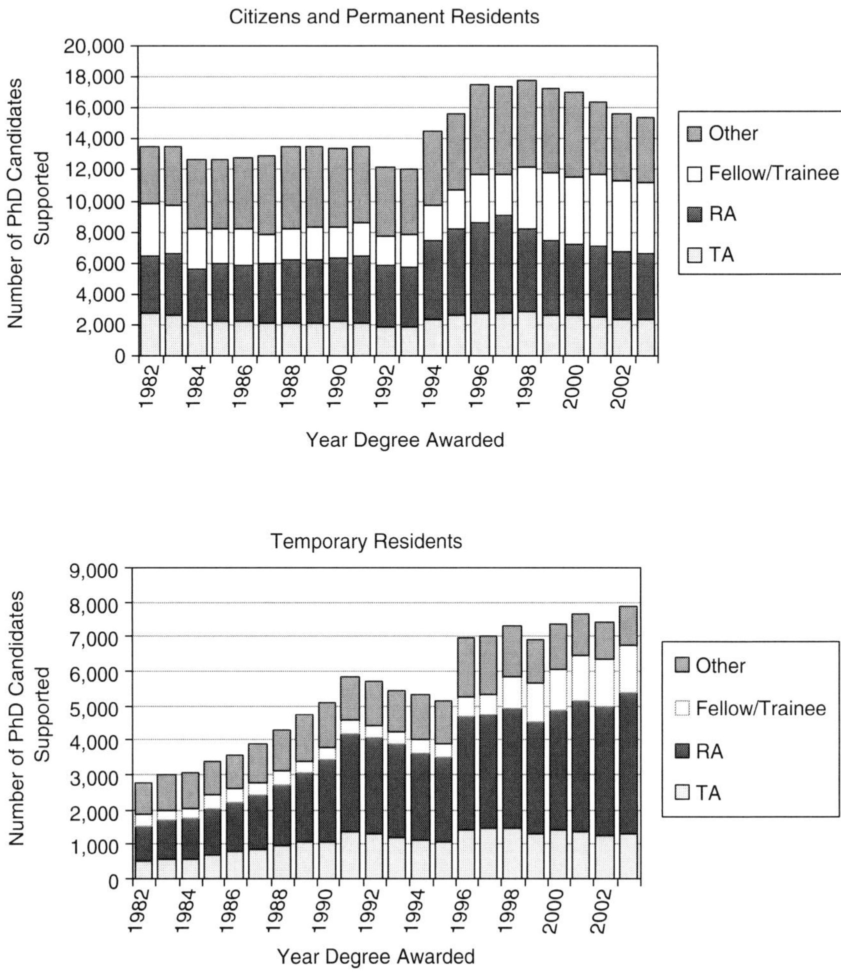

FIGURE HE-5 Number of PhD candidates supported, by support mechanism and citizenship status, 1982-2003.

NOTES: *Other:* Support from the student's or scholar's institution of higher education, state and local government, foreign sources, nonprofit institutions, or private industry; *traineeships:* educational awards given to students selected by the institution or by a federal agency; *research assistantships:* support for students whose assigned duties are primarily in research; *teaching assistantships:* support for students whose assigned duties are primarily in teaching.

SOURCE: National Science Foundation. *Survey of Earned Doctorates 2003.* Arlington, VA: National Science Foundation, 2004.

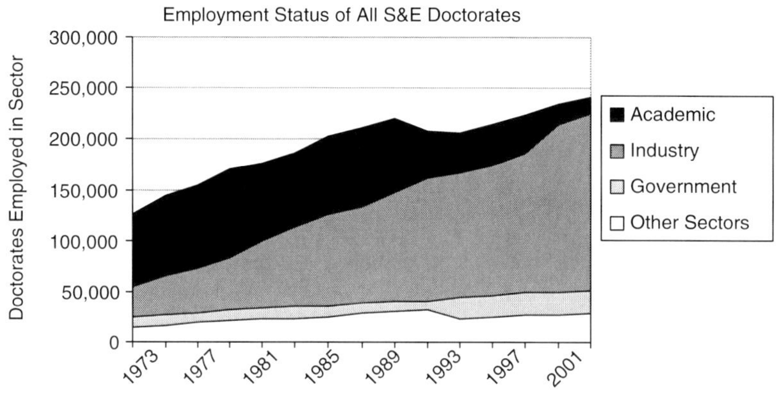

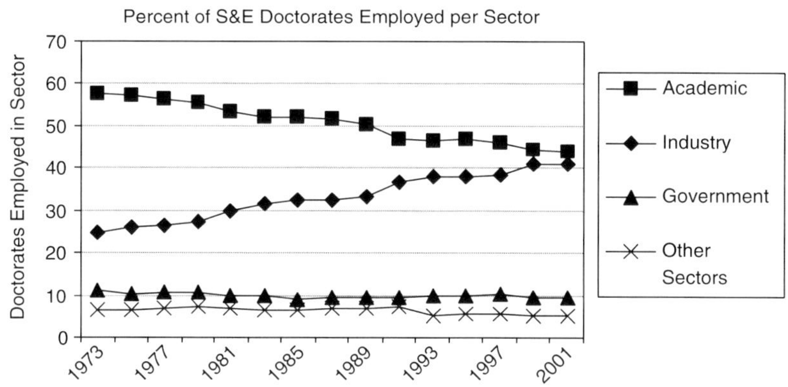

FIGURE HE-6 Number and percent of employment status of S&E doctorates, by sector, 1973-2001.
SOURCE: National Science Foundation. *Survey of Doctoral Recipients 2003*. Arlington, VA: National Science Foundation, 2005.

Options for the federal government include these:

• Establish education and traineeship grants to institutions focused on frontier research areas and multidisciplinary or innovation-oriented studies.[7]

• Eliminate the employer-employee stipulation in Office of Management Budget Circular A-21 to encourage the dual benefits to research and education of having graduate students serve as research assistants.[8]

---

[7]Ibid.

[8]Association of American Universities, Committee on Graduate Education. *Graduate Education*. Washington, DC: Association of American Universities, 1998.

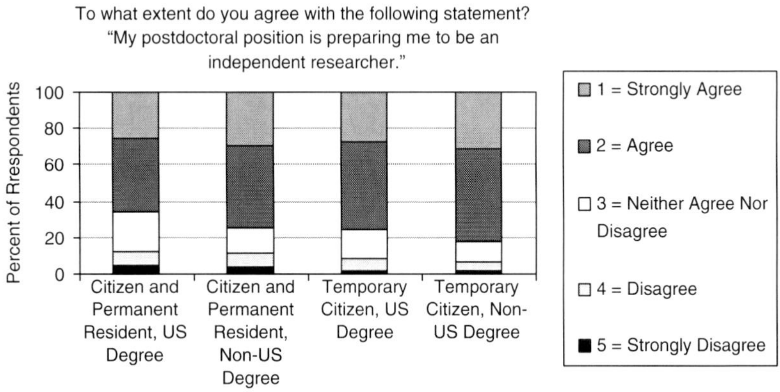

FIGURE HE-7 Response to postdoctoral survey question on preparation for independent research, by citizenship status and country of degree.
NOTE: 22,178 postdoctoral scholars at 46 institutions were contacted, including 18 of the 20 largest academic employers of postdoctoral scholars and National Institutes of Health. Postdoctoral status was confirmed by the institution. 8,392 (38%) responded; 6,775 (31%) of the respondents completed the entire survey, which included over 100 questions.
SOURCE: G. Davis. "Doctors Without Orders: Highlights of the Sigma Xi Postdoc Survey." *American Scientist* 93(3, supplement)(May-June 2005). Available at: http://postdoc.sigmaxi.org/results/.

- Require institutions applying for federal grants to report on the size, scope, and performance (student completion rates and career outcomes) of their graduate programs to determine whether these programs are meeting the interests of students in preparing them for diverse careers in academe, industry, government, and the nonprofit sector.[9]
- Provide graduate student stipends competitive with opportunities in other venues.[10]
- Direct the National Science Foundation to expand its data collection on S&E careers and its research into national and international workforce dynamics.[11]

Postdoctoral Training

For more than 2 decades, an increasing percentage of new PhD recipients have been pursuing postdoctoral study instead of employment after graduation. These experiences broaden and deepen the research and other skills that scientists and other highly trained professionals need to make major contributions to society (see Figure HE-7). Most postdoctoral schol-

[9]Ibid.
[10]National Science Foundation, National Science Board, 2003.
[11]NAS/NAE/IOM, 1995.

ars are funded by federal research grants (see Figure HE-8) and on average have stipends of under $35,000 per year.

However, mentors, institutions, and funding organizations have sometimes been slow to give postdoctoral fellows the status, recognition, and

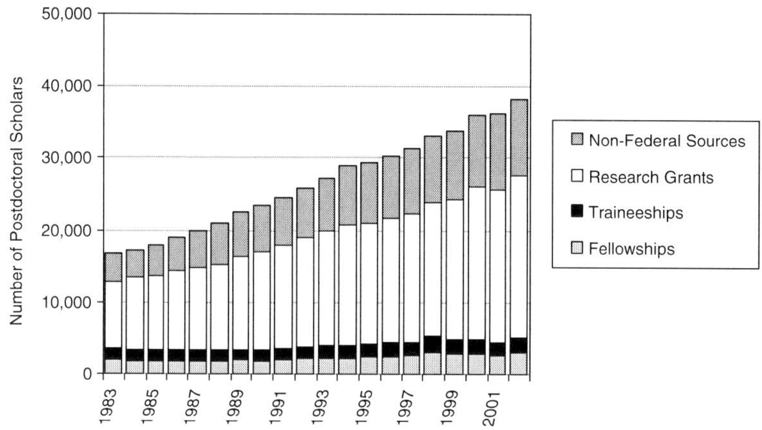

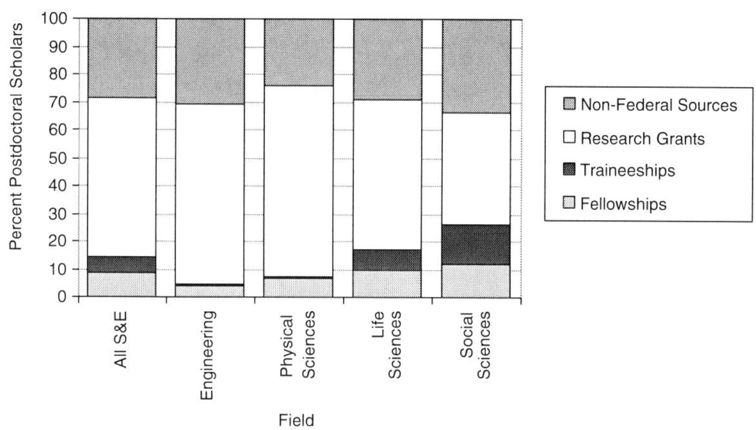

FIGURE HE-8 Number and percent of postdoctoral scholars, by funding mechanism and field, 1983-2002.
NOTE: *Non-federal support:* support may be from the institution of higher education, state, and local government, foreign sources, nonprofit institutions, or private industry; *research grants:* support from federal agencies to a principal investigator, under whom postdoctoral scholars work; *traineeships:* educational awards given to scholars selected by the institution or by a federal agency; *fellowships:* competitive awards given directly to scholars for financial support of their graduate or postdoctoral studies.
SOURCE: National Science Foundation. *Survey of Earned Doctorates 2002.* Arlington, VA: National Science Foundation, 2004.

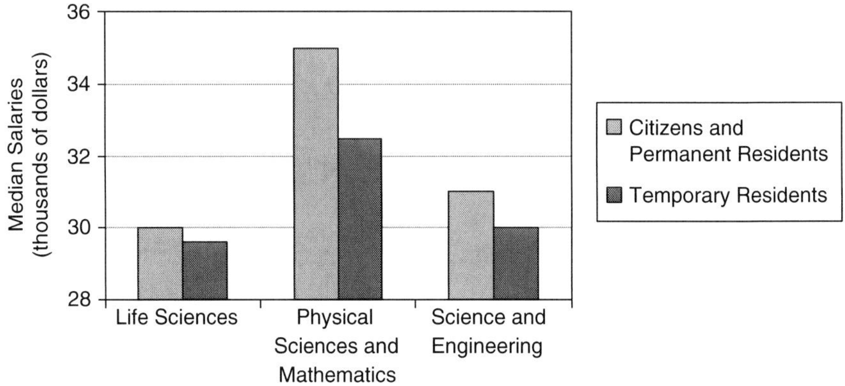

FIGURE HE-9 Median postdoctoral stipend, by field and citizenship status, 2001. SOURCE: National Science Foundation. *Survey of Earned Doctorates 2002*. Arlington, VA: National Science Foundation, 2004.

compensation that are commensurate with their skills and contributions to research (see Figure HE-9). Many postdoctoral scholars make substantial economic and familial sacrifices to pursue advanced training, yet they often do not have clearly defined rights, responsibilities, pay scales, access to benefits, or procedures for consideration of grievances.

To ensure a healthy research enterprise, the postdoctoral experience needs to be improved. The federal government should:

- Develop federal policies and standards for postdoctoral fellows supported on federal research grants, including letters of appointment, performance evaluations, benefits and leave, and stipend support. All postdoctoral scholars should have access to health insurance and to institutional services.[12]
- Help develop creative solutions to the problems faced by dual-career couples so that more US students opt to pursue research careers.
- Improve the quality and quantity of the data on postdoctoral working conditions, prospects, and careers.[13] Create standards for and require the submission of demographic information on postdoctoral scholars supported on federal research grants by investigators awarded such grants.

[12]NAS/NAE/IOM. *Enhancing the Postdoctorial Experience for Scientists and Engineers.* Washington, DC: National Academy Press, 2000.

[13]Ibid.

Implications of Changes in the Financing of Public Higher Education

SUMMARY

Public colleges and universities play a critical role in our nation's integrated system of education, research, and innovation. They educate the majority of undergraduates and constitute many of the nation's top research universities. They are training grounds for the people and ideas that drive innovation and improve our lives.

Yet even as public colleges and universities are becoming more important than ever in our knowledge-intensive society, many have come under intense financial pressure. Demographic changes in enrollments are driving up student enrollment in some places and reducing them in others, forcing institutions to adapt to new circumstances. The increasing costs of higher education have led to difficult tradeoffs affecting the quality of the education and services students receive. Extremely tight budgets in some states have reduced the relative appropriations to education in those states even as more students are looking to college as a means of personal advancement.

Though federal funding for student aid is up, more of this funding is going toward loans and tax benefits as opposed to student grants. Also, increases in funding have not been sufficient to match the needs of students.

This paper summarizes findings and recommendations from a variety of recently published reports and papers as input to the deliberations of the Committee on Prospering in the Global Economy of the 21st Century. Statements in this paper should not be seen as the conclusions of the National Academies or the committee.

The result has been a narrowing of educational choices for some students and concerns over deteriorating quality of public institutions.

Some organizations have proposed that the federal government take several important steps to improve the funding of public higher education and to increase student access to these institutions:

- Expand federal matching programs that encourage increased state appropriations for higher education.
- Reform the Medicaid program to slow the growth of state commitments that crowd out spending on higher education.
- Focus national resources on improving the purchasing power of Pell awards.
- Offer matching funds to states based on their funding of means-tested grant aid.

THE ROLE OF HIGHER EDUCATION IN THE KNOWLEDGE ECONOMY

Higher education has been central to the strength of the US economy over the last half-century. Broadened access for students has created social and economic opportunities for millions of Americans. The integration of education and research has become a key pillar of our research and innovation system. And the new knowledge generated has provided a strong engine for innovation and economic growth.

Public institutions are a particularly important component of America's higher education system. They enroll and educate one-quarter of all 4-year undergraduates (see Figure PHE-1). When community colleges are included, public schools account for more than 70% of all undergraduate enrollment (see Figures PHE-2A and B). Many of the nation's top research institutions, particularly in the Midwest and West, are public universities.

A strong system of higher education is more critical now than ever. Global competition in the knowledge economy is growing. Developed and developing countries are working to create high-quality educational institutions, often using American colleges and universities as a model. They are developing their own pool of knowledge workers and knowledge-sector firms.

For the United States to compete in this environment, American higher education needs to remain preeminent. It must continue to play a central role in the production of knowledge and innovation. It needs to create dynamic environments that will entice knowledge-based companies to locate in this country. The United States should facilitate world leadership of its higher education system by continuing to invest where it counts most.

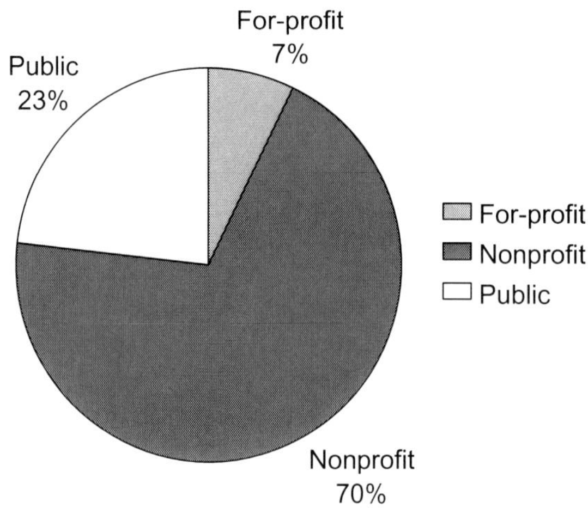

FIGURE PHE-1 Distribution of BA-granting institutions, by sector.
SOURCE: S. Turner. "Policy Implications of Changing Funding for Public Higher Education." Presentation to National Academies' Board on Higher Education and Workforce, April 2005.

STRESSES IN THE FINANCIAL STRUCTURE OF PUBLIC HIGHER EDUCATION

Public higher education is under severe financial pressures. The first source of pressure is increasing enrollments. The children of the baby boom are now reaching college age and will increase enrollments at some institutions over the coming decade (see Figures PHE-3A, B, and C). At the same time, the value of higher education as a means for students and society to achieve economic, social, and political goals also is boosting enrollments. Because public institutions typically do not charge students for the full cost of their education, the financial demands on these institutions are expected to grow significantly.[1]

A second stress on the system is the growing cost of higher education. Costs per student in higher education have grown consistently since the 1960s and steeply since the 1970s.[2] Both internal and external factors ap-

[1]R. C. Dickeson. *Collision Course: Rising College Costs Threaten America's Future and Require Shared Solutions.* Indianapolis, IN: Lumina Foundation for Education, Inc., 2004.

[2]J. L. Dionne and T. Kean. *Breaking the Social Contract: The Fiscal Crisis in Higher Education.* Report of the Commission on National Investment in Higher Education. New York: Council for Aid to Education, 1997.

	Distribution of Undergraduate Enrollment 1967	1996	Share of 1st time FT students in-state 1996
Community Colleges	21%	37%	92.7%
Other Public	38%	33%	81.5%
Flagship Public	13%	9%	72.0%
All Private	28%	21%	54.2%
Research I Private	3%	2%	23.6%
Liberal Arts Colleges	4%	2%	43.2%

FIGURE PHE-2A Distribution of undergraduate enrollment, by type of college, 1967 and 1996.
SOURCE: S. Turner. "Policy Implications of Changing Funding for Public Higher Education." Presentation to National Academies' Board on Higher Education and Workforce, April 2005.

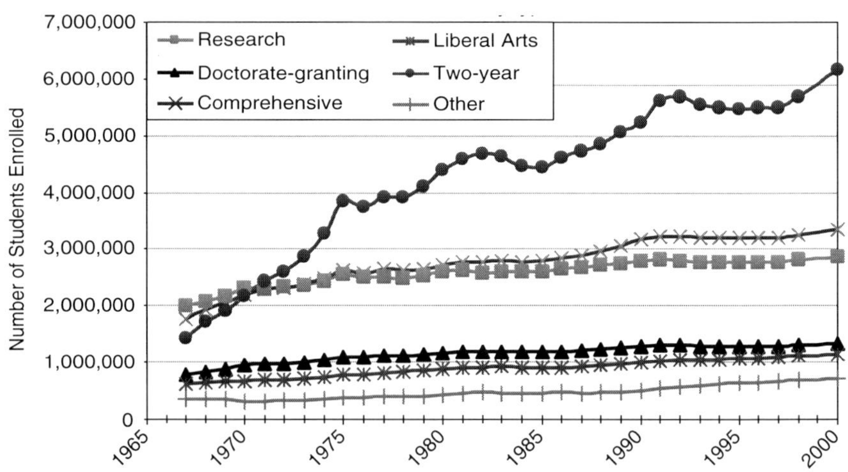

FIGURE PHE-2B Enrollment by type of institution, 1965-2000.
SOURCE: National Science Board. *Science and Engineering Indicators 2004*. NSB 04-01. Arlington, VA: National Science Foundation, 2004.

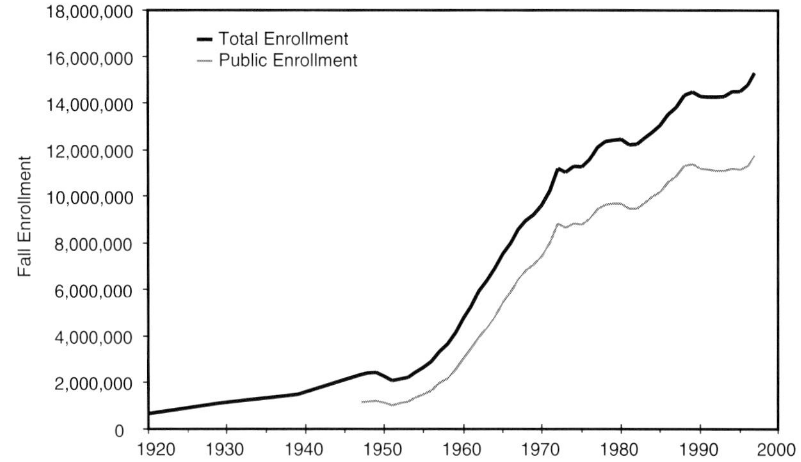

FIGURE PHE-3A Number of 18- to 24-year-olds enrolled in college, total and public colleges, 1920-2000.
SOURCE: S. Turner. "Policy Implications of Changing Funding for Public Higher Education." Presentation to National Academies' Board on Higher Education and Workforce, April 2005.

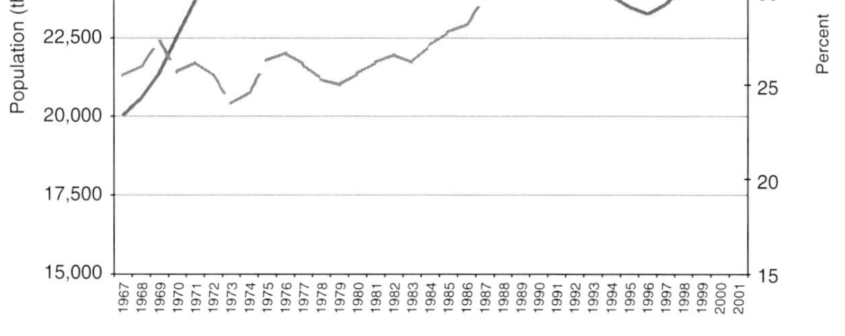

FIGURE PHE-3B Number and percent of 18- to 24-year-olds enrolled in college, 1967-2001.
SOURCE: T. J. Kane. "The Role of Federal Government in Financing Higher Education." Presentation to National Academies' Board on Higher Education and Workforce, March 21, 2005.

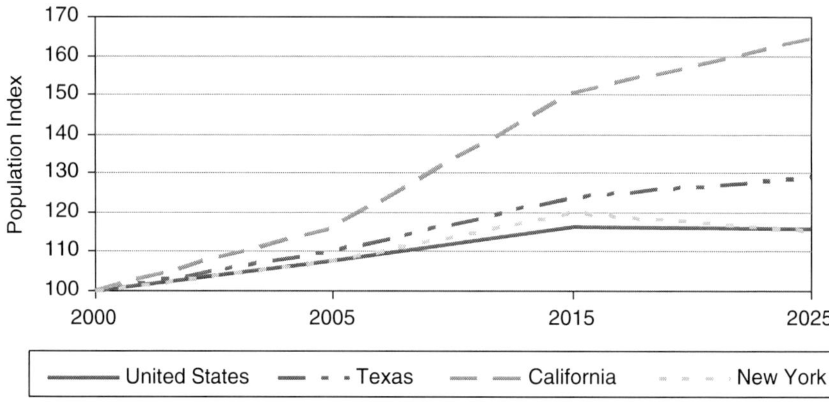

FIGURE PHE-3C Population growth projection for the United States, Texas, California, and New York, ages 18-24, 2000-2025.
NOTE: Calculations based on US Census Bureau, Population Projections.
SOURCE: T. J. Kane. "The Role of Federal Government in Financing Higher Education." Presentation to National Academies' Board on Higher Education and Workforce, March 21, 2005.

pear to be driving up costs. Universities need to compete for high-quality faculty, staff, and students. Computing services, information resources, and other services for students and faculty have added financial burdens (see Figure PHE-4). To cut costs in other areas, institutions have increased student:faculty ratios, shifted toward lower-cost part-time and nontenure-track faculty, encouraged early retirement, capped or postponed faculty salary increases, and outsourced noncritical missions[3] (see Figure PHE-5).

A third and perhaps the most important stress on public higher education has been a changing paradigm for public support at both the state and federal levels (see Figures PHE-6A, B, and C). Public colleges and universities—and even private ones that receive state support—have experienced strong competition for state resources over the last decade. Other state financial commitments—such as Medicaid payments—have continued to increase both in real dollars and as a percentage of state budget outlays, which has crowded out other spending priorities[4] (see Figure PHE-7).

[3]R. G. Ehrenberg and L. Zhang. *The Changing Nature of Faculty Employment*. Working Paper 44. Ithaca, NY: Cornell Higher Education Research Institute, 2004.

[4]T. J. Kane and P. R. Orszag. *Higher Education Spending: The Role of Medicaid and the Business Cycle*. Policy Brief #124. Washington, DC: The Brookings Institution, 2003.

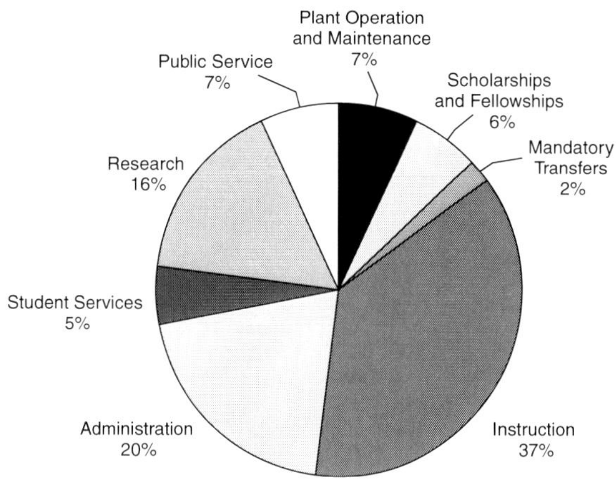

FIGURE PHE-4 Expenditure of all public institutions, by type of expense, 2001. SOURCE: National Center for Education Statistics. *Enrollment in Postsecondary Institutions, Fall 2001 and Financial Statistics, Fiscal Year 2001*. NCES 2004-155. Washington, DC: US Department of Education, December 23, 2003. Table 29. Available at: http://nces.ed.gov/pubsearch/pubsinfo.asp?pubid=2004155.

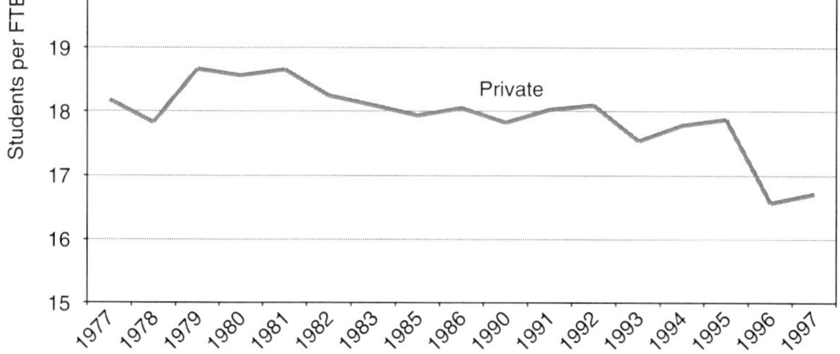

FIGURE PHE-5 Student-faculty ratios, by academic institution sector, 1977-1997. SOURCE: T. J. Kane. "The Role of Federal Government in Financing Higher Education." Presentation to National Academies' Board on Higher Education and Workforce, March 21, 2005.

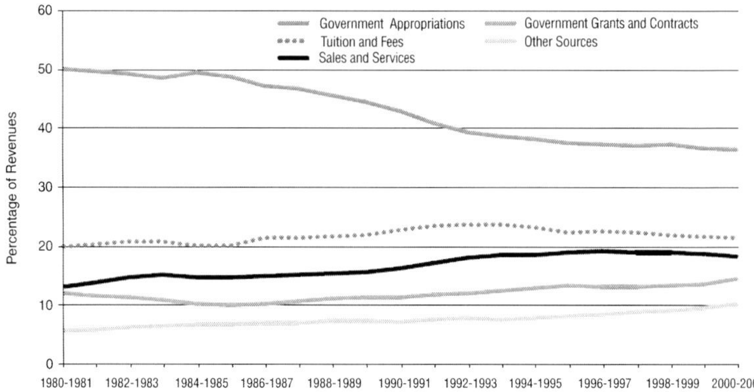

FIGURE PHE-6A Revenue sources for all public degree-granting institutions, 1980-1981 to 2000-2001.
SOURCE: College Board. *Trends in College Pricing, 2004*. Washington, DC: College Board, 2004. P. 20. Available at: http://www.collegeboard.com/prod_downloads/press/cost04/041264TrendsPricing2004_FINAL.pdf.

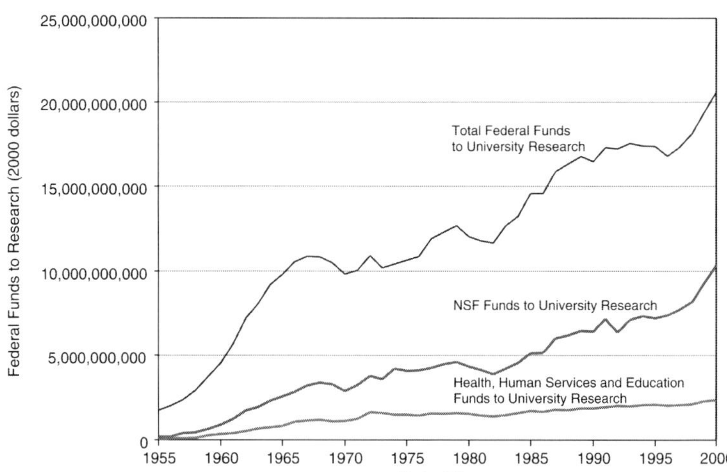

FIGURE PHE-6B Federal funds for university research, total and by select agency, 1955-2000.
SOURCE: S. Turner. "Policy Implications of Changing Funding for Public Higher Education." Presentation to National Academies' Board on Higher Education and Workforce, April 2005.

	Current Fund Revenues						Tuition	
	State & Local	Federal	Private	Endow.	Tuition	Aux. & Other	In-state	Out-of-State
Community Colleges	57.5%	11.7%	1.0%	0.1%	20.2%	9.5%	$1,814	$4,362
Other Public	36.3%	10.7%	4.0%	0.4%	18.3%	30.2%	$2,725	$6,981
Flagship Public	29.0%	14.8%	6.4%	1.3%	17.2%	31.4%	$3,493	$9,998
All Private	2.8%	10.3%	9.1%	5.1%	41.9%	30.8%		$12,881
Research I Private	2.3%	16.1%	9.5%	5.7%	22.9%	43.5%		$19,814
	1.4%	3.0%	9.1%	10.5%	55.5%	20.5%		$17,648

FIGURE PHE-6C Tuition and source of fund revenues, by academic institution type. SOURCE: S. Turner. "Policy Implications of Changing Funding for Public Higher Education." Presentation to National Academies' Board on Higher Education and Workforce, April 2005.

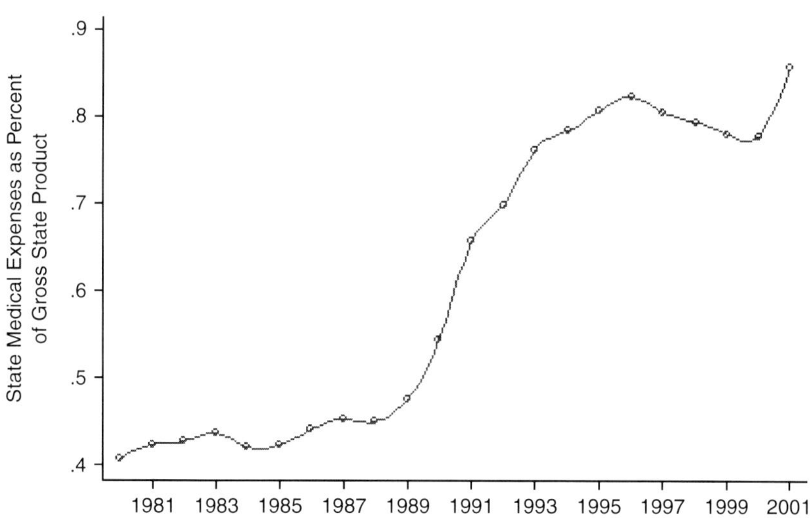

FIGURE PHE-7 Medicaid expenses as percent of gross state product, 1980-2001. SOURCE: T. J. Kane. "The Role of Federal Government in Financing Higher Education." Presentation to National Academies' Board on Higher Education and Workforce, March 21, 2005.

As a consequence of this financial pressure, education funding as a share of state spending, the percentage of education dollars directed to higher education, and the percentage of higher education dollars going to institutions (as opposed to students) all have declined[5] (see Figures PHE-8A and B). In brief, state support as a percentage of total revenue for public colleges and universities is down, and these institutions are adapting by restructuring costs and looking elsewhere (for example, to tuition) for financial support (see Figures PHE-9A and B).

At the federal level, spending for higher education appears on the surface to be strong. Spending on the Pell grant program, for example, increased 60% in real terms from 1999-2000 to 2003-2004[6] (see Figure PHE-10). However, hiding beneath the overall increases in federal support are important shifts in its distribution. The mix of federal support in 2003-2004 was 34% grants, 55% loans, and 5% tax benefits, the latter two of which have been growing as a percentage of federal support (see Figure PHE-11). Thus, there hasbeen a shift away from grants to other modes of support (for example, subsidized loans, tax credits, and tax-sheltered education accounts) and a shift from need-based to merit-based aid (see Figures PHE-12A, B, and C). Together, these changes have tended to shift subsidies away from students from lower-income families and toward the middle and upper-middle classes.

In addition, while there have been real increases in per student funding under the Pell grant program, they have not been adequate to offset larger increases in college prices. The size of the average grant has increased in real terms in recent years, but average tuition, fees, and room and board at public 4-year colleges and universities increased faster. As a result, the average Pell grant in 2003-2004 covered 23% of the charges at a public 4-year institution compared with 35% in 1980-1981[7] (see Figure PHE-13). Meanwhile, the Leveraging Education Assistance Partnerships (LEAP) program, which provides matching funds to states for providing need-based grant aid, has declined 31% in real terms over the last decade.[8]

IMPLICATIONS FOR AFFORDABILITY AND QUALITY

These developments have important implications both for access to higher education and for educational quality. As tuition increases, the array

[5]M. Rizzo. "State Preferences for Higher Education Spending: A Panel Data Analysis, 1977-2001." Paper presented at Cornell Higher Education Research Institute's Annual Conference, "Assessing Public Higher Education at the Start of the 21st Century." Ithaca, NY, May 22-23, 2005.

[6]College Board. *Trends in Student Aid 2004*. Washington, DC: College Board, 2004.

[7]Ibid.

[8]Ibid.

FIGURE PHE-8A Higher education appropriations share of state expenses, 1977-2002.
SOURCE: T. J. Kane. "The Role of Federal Government in Financing Higher Education." Presentation to National Academies' Board on Higher Education and Workforce, March 21, 2005.

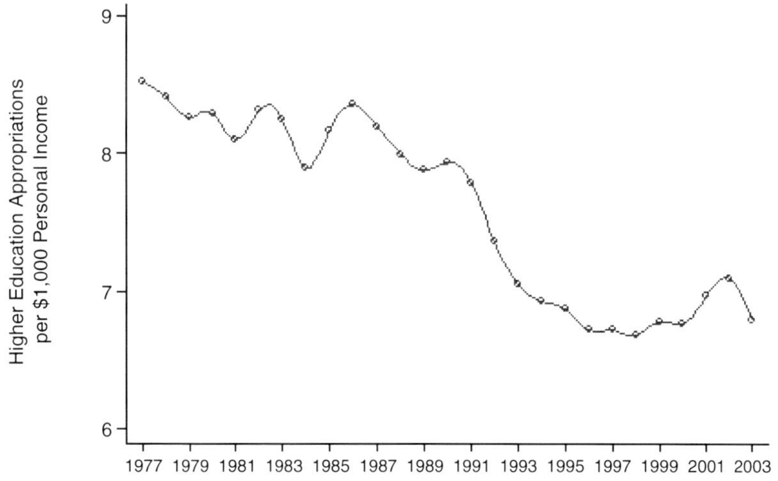

FIGURE PHE-8B Higher education appropriations relative to personal income, 1977-2003.
SOURCE: T. J. Kane. "The Role of Federal Government in Financing Higher Education." Presentation to National Academies' Board on Higher Education and Workforce, March 21, 2005.

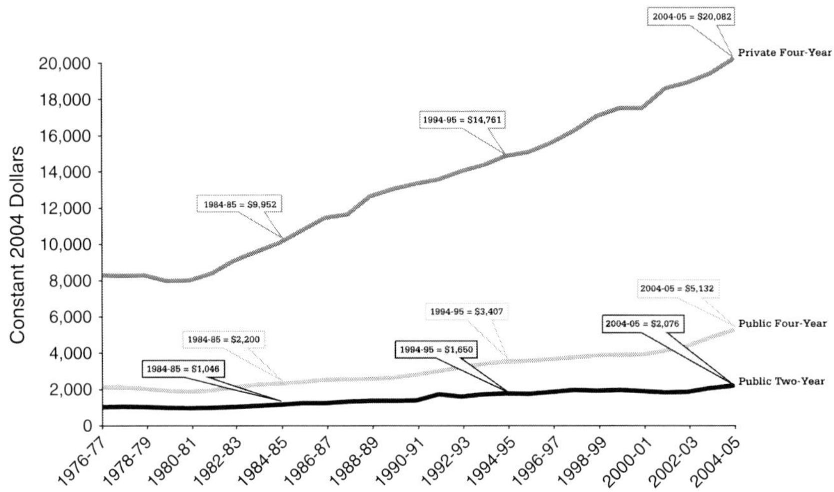

FIGURE PHE-9A Average published tuition and fee charges, enrollment-weighted, by institution type, in constant 2004 dollars, 1976-77 to 2004-05.
SOURCE: S. Baum. "Changes in Funding for Public Higher Education: College Prices and Student Aid." Presentation to National Academies' Board on Higher Education and Workforce, April 2005. Data are from College Board. "Trends in Higher Education Series 2004."

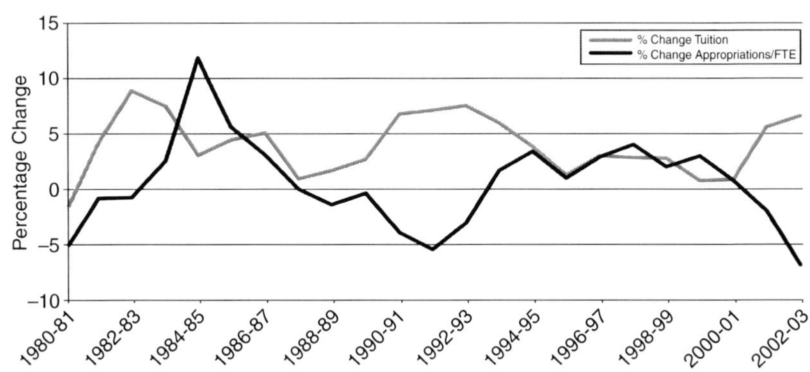

FIGURE PHE-9B Percent change in public 4-year institution tuition and appropriations, 1980-81 to 2002-03.
SOURCE: S. Baum. "Changes in Funding for Public Higher Education: College Prices and Student Aid." Presentation to National Academies' Board on Higher Education and Workforce, April 2005. Data are from College Board. "Trends in Higher Education Series 2004."

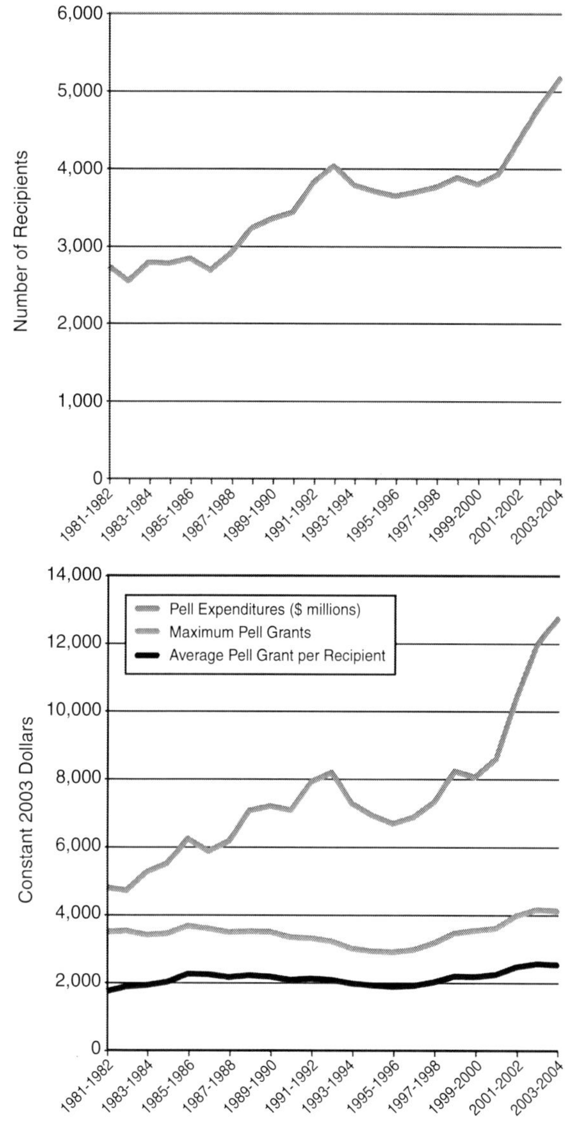

FIGURE PHE-10 Pell Grants: Number of recipients and total expenditures, maximum grant, and average grant, in constant 2003 dollars, 1981-1982 to 2003-2004.

SOURCE: S. Baum. "Changes in Funding for Public Higher Education: College Prices and Student Aid." Presentation to National Academies' Board on Higher Education and Workforce, April 2005. Data are from College Board. "Trends in Higher Education Series 2004."

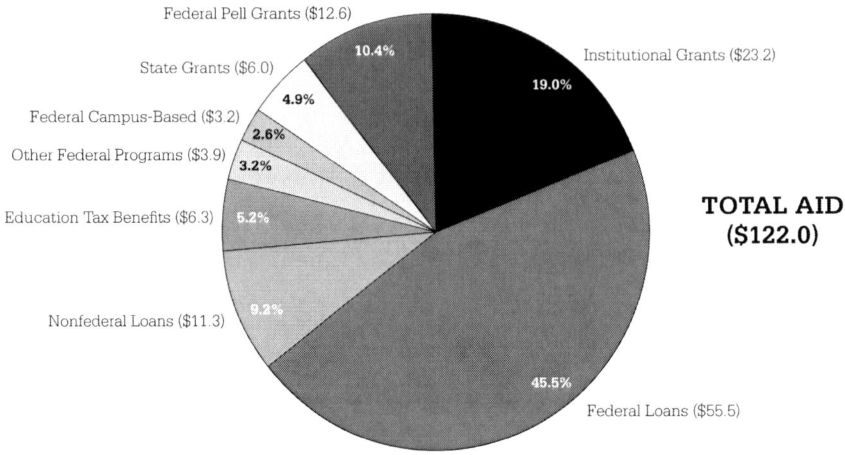

FIGURE PHE-11 Total higher education student aid, by source, in billions of current dollars, academic year 2003-2004.
SOURCE: S. Baum. "Changes in Funding for Public Higher Education: College Prices and Student Aid." Presentation to National Academies' Board on Higher Education and Workforce, April 2005. Data are from College Board. "Trends in Higher Education Series 2004."

of educational choices for students may be constrained unless the availability of financial aid can compensate. Especially for low-income students, the real and perceived cost increases for college education can limit access and lifetime opportunity (see Figures PHE-14A and B).

The second implication is for the quality of teaching and research. Reductions in funding for public education combined with constraints on tuition increases appear to be causing deterioration in the quality of public colleges and universities compared with private institutions.[9] Private universities benefit from larger endowments, have constrained enrollment growth to control costs, and have steadily increased tuition to offset inflation and provide new resources for qualitative improvement. Public institutions are less able to use these measures for fiscal control and as a result are falling behind private colleges and universities, in endowments, faculty salaries, student:faculty ratios, student services, and facilities (see Figure PHE-15). Also, to the extent that changes in faculty composition—such as increases in part-time and nontenure-track staff—affect the quality of teaching and mentoring and the availability of tenure-track faculty as role mod-

[9]J. Kissler. *Why It Is in the Interest to Address the Growing Gap Between Public and Private Universities*. Oakland, CA: University of California, 2005.

Federally Supported Programs (millions of constant 2003 dollars)	1993-1994	2003-2004
Grants		
Pell Grants	$7,196	$12,661
Supplemental Educational Opportunity Grants (SEOG)	742	760
Leveraging Educational Assistance Partnership Grant (LEAP)	91	64
Veterans	1,518	2,365
Military	515	981
Other Grants	245	353
Subtotal	10,308	17,184
Federal Work Study	982	1,218
Loans		
Perkins Loans	1,169	1,201
Subsidized Stafford	18,018	25,291
Unsubsidized Stafford	2,029	23,105
Parent Loan for Undergraduate Students (PLUS)	1,943	7,072
Supplemental Loans for Students (SLS)	4,415	—
Other Loans	580	125
Subtotal	28,780	56,794
Education Tax Benefits	—	6,298
Total Federal Aid	39,998	81,494
State Grant Aid	3,022	6,017
Institutional Grants	11,852	23,253
Total Federal, State Institutional	54,872	110,764
Nonfederal Loans	—	11,271

FIGURE PHE-12A Federal aid awarded to students (expenditures), in millions of constant 2003 dollars, 1993-1994 and 2003-2004.
SOURCE: S. Turner. "Policy Implications of Changing Funding for Public Higher Education." Presentation to National Academies' Board on Higher Education and Workforce, April 2005.

els, they may affect undergraduate persistence, graduation rates, and the propensity to continue to graduate school. The consequences include a more stratified, less dynamic society and a more limited workforce available for generating knowledge and innovation in the economy.

Issues of attainment also have come to the fore. With a growing number of postsecondary students starting out at community colleges and intending to transfer, 2- and 4-year institutions need to work to improve transfer and articulation agreements and processes to facilitate smooth transfers.[10] Colleges and universities must make a commitment to the stu-

[10]National Academy of Engineering and National Research Council. *Enhancing the Community College Pathway to Engineering Careers.* Washington, DC: The National Academies Press, 2005.

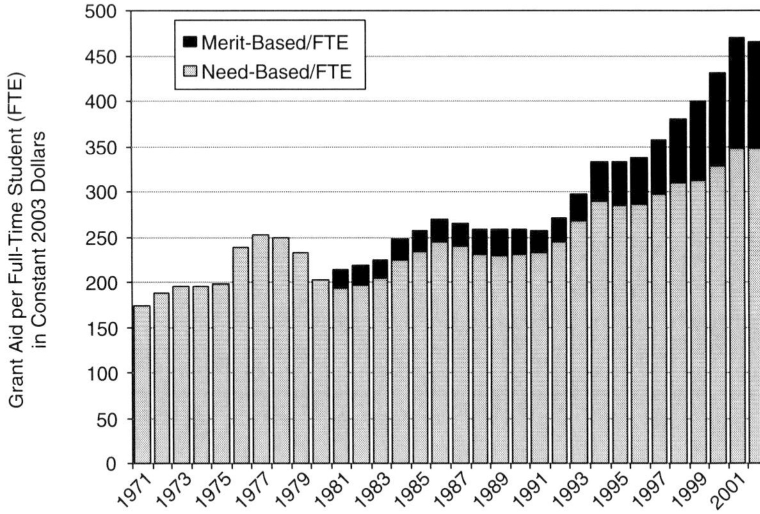

FIGURE PHE-12B Merit and need-based state grant aid per full-time student, by type of grant, 1971-2002.
SOURCE: S. Baum. "Changes in Funding for Public Higher Education: College Prices and Student Aid." Presentation to National Academies' Board on Higher Education and Workforce, April 2005. Data are from College Board. "Trends in Higher Education Series 2004."

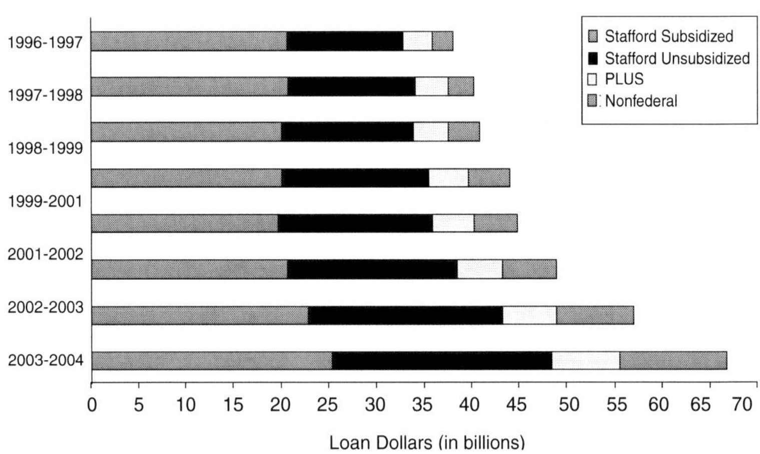

FIGURE PHE-12C Higher education loans, in billions of dollars, by type, 1996-1997 to 2003-2004.
SOURCE: S. Turner. "Policy Implications of Changing Funding for Public Higher Education." Presentation to National Academies' Board on Higher Education and Workforce, April 2005.

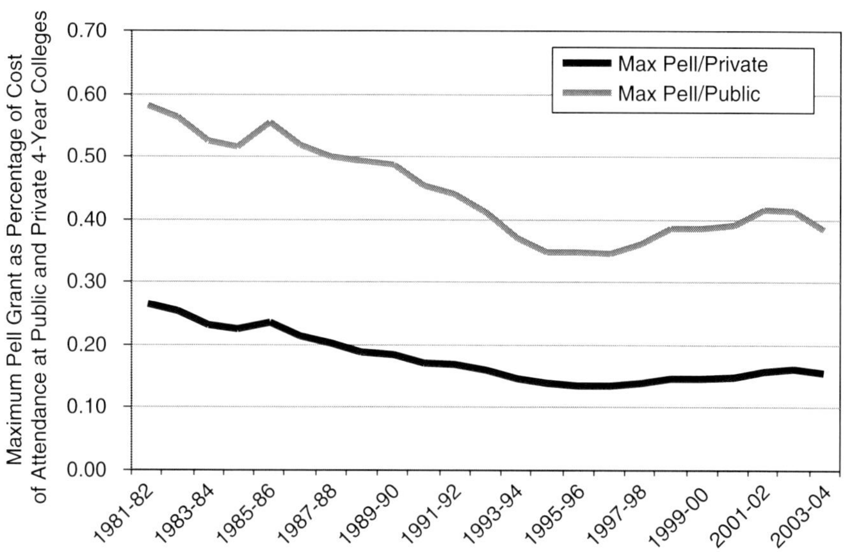

FIGURE PHE-13 Maximum Pell Grant as percentage of cost of attendance at public and private four-year colleges, 1981-82 to 2003-04.
SOURCE: S. Baum. "Changes in Funding for Public Higher Education: College Prices and Student Aid." Presentation to National Academies' Board on Higher Education and Workforce, April 2005. Data are from College Board. "Trends in Higher Education Series 2004."

	Total	Low Income (bottom 20%)	High Income (top 20%)
1972	49.2%	26.1%	63.8%
1981	53.9%	33.6%	67.6%
1996	65.0%	48.6%	78.0%

FIGURE PHE-14A Enrollment by income, 1972, 1981, and 1986.
SOURCE: S. Turner. "Policy Implications of Changing Funding for Public Higher Education." Presentation to National Academies' Board on Higher Education and Workforce, April 2005.

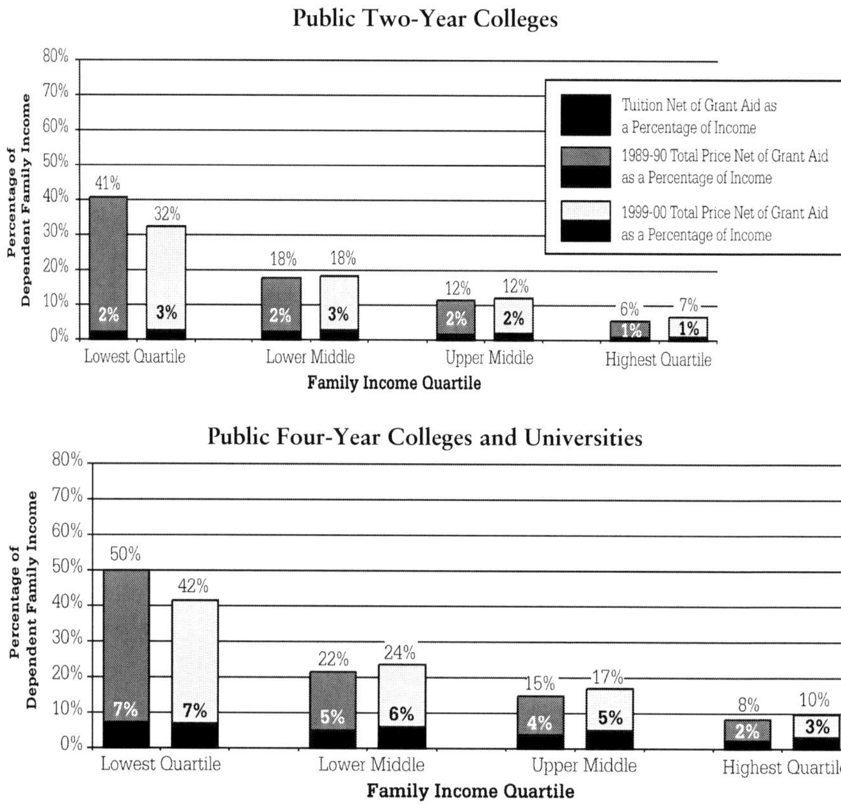

FIGURE PHE-14B Net tuition and fee and net cost of attendance as a percentage of family income, by family income quartile and institution type, 1989-1990 and 1999-2000.
SOURCE: S. Baum. "Changes in Funding for Public Higher Education: College Prices and Student Aid." Presentation to National Academies' Board on Higher Education and Workforce, April 2005. Data are from College Board. "Trends in Higher Education Series 2004."

dents they admit by supporting retention efforts so that students do not drop out of college with high debts and no degree.

ENSURING ADEQUATE FUNDING FOR PUBLIC HIGHER EDUCATION

The federal government has a number of options that could help public institutions receive revenues that reflect the true costs of higher education:

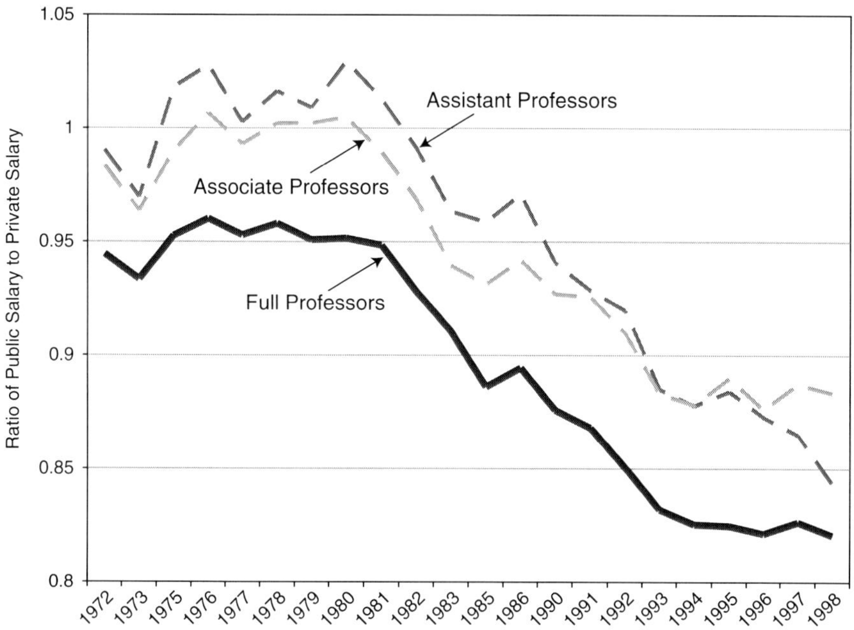

FIGURE PHE-15 Ratio of public institution salary to private institution salary, by faculty rank, 1972-1998.
SOURCE: T. J. Kane. "The Role of Federal Government in Financing Higher Education." Presentation to National Academies' Board on Higher Education and Workforce, March 21, 2005.

- Design or expand federal matching programs that encourage increased state appropriations for higher education. For example, to encourage states to expand means-tested grant aid, the federal government could offer matching funds to states based on their funding of such programs.
- Reform the Medicaid program to slow the growth of state commitments that crowd out spending on higher education.[11]
- Create "Learn Grant Universities" through a federal "Learn Grant Act" as significant as the Morrill Act of 1862 and the GI Bill of 1944.
- Enact a "Higher Education Millennium Partnership Act" that would integrate technology into the curriculum, create more flexible educational opportunities for part-time and nonresidential students, and develop new partnerships with schools, businesses, and local communities.[12]

---

[11]Kane and Orszag, 2003.

[12]J. J. Duderstadt and F. W. Womack. *Beyond the Crossroads: The Future of the Public University in America*. Baltimore, MD: Johns Hopkins University Press, 2003.

• Create a "Millennium Education Trust Fund" using the sale of unused communications spectrum over the next few years (with proceeds possibly greater than $18 billion) to provide students with the skills necessary for an age of innovation.

IMPROVING ACCESS TO HIGHER EDUCATION

In addition, the federal government can help the states improve access to higher education for all Americans through several actions:

• Focus national resources on improving the purchasing power of Pell awards.[13]
• Increase flexibility for states to buy more subsidized loan eligibility from the federal government.[14]
• Expand and restructure the LEAP program to allow private-sector matches from such organizations as Scholarship America and community foundations.[15]
• Institute a voucher program that would give more money to students from low-income homes.[16]
• Mandate that both public and private institutions use the average "net price" of attendance instead of the stated "sticker price" in all federal grant and loan programs to determine who qualifies for student-aid awards and how much they should be awarded. Using sticker prices as the official institutional "cost of attendance" misrepresents the actual average cost of attendance in most federal and state student-aid programs.[17]
• Consider eliminating the Free Application for Federal Student Aid. Changing laws to permit the use of Internal Revenue Service data to assess qualification for financial aid can simplify processes, save hundreds of millions of dollars, and remove bureaucratic barriers to postsecondary access.[18]

---

[13]Dickeson, 2004.
[14]Ibid.
[15]Ibid.
[16]R. Vedder. *Growing Broke by Degree: Why College Costs Too Much*. Washington, DC: AEI Press, 2004.
[17]A. F. King. "Policy Implications of Changes in Higher Education Finance." Presentation to the National Academies' Board on Higher Education and Workforce, April 21-22, 2005.
[18]Dickeson, 2004.

International Students and Researchers in the United States

SUMMARY

The United States has experienced a steadily growing influx of graduate students and postdoctoral scholars from throughout the world. International students now constitute more than a third of US science and engineering (S&E) graduate-school enrollments, up from less than a quarter in 1982. More than half the S&E postdoctoral fellows are temporary residents, half of whom earned a doctorate degree outside the United States. Including undergraduates, more than a half-million foreign citizens are studying at colleges and universities in the United States.

Many of the international students educated in this country choose to remain here after receiving their degrees. More than 70% of the foreign-born S&E doctorates who received their degrees in 2001 remained in the United States for more than 2 years, up from about half the 1989 doctorate recipients. These skilled migrants are an important source of innovation for the US economy.

The terrorist attacks of September 11, 2001, caused drops in the numbers of international students applying to and enrolling in US graduate programs. In addition, other countries are developing their own systems of graduate education to recruit and retain more highly skilled students and professionals. In this environment of increased competition and reduced

This paper summarizes findings and recommendations from a variety of recently published reports and papers as input to the deliberations of the Committee on Prospering in the Global Economy of the 21st Century. Statements in this paper should not be seen as the conclusions of the National Academies or the committee.

international mobility, the US education and research enterprise will have to readjust to be able to keep attracting the best students from home and abroad.

International exchanges of students and skilled professionals can benefit both the sending and receiving countries. Certainly, the United States S&E research enterprise depends critically on international students and scholars. Recommendations that various groups have made to maintain and enhance the ability of the United States to attract these highly skilled people include the following:

- Create new nonimmigrant visa categories exempted from the 214b provision for doctoral-level graduate students and postdoctoral scholars.
- Extend the validity of Visas Mantis security clearances for international students and scholars from the current 2-year limit to the duration of their academic appointments.
- Allow international students, scholars, scientists, and engineers to renew their visas in the United States.
- Implement a points-based immigration policy, similar to that of Canada or the United Kingdom, in which graduate education and S&E skills count toward obtaining citizenship.

SCIENCE AND ENGINEERING GRADUATE ENROLLMENTS AND DEGREES

The exchange of people and ideas across borders, accelerated in the last two decades by *perestroika* and the emergence of East Asia as a world economic power, has transformed institutions and individuals. Most countries today send bright young people to study abroad.[1] Many of them stay and contribute in lasting ways to their adopted countries. And whether they stay, return home, or move on to a third country, they become part of a global network of researchers, practitioners, and educators that provides cultural and intellectual support for students and scholars whatever their origins.

Since World War II, the United States has been the most popular destination for S&E graduate students and postdoctoral scholars choosing to study abroad. With about 6% of the world's population, the United States has been producing over 20% of S&E PhD degrees.[2] International graduate students and postdoctoral researchers, many of whom stay in the United

---

[1]T. M. Davis. *Atlas of Student Mobility*. New York: Institute of International Education, 2003.

[2]National Science Board. *Science and Engineering Indicators 2004*. NSB 04-01. Arlington, VA: National Science Foundation, 2004.

States after completing their studies, make substantial contributions to our society by creating and applying new knowledge.

The total number of S&E graduate students in US institutions has grown consistently over the last several decades, with an acceleration during the 1990s.[3] These increases have taken place despite evidence that US graduate schools give preference to domestic applicants.[4] Since the 1970s, the strongest inflow of graduate students has been from Asian countries. From 1985 to 2001, students from China, Taiwan, India, and South Korea earned more than half of the 148,000 US science and engineering doctoral degrees awarded to foreign students, four times the number awarded to students from Europe.

The percentage of international students in US graduate schools has risen from 23.4% in 1982 to 34.5% in 2002 (see Figure IS-1). In 2002, international students received 19.5% of all doctorates awarded in the social and behavioral sciences, 18.0% in the life sciences, 35.4% in the physical sciences, and 58.7% in engineering.[5] For doctorate-granting institutions, total enrollment of international S&E graduate students increased dramatically between 2000 and 2002. In 2002, 55.5% of international S&E graduate students were enrolled at Research I (R1) universities; R1s also enroll the highest proportion (26.0%) of international students (see Figure IS-2). Today, the total number of foreign citizens studying in US universities (including undergraduates) has passed the half-million mark.

A recent study further delineates the changing demographics of graduate students in US institutions.[6] In 1966, US-born males accounted for 71% of S&E PhD graduates, and 6% were awarded to US-born females; 23% of doctorate recipients were foreign-born. In 2000, 36% of doctorate recipients were US-born males, 25% US-born females, and 39% foreign-born. Among postdoctoral scholars, the participation rate of temporary residents has increased from 37.4% in 1982 to 58.8% in 2002 (see Figure IS-3). Similarly, the share of foreign-born faculty who earned their doctoral degrees at US universities has increased from 11.7% in 1973 to 20.4% in

---

[3]Ibid.

[4]G. Attiyeh and R. Attiyeh. "Testing for Bias in Graduate School Admissions." *Journal of Human Resources* 32(1997):524-548.

[5]National Science Foundation. *Survey of Graduate Students and Postdoctorates in Science and Engineering 2002*. Arlington, VA: National Science Foundation, 2004. Life sciences include biological sciences, agricultural sciences, and health fields; social sciences include psychology; and physical sciences include physics, chemistry, mathematics, computer science, and earth sciences.

[6]R. B. Freeman, E. Jin, and C.-Y. Shen. *Where Do New US-Trained Science-Engineering PhDs Come From?* Working Paper Number 10544. Cambridge, MA: National Bureau of Economics Research, 2004.

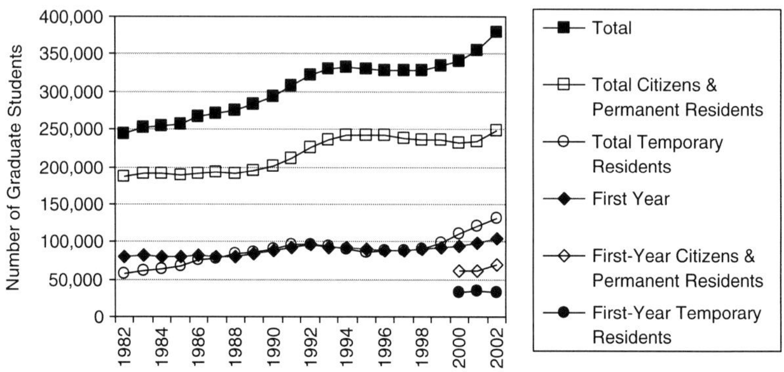

FIGURE IS-1 Number of graduate students enrolled, by citizenship status, 1982-2002.
NOTE: Enrollment numbers include medical fields.
SOURCE: National Science Foundation. *Survey of Graduate Students and Postdoctorates in Science and Engineering 2002*. Arlington, VA: National Science Foundation, 2004.

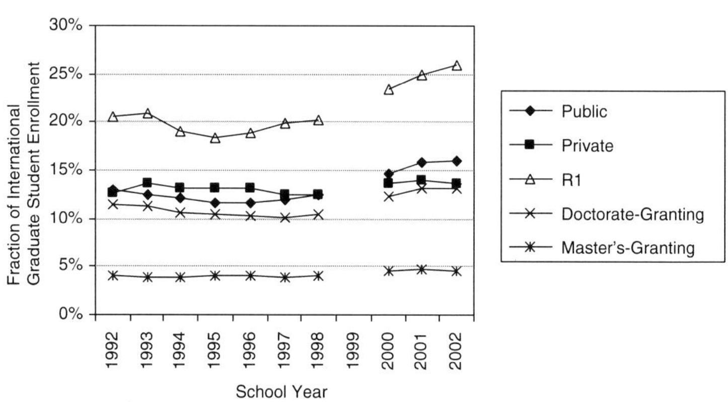

FIGURE IS-2 Fraction of international graduate student enrollment, by institution type, 1992-2002.
NOTE: The CGS enrollment numbers include all major S&E fields, as business, education, humanities and arts, and public administration and services. The non-S&E fields have 3% and 17% enrollment of international students. CGS states, "Institution type was a major differentiating variable in the enrollment of non-US students, reflecting the concentration of international students in doctoral programs in science and engineering."
SOURCE: The Council of Graduate Schools. "CGS/GRE Graduate Enrollment and Degrees: Annual Surveys from 1992-2002." Washington, DC. Available at: http://www.cgsnet.org/VirtualCenterResearch/graduateenrollment.htm.

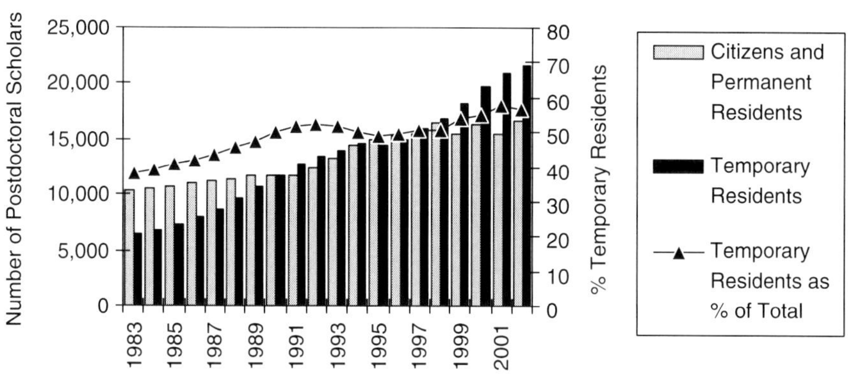

FIGURE IS-3 Total postdoctoral pool, by US residency status, 1983-2002.
NOTE: Medical fields are included, but postdoctoral scholars with medical degrees (presumably acting as physicians) are excluded from the analysis.
SOURCE: National Science Foundation. *Survey of Graduate Students and Postdoctorates in Science and Engineering 2002.* Arlington, VA: National Science Foundation, 2004.

1999. In engineering fields, the share increased from 18.6 to 34.7% in the same period.[7]

Stay Rates of International Graduate Students and Scholars

Representation of foreign-born scientists and engineers in US S&E occupations varies by field, country of origin, economic conditions in the sending country, and when the PhD was awarded. In total, foreign-born scientists and engineers were 22.7% of the US S&E labor force in 2000, an increase from 12.7% in 1980. Foreign-born doctorates were 37.3% of the US S&E labor force, an increase from 23.9% in 1990.

One study found that 45% of international students from developing countries planned to enter the US labor market for a time, and 15% planned to stay permanently; another 15% planned to go to a third country.[8] Another study showed that the stay rate of international doctorate scientists

[7]National Science Board. *Science and Engineering Indicators 2004.* NSB 04-01. Arlington, VA: National Science Foundation, 2004. Appendix Table 5-24. Available at: http://www.nsf.gov/sbe/srs/seind02/append/c5/at05-24.xls.

[8]N. Aslanbeigui and V. Montecinos. "Foreign Students in US Doctoral Programs." *Journal of Economic Perspectives* 12(1998):171-182.

and engineers has increased steadily and substantially in the last decade.[9] The proportion of foreign-born doctorates remaining in the United States for at least 2 years after receiving their degrees increased from 49% for the 1989 cohort to 71% for the larger 2001 cohort.[10]

Stay rates were highest among engineering, computer-science, and physical-science graduates. Stay rates also varied dramatically among graduate students from the top source countries—China (96%), India (86%), Taiwan (40%), and Korea (21%). Decisions to stay in the United States appear to be strongly affected by conditions in the students' home countries, primarily the unemployment rate, the percentage of the labor force that works in agriculture, and per capita GDP.[11]

COSTS AND BENEFITS OF INTERNATIONAL MOBILITY

Skilled migrants contribute to the US economy as technicians, teachers, and researchers and in other occupations in which technical training is desirable (see Table IS-1). Some research suggests that they generate economic gains by contributing to industrial and business innovation, resulting in a net increase in real wages for both citizen and immigrant workers. One study, for example, found that the immigration of skilled workers added to local skills rather than substituting for them.[12] The authors' econometric analyses suggest that a 10% increase in the number of international graduate students would raise university patent grants by 6% and nonuniversity patent grants by 4%. The authors concluded that bureaucratic hurdles in obtaining student visas may impede innovation if they decrease the inflow of international graduate students.

Foreign-born and foreign-educated scientists and engineers have made a disproportionate number of "exceptional" contributions to the S&E en-

---

[9]Although *international student* is usually taken to mean a student on a temporary visa, the figures sometimes include students on both temporary and permanent visas to compensate for the large number of Chinese students in the 1990s who became permanent residents by special legal provisions. This issue is discussed in greater detail by Finn (see next footnote), who finds the stay rate for those on temporary and permanent visas almost the same.

[10]M. G. Finn. *Stay Rates of Foreign Doctorate Recipients from US Universities, 2001*. Oak Ridge, TN: Oak Ridge Institute for Science and Education, 2003. The stay rate was defined as remaining in the United States for at least 2 years after receipt of the doctorate, but Finn estimates that these rates do not fall appreciably during the first 5 years after graduation.

[11]D. L. Johnson. *Relationship Between Stay Rates of PhD Recipients on Temporary Visas and Relative Economic Conditions in Country of Origin*. Working Paper. Oak Ridge, TN: Oak Ridge Institute for Science and Education, 2001.

[12]G. Chelleraj, K. E. Maskus, and A. Mattoo. *The Contribution of Skilled Immigration and International Graduate Students to US Innovation*. Working Paper 04-10. Boulder, CO: University of Colorado, 2004.

TABLE IS-1 Number of Foreign Born in US S&E Occupations, by Degree and Field, 2000

	Number of Foreign-Born in US S&E Occupations, 2000					
	All S&E	Engineering	Life Sciences	Mathematics and Computer Sciences	Physical Sciences	Social Sciences
All college-educated	816,000	265,000	52,000	370,000	92,000	37,000
Bachelor's degree	365,000	132,000	6,000	197,000	21,000	9,000
Master's degree	291,000	100,000	10,000	146,000	21,000	14,000
Professional degree	25,000	5,000	8,000	6,000	4,000	2,000
Doctoral degree[a]	135,000	28,000	28,000	21,000	46,000	12,000

[a]In 2001, 57% of those who were foreign-born S&E doctorate holders were US citizens.
NOTE: Data are from US Census 2000 5% Public Use Microdata Samples (PUMS) and include all S&E occupations other than postsecondary teachers, because field of instruction was not included in occupation coding for the 2000 census.
SOURCE: The National Academies. *Policy Implications of International Graduate Students and Postdoctoral Scholars in the United States.* Washington, DC: The National Academies Press, 2005. Table 1-5.

terprise of the United States.[13] Since 1990, almost half the US Nobel laureates in science fields were foreign-born; 37% received their graduate education abroad. The large number of foreign-born scientists and engineers working in the United States who were educated abroad suggests that the United States has benefited from investments in education made by other countries.

Many people believe that emigration of technically skilled individuals—often called a "brain drain"—is detrimental to the country of origin. However, the concept of brain drain may be too simplistic inasmuch as it ignores the many benefits of emigration, including remittances, international collaborations, the return of skilled scientists and engineers, diaspora-facilitated international business, and a general investment in skills caused by the prospect

[13]P. E. Stephan and S. G. Levin. Foreign Scholars in US Science: Contributions and Costs. In R. Ehrenberg and P. Stephan, eds. *Science and the University.* Madison, WI: University of Wisconsin Press, 2005. The authors use six criteria to indicate "exceptional" contributions (not all contributions) in S&E: individuals elected to the National Academy of Sciences (NAS) and/or National Academy of Engineering (NAE), authors of citation classics, authors of hot papers, the 250 most cited authors, authors of highly cited patents, and scientists who have played a key role in launching biotechnology firms.

of emigration.[14] As the R&D enterprise becomes more global, some observers propose that "brain drain" be recast as "brain circulation"[15] and include the broader topics of the international circulation of thinkers, knowledge workers, and rights to knowledge.[16] Such a discussion would include issues of local resources; many countries lack the educational and technical infrastructure to support advanced education, so aspiring scientists and engineers have little choice but to seek at least part of their training abroad, and in many instances such travel is encouraged by governments. Supporting the concept of brain circulation is the finding that ethnic networks developed in the United States by international students and scholars help to support knowledge transfer and economic development in both the United States and the sending country.[17]

In other countries, migration for employment, particularly for highly skilled workers, remains a core concern.[18] European Union (EU) countries, especially those with developed S&E capacity, have implemented strategies to facilitate retention and immigration of the technically skilled. Several Organisation for Economic Co-operation and Development (OECD) countries have relaxed their immigration laws to attract high-skilled students and workers.[19] Some are increasing growth in their international student populations and are encouraging these students to apply for resident status.

Point-based immigration systems for high-skilled workers, while not widespread, are starting to develop.[20] Canada, Australia, and New Zealand use

---

[14]D. Kapur and J. McHale. Sojourns and Software: Internationally Mobile Human Capital and High-Tech Industry Development in India, Ireland, and Israel. In A. Arora and A. Gambardella, eds. *From Underdogs to Tigers: The Rise and Growth of the Software Industry in Israel, Ireland and India.* Oxford, UK: Oxford University Press, 2005.

[15]Organisation for Economic Co-operation Development. *International Mobility of the Highly Skilled.* Policy Brief 92 2002 01 1P4. Washington, DC: OECD, 2002. Available at: http://www.oecd.org/dataoecd/9/20/1950028.pdf.

[16]B. Jewsiewicki. *The Brain Drain in an Era of Liberalism.* Ottawa, ON: Canadian Bureau for International Education, 2003.

[17]W. Kerr. "Ethnic Scientific Communities and International Technology Diffusion." Working Paper. 2004. Available at: http://econ-www.mit.edu/faculty/download_pdf.php?id=994.

[18]OECD members countries include Australia, Austria, Belgium, Canada, the Czech Republic, Denmark, Finland, France, Germany, Greece, Hungary, Iceland, Ireland, Italy, Japan, Korea, Luxembourg, Mexico, The Netherlands, New Zealand, Norway, Poland, Portugal, the Slovak Republic, Spain, Sweden, Switzerland, Turkey, the United Kingdom, and the United States.

[19]K. Tremblay. "Links Between Academic Mobility and Immigration." Symposium on International Labour and Academic Mobility: Emerging Trends and Implications for Public Policy, Toronto, October 22, 2004.

[20]Organisation for Economic Co-operation Development. *Trends in International Migration: 2004 Annual Report.* Paris: OECD, 2005. See http://www.workpermit.com for more information on immigration policies in English-speaking countries and the European Union.

such systems to recruit highly skilled workers. The United Kingdom has been doing so since 2001, and the Czech Republic set up a pilot project that started in 2004. In 2004, the European Union Justice and International Affairs council adopted a recommendation to facilitate the immigration of researchers from non-EU countries, asking member states to waive requirements for residence permits or to issue them automatically or through a fast-track procedure and to set no quotas that would restrict their admission. Also, the European Commission has adopted a directive for a special admissions procedure for third-world nationals coming to the EU to perform research.

RECENT TRENDS IN GRADUATE SCHOOL ENROLLMENT

Declines in international student applications for entry to US graduate schools have stimulated considerable discussion and more than a few warnings that our national S&E capacity may have begun to weaken. In 2002, the National Science Foundation noted a decrease in first-time full-time S&E graduate enrollments among temporary residents, by about 8% for men and 1% for women.[21] At the same time, first-time full-time S&E graduate-student enrollment increased by almost 14% for US citizens and permanent residents—15% for men and more than 12% for women (see Figure IS-1).

More recent surveys by the Council on Graduate Schools showed dramatic decreases in applications among international students for the 2003 academic year but much smaller decreases in admissions. Applications and admissions for domestic students did not change appreciably during this period, whereas enrollments decreased by 5%. There appear to be much smaller effects on applications for the 2004 academic year (see Table IS-2).

These declines were partly in response to the terrorist attacks of September 11, 2001, after which it became clear to everyone that the issuance and monitoring of visas are as important to graduate education as the training experience. Even more so, however, the declines reflect increasing global competition for graduate students amid the globalization of S&E education and research.

RISING GLOBAL CAPACITY FOR HIGHER EDUCATION

Given the fast-rising global tide of S&E infrastructure and training, it would be surprising if the S&E education and research enterprise currently dominated by the United States did not begin to change into a more global

---

[21]National Science Foundation. *Graduate Enrollment in Science and Engineering Fields Reaches New Peak; First-Time Enrollment of Foreign Students Declines.* NSF 04-326. Arlington, VA: National Science Foundation, 2004.

TABLE IS-2 Applications, Admissions, and Enrollments of International Graduate Students, by Field, 2002-2003

	Total	Engineering	Life Sciences	Physical Sciences
Applications	−28% (−5%)[a]	−36% (−7%)	−24% (−1%)	−26% (−3%)
Admissions	−18%	−24%	−19%	−17%
Enrollments	−6%	−8%	−10%	+6%

[a]Available data for the 2005 academic year are shown in parentheses.
SOURCE: H. Brown. *Council of Graduate Schools Finds Decline in New International Graduate Student Enrollment for the Third Consecutive Year.* Washington, DC: Council of Graduate Schools, November 4, 2004.

network of scientific and economic strength. Indeed, there is considerable evidence that that process has begun. Students have been leaving their home countries in search of academic opportunities abroad for thousands of years.[22] For scientists and engineers, the trend gained importance with the rise of universities and the need for formal training unavailable at home. As early as the late 19th century, many Americans were drawn abroad to German universities to gain expertise in fast-growing new technical fields.[23] In the following decades, that trend gradually reversed as US universities gained technical strength and attracted both faculty and students. US universities also benefited from an influx of educated refugees fleeing war-torn Europe during and after World War II.

Now, even while the United States can boast of 17 of the world's top 20 universities,[24] the US share of the world's S&E graduates is declining rap-

[22]W. I. Cohen. *East Asia at the Center: Four Thousand Years of Engagement with the World.* New York: Columbia University Press, 2001.

[23]D. E. Stokes. *Pasteur's Quadrant: Basic Science and Technological Innovation.* Washington, DC: Brookings Institution, 1997. Pp. 38-41. Stokes explains the effect of this export and re-importation of S&E talent on US universities: "This tide, which was at a flood in the 1880's, reflected the lack of an American system of advanced studies adequate to the needs of a rising industrial nation, and was a standing challenge to create one. The efforts to fill this gap in American higher education were generously supported by America's economic expansion, particularly by the private individuals who had acquired great wealth in the decades after the Civil War, many of whom had gained a vision of what might be done from their studies in the German universities."

[24]Shanghai's Jiao Tong University Institute of Higher Education. "Academic Ranking of World Universities." 2004. Available at: http://ed.sjtu.edu.cn/rank/2004/2004Main.htm. The ranking emphasizes prizes, publications, and citations attributed to faculty and staff, as well as the size of institutions. The *Times Higher Education* supplement also provides international comparisons of universities.

idly. European and Asian universities have increased degree production while the number of students obtaining US graduate degrees has stagnated (see Figure IS-4). Other interesting notes:

• The percentage of foreign students on OECD campuses rose by 34.9% on average between 1998 and 2002 and by 50% or more in the Czech Republic, Iceland, Korea, New Zealand, Norway, Spain, and Sweden. In absolute terms, more than 450,000 new individuals crossed borders to study in an OECD country during this short period, raising the number of foreign students enrolled on OECD campuses to 1,781,000. K. Tremblay. "Links Between Academic Mobility and Immigration." Symposium on International Labour and Academic Mobility: Emerging Trends and Implications for Public Policy, Toronto, October 22, 2004.

• In 2000, the EU was ahead of the United States and Japan in the production of S&E graduates. As a proportion of PhDs per 1,000 population aged 25-34 years, the EU-15 had an average of 0.56, the United States had 0.48 and Japan had 0.24. However, the emigration of EU-15 S&E graduates is creating a restriction for European R&D. In the late 1990s, the European S&E workforce accounted for 5.4 per thousand workers vs 8.1 per thousand in the United States and 9.3 in Japan. European Commission. *Towards a European Research Area. Science, Technology, and Innovation, Key Figures 2002*. Brussels: European Commission, 2002. Pp. 36-38. Available at: ftp://ftp.cordis.lu/pub/indicators/docs/ind_kf2002.pdf.

• Two independent estimates indicate that of the 60% of academic postdoctoral scholars who hold temporary visas, about four-fifths have non-US doctorates, which means that half of all US academic postdoctoral scholars have non-US doctorates.[25] Of postdoctoral scholars on temporary visas, almost 80% had earned their PhDs outside the United States. Of those with non-US PhDs, the highest number came from China (25%), followed by India (11%), Germany (7%), South Korea (5%), Canada (5%), Japan (5%), the UK (4%), France (4%), Spain (2%), and Italy (2%). The United States is benefiting from an inflow of postdoctoral scholars who have received graduate support and training elsewhere.

As countries develop knowledge-based economies, they seek to reap more of the benefits of international educational activities, including strong positive effects on gross domestic product (GDP) growth.[26] Emerging econo-

---

[25]Estimates based on the NSF Survey of Doctorate Recipients 2001, the NSF Survey of Graduate Students and Postdocs 2001, and the 2004 Sigma Xi National Postdoctoral Survey. Available at: http://postdoc.sigmaxi.org.

[26]The Conference Board of Canada. *The Economic Implications of International Education for Canada and Nine Comparator Countries: A Comparison of International Education*

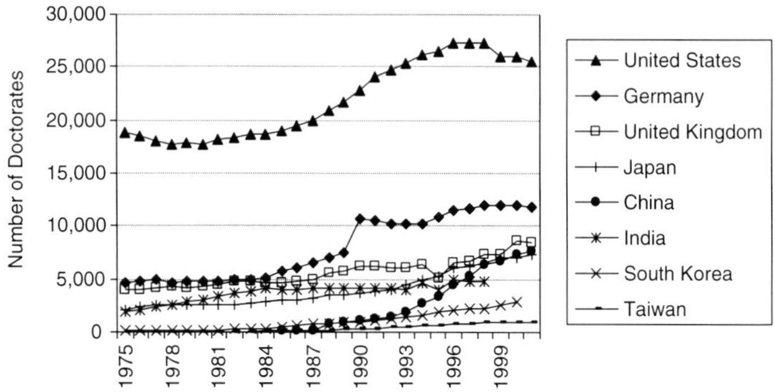

FIGURE IS-4 S&E doctorate production, by selected country, 1975-1999.
SOURCE: Based on National Science Board. *Science and Engineering Indicators 2004*. NSB 04-01. Arlington, VA: National Science Foundation, 2004. Appendix Tables 2-38 and 2-39.

mies have coupled education-abroad programs with strategic investments in S&E infrastructure—in essence pushing students away to gain skills and creating jobs to draw them back. Other countries, particularly in Europe, are trying to retain their best students and also to increase quality and open international access to their own higher educational institutions.

VISA AND IMMIGRATION POLICY

A growing challenge for policy-makers is to reconcile the flow of people and information with security needs. Policies and regulations, particularly those governing visas and immigration, can disrupt the global movement of individuals and therefore the productivity of scientists and engineers. In turn, this can affect a nation's economic capabilities.

The repercussions of the terror attacks of September 11, 2001, have included security-related changes in federal visa and immigration policy. Other immigration-related policies relevant to international student flows are international reciprocity agreements and deemed-export policies. Policy changes intended to restrict the illegal movements of an extremely small

Activities and Economic Performance. Ottawa, ON: Department of Foreign Affairs and International Trade, 1999. Also see A. Saxenian. *Silicon Valley's New Immigrant Entrepreneurs*. San Francisco: Public Policy Institute, 1999. Available at: http://www.ccis-ucsd.org/PUBLICATIONS/wrkg15.PDF.

population have had a substantial effect on international graduate students and postdoctoral scholars already in the United States or contemplating a period of study here.

Changes in visa and immigration policies and structures had a rapid and adverse effect on student mobility. Nonimmigrant-visa issuance rates decreased, particularly for students (see Figure IS-5). Implementation of the student-tracking system, the Student and Exchange Visitor Information System (SEVIS), and enhanced Visas Mantis security screening led to closer scrutiny and longer times for visa processing, in some cases causing students to miss classes or to turn to other countries for their graduate training.[27] After intense discussions between the university community and government agencies,[28] some of these policies have been adjusted to reduce effects on student mobility (see Figure IS-6 and Box IS-1). However, unfavorable perceptions remain, and international sentiment regarding the United States and its visa and immigration processes is a lingering problem for the recruitment of international students and scholars.

RECOMMENDATIONS

To maintain its leadership in S&E research, the United States must be able to recruit the most talented people worldwide for positions in academe, industry, and government.[29] The United States therefore must work to attract the best international talent while seeking to improve the mentoring, education, and training of its own S&E students, including women and members of underrepresented minority groups. This dual goal is especially important in light of increasing global competition for the best S&E students and scholars.

Federal actions that have been recommended include the following:

- Create new nonimmigrant-visa categories for doctoral-level graduate students and postdoctoral scholars, whether they are coming to the United States for formal educational or training programs or for short-term research collaborations or scientific meetings.[30] The categories should be exempted

---

[27]See, among many examples: "A Visa System Tangled in Red Tape and Misconceived Security Rules Is Hurting America." *The Economist*, May 6, 2004; C. Alphonso. "Facing Security Hurdles, Top Students Flock to Canada." *The Globe and Mail*, February 22, 2005.

[28]"Statement and Recommendations on Visa Problems Harming America's Scientific, Economic, and Security Interests," February 11, 2004, signed by 22 scientific, engineering, and academic leaders.

[29]The National Academies. *Policy Implications of International Graduate Students and Postdoctoral Scholars in the United States*. Washington, DC: The National Academies Press, 2005.

[30]Ibid.

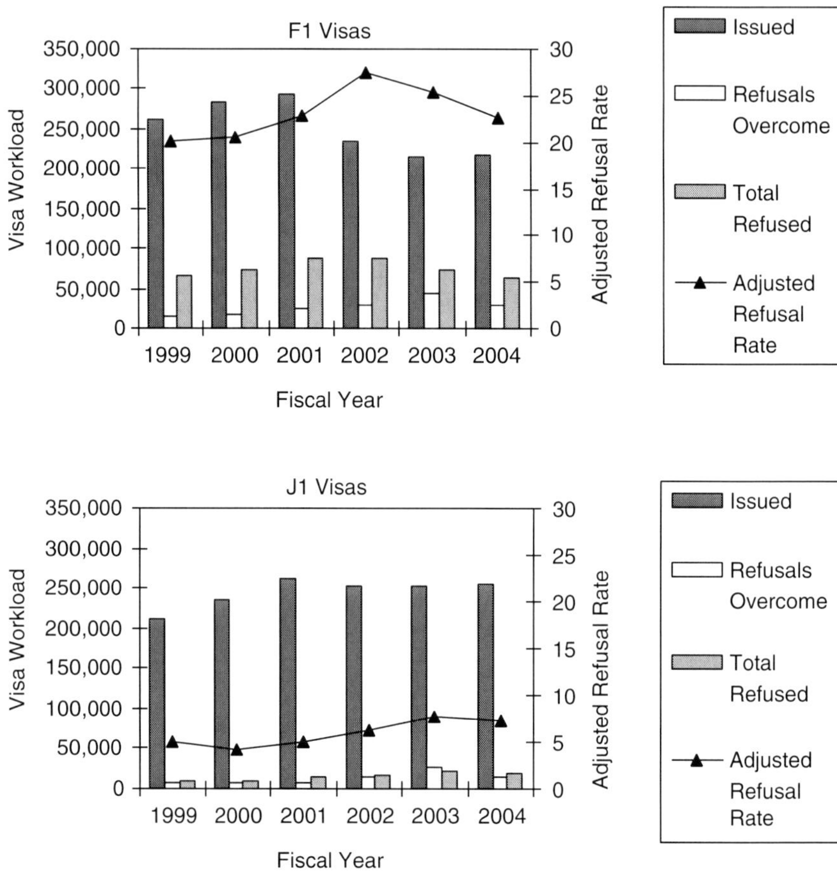

FIGURE IS-5 Visa workload and outcomes, by visa type, 1999-2004.
NOTE: *Report of the Visa Office* is an annual publication of the US Department of State, published by the Bureau of Consular Affairs. Recent editions are available at: http://travel.state.gov/visa/report.html. The adjusted refusal rate is calculated with the following formula: (Refusals – Refusals Overcome/Waived)/(Issuances + Refusals – Refusals Overcome/Waived).

A steep decline in visa issuances began in 2001 and continued through 2003. J-visa issuances, mostly to Europeans, followed roughly the same pattern, with a larger rise in the 1990s and a smaller downturn after 2001. To date, the downturn has reflected an increased denial rate more than a decreased application rate. As seen in the figure, the refusal rate for J-visa applicants rose steadily from 2000 through 2003. The adjusted refusal rate for F-visa applicants peaked in 2002. In 2004, denial rates had decreased considerably and were approaching 1999 levels.
SOURCE: United States Department of State, Bureau of Consular Affairs. *Report of the Visa Office: Multi Year Reports (1992-2004)*. Washington, DC: US Department of State, 2004. Available at: http://travel.state.gov/visa/report.html.

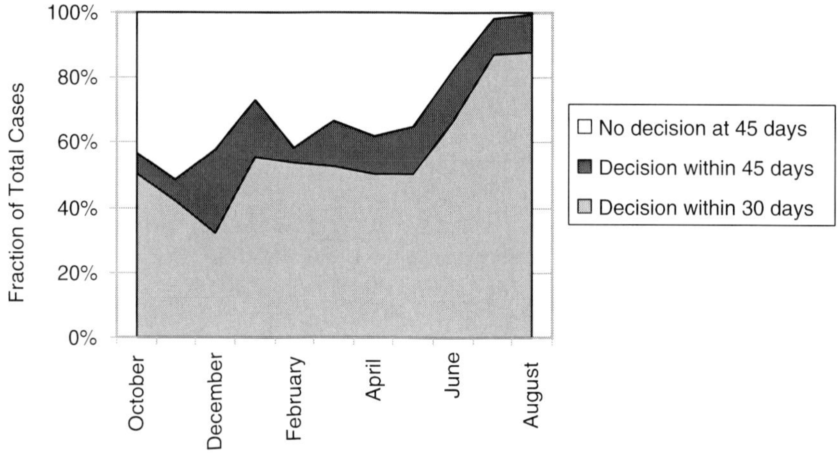

FIGURE IS-6 Visas Mantis Security Advisory Opinion (SAO) workload, FY 2004.
SOURCE: Data presented to Committee on Science, Engineering, and Public Policy's Committeee on Policy Implications of International Graduate Students and Postdoctoral Scholars in the United States on October 12, 2004, by Janice Jacobs, deputy assistant secretary of visas affairs, US Department of State.

from the 214b provision whereby applicants must show that they have a residence in a foreign country that they have no intention of abandoning.

- Allow international students, scholars, scientists, and engineers to renew their visas in the United States.[31]
- Negotiate visa reciprocity agreements between the United States and key sending countries, such as China, to extend visa duration and to permit multiple entries.[27,28]
- In the case of deemed-export controls, clear students and scholars to conduct research and use equipment required for such research through the visa process.[32]
- Implement a points-based immigration policy, similar to that of Canada or the United Kingdom, in which US graduate education and S&E skills count toward obtaining US citizenship.[33]

---

[31]"Recommendations for Enhancing the US Visa System to Advance America's Scientific and Economic Competitiveness and National Security Interests," May 18, 2005, signed by the National Academies presidents and 38 higher education and business organizations.

[32]Association of American Universities. "Revision and Clarification of Deemed Export Regulatory Requirements," submitted to the Bureau of Industry and Security, US Department of Commerce, June 27, 2005.

[33]Organisation for Economic Co-operation and Development. *Trends in International Migration: 2004 Annual Report*. Paris: OECD, 2005. See appendix for information on existing immigration policies.

BOX IS-1
VISA UPDATE

In 2002, a new antiterrorist screening process called Visas Condor was added for nationals of US-designated state sponsors of terrorism[a] that initially overloaded the Security Advisory Opinion (SAO) interagency process and slowed Mantis clearances.[b] The problem of extended waiting times for clearance of nonimmigrant visas flagged by Mantis has for the most part been addressed successfully.[c] By August 2004, the proportion of Visas Mantis visitors cleared within 30 days had risen substantially, and fewer than 15% took more than 30 days. The Visas Mantis process[d] is triggered when a student or exchange-visitor applicant intends to study a subject covered by the Technology Alert List (TAL). The express purpose of the TAL, originally drawn up as a tool for preventing proliferation of weapons technology, is to prevent the export of "goods, technology, or sensitive information" through such activities as "graduate-level studies, teaching, conducting research, participating in exchange programs, receiving training or employment."[e] Initially, Mantis procedures were applied on entry and each re-entry to the United States for persons studying or working in sensitive fields. In 2004, SAO clearance was extended to 1 year for those who were returning to a US government-sponsored program or activity and performing the same duties or functions at the same facility or organization that was the basis for the original Mantis authorization.[f] In 2005, the US Department of State extended the validity of Mantis clearances for F-, J-, H-, L-, and B-visa categories. Clearances for F-visas are valid for up to 4 years unless the student changes academic positions. H, J, and L clearances are valid for up to 2 years unless the visa holder's activity in the United States changes.[g]

[a]Countries designated section 306 in 2005: Iran, Syria, Libya, Cuba, North Korea, and Sudan. See http://travel.state.gov/visa/temp/info/info_1300.html.

[b]Government Accountability Office. Border Security: Improvements Needed to Reduce Time Taken to Adjudicate Visas for Science Students and Scholars. GAO-04-371. Washington, DC: Government Accountability Office, 2004. In April-June 2003, applicants waited an average of 67 days for completion of security checks associated with visa applications.

[c]Government Accountability Office. Border Security: Streamlined Visas Mantis Program Has Lowered Burden on Science Students and Scholars, but Further Refinements Needed. GAO-05-198. Washington, DC: Government Accountability Office, 2005.

[d]The Visa Mantis program was established in 1998 and applies to all nonimmigrant visas, including student (F), exchange-visitor (J), temporary-worker (H), intracompany-transferee (L), business (B-1), and tourist (B-2).

[e]See http://travel.state.gov/visa/testimony1.html for an overview of the Visas Mantis and Condor programs.

[f]See Department of State cable, 04 State 153587, No. 22: Revision to Visas Mantis Clearance Procedure. Available at: http://travel.state.gov/visa/state153587.html.

[g]"Extension of Validity for Science-related Interagency Visa Clearances." Media Note 2005/182. US Department of State, February 11, 2005. Available at: http://www.state.gov/r/pa/prs/ps/2005/42212.htm.

ANNEX 1

Existing High-Skilled Immigration Policies in OECD Countries[34]

Migration for employment, particularly for high-skilled workers, remains a core concern for OECD member countries.[35] EU countries, especially those with developed S&E capacity, have implemented strategies to facilitate retention and immigration of the technically skilled. Several OECD countries have relaxed their immigration laws to attract high-skilled students and workers. Some are increasing growth in their international-student populations and encouraging these students to apply for resident status.[36]

(1) Points-Based Immigration for High-Skilled Workers

Points systems, while not widespread, are starting to develop. Canada, Australia, New Zealand, and the United Kingdom use such systems to recruit highly skilled workers. The Czech Republic set up a pilot project that started in 2004. In 2004, the EU Justice and International Affairs council adopted a recommendation to facilitate researchers from non-EU countries, which asks member states to waive requirements for residence permits or to issue them automatically or through a fast-track procedure and to set no quotas that would restrict their admission. Permits should be renewable and family reunification facilitated. The European Commission has adopted a directive for a special admissions procedure for third-world nationals coming to the EU to perform research. This procedure will be in force in 2006.

- *Canada* has put into place a points-based program aimed at fulfilling its policy objectives for migration, particularly in relation to the labor-market situation. The admission of skilled workers depends more on human capital (language skills and diplomas, professional skills, and adaptability) than on specific abilities.[37] Canada has also

[34]Unless otherwise noted, policies listed are from an overview presented in: Organisation for Economic Co-operation and Development. *Trends in International Migration: 2004 Annual Report.* Paris: OECD, 2005.

[35]OECD members countries include Australia, Austria, Belgium, Canada, the Czech Republic, Denmark, Finland, France, Germany, Greece, Hungary, Iceland, Ireland, Italy, Japan, Korea, Luxembourg, Mexico, The Netherlands, New Zealand, Norway, Poland, Portugal, the Slovak Republic, Spain, Sweden, Switzerland, Turkey, the United Kingdom, and the United States.

[36]K. Tremblay. "Links Between Academic Mobility and Immigration." Symposium on International Labour and Academic Mobility: Emerging Trends and Implications for Public Policy, Toronto, October 22, 2004.

[37]Applicants can check online their chances to qualify for migration to Canada as skilled workers. A points score is automatically calculated to determine entry to Canada under the Skilled Worker category. See Canadian Immigration Points Calculator Web site at http://www.workpermit.com/canada/points_calculator.htm.

instituted a business-immigrant selection program to attract investors, entrepreneurs, and self-employed workers.

- *Germany* instituted a new immigration law on July 9, 2004. Among its provisions, in the realm of migration for employment, it encourages settlement by high-skilled workers, who are eligible immediately for permanent residence permits. Family members who accompany them or subsequently join them have access to the labor market. Like Canada, Germany encourages the immigration of self-employed persons, who are granted temporary residence permits if they invest a minimum of 1 million euros and create at least 10 jobs. Issuance of work permits and residence permits has been consolidated. The Office for Foreigners will issue both permits concurrently, and the Labor Administration subsequently approves the work permit.
- *UK*[38] The UK Highly Skilled Migrant Programme (HSMP) is an immigration category for entry to the UK for successful people with sought-after skills. It is in some ways similar to the skilled migration programs for entry to Australia and Canada. The UK has added an MBA provision to the HSMP. Eligibility for HSMP visas is assessed on a points system with more points awarded in the following situations:
  - Preference for applicants under 28 years old.
  - Skilled migrants with tertiary qualifications.
  - High-level work experience.
  - Past earnings.
  - In a few rare cases, HSMP points are also awarded if one has an achievement in one's chosen field.
  - One may also score bonus points if one is a skilled migrant seeking to bring a spouse or partner who also has high-level skills and work experience.
- *Australia* encourages immigration of skilled migrants, who are assessed on a points system with points awarded for work experience, qualifications, and language proficiency.[39] Applicants must demonstrate skills in specific job categories.

(2) Business Travel

- *Asia-Pacific Economic Cooperation (APEC)* has instituted the Business Travel Card Scheme designed to liberalize trade and stimulate economic growth. The scheme facilitates travel for business people

---

[38]The UK Highly Skilled Migrant Programme Web page also has a points calculator. See http://www.workpermit.com/uk/highly_skilled_migrant_program.htm.

[39]See points calculator at: http://www.workpermit.com/australia/point_calculator.htm.

traveling for short periods to participating countries (in 2004, APEC had 16 member countries, including China). Travel is possible between participating countries after submission of a single application, which is filtered by the applicant's home country and forwarded to all the participating countries for precertification. Cardholders are checked against police records in their own country as well as against warning lists in participating countries. Approved travelers get cards valid for 3 years that provide special access to fast-track lanes at airports. In 2004, there were over 5,000 cards in circulation.

(3) **Student Visas** Many OECD countries are determined to attract a larger number of international students. In addition to developing special programs and streamlining application processes, some countries have signed bilateral agreements while others have decided to offer job opportunities to graduates.

- *Canada* Students no longer require study permits for stays of less than 6 months.
- *France* Since 1999, it has been possible to obtain a 3- to 6-month visa for short-term studies without registration.

(4) **Work Permits for International Students and Spouses**

- *Canada*[40] A new off-campus work program allows international students at public postsecondary institutions to work off campus, extending the previous policy enacted earlier in 2005 that allowed students to work on campus while in Canada on a student visa.
- *Germany* Since 2003, international students have been allowed to work 180 half-days per year without a work permit.
- *Austria* Since 2003, students can work half-time to finance their studies.

(5) **Permit to Stay After Graduation to Find a Job**

- *Canada*[41] As of May 16, 2005, a new policy allows certain students to work in their field of study for up to 2 years after graduation. Previously, international students were allowed to stay only 1 year after graduation to work in Canada.

---

[40]Office of Science and Technology. "Canada: Immigration Policy Change Widens Door for Foreign Students and Scholars." *Bridges* 6(July 13, 2005). Available at: http://bridges.ostina.org.

[41]Ibid.

- *Germany* International students may remain in Germany for 1 year after the end of their studies to seek employment.
- *UK*[42] Foreign students at UK universities graduating from specific engineering, physical-sciences and mathematics courses are now permitted to stay in the UK for 1 year after graduation to take up employment.[43] The Science and Engineering Graduate Scheme was launched on October 25, 2004, and is now fully operational. This new immigration category allows non-European Economic Area nationals who have graduated from UK higher or further education establishments in certain mathematics, physical-sciences, and engineering subjects with a 2.2 degree or higher to remain in the UK for 12 months after their studies to pursue a career. Only those who have studied approved programs are eligible to apply to remain under the scheme. The scheme was first announced in the UK 2003 budget as an incentive to encourage foreign students to study in these fields in the UK and to be an asset to the workplace after graduation by relieving the shortages of engineering, physical-sciences, and mathematics graduates in the UK. Applicants must
  - Have successfully completed a degree course with second-class honors (2.2) or higher, a master's course or PhD on the relevant list of Department for Education or skills-approved physical-sciences, mathematics, and engineering courses at a UK institution of higher or further education.
  - Intend to work during the period of leave granted under the scheme.
  - Be able to maintain and accommodate themselves and any dependents without recourse to public funds.
  - Intend to leave the UK at the end of their stay (unless granted leave as a work-permit holder, high-skilled migrant, business person, or innovator).

---

[42]UK Home Office "Working in the UK" Web page. Available at: http://www.workingintheuk.gov.uk/working_in_the_uk/en/homepage/schemes_and_programmes/graduate_students.html.

[43]The scheme was highlighted in Sir Gareth Roberts' review, "The Supply of People with Science, Technology, Engineering and Mathematics Skills" (see http://www.kent.ac.uk/stms/research-gc/roberts-transferable-skills/roberts-recommendations.doc), that the UK was suffering from a shortage of engineering, mathematics, and physical sciences students at university and skilled workers in the labor market. This shortage could do serious damage to the UK's future economical growth. There is currently a reported shortage in sectors such as research and development and financial services for mathematics, science, and engineering specialists.

Achieving Balance and Adequacy in Federal Science and Technology Funding

SUMMARY

The complementary goals of *balance* and *adequacy* in federal funding for science and technology require both diversity and cohesion in the nation's R&D system. Diversity fosters creativity, creates competition among people and ideas, brings new perspectives to problems, and fosters linkages among sectors. Cohesion helps ensure that basic research is not squeezed out by more immediate needs and that the highest quality research is supported.

Federal actions that could improve the balance of federal science and technology (FS&T) funding include the following:

- Create a process in Congress that examines the entire FS&T budget before the total federal budget is aggregated into allocations to appropriations committees and subcommittees.
- Establish a stronger coordinating and budgeting role for the Office of Science and Technology Policy to promote cohesion among federal R&D agencies.
- Maintain the diversity of FS&T funding in terms of sources of funding, performers, time horizons, and motivations.
- Balance funding between basic and applied research and across fields of research to stimulate innovative cross-disciplinary thinking.

This paper summarizes findings and recommendations from a variety of recently published reports and papers as input to the deliberations of the Committee on Prospering in the Global Economy of the 21st Century. Statements in this paper should not be seen as the conclusions of the National Academies or the committee.

- Protect funding for high-risk research by setting aside a portion of the R&D budgets of federal agencies for this purpose.
- Maintain a favorable economic and regulatory environment for capitalizing on research—for example, by using tax incentives to build stronger partnerships among academe, industry, and government.
- Encourage industry to boost its support of research conducted in colleges and universities from 7 to 20% of total academic research over the next 10 years.

Two important goals can help policy-makers judge the adequacy of federal funding for FS&T. First, the United States should be among the world leaders in all major areas of science. Second, the United States should maintain clear leadership in some areas of science. The recent doubling of the budget of the National Institutes of Health—and other recent increases in R&D funding—acknowledge the tremendous opportunities and national needs that can be addressed through science and technology. Similar opportunities exist in the physical sciences, engineering, mathematics, computer science, environmental science, and the social and behavioral sciences—fields in which federal funding has been essentially flat for the last 15 years.

Among the steps that the federal government could take to ensure that funding for science and technology is adequate across fields are these:

- Increase the budget for mathematics, the physical sciences, and engineering research by 12% a year for the next 7 years within the research accounts of the Department of Energy, the National Science Foundation, the National Institute for Standards and Technology, and the Department of Defense.
- Return federal R&D funding to at least 1% of US gross domestic product.
- Make the R&D tax credit permanent to promote private support for research and development, as requested by the Administration in the fiscal year (FY) 2006 budget proposal.

Support for a new interdisciplinary field of quantitative science and technology policy studies could shed light on the complex effects that scientific and technologic advances have on economic activities and social change.

A Century of Science and Technology

In 1945, in his report *Science—The Endless Frontier*, Vannevar Bush proposed an idea that struck many people as far-fetched.[1] He wrote that the

[1] V. Bush. *Science—The Endless Frontier*. Washington, DC: US Government Printing Office, 1945.

federal government should fund the research of scientists without knowing exactly what results the research would yield—an idea that flatly contravened the US government's historical practice.[2]

Despite the misgivings of many policy-makers, the US government eventually adopted Bush's idea. The resulting expansion of scientific and technological knowledge helped produce a half-century of unprecedented technologic progress and economic growth. New technologies based on increased scientific understanding have enhanced our security, created new industries, advanced the fight against disease, and produced new insights into ourselves and our relationship with the world. If the 20th century was America's century, it also was the century of science and technology.

Since 1950, the federal government's annual support for research and development (R&D) has grown from less than $3 billion to more than $130 billion—more than a 10-fold expansion in real terms.[3] Today, about 1 in every 7 dollars in the federal discretionary budget goes for R&D. Performers of federal R&D include hundreds of colleges and universities and many thousands of private companies, federal laboratories, and other nonprofit institutions and laboratories. These institutions produce not only new knowledge but also the new generations of scientists and engineers who are responsible for a substantial portion of the innovation that drives changes in our economy and society.

Major priorities within the federal R&D budget have shifted from the space race in the 1960s to energy independence in the 1970s to the defense buildup of the 1980s to biomedical research in the 1990s. In the 1990s, the nation's R&D system also began to encounter challenges that it had not faced before. The end of the Cold War, an acceleration of economic globalization, the rapid growth of information technologies, new ways of conducting research, and very tight federal budgets led to thorough re-evaluations of the goals of federal R&D. Though Vannevar Bush's vision remains intact, the R&D system today is much more complex, diversified, and integrated into society than would have been imagined 60 years ago.

In this decade, the challenges to the R&D system have intensified. International competitors are now targeting service sectors, including R&D, just as they have targeted manufacturing sectors in the past. Global development and internationalization, new trade agreements, and the rapid flow of capital are reshaping industries so quickly that policy-makers barely have time to respond. Similarly, workplace technologies and demands change so quickly that workers must be periodically retrained to remain competitive.

---

[2]A. H. Dupree. *Science in the Federal Government: A History of Policies and Activities*, 2nd ed. Baltimore, MD: Johns Hopkins University Press, 1986.

[3]National Science Foundation, National Science Board. *Science and Engineering Indicators 2000*. Arlington, VA: National Science Foundation, 2000.

Throughout modern economies, advantages accrue to individuals, governments, and companies that are adaptable, forward-looking, knowledgeable, and innovative.

At the beginning of the 21st century, the United States stands at a crossroads. The only way for this nation to remain a high-wage, high-technology country is to remain at the forefront of innovation. Achieving this goal will require that the nation remain a leader in the scientific and technological research that contributes so heavily to innovation.

ACHIEVING BALANCE IN FEDERAL SCIENCE AND TECHNOLOGY FUNDING

Federal funding for science and technology in the United States historically has been balanced along several dimensions—between research and development, between defense and nondefense R&D, between academic and nonacademic R&D performers, and so on. Much of this balance arises in a de facto manner from the independent actions of a wide range of array supporters and performers. But some is the consequence of explicit policy decisions by the executive and legislative branches.

In the 1995 report *Allocating Federal Funds for Science and Technology*, a committee of the National Research Council laid out five broad principles designed in part to help the federal government achieve the proper balance of R&D funding:[4]

- Make the allocation process more coherent, systematic, and comprehensive.
- Determine total federal spending for federal science and technology based on a clear commitment to ensuring US leadership.
- Allocate funds to the best projects and people.
- Ensure that sound scientific and technical advice guides allocation decisions.
- Improve federal management of R&D activities.

The report recommended that

- The President present an annual comprehensive FS&T budget, including areas of increased and reduced emphasis. The budget should be sufficient to serve national priorities and foster a world-class scientific and technical enterprise.
- Departments and agencies make FS&T allocations based on clearly

---

[4]National Research Council, Committee on Criteria for Federal Support of Research and Development. *Allocating Federal Funds for Science and Technology*. Washington, DC: National Academy Press, 1995.

articulated criteria that are congruent with those used by the Executive Office of the President and by Congress.

- Congress create a process that examines the entire FS&T budget before the total federal budget is disaggregated into allocations to appropriations committees and subcommittees.
- The President and Congress ensure that the FS&T budget is sufficient to allow the United States to achieve preeminence in a select number of fields and perform at a world-class level in other major fields.

The Executive Branch responded by providing, as part of the President's budget submission, an analysis of the FS&T budget that encompasses federal funds spent specifically on scientific and technological research programs, the development and maintenance of the necessary research infrastructure, and the education and training of scientists and engineers. In addition, the White House Office of Management and Budget (OMB) and Office of Science and Technology Policy (OSTP) issue a joint budget memorandum that articulates the President's goals for the upcoming budget year to aid them in the preparation of agency budgets before submission to OMB.

Analysis of this budget reveals trends in the support of scientific and technologic research that the broader category of R&D obscures. For example, in the president's FY 2006 budget request, federal R&D would be up 1% from $131.5 billion to $132.3 billion. But FS&T would be down 1%, from $61.7 billion to $60.8 billion (see Figures R&D-1 and R&D-2).[5] (The director of OSTP has pointed out that it can be misleading to compare proposed budgets with enacted budgets because the latter can contain funds specified by Congress for research projects that were not included in the President's budget.[6])

Congress has not yet adopted a process that entails an overall consideration of the scientific and technological research supported by the federal government.[7] Subcommittees in both the House and Senate still consider portions of the federal R&D budget separately without deliberations or hearings on the broad objectives of S&T spending. At a minimum, the use of a common budget classification code could allow Congress more easily to address science and technology programs in a unified manner.

Overall consideration of the FS&T budget could reiterate the importance of basic research and of diversity among research supporters and performers.

[5]Office of Management and Budget. *Budget of the United States Government, Fiscal Year 2006*. Washington, DC: US Government Printing Office, 2005.

[6]John Marburger, speech to the 20th Annual AAAS Forum on Science and Technology Policy, April 21, 2005.

[7]J. Bingaman, R. M. Simon, and A. L. Rosenberg. "Needed: A Revitalized National S&T Policy." *Issues in Science and Technology* (Spring 2004):21-25.

	2004 Actual	2005 Estimate	2006 Proposed	Dollar Change: 2005 to 2006	Percent Change: 2005 to 2006
By Agency					
Defense	65,462	70,422	70,839	417	1%
Health and Human Services	28,047	28,752	28,807	55	....
NASA	10,574	10,990	11,527	537	5%
Energy	8,779	8,629	8,528	–101	–1%
National Science Foundation	4,160	4,082	4,194	112	3%
Agriculture	2,222	2,415	2,039	–376	–16%
Homeland Security	1,053	1,185	1,467	282	24%
Commerce	1,137	1,134	1,013	–121	–11%
Transportation	661	748	808	60	8%
Veterans Affairs	866	784	786	2	....
Interior	627	615	582	–33	–5%
Environmental Protection Agency	661	572	569	–3	–1%
Other	1,089	1,243	1,145	–98	–8%
Total	**125,338**	**131,571**	**132,304**	**733**	**1%**
Basic Research					
Defense	1,358	1,513	1,319	–194	–13%
Health and Human Services	14,780	15,124	15,246	122	1%
NASA	2,473	2,368	2,199	–169	–7%
Energy	2,847	2,887	2,762	–125	–4%
National Science Foundation	3,524	3,432	3,480	48	1%
Agriculture	829	851	788	–63	–7%
Homeland Security	68	85	112	27	32%
Commerce	43	58	71	13	22%
Transportation	20	38	41	3	8%
Veterans Affairs	347	315	315	....	....
Interior	37	36	30	–6	–17%
Environmental Protection Agency	113	66	70	4	6%
Other	149	155	175	20	13%
Subtotal	**26,588**	**26,928**	**26,608**	**–320**	**–1%**

FIGURE R&D-1 Federal research and development spending, in millions of dollars, for all R&D and for basic research, by agency, 2004-2006.

SOURCE: Executive Office of the President. *Budget of the United States Government, Fiscal Year 2006, Part Two: Analytical Perspectives*. Washington, DC: US Government Printing Office, 2005. P. 66. Available at: http://www.ostp.gov/html/budge/2006/FY06RDChapterFinal.pdf.

	2004 Actual	2005 Estimate	2006 Proposed	Dollar Change: 2005 to 2006	Percent Change: 2005 to 2006
Applied Research					
Defense	4,351	4,851	4,139	−712	−15%
Health and Human Services	13,007	13,274	13,410	136	1%
NASA	3,006	2,497	3,233	736	29%
Energy	2,693	2,760	2,709	−51	−2%
National Science Foundation	266	279	276	−3	−1%
Agriculture	1,055	1,093	942	−151	−14%
Homeland Security	247	346	399	53	15%
Commerce	828	825	763	−62	−8%
Transportation	349	423	494	71	17%
Veterans Affairs	476	430	433	3	1%
Interior	538	530	495	−35	−7%
Environmental Protection Agency	423	365	386	21	6%
Other	599	562	553	−9	−2%
Subtotal	**27,838**	**28,235**	**28,232**	**−3**	...
Development					
Defense	59,701	63,903	65,331	1,428	2%
Health and Human Services	41	54	28	−26	−48%
NASA	3,189	3,727	3,511	−216	−6%
Energy	1,992	1,846	1,959	113	6%
National Science Foundation	...	...	...	...	...
Agriculture	159	157	146	−11	−7%
Homeland Security	481	599	746	147	25%
Commerce	152	149	90	−59	−40%
Transportation	279	269	254	−15	−6%
Veterans Affairs	43	39	38	−1	−3%
Interior	49	46	54	8	17%
Environmental Protection Agency	125	141	113	−28	−20%
Other	324	495	396	−99	−20%
Subtotal	**66,535**	**71,425**	**72,666**	**1,241**	**2%**
Facilities and Equipment					
Defense	52	155	50	−105	−68%
Health and Human Services	219	300	123	−177	−59%
NASA	1,906	2,398	2,584	186	8%
Energy	1,247	1,136	1,098	−38	−3%
National Science Foundation	370	371	438	67	18%
Agriculture	179	314	163	−151	−48%
Homeland Security	257	155	210	55	35%
Commerce	114	102	89	−13	−13%
Transportation	13	18	19	1	...
Veterans Affairs	...	...	...	...	N/A
Interior	3	3	3	...	...
Environmental Protection Agency	...	...	...	...	N/A
Other	17	31	21	−10	−32%
Subtotal	**4,377**	**4,983**	**4,798**	**−185**	**−4%**

FIGURE R&D-2 Federal research and development spending, in millions of dollars, by agency, for applied research, development, facilities, and equipment, 2004-2006. SOURCE: Executive Office of the President. *Budget of the United States Government, Fiscal Year 2006, Part Two: Analytical Perspectives*. Washington, DC: US Government Printing Office, 2005. P. 67. Available at: http://www.ostp.gov/html/budget/2006/FY06RDChapterFinal.pdf.

Especially when budgets are tight, basic research can be displaced by the more immediate needs of applied research and technology development. In fact, less than half of all federal R&D funding is allocated for basic and applied research (see Figure R&D-3). The FS&T budget has increased since 2000, but these increases are primarily due to increases in funding of the National Institutes of Health (NIH). Nondefense-related R&D funding has

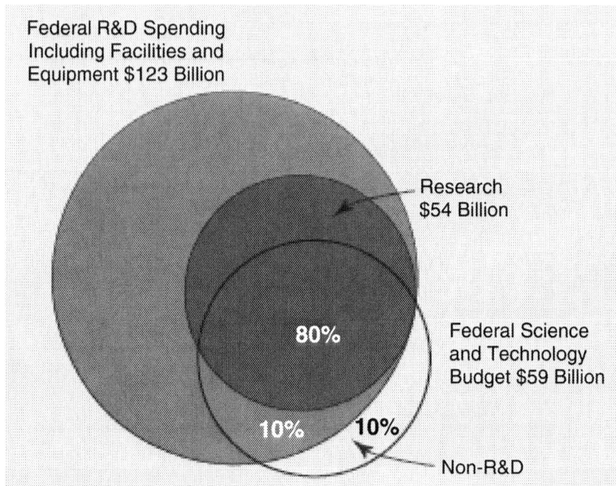

FIGURE R&D-3 Funding concepts in FY 2004 budget proposal.
SOURCE: National Science Board. *Science and Engineering Indicators 2004.* NSB 04-01. Arlington, VA: National Science Foundation, 2004. Figure 4-12.

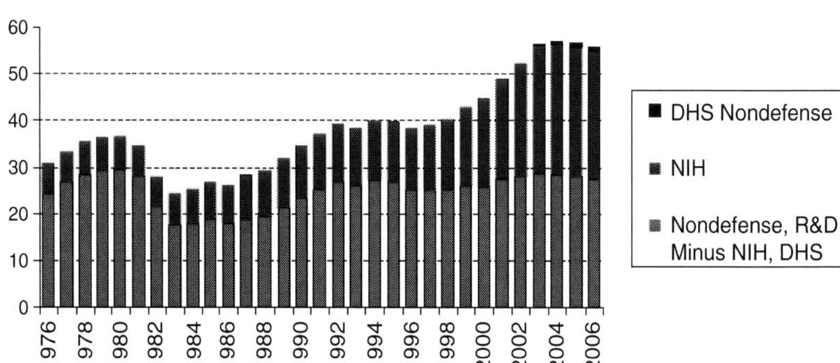

FIGURE R&D-4 Selected trends in nondefense R&D, FY 1976-FY 2006, in billions of constant FY 2005 dollars.
SOURCE: American Association for the Advancement of Science. *Chart: Selected Trends in Nondefense R&D: FY 1976-2006.* Washington, DC: American Association for the Advancement of Science, 2005. Available at: http://www.aaas.org/spp/rd/trnon06c.pdf.

been stagnant in recent years (see Figure R&D-4). Recently, the FS&T budget has been declining since the charge to double NIH funding has been completed (see Figure R&D-5). Recent Department of Defense (DOD) budgets offer another example—ever the last decade, the resources provided for basic research by the DOD have declined substantially.[8] Recent trends show that while defense R&D budgets have been increasing overall, the amount of resources allocated to science research in DOD is decreasing (see Figures R&D-6A and B). This lack of support for basic research could have major consequences for the development of necessary future military capabilities.

Allocating Federal Funds for Science and Technology also recommended that:

- R&D conducted in federal laboratories focus on the objectives of the sponsoring agency and not expand beyond the assigned missions of the laboratories. The size and activities of each laboratory should correspond to changes in mission requirements.
- FS&T funding generally favor academic institutions because of their flexibility and inherent quality control and because they link research to education and training in science and engineering.
- FS&T budget decisions give preference to funding projects and people rather than institutions. That approach will increase the flexibility in responding to new opportunities and changing conditions.
- Competitive merit review, especially that involving external reviewers, be the preferred way to make awards, because competition for funding is vital to maintain the high quality of FS&T programs.
- Evaluations of R&D programs and of those performing and sponsoring the work also incorporate the views of outside evaluators.
- R&D be well managed and accountable but not micromanaged or hobbled by rules and regulations that have little social benefit.

Diversity cannot be an excuse for mediocrity. People, projects, and institutions need to be reviewed to ensure that they are meeting national needs in science and technology. Open competition involving evaluation of merit by peers is the best-known mechanism to maintain support for the highest-quality projects and people. Quality also can be maintained by knowledgeable program managers who have established external scientific and technical advisory groups to help assess quality and to help monitor whether agency needs are being met.

Possible actions for the federal government to maintain the diversity

[8] National Research Council, Committee on Department of Defense Basic Research. *Assessment of Department of Defense Basic Research*. Washington, DC: The National Academies Press, 2005.

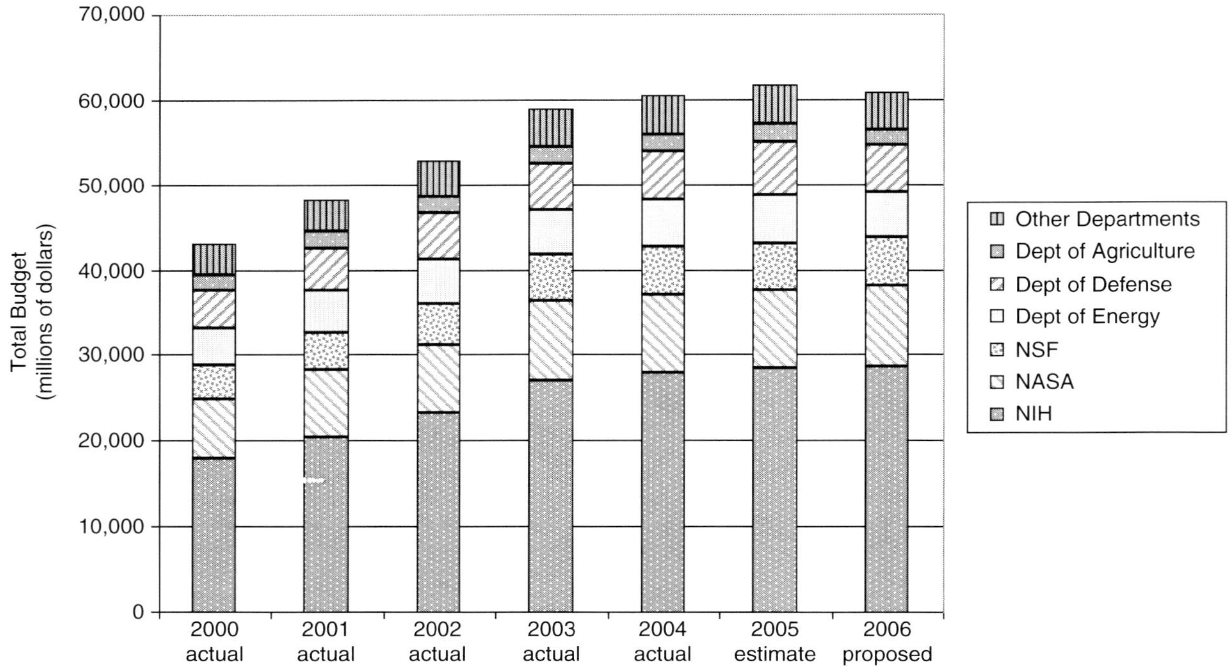

FIGURE R&D-5 Federal science and technology (FS&T) budget, in millions of dollars, FY 2000-FY 2006.
SOURCE: Based on data in several editions of Executive Office of the President. *Budget of the United States Government, Part Two: Analytical Perspectives*. Washington, DC: US Government Printing Office, 2005. Chapter 5. For research and development in the FY 2006 budget, see Table 5-3. Available at: http://www.gpoaccess.gov/usbudget.fy06/browse.html.

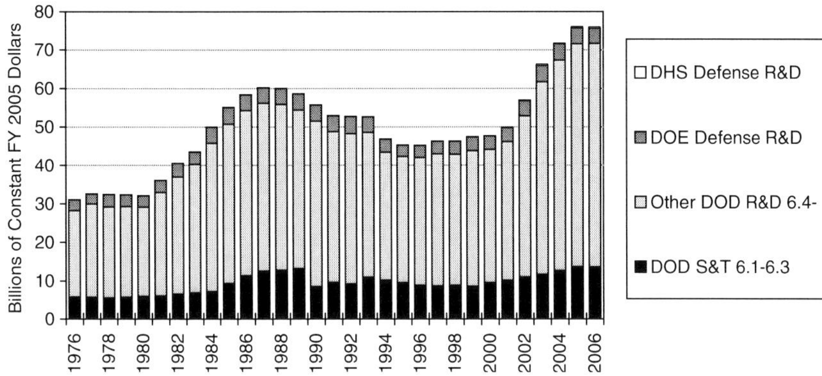

FIGURE R&D-6A Trends in defense R&D, FY 1976-FY 2006, in billions of constant fiscal year (FY) 2005 dollars, by agency.
SOURCE: American Association for the Advancement of Science. *Chart: Trends in Defense R&D: FY 1976-2006.* Washington, DC: American Association for the Advancement of Science, February 2005. Available at: http://www.aaas.org/spp/rd/trdef06c.pdf.

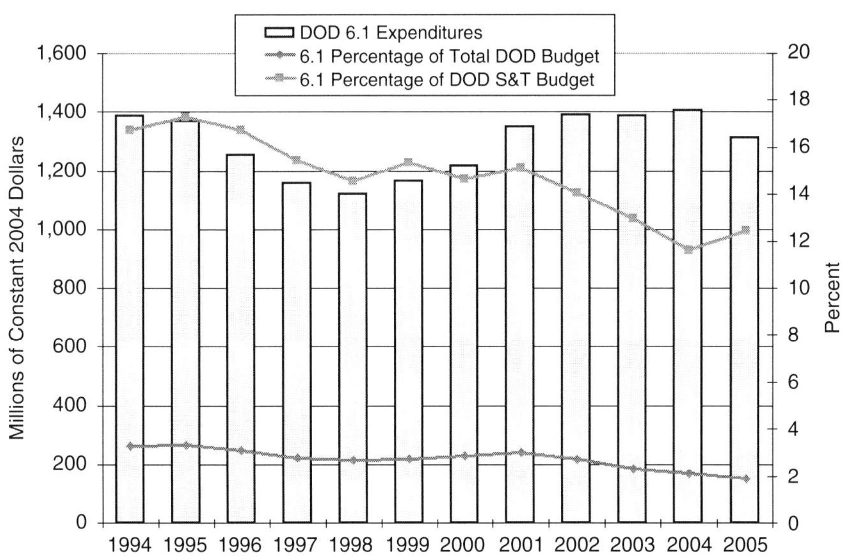

FIGURE R&D-6B Trends in Department of Defense (DOD) 6.1 R&D, FY 1994-FY 2005, in millions of constant FY 2004 dollars.
SOURCE: National Science Board. *Science and Engineering Indicators 2004.* NSB 04-01. Arlington, VA: National Science Foundation, 2004.

and balance of federal funding for science and technology include the following:

- Create a process in Congress that examines the entire FS&T budget before the total federal budget is aggregated into allocations to appropriations committees and subcommittees.[9]
- Establish a stronger coordinating and budgeting role for OSTP to promote cohesion among federal R&D agencies.[10]
- Maintain the diversity of FS&T funding in terms of sources of funding, performers, time horizons, and motivations.[11]
- Balance funding between basic and applied research and across fields of research to stimulate innovative cross-disciplinary thinking.[12]
- Protect funding for high-risk research by setting aside a portion of the R&D budgets of federal agencies for this purpose.[13]
- Maintain a favorable economic and regulatory environment for capitalizing on research—for example, by using tax incentives to build stronger partnerships among academe, industry, and government.[14]
- Encourage industry to boost its support of research conducted in colleges and universities from 7 to 20% of total academic research over the next 10 years.[15]

ACHIEVING ADEQUACY IN FEDERAL SCIENCE AND TECHNOLOGY FUNDING

Given the importance of maintaining balance and diversity in the FS&T budget, the next logical question is, What is the appropriate magnitude of federal support for science and technology?

In 1993, the Committee on Science, Engineering, and Public Policy

---

[9]Committee on Criteria for Federal Support of Research and Development, 1995.

[10]National Research Council, Board on Science, Technology, and Economic Policy. *Trends in Federal Support of Research and Graduate Education*. Washington, DC: National Academy Press, 2001.

[11]NAS/NAE/IOM. *Capitalizing on Investments in Science and Technology*. Washington, DC: National Academy Press, 1999.

[12]National Academy of Engineering, Committee on the Impact of Academic Research on Industrial Performance. *The Impact of Academic Research on Industrial Performance*. Washington, DC: The National Academies Press, 2003.

[13]Council on Competitiveness. *Innovate America*. Washington, DC: Council on Competitiveness, 2004.

[14]NAS/NAE/IOM. *Capitalizing on Investments in Science and Technology*. Washington, DC: National Academy Press, 1999.

[15]National Research Council, Office of Special Projects. *Harnessing Science and Technology for America's Economic Future: National and Regional Priorities*. Washington, DC: National Academy Press, 1999.

(COSEPUP) of the National Academy of Sciences, the National Academy of Engineering, and the Institute of Medicine established two broad goals to guide federal investments in science and technology:[16]

- The United States should be among the world leaders in all major areas of science. Achieving this goal would allow this nation quickly to apply and extend advances in science wherever they occur.
- The United States should maintain clear leadership in some areas of science. The decision to select a field for leadership would be based on national objectives and other criteria external to the field of research.

These goals provide a way of assessing the adequacy of federal funding for science and technology. Being world class across fields requires that the United States have the funding, infrastructure, and human resources for researchers to work at the frontiers of research. Preeminence in fields relevant to national priorities requires that policy-makers choose specific areas in which to invest additional resources.

An important way of measuring leadership and preeminence in fields and subfields of research is benchmarking of US research efforts against those in other countries. Experiments with benchmarking have demonstrated that data can be gathered fairly readily for analysis.[17] Benchmarking analyses then can be converted into funding guidance that takes into account the activities of other research performers (including industry and other countries) and the inherent uncertainties of research.

Responding to abundant opportunities and national priorities in science and technology, the federal government has increased R&D funding substantially in recent years. From 1990 to 2002, inflation-adjusted investment by the federal government in academic research went up 66%.[18] Increases in total R&D have been especially dramatic in the last few years because of increases for defense weapons development, the creation of homeland-security R&D programs, and the effort to double the budget of NIH.

However, as a percentage of gross domestic product (GDP), R&D has fallen from 1.25% in 1985 to about 0.75% today, and a continuation of current trends will extend this decline into the future (see Figure R&D-7). Compared with the European Union, the Organisation for Economic Co-operation and Development, and Japan, US federal R&D expenditures as a

---

[16]NAS/NAE/IOM. *Science, Technology, and the Federal Government: National Goals for a New Era*. Washington, DC: National Academy Press, 1993.

[17]NAS/NAE/IOM. *Experiments in International Benchmarking of US Research Fields*. Washington, DC: National Academy Press, 2000.

[18]National Science Board. *Science and Engineering Indicators 2004*. NSB 04-01. Arlington, VA: National Science Foundation, 2004.

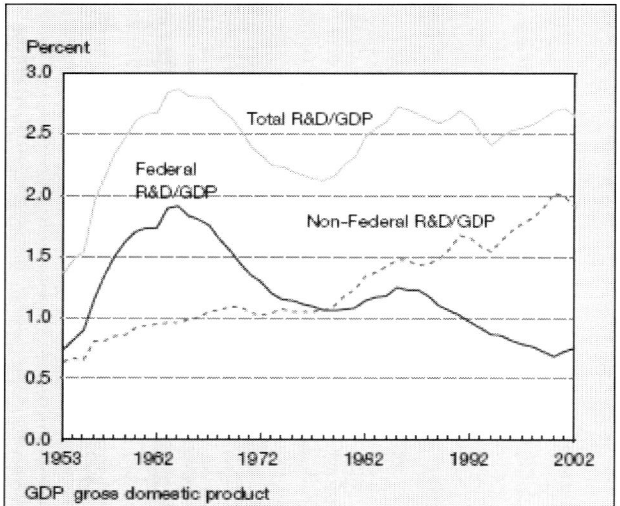

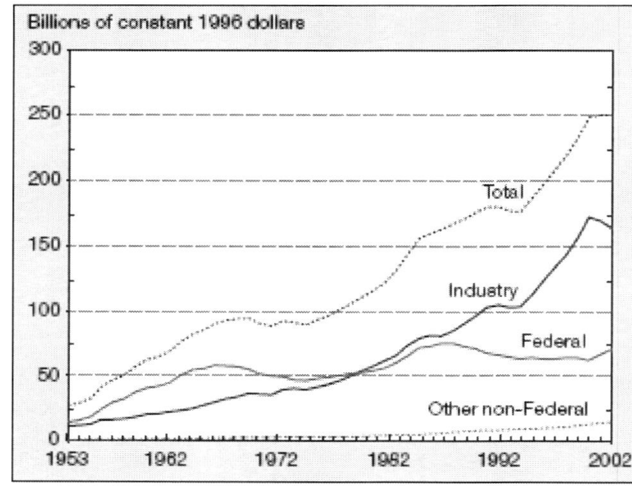

FIGURE R&D-7 R&D share of GDP, 1953-2002.
SOURCE: National Science Board. *Science and Engineering Indicators 2004*. NSB 04-01. Arlington, VA: National Science Foundation, 2004. Figures 4-3 and 4-5.

share of GDP are declining (see Figure R&D-8). Sweden, Finland, Japan, and Korea all invest a larger percentage of their GDP in R&D than the United States (see Figure R&D-9). In the president's FY 2006 budget request, most R&D programs would drop in real terms, and overall expenditures for R&D would fail to keep pace with inflation for the first time in more than a decade.[19] Funding for all three multiagency R&D initiatives—Networking and Information Technology R&D, the National Nanotechnology Initiative, and the Climate Change Science Program—would drop in FY 2006. Furthermore, with record-breaking budget deficits and new federal obligations ranging from the war in Iraq to the expansion of Medicare to pay for prescription drugs, prospects for outyear increases in R&D are dim.

The doubling of the NIH budget from 1998 to 2003 implicitly acknowledged that the rate of return on additional federal investments in science and technology is very high. Similar opportunities exist in the physical sciences, engineering, mathematics, computer science, environmental science, and the social and behavioral sciences—fields in which federal funding has been essentially flat for the last 15 years (see Figure R&D-10). Microelectronics, biotechnology, information technology, systems analysis, alternative fuels, robotics, nanotechnology, and many other research areas all have the potential to transform entire industries. Even such seemingly esoteric fields as cosmology and elementary particle physics could reveal new aspects of matter that not only could have practical implications but will inspire future generations of scientists, engineers, and mathematicians.

In addition, increases in funding of fields outside the biomedical sciences can pay dividends by complementing the tremendous advances occurring in molecular biology. Much of the recent progress in the health sciences has been underpinned by earlier achievements in mathematics, the physical sciences, and engineering. Deciphering the human genome, for example, was heavily dependent on advancements in robotics and computers. The development of modern imaging machines was made possible to a great extent by advances in engineering and mathematics.

The federal government could take several steps to ensure that funding for science and technology is adequate across fields:

- Increase the budget for mathematics, the physical sciences, and engineering research by 12% a year for the next 7 years in the research accounts of the Department of Energy, the National Science Foundation, the National Institute for Standards and Technology, and the Department of Defense.[20]

---

[19]American Association for the Advancement of Science. *AAAS Analysis of R&D in the FY 2006 Budget*. Washington, DC: American Association for the Advancement of Science, 2006.

[20]Alliance for Science & Technology Research in America. "Basic Research: Investing in America's Innovation Future." Presentation for the House Republican High-Tech Working Group, March 31, 2004.

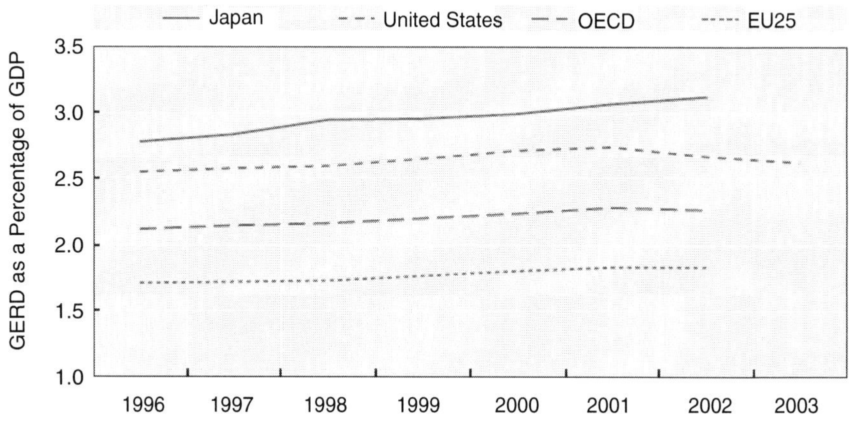

FIGURE R&D-8 Trends in R&D intensity, in United States, Japan, OECD, and EU, 1996-2003. R&D intensity is the gross domestic expenditure on R&D (GERD) as a percentage of GDP.
SOURCE: Organisation for Economic Co-operation and Development. *Main Science and Technology Indicators 2004*. Paris: OECD, June 2004.

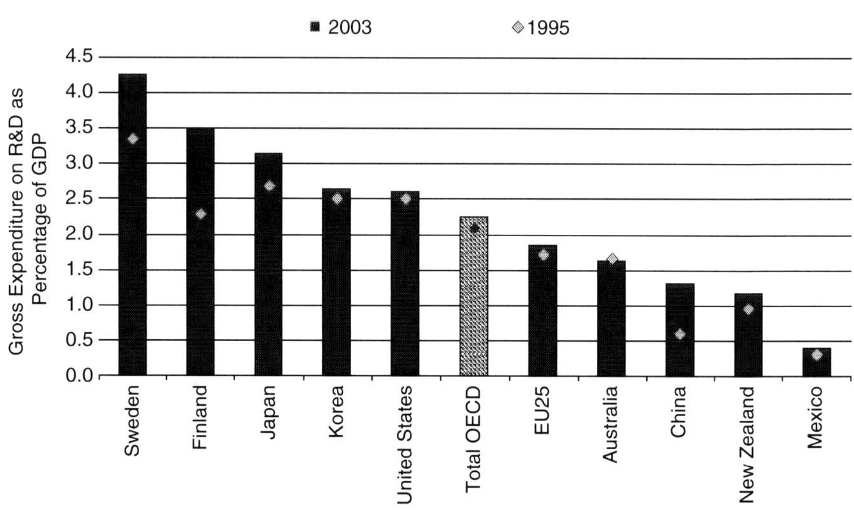

FIGURE R&D-9 Gross expenditure on R&D investments as a percentage of GDP, for select countries, OECD, and EU, 1995 and 2003.
SOURCE: Organisation for Economic Co-operation and Development. *Main Science and Technology Indicators 2005*. Paris: OECD, June 2005. Available at: http://www.oecd.org/document/26/0,2340,en_2649_34451_1901082_1_1_1_1,00.html.

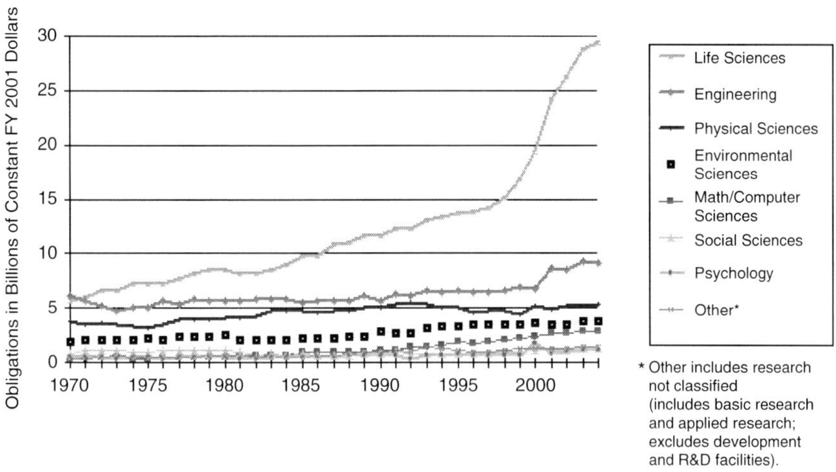

FIGURE R&D-10 Trends in federal research by field, FY 1970-FY 2004, in obligations of billions of constant FY 2004 dollars.
SOURCE: American Association for the Advancement of Science. *Chart: Trends in Federal Research by Discipline: FY 1970-2004.* Washington, DC: American Association for the Advancement of Science, February 2005. Available at: http://www.aaas.org/spp/rd/discip04.pdf.

- Return federal R&D funding to at least 1% of the US GDP.[21]
- Minimize earmarks in science and technology funding because these types of research requests diminish the funding available for competitive merit-reviewed research.[22]
- Provide a tax credit to corporations that fund basic research in science and technology at our nation's universities.
- Make the R&D tax credit permanent to promote private support of R&D, as requested by the Administration in the FY 2006 budget proposal.

LEARNING MORE ABOUT THE EFFECTS OF RESEARCH

Innovation has become more important than capital or labor in boosting economic productivity, but the course and effects of innovation are much harder to predict and understand. New technologies can spread rap-

---

[21]Council on Competitiveness. *Innovate America.* Washington, DC: Council on Competitiveness, 2004.
[22]NAS/NAE/IOM, 2003.

idly through a society, transforming multiple areas of economic activity and in turn triggering further innovations. The prime example is information technology, which has had a dramatic and accelerating influence on manufacturing, the provision of services, and other economic activities.

Intensive study of innovation as an engine of economic growth and social change in an extremely complex social context could provide guidance for policy-makers and other leaders. For example, is the current federal support of science and technology appropriately balanced across fields? What would be the effects if federal R&D were returned to its historical high as a share of GDP?

Another important topic for research is the organization of the federal agencies that support R&D. New organizational models could be explored, performance metrics developed, and approaches tested.

Options for the federal government include the following:

- Support the development of a new interdisciplinary field of quantitative science and technology-policy studies that could work to predict the effect of specific science and technology projects on the world's economies and workforces.[23]
- Support research to examine the organization models of R&D agencies and potential changes in practices and structures.

[23]Marburger, 2005.

The Productivity of Scientific and Technological Research

SUMMARY

Innovation—the process of converting inventions, ideas, or concepts into commercial products or processes—has always been a convoluted process, but today it is becoming even more difficult to understand and predict. Seemingly minor developments can have major consequences, producing a nonlinearity that defies forecasting. Developments in one field can heavily influence other fields, creating multidisciplinary networks of cause and effect. New ideas can come from anywhere in the production process, not just from the basic research that traditionally has been seen as the driver of innovation. In such a fluid, interconnected system, policy-makers need to create the optimal environment for innovation and then stand back and let the system do its job.

The effectiveness of scientific and technologic innovation depends on many factors in research organizations, including the management and review of research programs, the policies and procedures that apply to those programs, and the broader environment and culture of research. Federal options to improve this effectiveness include the following:

This paper summarizes findings and recommendations from a variety of recently published reports and papers as input to the deliberations of the Committee on Prospering in the Global Economy of the 21st Century. Statements in this paper should not be seen as the conclusions of the National Academies or the committee.

The Research Environment and Culture

- Increase the size and duration of project awards so that researchers spend more time doing research and less time ensuring that their research is supported.
- Increase the diversity of the individuals and organizations doing research.
- Fund risky projects that could dramatically advance an area of research or open new research frontiers.
- Develop a new digital cyberinfrastructure to make the best use of rapidly expanding databases and multidisciplinary collaborations.
- Expand funding for merit-reviewed, cross-disciplinary, collaborative research centers.

Program Management and Review

- Ensure that federal agencies include research programs in their strategic plans and that they evaluate the success of those programs in performance reports.
- Evaluate research in terms of quality, relevance, and leadership. For basic research, include assessments of the historical value of basic research in contributing to national goals.
- Evaluate how well research programs develop human resources and the quality, relevance, and leadership of the programs.
- Establish a formal process to identify and coordinate areas of research that are supported by multiple agencies, and designate a lead agency for each such field.

Administrative Policies and Procedures

- Develop a new framework for the development of policies, rules, regulations, and laws affecting the partnership between the federal government and the institutions that perform research.
- Raise the cap on reimbursement of indirect costs to reflect the costs to universities of conducting research.
- Expand and enhance the Federal Demonstration Partnership to enroll more institutions and heighten the visibility of this important initiative.

THE RESEARCH ENVIRONMENT AND CULTURE

Because innovation does not have a single obvious pathway to success, much depends on the environment and culture that make innovation possible. These factors range widely across social, administrative, and technological dimensions. The social factors include such considerations as commitment, collaboration, communication, the treatment of multiple viewpoints, workplace diversity, and the willingness to take risks. Administrative factors include salaries, benefits, workplace conditions, the availability of sabbaticals, and travel funding. Technological factors include technical support, training, access to high-speed computing and communications, information services, and so on.

Each of these environmental and cultural dimensions can itself be the subject of innovation. This is most obvious with regard to information technology. To take just one example, a Web site called InnoCentive (www.innocentive.com) now allows companies to post R&D problems online and offer scientists financial rewards for solutions.

The consequences of innovation extend into the social and administrative spheres. For example, increasing the number of women in the biomedical sciences helped focus attention on women's health issues, with corresponding increases in research in these areas. Similarly, funding researchers at different stages in their careers and at different types of institutions can expand the range of viewpoints brought to bear on a problem.

The federal initiatives that could improve the research environment and culture are unlimited. Among those suggested are the following:[1]

- Increase the size and duration of project awards so that researchers spend more time doing research and less time ensuring that their research is supported (see Figures RP-1 and RP-2).
- Increase the diversity of the individuals and organizations doing research.
- Fund risky projects that could dramatically advance an area of research or open new research frontiers.
- Develop a new digital cyberinfrastructure to make the best use of rapidly expanding databases and multidisciplinary collaborations.
- Expand funding for merit-reviewed, cross-disciplinary, collaborative research centers.
- Collect the best practices and attributes of federal agencies and research performers and disseminate this information widely.
- Develop a common electronic grant-application system that combines the best features of current systems and can be used by all researchers and all federal agencies.

[1] National Science and Technology Council, Business Models Subcommittee. "Comments from the Request for Information." 2003. Available at: http://rbm.nih.gov/fed_reg_20030906/index.htm.

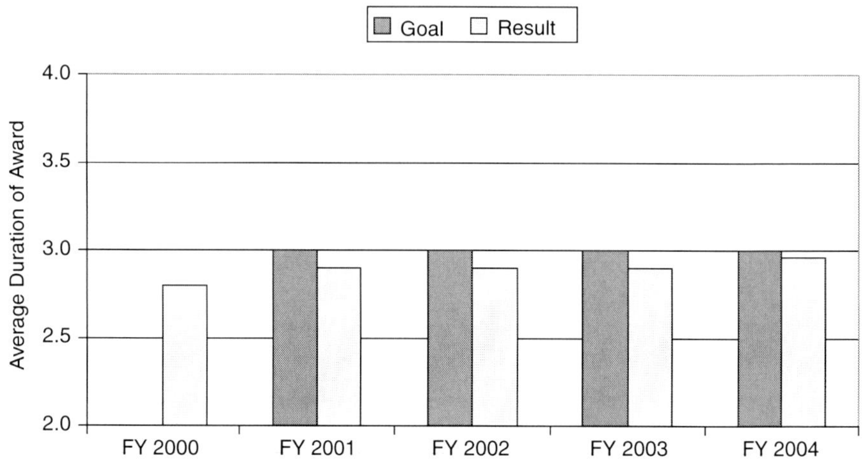

FIGURE RP-1 Average duration of research grant award at NSF, FY 2000-FY 2004. SOURCE: National Science Foundation. *FY 2004 Performance and Accountability Report*. Arlington, VA: National Science Foundation, 2004.

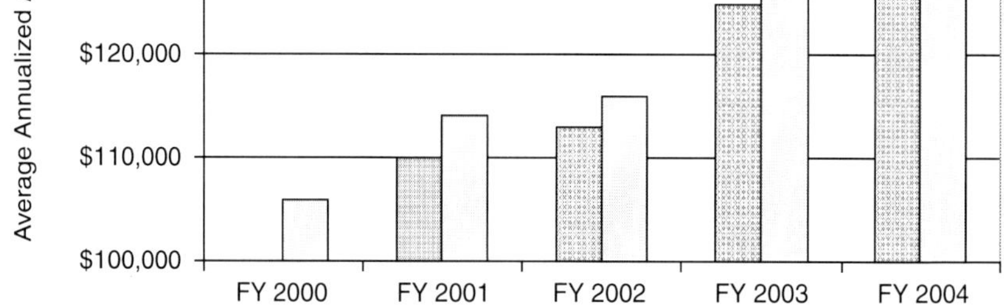

FIGURE RP-2 Average annualized award size at NSF, FY 2000-FY 2004. SOURCE: National Science Foundation. *FY 2004 Performance and Accountability Report*. Arlington, VA: National Science Foundation, 2004.

PROGRAM MANAGEMENT AND REVIEW

In an era of innovation, the innovation process itself needs to be the subject of research and development. Federal policies that influence scientific and technological research and the commercialization of that research need to be continually re-examined and improved. Valuable sources of insight include international comparisons, the results of small-scale experiments, lessons from other sectors of the economy, and clear, data-based thinking.

One useful way to improve the effectiveness of research programs is by setting goals for those programs and then monitoring the ability of programs to achieve those goals. This was one of the aims of the 1993 Government Performance and Results Act (GPRA), which was designed to encourage greater efficiency, effectiveness, and accountability in federal programs and spending. The act required federal agencies to set strategic goals for at least a 5-year period and then measure their success annually in meeting those goals.

For agencies that support research activities, implementing GPRA has presented many challenges.[2] Applied-research programs, whether conducted by federal agencies or private companies, have desired outcomes that are directly related to agency or company missions. Evaluating such programs is therefore relatively straightforward. A series of milestones that should be achieved by particular times can be established, and periodic reporting can indicate progress toward those milestones.

But the usefulness of new basic research is inherently unpredictable. Though history abundantly demonstrates the tremendous value of basic research, the practical outcomes of such research can seldom be identified while the research is in progress. Furthermore, misuse of measurements for basic research could lead to strongly negative results. Measuring this research on the basis of short-term relevance, for example, could be very destructive to quality work.

For both basic and applied research, there are meaningful measures of quality, relevance to agency goals and intended users, and contributions to world leadership in the relevant fields. These measures can be regularly reported, and they represent a sound way to ensure that the country is getting a good return on its research investments. A full description of an agency's goals and results should contain an evaluation of all research activities and their relevance to an agency's mission.

Evaluating basic research requires substantial scientific or engineering knowledge. Evaluating applied research requires, in addition, the ability to

---

[2]NAS/NAE/IOM. *Evaluating Federal Research Programs: Research and the Government Performance and Results Act*. Washington, DC: National Academy Press, 1999.

recognize its potential applicability to practical problems, which typically requires input from potential users. Expert review should be used to assess both basic-research and applied-research programs. A balance must be achieved between having the most knowledgeable and the most independent individuals serve as reviewers.

Pluralism is a major strength of the US research enterprise. But better communication among agencies would enhance opportunities for collaboration, keep important questions from being overlooked, and reduce inefficient duplication of effort. Identifying a single agency to serve as the focal point for particular fields of research could bring needed cohesion to the federal research effort. In some cases, it may make sense to adopt the model used at the Defense Advanced Research Projects Agency (DARPA), in which the desired end product or technology is defined before research begins, so that research teams can coordinate their efforts to solve the problem.

To improve the effectiveness of federal research and development programs, the federal government could:

- Ensure that federal agencies include research programs in their strategic plans and that they evaluate the success of those programs in performance reports.[3]
- Evaluate research in terms of quality, relevance, and leadership. For basic research, include assessments of the historical value of basic research in contributing to national goals.
- Evaluate how well research programs develop human resources and the quality, relevance, and leadership of the programs. If federal research activities do not continue to produce a flow of well-educated scientists and engineers, the capability of an agency to fulfill its mission will be compromised and the knowledge learned and technology developed will be lost.
- Establish a formal process to identify and coordinate areas of research that are supported by multiple agencies. A lead agency should be identified for each such field, and that agency should be responsible for ensuring that coordination occurs among the agencies.
- Investigate and experiment with innovative ways of managing research, such as establishment of long-term research goals, very flat management structures, multidisciplinary teams, and a focus on technology transfer (these are some of the approaches that have met with considerable success at DARPA).[4]

---

[3]Ibid.

[4]L. H. Dubois. DARPA's Approach to Innovation and Its Reflection in Industry. In *Reducing the Time from Basic Research to Innovation in the Chemical Sciences: A Workshop Report to the Chemical Sciences Roundtable*. Washington, DC: The National Academies Press, 2003. Pp. 37-48.

ADMINISTRATIVE POLICIES AND PRACTICES

The performers of research sponsored by the federal government operate under an increasing number and variety of administrative requirements. Examples include rules for human subjects, animal welfare, conflicts of interest, costing and administration, agency-specific requirements, and indirect costs. While each rule has its own history and justifications, the combination of often poorly coordinated requirements imposes a significant burden on research performers.

Two publications from the Office of Management and Budget (OMB)—Circular A-21, *Cost Principles for Educational Institutions*, and Circular A-110, *Uniform Administrative Requirements for Grants and Other Agreements with Institutions of Higher Education, Hospitals, and Other Non-Profit Organizations*—form the framework for current cost and administrative regulations. Both are in need of revision. In 1999, the National Science and Technology Council (NSTC) released a report titled *Renewing the Federal Government-University Research Partnership for the 21st Century*, which laid out a set of guiding principles to provide a framework for the development of new policies, rules, regulations, and laws. These principles could be used to define acceptable standards for the conduct of research that could identify areas of deficiency and foster an appropriate balance between compliance with regulations and administrative flexibility.

A particularly contentious issue for college and university researcher performers has been the 26% cap on reimbursement of administrative costs imposed by the federal government in 1991.[5] Currently, about a quarter of federal funds spent on research at universities reimburses indirect costs. The two major components of indirect costs are for the construction, maintenance, and operation of facilities used for research and for supporting administrative expenses, such as financial management, institutional review boards, and environment, health, and safety management.

As the administrative demands on universities have increased, these institutions have had to pay for an increasing percentage of indirect costs that are not covered under the 26% cap. As a result, universities have had to shift funds to cover administrative costs from other sources, including tuition, endowments, or state appropriations. Eventually, this cost shifting will be detrimental to the health of these institutions, resulting either in less research, higher tuitions, or reduced services to students.

A more flexible and responsive relationship between federal agencies and universities could help control the administrative costs of research. In 1986, the program now known as the Federal Demonstration Partnership

---

[5]Office of Science and Technology Policy. *Analysis of Facilities and Administrative Costs at Universities*. Washington, DC: Executive Office of the President, 2000.

(FDP) was established to examine, streamline, and reduce the burdens of grant administration. The goals of the FDP are to standardize terms and conditions across federal agencies, simplify the prior-approval process, and streamline award distribution—for example, the FDP is doing a long-term study of institutional burdens related to the OMB circulars. Extending the FDP to colleges with less involvement in federal research awards would help disseminate best practices among federal agencies and institutions of higher education.

Among the actions the federal government could take to reduce the administrative burden on the performers of research are the following:

- Use the "Principles of the Federal Partnership with Universities in Research" developed by the NSTC to provide a framework for the development of new policies, rules, regulations, and laws affecting the government-university partnership.
- Raise the cap on reimbursement of indirect costs to reflect the costs to universities of conducting research.
- Expand and enhance the FDP to enroll more institutions and heighten the visibility of this important initiative.
- Streamline and align the grant-administration process across agencies to the extent that is consistent with agency needs; all agencies should use uniform terms and conditions for all research and research-related project grants.

Investing in High-Risk and Breakthrough Research

SUMMARY

If processes for awarding research grants are too risk-averse, innovative research projects that could lead to future breakthroughs in science and technology may never be funded. To avoid over-cautious R&D funding, recent reports and new programs have focused on three critical areas: adequate funding for basic, discovery-oriented research; independent research funding for young investigators; and funding for individuals who propose visionary research.

Among the federal actions that have been proposed to encourage high-risk research are the following:

- Reallocate 3% of all federal-agency R&D budgets toward grants that invest in novel, high-risk, and exploratory research.
- Establish a program at the National Institutes of Health (NIH) to promote the conduct of innovative research by scientists transitioning into their first independent positions.
- Within NIH, continue to explore programs, such as the Pioneer Awards, to increase funding for high-risk, high-benefit biomedical research.

This paper summarizes findings and recommendations from a variety of recently published reports and papers as input to the deliberations of the Committee on Prospering in the Global Economy of the 21st Century. Statements in this paper should not be seen as the conclusions of the National Academies or the committee.

SUPPORT HIGH-RISK RESEARCH

Besides favoring older investigators, the current peer-review system can tend to drive award decisions toward conservative research that is based on precedent and is consensus-oriented. As a result, public funding for research can gradually shift from investments in bold, transformational discovery to much more incremental research.

The Council on Competitiveness proposes in the 2004 report *Innovate America* that the nature of discovery-focused research creates a need for government support. However, federal research support since the Cold War has become more conservative, focusing on short-term, incremental, low-risk goals. Outside the government, the council believes that risk-based investments are also needed to promote innovation. Investors tend to focus on short-term profits and are unwilling to accept the risks that come with investing in a long-term research project (see Figure HRR-1).[1] The report recommends the following:

- Reallocate 3% of all federal-agency R&D budgets toward grants that invest in novel, high-risk, and exploratory research.
- Provide a 25% tax credit for early-stage investments of at least $50,000 through qualified angel funds.[2]

In the United States, NIH has, through its Roadmap initiative, also begun to seed more innovative, high-risk research. "The past two decades have brought tremendous scientific advances that can greatly benefit medical research," the Roadmap argues. "While progress will continue into the foreseeable future, human health and well-being would benefit from accelerating the current pace of discovery. One way to achieve this goal is to support scientists of exceptional creativity who propose highly innovative approaches to major contemporary challenges in biomedical research. NIH has traditionally supported research projects, not individual investigators. However, complementary means might be necessary to identify scientists with ideas that have the potential for high impact, but that may be too novel, span too diverse a range of disciplines, or be at a stage too early to fare well in the peer review process." As part of this initiative, NIH has created the NIH Director's Pioneer Awards "to encourage creative, outside-the-box thinkers to pursue exciting and innovative ideas about biomedical research." The first Pioneer Awards were granted in 2004.[3]

---

[1]Council on Competitiveness. *Innovate America*. Washington, DC: Council on Competitiveness, 2004.

[2]Ibid.

[3]National Institutes of Health, NIH Roadmap. "High Risk Research." 2005. Available at: http://nihroadmap.nih.gov/highrisk/.

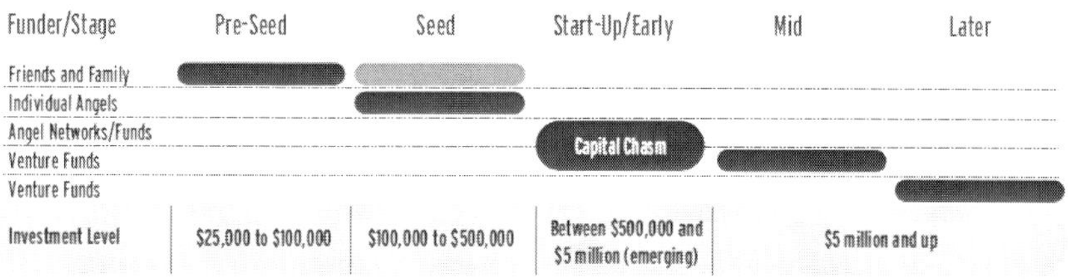

FIGURE HRR-1 Funding for innovation, by funder and investment stage.
SOURCE: Council on Competitiveness. *Innovate America*. Washington, DC: Council on Competitiveness, 2004. P. 36. Figure 6.

To revitalize frontier research capable of providing breakthroughs, the federal government could

- Within NIH, continue to explore programs, such as the Pioneer Awards, to increase funding for high-risk, high-benefit biomedical research.

The National Science Board, at the National Science Foundation (NSF), is also discussing this issue. In 2004, an ad hoc Task Group on High-Risk Research was formed, which recommended that a formal Task Force on Transformative Research be established under the Committee on Programs and Plans. Additionally, the ad hoc Task Group noted that there is no formal definition of "high-risk" or "transformative" research, so there is no way to adequately determine how much support NSF is providing to such projects, but there are several reasons to begin doing so. The formal committee is researching these and other questions, and a report is expected within 2 years.[4]

The European Commission (EC), meanwhile, has focused part of its R&D funding on seeding high-risk research. Under its Sixth Framework Programme (FP6), the EC has established a New and Emerging Science and Technology (NEST) program at €215 million to "support unconventional and visionary research with the potential to open new fields for European science and technology, as well as research on potential problems uncovered by science."[5]

---

[4]National Science Board. "Committee on Programs and Plans, Charge to the Task Force on Transformative Research." Available at: http://www.nsf.gov/nsb/committees/cpptrcharge.htm.

[5]European Commission, Enterprise and Industry Directorate-General. "New and Emerging Science and Technology (NEST) Programme." 2005. Available at: http://www.cordis.lu/nest/home.html.

FOSTER INNOVATION THROUGH YOUNG INVESTIGATORS

While peer review provides a high-integrity process sheltered from political forces, evidence suggests that it tends to favor both established investigators and investigators, new or continuing, who build on established research lines.[6] As a result, young investigators have difficulty establishing themselves as independent researchers, which can have a variety of negative consequences for establishing careers, ensuring an adequate research workforce, and bringing fresh insights and ideas to the research enterprise. Indeed, recent research indicates that the age at which great innovations are produced has increased by about 6 years over the 20th century, and the loss of productivity at earlier ages is not compensated for by increased productivity after early middle age[7] (see Figures HRR-2A and B). The risk is that competence and productivity can be honored to the point where they become the "enemies of greatness."

The current system tends to emphasize the number of papers published and can overlook whether important problems are being tackled. Because requests for grant funds from new investigators are evaluated on the basis of "preliminary results," most funded research becomes constrained to well-worn research paths, which for new investigators often means the research they previously pursued when they were postdoctoral fellows in established laboratories. In short, innovation can become the victim of a system that has become too risk-averse.

Because of the difficulties facing new investigators, the median age at which investigators receive their first research grant from NIH, for example, had crept up to 42 years in 2002. This raises the concern that new investigators are being driven to pursue more conservative research projects instead of high-risk, high-reward research that can significantly advance science. Also, young investigators can end up focusing much of their attention on others' research, forfeiting the special creativity that they may bring to their own work (see Figures HRR-3A, B, and C).[8]

The same considerations apply to work funded by the Department of Defense (DOD). The need for new discoveries and innovation argues for substantial involvement of university researchers. Yet some younger university researchers in the expanded fields of interest to the DOD are discouraged by difficulty in acquiring research support from the department.[9]

---

[6]National Research Council. *Bridges to Independence: Fostering the Independence of New Investigators in Biomedical Research*. Washington, DC: The National Academies Press, 2005.

[7]B. Jones. *Age and Great Innovation*. Working Paper 11359. Cambridge, MA: National Bureau of Economic Research, 2005. Available at: http://www.nber.org/papers/w11359.

[8]National Research Council. *Bridges to Independence: Fostering the Independence of New Investigators in Biomedical Research*. Washington, DC: The National Academies Press, 2005.

[9]National Research Council. *Assessment of Department of Defense Basic Research*. Washington, DC: The National Academies Press, 2005.

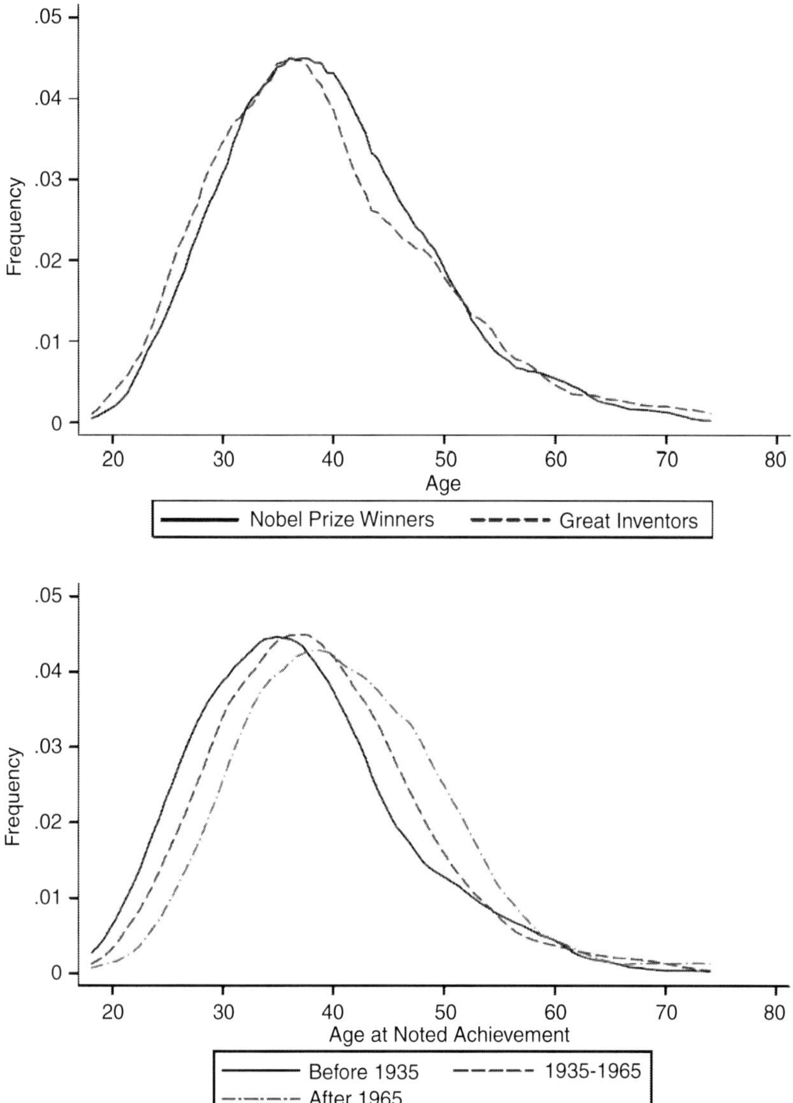

FIGURE HRR-2A Frequency distribution of age of Nobel Prize winners and great inventors at time of noted achievement.
SOURCE: B. Jones. *Age and Great Innovation.* Working Paper 11359. Cambridge, MA: National Bureau of Economic Research, 2005. Available at: http://www.nber.org/papers/w11359/.

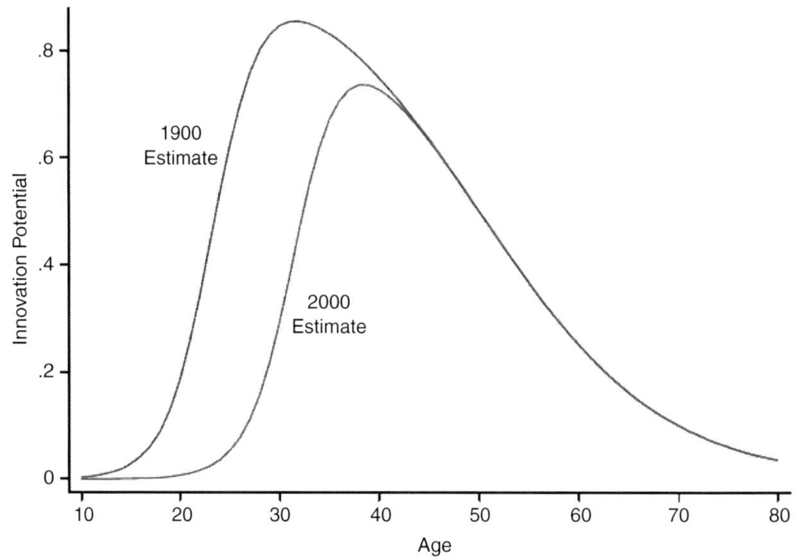

FIGURE HRR-2B Maximum likelihood estimates for potential to produce great innovations as a function of age.
SOURCE: B. Jones. *Age and Great Innovation.* Working Paper 11359. Cambridge, MA: National Bureau of Economic Research, 2005. Available at: http://www.nber.org/papers/w11359/.

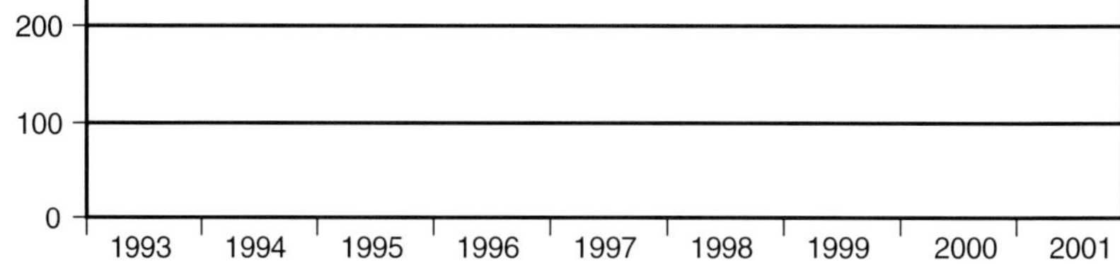

FIGURE HRR-3A Number of federal awards received by those 35 and under, 1993-2001.
SOURCE: P. Stephan. Presentation at Bridges to Independence Workshop. Board on Life Sciences, The National Academies, June 16, 2004. Available at: http://dels.nas.edu/bls/bridges/Stephan.pdf. Data are drawn from the National Science Foundation's Survey of Doctorate Recipients.

Success rate of competing new R01 and R29 grant application by age of principal investigator.

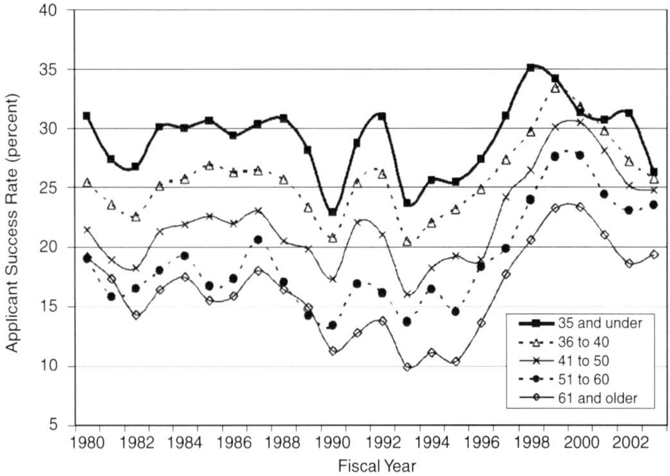

Number of R01, R23, R29, or R37 applicants by age cohort.

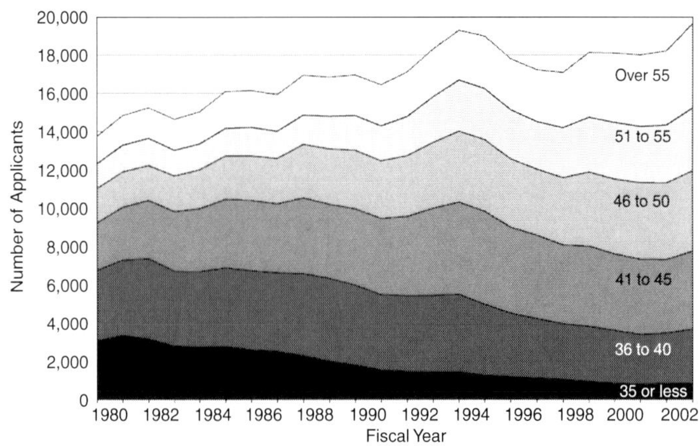

FIGURE HRR-3B Success rate of competing new R01 and R29 grant application by age of principal investigator and number of R01, R23, R29, or R37 applicants by age cohort.

NOTE: Data are from the National Institutes of Health, Office of Extramural Research. Available at: http://grants.nih.gov/grants/oer.htm.

SOURCE: National Research Council. *Bridges to Independence: Fostering the Independence of New Investigators in Biomedical Research*. Washington, DC: The National Academies Press, 2005.

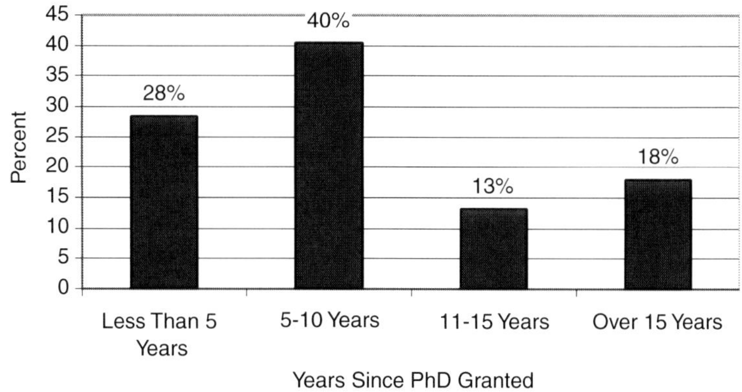

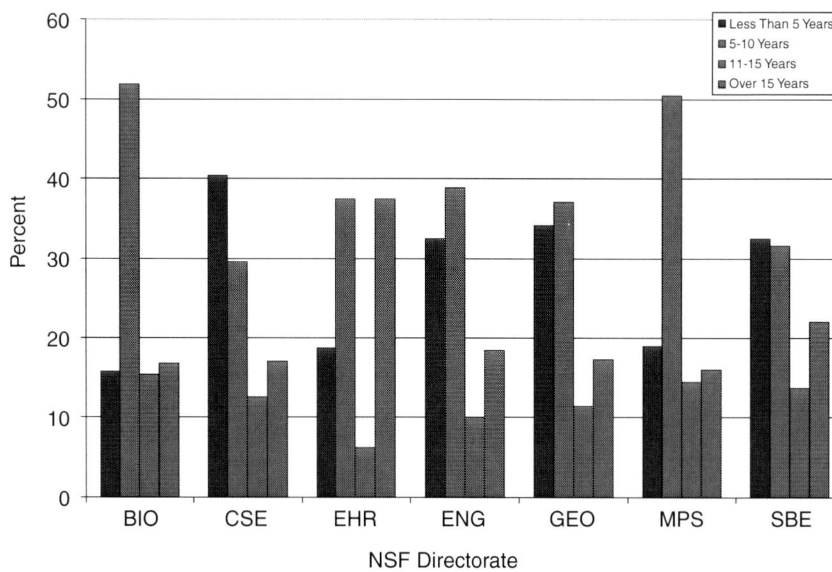

FIGURE HRR-3C Percent of fiscal year 2003 awards to new principal investigators, versus year since PhD, by field and NSF directorate.

NOTE: BIO = Biological Sciences; CSE = Computer, Information Sciences, and Engineering; EHR = Education and Human Resources; ENG = Engineering; GEO = Geosciences; MPS = Mathematical and Physical Sciences; SBE = Social, Behavioral, and Economic Sciences.

SOURCE: M. Clutter. Presentation at Bridges to Independence Workshop. Board on Life Sciences, The National Academies, June 16, 2004. Available at: http://dels.nas.edu/bls/bridges/Clutter.pdf.

To address these needs, the federal government could:

• Establish a program at NIH to promote the conduct of innovative research by scientists transitioning into their first independent positions. These research grants would replace the existing collection of K22 awards and would provide sufficient funding and resources for promising scientists to initiate independent research programs and allow for increased risk-taking during the final phase of these efforts. The program should make 200 grants annually of $500,000 each, payable over 5 years. Each award would provide funding for 2 years of postdoctoral training support while the awardee develops an independent research program and 3 years of support as a fully independent researcher.[10]

• Establish and implement uniformly across all the NIH institutes a New Investigator R01 grant. The "preliminary results" section of the application should be replaced with "previous experience" to be appropriate for new investigators and to encourage higher-risk proposals or scientists branching out into new areas. This award should include a full budget and have a 5-year term. NIH should track New Investigator R01 awardees in a uniform manner, including their success on future R01 applications.[11]

• Encourage, through DOD funding and policies for university research, participation by younger researchers as principal investigators.[12]

---

[10]National Research Council. *Bridges to Independence*, 2005.

[11]Ibid.

[12]National Research Council. *Assessment of Department of Defense Basic Research*, 2005.

Ensuring That the United States Is at the Forefront in Critical Fields of Science and Technology

SUMMARY

As concerns over the declining competitiveness of some US industries emerged in the 1980s, policies and programs were put into place with the goal of enabling new ideas—particularly those created through federal support—to be commercialized more quickly.

These policies and programs have taken a number of forms. They have included support for R&D partnerships among companies and between industry and government, support for R&D activities in small companies, programs to support academic research in areas of interest to industry, policies to encourage commercialization of inventions made by federal laboratories and those made by academic researchers with federal support, initiatives to coordinate federal R&D in areas of interest to several agencies, and the creation of private-sector advisory committees concerned with the future international competitiveness of particular industries.

Some of these programs have attracted controversy. For example, the Advanced Technology Program (ATP), having survived several attempts to eliminate it, was not appropriated funds for new awards in fiscal year (FY) 2005.[1] Others have continued and expanded or have made a variety of transitions—for example, from government-supported to privately funded.

This paper summarizes findings and recommendations from a variety of recently published reports and papers as input to the deliberations of the Committee on Prospering in the Global Economy of the 21st Century. Statements in this paper should not be seen as the conclusions of the National Academies or the committee.

[1]See ATP Web site's "Update for 2005." Available at: http://www.atp.nist.gov/atp/05comp.htm.

Federal actions that have been proposed include the following:

New Policies and Initiatives

- Create interdisciplinary discovery-innovation institutes to bring together research, education, and practice around the solution of major societal problems.
- Create a program of "Innovation Acceleration" grants to stimulate high-risk research through a set aside of 3% of agency R&D budgets.
- Create a National Institute of Innovation to provide venture capital for innovative startups.
- Expand industry-led roadmaps for R&D priorities.
- Launch a large new initiative to develop the computational science base and the necessary broad infrastructure (such as networks) and domain-specific tools for research and education enabled by information technology across the various fields of science, engineering, and medicine.
- Establish centers for production excellence and Innovation Extension Centers to improve the capabilities of small and medium-sized enterprises.

Modifications of Existing Policies and Programs

- Make improvements to the Small Business Innovation Research program, including bridges between phase I and phase II funding, increased phase II funding relative to phase I funding, and regular assessments across agencies.
- Restore ATP funding—including the ability to support new awards—to the average level of recent years.
- Make improvements in ATP, including streamlining the application process and widening the window for funding, better integrating ATP with other programs, and focusing some funding in thematic areas.
- Have such agencies as the Securities and Exchange Commission, the Federal Communications Commission, and the Internal Revenue Service consider launching industry–university collaborative research centers to benefit the services industries.
- Re-examine and amend the Bayh–Dole Act to encourage collaboration among university licensing offices, thereby promoting economic development.

THE FEDERAL GOVERNMENT AS VENTURE CAPITALIST

The Small Business Innovation Research (SBIR) and Small Business Technology Transfer (STTR) programs have sought to encourage the innovative activities of small businesses. SBIR was established in 1982 and sets aside 2.5% of the extramural R&D budgets of the largest federal science agencies for funding R&D by small businesses; it currently runs at over $1 billion per year.[2] Table EL-1 shows the overall trend. SBIR encompasses three phases: feasibility, development, and commercialization. SBIR has been reviewed and evaluated a number of times over the course of its existence.[3] The National Research Council is currently undertaking a new assessment of the program.[4]

STTR was established in 1992 to encourage small businesses to partner with research institutions in R&D and commercialization.[5]

Although there has been debate over the years about the impacts of these programs and the appropriate evaluation metrics, past assessments have been positive overall. Political support also has been very strong, with a number of technical changes having been recommended and enacted over the years.

Possible federal actions to improve and extend these programs include the following:

- Bridge the funding gap between phase I and phase II awards provided by the SBIR program.[6]
- Increase the number of phase II SBIR awards at the expense of phase I awards.[7]
- Regularly assess SBIR program results and compare with the Department of Defense (DOD) Fast Track results, and assess the costs and benefits of better integrating SBIR awards in the development of "clusters" around universities and technology parks.[8]
- Create a National Institute of Innovation that would provide venture capital for innovative startup companies to smooth the peaks and valleys of private-sector venture-capital flows.[9] A similar idea, called the Civil-

---

[2]National Research Council. *SBIR: Program Diversity and Assessment Challenges, Report of a Symposium*. Washington, DC: The National Academies Press, 2004.

[3]National Research Council. *SBIR: An Assessment of the Department of Defense's Fast Track Initiative*. Washington, DC: National Academy Press, 2000; National Research Council. *SBIR: Challenges and Opportunities*. Washington, DC: National Academy Press, 1999.

[4]National Research Council. *An Assessment of the Small Business Innovation Research Program: Project Methodology*. Washington, DC: The National Academies Press, 2004.

[5]US General Accounting Office. "Contributions to and Results of the Small Business Technology Transfer Program." Statement by Jim Wells. GAO-01-867T. Washington, DC: General Accounting Office, 2001.

[6]National Research Council. *The Small Business Innovation Research Program: Challenges and Opportunities*. Washington, DC: National Academy Press, 1999.

[7]Ibid.

[8]National Research Council, 2000.

[9]K. Hughes. "Facing the Global Competitiveness Challenge." *Issues in Science and Technology* 21(Summer 2005).

TABLE EL-1 Small-Business Innovation Research Award Funding, by Type of Award: FY 1983-FY 2001

	All Agencies		
FY	Total	Phase I (feasibility)	Phase II (main phase)
1983	45	45	0
1984	108	48	60
1985	199	69	130
1986	298	99	199
1987	351	110	241
1988	389	102	285
1989	432	108	322
1990	461	118	342
1991	483	128	336
1992	508	128	371
1993	698	154	491
1994	718	220	474
1995	835	232	602
1996	916	229	646
1997	1,107	278	789
1998	1,067	262	804
1999	1,097	300	797
2000	1,190	302	888
2001	1,294	317	977

SOURCE: National Science Board. *Science and Engineering Indicators 2004*. NSB 04-01. Arlington, VA: National Science Foundation, 2004. Appendix Table 4-39.

ian Technology Corporation, was proposed by a National Academies committee some years ago.[10]

THE ADVANCED TECHNOLOGY PROGRAM AND OTHER CONSORTIA

Partly as a response to Japan's success in benefiting from industrial consortia in such areas as steel and semiconductors, Congress passed the National Cooperative Research Act in 1984. This legislation limited potential antitrust liabilities in order to encourage corporate R&D consortia.

[10]NAS/NAE/IOM. *The Government Role in Civilian Technology*. Washington, DC: National Academy Press, 1992.

With the launch of SEMATECH in 1987, the US government moved to actual financial support for collaborative industrial R&D. SEMATECH was founded as a partnership between US semiconductor companies and the DOD. In the succeeding years, as the US semiconductor industry regained competitive strength, the federal contribution to SEMATECH was gradually reduced and then eliminated.[11] The consortium, now named International SEMATECH, includes countries based in Europe, Korea, and Taiwan in addition to those based in the United States.

ATP was established in 1988 as a program of the National Institute of Standards and Technology (NIST). ATP supports collaborative research among companies. The program has operated at a level of $150 million to $200 million per year in recent years. As mentioned above, the FY 2005 budget included funds to continue existing projects but no money to fund new proposals. Figure EL-1 shows how ATP funding has fluctuated over the years. ATP also supports an extensive program of evaluation and research, which has supported work at the National Academies and the National Bureau of Economic Research.[12]

Possible federal actions to derive advantage from government–industry partnerships and industrial consortia include the following:

- Create "Innovation Acceleration" grants to stimulate high-risk research.[13] These grants would be supported through a set aside of 3% of agency R&D budgets.
- Restore the support of ATP and its ability to fund new projects to the level of recent years.
- Streamline and shorten the ATP application process and timeline.[14]
- Give applications from single companies parity with those from joint ventures or consortia.[15]
- Extend the window for ATP award applications, accelerate the decision-making process for awards, and extend the period in which awards can be made.[16]

---

[11]National Research Council. *Securing the Future: Regional and National Programs to Support the Semiconductor Industry*. Washington, DC: The National Academies Press, 2003. See also the "History" page on the International SEMATECH Web site, http://www.sematech.org/corporate/history.htm.

[12]See the ATP Web site. Available at: http://www.atp.nist.gov/factsheets/1-a-1.htm.

[13]Council on Competitiveness. *Innovate America*. Washington, DC: Council on Competitiveness, 2004.

[14]National Research Council. *The Advanced Technology Program: Challenges and Opportunities*. Washington, DC: National Academy Press, 1999.

[15]Ibid.

[16]National Research Council. *The Advanced Technology Program: Assessing Outcomes*. Washington, DC: National Academy Press, 2001.

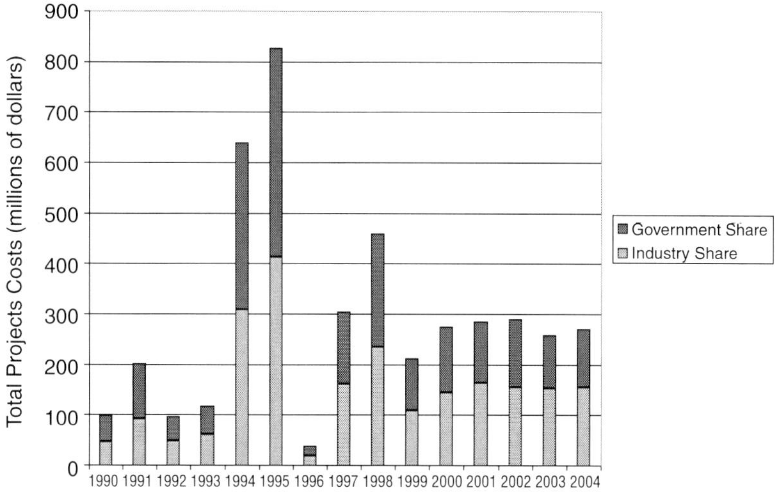

FIGURE EL-1 Summary of ATP awards, by source of funding, 1990-2004.
SOURCE: Advanced Technology Program. "ATP Factsheet: 3.A.3: ATP Awards Summary Data-Funding ($ Millions)." September 2004. Available at: http://www.atp.nist.gov/factsheets/3-a-3.htm.

- Retain the debriefing process for unsuccessful ATP applicants.[17]
- Concentrate a significant portion of ATP awards in selected thematic areas.[18]
- Coordinate ATP with SBIR and national initiatives.[19]
- Establish a regular outreach program within NIST to coordinate ATP awards with matching grants by states.[20]
- Pass legislation that would allow industries to form self-organizing investment boards that would raise funds through a "tax" on sales of their products in order to support R&D on common problems.[21]

---

[17]Ibid.

[18]Ibid.

[19]Ibid.

[20]Ibid.

[21]P. Romer. Implementing a National Technology Strategy with Self-Organizing Industry Investment Boards. In M. N. Baily, P. C. Reiss, and C. Winston, eds. *Brookings Papers on Economic Activity: Microeconomics* (2). Washington, DC: Brookings Institution, 1993. Pp. 345-399.

UNIVERSITY-BASED CENTERS

Federally supported university-based centers constitute a category of programs that support collaborative (usually interdisciplinary) research between universities and industries. These include such programs as the Engineering Research Centers (ERCs), Science and Technology Centers (STCs), and Industry–University Cooperative Research Centers (I/UCRCs) of the National Science Foundation (NSF). Other agencies, such as the Department of Transportation and the Department of Energy (DOE), also support university-based centers. These programs are generally awarded on a continuing basis with renewal reviews at fixed periods. NSF support for individual STCs phases out after 11 years, while other center programs are funded longer. Leveraged support from industry is generally required, the level of which varies by program.

The NSF efforts have the longest track record. For example, the ERCs program was established in 1985.[22] The program itself is occasionally evaluated internally and by an external contractor using surveys, bibliometric analysis, and other methods.[23] These evaluations generally show that a large percentage of industry participants derive benefits from participation, including knowledge transfer and the ability to hire students. At the time when the STCs program was being considered for renewal, a National Academies committee recommended that the program continue.[24] Figure EL-2 shows how the various NSF centers programs fit into the overall funding picture.

Options for federal action include the following:

- Establish a new, large, multi-agency centers program. In a preliminary report released for public comment earlier this year, a committee of the National Academy of Engineering proposed to create a program of interdisciplinary discovery-innovation institutes on research-university campuses. The institutes would bring together research, education, and practice around the solution of major societal problems.[25] Multi-agency federal support for the institutes would build to several billion dollars per year, to be supplemented by support from industry, states, and nonprofits.

---

[22]L. Parker. *The Engineering Research Centers (ERC) Program: An Assessment of Benefits and Outcomes*. Arlington, VA: National Science Foundation, 1997. Available at: http://www.nsf.gov/pubs/1998/nsf9840/nsf9840.htm.

[23]J. D. Roessner, D. Cheney, and H. R. Coward. *Impact of Industry Interactions with Engineering Research Centers—Repeat Study*. Arlington, VA: SRI International, 2004.

[24]NAS/NAE/IOM. *An Assessment of the National Science Foundation's Science and Technology Centers Program*. Washington, DC: National Academy Press, 1996.

[25]National Academy of Engineering. *Assessing the Capacity of the US Research Enterprise*. Preliminary Report for Public Comment. Washington, DC: The National Academies Press, 2005.

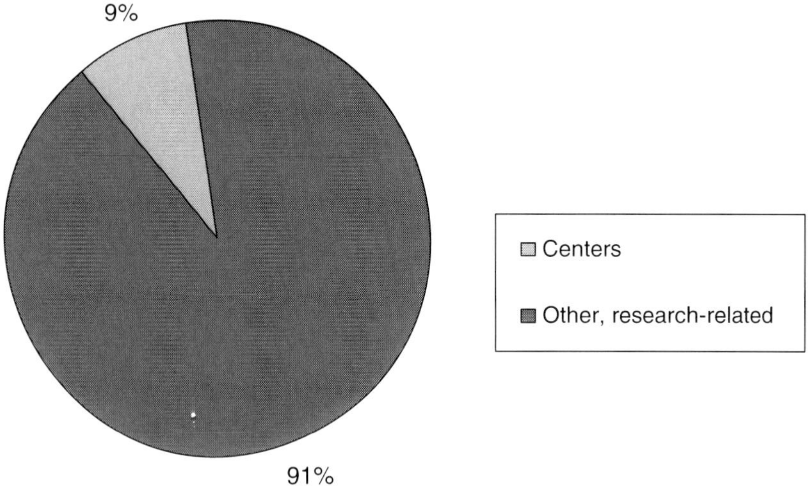

FIGURE EL-2 Centers as a percentage of the NSF research and related account. SOURCE: Based on data in the National Science Foundation. *FY 2005 Performance and Accountability Report.* Arlington, VA: National Science Foundation, 2005.

- Establish centers in agencies that have not supported centers in the past. Federal mission and regulatory agencies with primary responsibility for the services industries—such as the Securities and Exchange Commission, the Internal Revenue Service, the Federal Communications Commission, and the Department of Health and Human Services (DHHS)—should consider funding academic research in ways that encourage greater participation by the services industries.[26]

COLLABORATIVE RESEARCH AND DEVELOPMENT AGREEMENTS

Another mechanism for government–industry collaboration is a collaborative and development agreement (CRADA). The Stevenson–Wydler Technology Innovation Act of 1986 allowed federal laboratories to enter into CRADAs with private companies. The legislation has been amended several times and covers most agencies.The National Aeronautics and Space

[26]National Academy of Engineering. *The Impact of Academic Research on Industrial Performance.* Washington, DC: The National Academies Press, 2003.

Administration has a separate authority under the 1958 Space Act and the 1989 National Space Policy.[27]

As of FY 2001, there were 3,603 active CRADAs, 80% of which involved DOD, DOE, or the Department of Health and Human Services.[28]

CRADAs can range from focused collaboration on a specific technology to large programs, such as FreedomCAR, a successor to the Partnership for a New Generation of Vehicles (PNGV) CRADA between DOE and the big three automakers.[29] PNGV was reviewed by a standing National Academies committee.[30] Although the research made impressive technological progress, only with the recent rapid rise in gasoline prices are advanced technologies for high-fuel-economy vehicles becoming a competitive factor in the marketplace.

THE BAYH–DOLE ACT

The Bayh–Dole Act of 1980, which allowed universities to own and license patents of university inventions (even inventions supported by federal funds), ushered in an explosion of university patenting and licensing activity.[31] There is broad recognition that Bayh–Dole has encouraged a variety of university–industry collaborations and small-firm startups. Figures EL-3 and EL-4 show how industry support for university research and university licensing income has gone up. There has been continuing research and debate on the ultimate impacts.[32]

Calls to amend or rethink Bayh–Dole have come from several quarters in recent years. Some companies and universities have found it difficult to work out the intellectual-property aspects of collaboration.[33] There also have been cases in which university intellectual-property rights might have

---

[27]National Aeronautics and Space Administration. *Space Act Manual.* Washington, DC: National Aeronautics and Space Administration, 1998. Available at: http://nodis3.gsfc.nasa.gov/1050-1.html.

[28]National Science Board. *Science and Engineering Indicators 2004.* NSB 04-01. Arlington, VA: National Science Foundation, 2004. See summary points for Chapter 4 at: http://www.nsf.gov/sbe/srs/seind04/c4/c4h.htm.

[29]US General Accounting Office. "Lessons Learned from Previous Research Could Benefit FreedomCAR Initiative." Statement of Jim Wells. GAO-02-810T. Washington, DC: General Accounting Office, 2002.

[30]National Research Council. *Review of the Research Program of the Partnership for a New Generation of Vehicles.* Washington, DC: National Academy Press, 2001.

[31]Council on Government Relations. *The Bayh-Dole Act: A Guide to the Law and Implementing Regulations.* Washington, DC: Council on Government Relations, 1999. Available at: www.ucop.edu/ott/bayh.html.

[32]D. C. Mowery and A. A. Ziedonis. Numbers, Quality and Entry: How Has the Bayh-Dole Act Affected US University Patenting and Licensing? In A. B. Jaffe, J. Lerner, and S. Stern, eds. *Innovation Policy and the Economy, Volume 1.* Cambridge, MA: MIT Press, 2001.

[33]S. Butts and R. Killoran. "Industry-University Research in Our Times: A White Paper." 2003. Available at: http://www7.nationalacademies.org/guirr/IP_background.html.

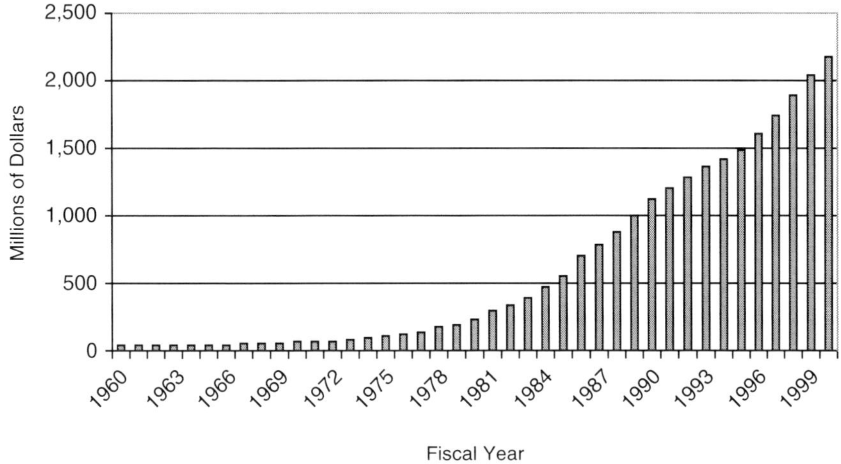

FIGURE EL-3 Industry support of science and engineering research at US colleges and universities, in millions of dollars, 1960-1999.
SOURCE: R. Killoren and S. Butts. *Industry-University Research in Our Times*. Background paper for *Re-Engineering Intellectual Property Rights Agreements in Industry-University Collaborations*. Government-University-Industry Research Roundtable, National Academies, June 26, 2003. Available at: http://www7. national academies.org/guirr/IP_background.html.

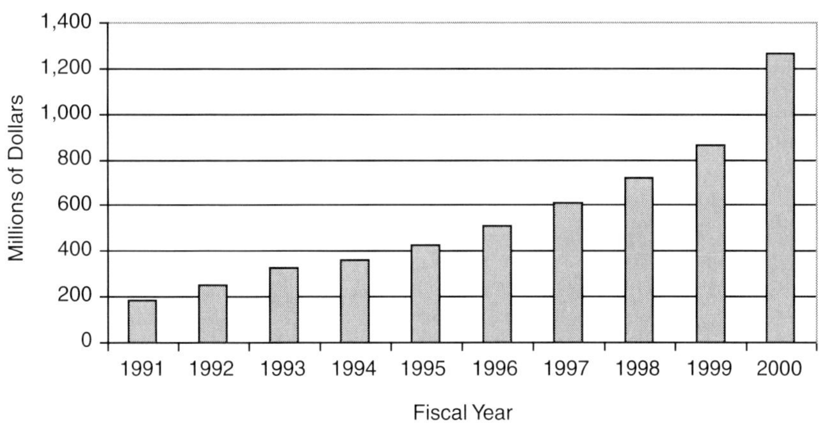

FIGURE EL-4 License income to North American universities and research institutes, in millions of dollars, 1991-2000.
SOURCE: R. Killoren and S. Butts. *Industry-University Research in Our Times*. Background paper for *Re-Engineering Intellectual Property Rights Agreements in Industry-University Collaborations*. Government-University-Industry Research Roundtable, National Academies, June 26, 2003. Available at: http://www7. nationalacademies.org/guirr/IP_background.html.

impeded the flow of a superior medical treatment to the market, to the detriment of public health.[34]

Possible options for federal action include the following:

- Evaluate and amend the Bayh–Dole Act to promote collaborations between university technology-transfer offices, local community colleges, local economic-development planning agencies, federal laboratories, select managers of venture funds, and industry leaders. This would respond to the increasing pressure on university technology-transfer specialists to become stewards of their regional economic development. Cooperative Economic Development Agreements (CEDAs) can accomplish this goal.[35]

COMMISSIONS AND COUNCILS ON SPECIFIC INDUSTRIES AND TECHNOLOGIES

Over the years, a number of national advisory bodies have been set up to develop policy ideas and recommendations affecting specific industries. These bodies have sometimes taken on science and engineering issues as a central part of their work. The National Advisory Committee on Semiconductors, which operated in the late 1980s and early 1990s, is one example. A more recent example is the Commission on the Future of the United States Aerospace Industry.[36] A followup effort, the National Aerospace Initiative, has sought to involve the relevant agencies in the development of technology roadmaps for the industry.[37]

The President's Information Technology Advisory Committee, which was disbanded in June 2005, issued a final report recommending that federal agencies change the way they fund computational science and calling on the National Academies to lead a roadmapping effort.[38] Several years ago, an advisory committee to NSF recommended the launch of an effort to boost cyberinfrastructure for research enabled by information technology.[39]

---

[34]A. B. Shalom and R. Cook-Deegan. "Patents and Innovation in Cancer Therapeutics: Lessons from CellPro." *The Milbank Quarterly* 80(December 2002):iii-iv, 637-676.

[35]C. Hamilton. "University Technology Transfer and Economic Development: Proposed Cooperative Economic Development Agreements Under the Bayh-Dole Act." *John Marshall Law Review* (Winter 2003).

[36]Commission on the Future of the United States Aerospace Industry. *Final Report*. Arlington, VA: Commission on the Future of the United States Aerospace Industry, 2002. Available at: http://www.ita.doc.gov/td/aerospace/aerospacecommission/AeroCommissionFinalReport.pdf.

[37]National Research Council. *Evaluation of the National Aerospace Initiative*. Washington, DC: The National Academies Press, 2004.

[38]President's Information Technology Advisory Committee. *Computational Science: Ensuring America's Competitiveness*. Washington, DC: National Coordination Office for Information Technology Research and Development (NCO/ITR&D), 2005.

[39]Blue-Ribbon Advisory Panel on Cyberinfrastructure. *Revolutionizing Science and Engineering Through Cyberinfrastructure*. Arlington, VA: National Science Foundation, 2003.

Possible options for federal action include the following:

- Make coordinated, fundamental, structural changes that affirm the integral role of computational science in addressing the 21st century's most important problems, which are predominantly multidisciplinary, multi-agency, multisector, and collaborative. To initiate the required transformation, the federal government, in partnership with academe and industry, must create and execute a multidecade roadmap directing coordinated advances in computational science and its applications in science and engineering disciplines.
- Commission the National Academies to convene one or more task forces to develop and maintain a multidecade roadmap for computational science and the fields that require it, with a goal of ensuring continuing US leadership in science, engineering, the social sciences, and the humanities.
- Direct NSF to establish and lead a large-scale, interagency, and internationally coordinated Advanced Cyberinfrastructure Program to create, deploy, and apply cyberinfrastructure in ways that radically empower all scientific and engineering research and allied education. Sustained new NSF funding of $1 billion per year is required to achieve "critical mass" and to leverage the necessary coordinated coinvestment from other federal agencies, universities, industry, and international sources required to empower a revolution.[40]

MANUFACTURING AND INNOVATION EXTENSION

The Manufacturing Extension Partnership (MEP) program of NIST was established in 1989 and now comprises about 350 nonprofit MEP centers that collectively receive a little over $100 million annually from NIST.[41] The centers have been successful in attracting support from states, industry, and other entities.

Several recent recommendations for federal action are related to manufacturing technology and extension services:

- Establish a program of Innovation Extension Centers to enable small and medium-sized enterprises to become first-tier manufacturing partners.[42]
- Create centers for production excellence that include shared facilities and consortia.[43]

---

[40]Ibid.

[41]See the NIST Web site. Available at: http://www.mep.nist.gov/about-mep/about.html.

[42]Council on Competitiveness. *Innovate America*. Washington, DC: Council on Competitiveness, 2004.

[43]Ibid.

Understanding Trends in Science and Technology Critical to US Prosperity

SUMMARY

Sound policies rest on a solid foundation of information and analysis. The collection and analysis of data have become key components of the innovation system.

During the late 1980s and early 1990s, policy-makers expressed a growing interest in assessments and international comparisons of critical technologies. This interest was prompted by the rapid (and unexpected) emergence during the 1980s of Japanese companies in high-technology fields, such as microelectronics, robotics, and advanced materials. Policy-makers proposed that regular efforts to identify the technologies likely to underlie future economic growth and to assess the relative international standing of the United States in those technologies would yield information useful for making investment decisions.

Today, a number of government and private groups undertake a variety of technology assessments that enhance our understanding of America's relative standing in specific science and engineering fields. More detailed and innovative measures could provide important additional information on the status and effects of scientific and technological research.

Recommendations for federal actions in these areas include the following:

This paper summarizes findings and recommendations from a variety of recently published reports and papers as input to the deliberations of the Committee on Prospering in the Global Economy of the 21st Century. Statements in this paper should not be seen as the conclusions of the National Academies or the committee.

International Benchmarking of US Research Fields

• Establish a system to conduct regular international benchmarking assessments of US research to provide information on the world leadership status of key fields and subfields of scientific and technologic research.

Critical Technologies

• Establish a federal office that would coordinate ongoing private and public assessments of critical technologies and initiate additional assessments where needed.

Data Collection and Dissemination

• Mandate that the White House Office of Science and Technology Policy prepare a regular report on innovation that would be linked to the federal budget cycle.

• Provide the National Science Foundation (NSF) Division of Science Resources Statistics (SRS) with resources to launch a program of innovation surveys.

• Ensure that research and innovation survey programs, such as the NSF R&D survey, incorporate emerging, high-growth, technology-intensive industries, such as telecommunications and biotechnology, and industries across the service sector—financial services, transportation, and retailing, among others.

SCIENCE AND TECHNOLOGY BENCHMARKING

As part of the technology and international-competitiveness debates of the 1980s and 1990s, several initiatives were launched to assess national capabilities in specific fields of science and engineering. Many of the early assessments looked at Japanese capabilities and were performed by US or international panels.[1] In the late 1980s, the Japan Technology Evaluation Center started as an interagency federal initiative managed by SAIC; it evolved into an NSF-contracted center at Loyola College of Maryland and is now an independent nonprofit known as WTEC, Inc.[2] WTEC assessments cover a variety of countries and fields and are undertaken on an ad hoc basis. They are funded by the federal agencies most interested in the specific field being assessed.

---

[1]National Research Council, National Materials Advisory Board. *High-Technology Ceramics in Japan.* Washington, DC: National Academy Press, 1984.

[2]See the WTEC, Inc., Web site. Available at: http://www.wtec.org/welcome.htm.

A 1993 National Academies report recommended that the world leadership status of research fields be evaluated through international benchmarking.[3] A followup report that reviewed three benchmarking experiments (mathematics, immunology, and materials science and engineering) concluded that the approach of using expert panels could yield timely, accurate "snapshots" of specific fields.[4] The report also suggested that benchmarking assessments be conducted every 3-5 years to capture changes in the subject fields. Figure UT-1 illustrates one such assessment.

The factors considered most important in determining US leadership status, on the basis of all the international benchmarking experiments, were human resources and graduate education, funding, innovation process and industry, and infrastructure.

In addition, the Bureau of Industry and Security of the US Department of Commerce undertakes assessments of the US industrial and technology base in areas considered important for national defense.[5] These assessments often take into account international competitiveness.

Possible federal action includes the following:

- Establish a system to conduct regular international benchmarking assessments of US research to provide information on the world leadership status of key fields and subfields of scientific and technological research.

An example of the potential utility of this information is shown in Figures UT-2 to UT-5 which show funding and innovation process metrics for nanotechnology.

CRITICAL TECHNOLOGIES

In 1990, Congress mandated that a biennial review be conducted of America's commitment to critical technologies deemed essential for "maintaining economic prosperity and enhancing the competitiveness of the US research enterprise." The legislation required that the number of technologies identified in the report not exceed 30 and include the most economically important civilian technologies expected after the decade following the report's release with the estimated current and future size of the domes-

---

[3]NAS/NAE/IOM. *Science, Technology, and the Federal Government*. Washington, DC: National Academy Press, 1993.

[4]NAS/NAE/IOM. *Experiments in International Benchmarking of U.S. Research Fields*. Washington, DC: National Academy Press, 2000.

[5]See http://www.bis.doc.gov/defenseindustrialbaseprograms/osies/DefMarketResearchRpts/Default.htm.

Sub-Subfield	Current Position					Likely Future Position					Comments
	1 Fore-front	2	3 Among world leaders	4	5 Behind world leaders	1 Gaining/ Extending	2	3 Main-taining	4	5 Losing	
Tissue engineering	•					•					Clear US leadership; tremendous worldwide interest.
Molecular architecture			•					•			Strong US competition from Germany and Japan.
Protein analogs	•					•					US dominates, driven by a basic-science approach.
Biomimetics			•					•			Strong players in North America, UK, Japan.
Contemporary diagnostic systems			•						•		Large European Community investments in biosensors research could lower US ranking.
Advanced controlled-release systems		•					•				US leads; extremely high worldwide interest could change this.
Bone biomaterials			•					•			Important developments in Europe and Japan.

FIGURE UT-1 Example of international benchmarking for several materials science and engineering subfields.
SOURCE: NAS/NAE/IOM. *Experiments in International Benchmarking of US Research Fields*. Washington, DC: National Academy Press, 2000.

tic and international markets for products derived from the identified technologies. However, the exact definition of critical technologies was not included in the legislation.

The Office of Science and Technology Policy (OSTP) prepared National Critical Technologies Reports (NCTR) to Congress in 1991,[6] 1993,[7] 1995,[8] and 1998.[9] The content of and methods used to prepare the NCTRs varied

[6]National Critical Technologies Panel. *Report of the National Critical Technologies Panel*. Washington, DC: US Government Printing Office, 1991.

[7]National Critical Technologies Panel. *The Second Biennial Report of the National Critical Technologies Panel*. Washington, DC: US Government Printing Office, 1993.

[8]National Critical Technologies Panel. *The National Critical Technologies Report*. Washington, DC: US Government Printing Office, 1995.

[9]S. W. Popper, C. S. Wagner, and E. V. Larson. *New Forces at Work: Industry Views Critical Technologies*. Santa Monica, CA: RAND, 1998.

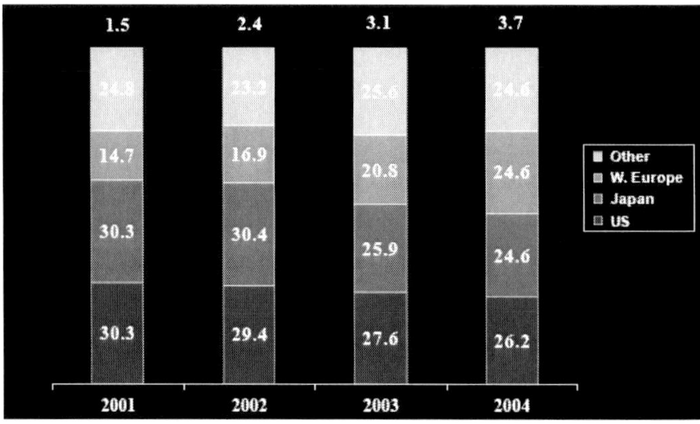

FIGURE UT-2 Share of total government investment for nanotechnology, in billions of dollars.
SOURCE: S. Murdock. Testimony before the Research Subcommittee of the Committee on Science of the United States House of Representatives. Hearing on "Nanotechnology: Where Does the US Stand?" June 29, 2005.

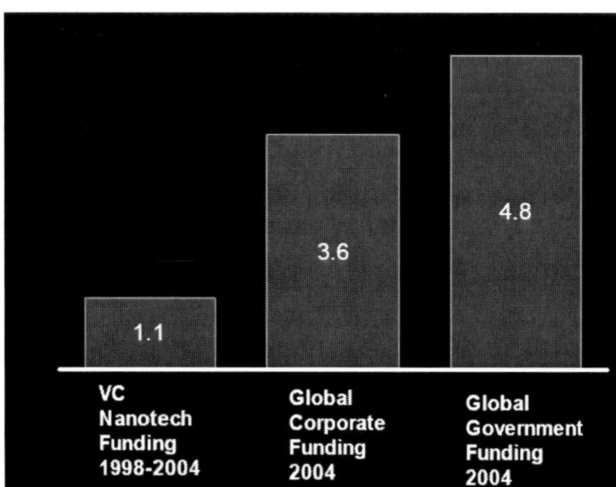

FIGURE UT-3 Venture capital, global corporate, and global government nanotechnology funding, in billions of dollars.
SOURCE: S. Murdock. Testimony before the Research Subcommittee of the Committee on Science of the United States House of Representatives. Hearing on "Nanotechnology: Where Does the US Stand?" June 29, 2005.

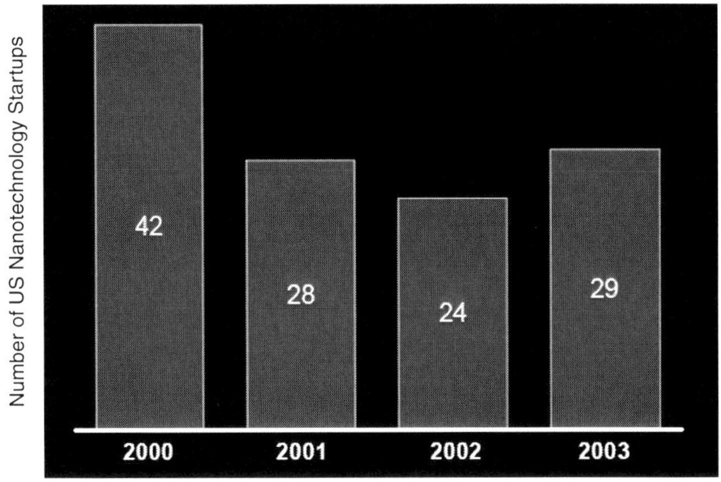

FIGURE UT-4 Number of US nanotechnology startups, 2000-2003.
SOURCE: S. Murdock. Testimony before the Research Subcommittee of the Committee on Science of the United States House of Representatives. Hearing on "Nanotechnology: Where Does the US Stand?" June 29, 2005.

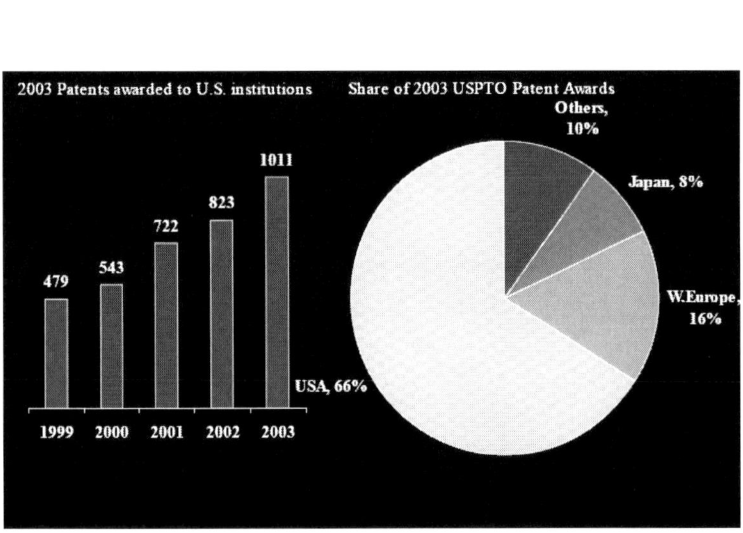

FIGURE UT-5 US patents awarded to US institutions, 2003.
SOURCE: S. Murdock. Testimony before the Research Subcommittee of the Committee on Science of the United States House of Representatives. Hearing on "Nanotechnology: Where Does the US Stand?" June 29, 2005. This figure was based on an analysis done by Jim Murday and Mike Roco of the Nano Business Alliance.

throughout the decade.[10] The 1995 report, for example, identified seven "technology categories" (energy, environmental quality, information and communication, living systems, manufacturing, materials, and transportation), which were divided into 27 "technology areas." Figure UT-6 illustrates the NCTR analyses for materials research. Each of the 27 areas was identified on a competitive scale ranging from lagging to leading, and each area was then compared with Europe and Japan.[11]

Over the 1990s, the RAND Corporation played an increasingly important role in the preparation of the NCTRs. RAND assisted with the background research for the 1993 report and was a co-author of the 1995 report with OSTP.[12] The 1998 critical-technologies report was prepared by RAND with little involvement of OSTP.[13] This report, which refocused the study specifically on input from the private sector, identified five critical sectors of technology: software, microelectronics and telecommunications technologies, advanced manufacturing, materials, and sensor and imaging technologies.[14] After the release of the 1998 report, the legal requirement for OSTP to prepare the NCTR was removed.

Those involved in the NCTR process point out that federal agencies and state and local governments used the reports as a basis for policy-making. However, the NCTRs do not appear to have had a formal effect on US federal policy toward technology development.[15] For example, the NCTRs did not lead to the creation of any large cross-agency technology initiative. Nanotechnology was not a focus of the final 1998 NCTR, but OSTP started work around that time on discussions that would culminate in the creation of the National Nanotechnology Initiative several years later.[16]

In addition to the NCTRs, several other public and private efforts to identify critical technologies in both the defense and civilian arenas were undertaken during the 1990s by such groups as the US Department of Defense[17] and the Council on Competitiveness.[18] More recently, several government agencies have expressed interest in assessing international capabilities in

---

[10]C. S. Wagner and S. W. Popper. "Identifying Critical Technologies in the USA." *Journal of Forecasting* 22(2003):113-128.

[11]National Critical Technologies Panel, 1995.

[12]Wagner and Popper, 2003, p. 120.

[13]Ibid.

[14]Popper, Wagner, and Larson, 1998.

[15]Wagner and Popper, 2003, p. 123.

[16]N. Lane and T. Kalil. "The National Nanotechnology Initiative: Present at the Creation." *Issues in Science and Technology* 21(Summer 2005):49-54.

[17]See the Militarily Critical Technologies Web site. Available at: http://www.dtic.mil/mctl.

[18]Council on Competitiveness. *Gaining New Ground: Technology Priorities for America's Future*. Washington, DC: Council on Competitiveness, 1991.

Materials

Technology Area	Technology Sub-Area	Specific Technologies	EP	NS	Sample Applications to Which Specific Techs Contribute
Materials	**alloys**	light weight structural alloys	●	●	lighter, stiffer airframes, automobile frames
		intermetallic alloys (Ti/Ni/Fe aluminides)	●	●	high-termp structural applications in aerospace
	ceramic materials	ceramic IC packaging	●		improved interconnection/data rates for multi-IC assemblies
		ceramic matrix composites (see "composites")	●	●	
		ceramic coatings	●	●	improved wear characteristics in high-speed moving parts: turbine engines, cutting tools
	composites	polymer matrix composities	●	●	economical, bulk PMCs for jet engines, automobiles, high-T aerodynamic surf
		ceramic matrix composites	●	●	high-perf aircraft engines, automobile engines, advanced armor
		metal-matrix composites	●	●	space applic. for null-outgassing, rad-resist important, hypersonic aircraft, HSCT, armor
		carbon-carbon composites	●	●	ultra-high temperature aerospace applications (rocket nozzles, thermal surfaces)
		molecular-scale composites	●	●	
	electronic materials	GaAs	●	●	superior, economic GaAs-based ICs (faster and/or hardened electronics)
		thin-film dielectric materials	●	●	miniaturization of microwave communication devices
	photonic materials	semiconductor lasers/laser arrays	●	●	laser/detector techs to support communications
		advanced detectors	●	●	and optical data processing, with eventual goal of OEICs
	high energy-density materials	advanced solid and liquid propellants		●	space launch vehicles, longer-range artillery
		explosives		●	improved conventional weapons
	highway/infrastructure mater	paving	●		higher-durability, more economical materials to maintain surface transport infrastructure
		repair materials	●		
		polymer matrix composites	●		retrofit reinforcement for earthquake damage prevension
	biocompatible materials		●	●	See "Living Systems"
	stealth materials	radar-absorbing materials/coatings		●	increased abilities for military aircraft to operate unobserved by radar
	superconductors	high temperature superconductors	●	●	sensors, low-power electronics, power transmission, energy storage,
		advanced low-temp superconductors	●	●	powerful magnets for research, medical diagnostics, maglev technology

FIGURE UT-6 Example of critical technologies list for materials.
NOTE: EP = Economic Prosperity, NS = National Security.
SOURCE: Office of Science and Technology Policy. "National Critical Technologies List, March 1995." Available at: http://clinton1.nara.gov/White_House/EOP/OSTP/CTIformatted/AppA/appa.html.

militarily critical technologies.[19] Also, a number of countries are engaged in periodic assessments of critical technologies and international capabilities.

Possible federal actions include the following:

- Establish a federal office that would coordinate ongoing private and public assessments of critical technologies and initiate additional assessments where needed.
- Analyze the technology forecasting and foresight activities of other countries to identify where such activities can provide useful input to policy processes.

DATA ON RESEARCH AND INNOVATION

The adequacy of measures and statistical data to inform policy-making remains a concern of the science and technology policy community. For example, during the 1990s, information technologies were widely deployed throughout the US economy and played a major role in a surge of US innovation, yet this process was captured poorly, if at all, by traditional indicators of research and innovation. Except for statistics on formal R&D spending, patents, and some aspects of science and engineering education, innovation-related data are extremely limited.[20]

Among the steps the federal government could take to improve data collection and analysis are the following:

- Mandate that OSTP prepare a regular report on innovation that would be linked to the federal budget cycle.[21] The goal of the report would be to give the government and the public a clear sense of how federal support for R&D fits into the larger national economic system and how both are linked to an increasingly international process of innovation.
- Provide the NSF SRS with resources to launch a program of innovation surveys.[22] SRS should work with experts in universities and public institutions that have expertise in a broad spectrum of related issues. In some cases, it may be judicious to commission case studies. NSF also should

---

[19]National Research Council, Division on Engineering and Physical Sciences. *Avoiding Surprise in an Era of Global Technology Advances*. Washington, DC: The National Academies Press, 2005.

[20]National Research Council, Committee on National Statistics. *Measuring Research and Development Expenditures in the U.S. Economy*. Washington, DC: The National Academies Press, 2004.

[21]K. Hughes. "Facing the Global Competitiveness Challenge." *Issues in Science and Technology* 21(Summer 2005):72-78.

[22]National Research Council, 2004.

build an internal capacity to resolve the methodologic issues related to collecting innovation-related data.

- Ensure the collection of information needed to construct data series of federal science and technology (FS&T).[23] NSF needs to continue to collect the additional data items that are readily available in the defense agencies and expand collection of civilian data that would permit users to construct data series on FS&T expenditures in the same manner as the FS&T presentation in the president's budget documentation.
- Overhaul the field-of-science classification system to take account of changes in academic research, including interdisciplinary and multidisciplinary research.[24] It has been some three decades since the field-of-science classification system has been updated, and the current classification structure no longer adequately reflects the state of science and engineering fields. The Office of Management and Budget needs to initiate a review of the *Classification of Fields of Science and Engineering*, last published as Directive 16 in 1978. The SRS could serve as the lead agency for an effort that must be conducted on a governmentwide basis. NSF should engage in a program of outreach to the disciplines to begin to develop a standard concept of interdisciplinary and multidisciplinary research, and on an experimental basis it should initiate a program to collect information from a subset of academic and research institutions.
- Redesign NSF's industrial R&D survey.[25] The redesign should begin by assessing the US survey against the international "standard"—the definitions promulgated through the *Frascati Manual* from the Organisation for Economic Co-operation and Development. The redesign also should update the industry *questionnaire* to facilitate an understanding of new and emerging R&D issues, enhance the program of *data analysis and publication*, revise the *sample* to enhance coverage of growing sectors, and improve the *collection procedures* to better involve and educate the respondents.
- Ensure that research and innovation survey programs, such as NSF's R&D survey, incorporate emerging, high-growth, technology-intensive industries, such as telecommunications and biotechnology, and industries across the service sector—financial services, transportation, and retailing, and others.[26] Also, survey programs should collect information at the business-unit level of corporate activity rather than on a firm as a whole, and geographic location detail should be collected.

---

[23]Ibid.
[24]Ibid.
[25]Ibid.
[26]National Research Council, Board on Science, Technology, and Economic Policy. *Industrial Research and Innovation Indicators*. Washington, DC: National Academy Press, 1997.

• NSF should increase the analytic value of its data by improving comparability and linkages among its data sets and between its data and data from other sources, such as the US census.[27]

• SRS should develop a long-term plan for its *Science and Engineering Indicators* publication so that it is smaller, more policy-focused, and less duplicative of other SRS publications.[28] SRS also should substantially reduce the time between the reference date and data release of each of its surveys to improve the relevance and usefulness of its data.

---

[27]Committee on National Statistics, 2004.

[28]Ibid.

Ensuring That the United States Has the Best Environment for Innovation

SUMMARY

A number of recent reports have raised concerns about the United States' long-term ability to sustain its global science and engineering (S&E) leadership.[1] They argue that erosion of this leadership threatens our ability to reap the rewards of innovation in the form of higher incomes and living standards, better health, a cleaner environment, and other societal benefits.

Certainly, the leadership position the United States has maintained in research and the creation of new knowledge since World War II has been an important contributor to economic growth and other societal rewards. However, a look at US history and some contemporary international examples shows that leadership in research is not a sufficient condition for gaining the lion's share of benefits from innovation. A favorable environment for innovation is also necessary. The environment for innovation includes such elements

---

This paper summarizes findings and recommendations from a variety of recently published reports and papers as input to the deliberations of the Committee on Prospering in the Global Economy of the 21st Century. Statements in this paper should not be seen as the conclusions of the National Academies or the committee.

[1]American Electronics Association. *Losing the Competitive Advantage? The Challenge for Science and Technology in the United States*. Washington, DC: American Electronics Association, 2004; Council on Competitiveness. *Innovate America*. Washington, DC: Council on Competitiveness, 2004; R. B. Freeman. *Does Globalization of the Scientific/Engineering Workforce Threaten US Economic Leadership?* NBER Working Paper 11457. Cambridge, MA: National Bureau of Economic Research, 2005; Task Force on the Future of American Innovation. *The Knowledge Economy: Is America Losing Its Competitive Edge?* Washington, DC: The Task Force on the Future of American Innovation, 2005.

as the market and regulatory environment, trade policy, intellectual-property policies, policies that affect the accumulation of human capital, and policies affecting innovation environments in specific regions. In addition, grand challenges issued by the president (such as the reaction to Sputnik and the call for the Apollo project) can mobilize resources and the national imagination in pursuit of important innovation-related goals.

How can the United States sustain and improve the environment for innovation even in a future where its relative share of global S&E inputs to the innovation process (such as R&D spending, S&E personnel, and the quantity and quality of scientific literature) declines?

Many approaches to improving the innovation environment have been suggested. On some issues, including the offshoring of service-industry jobs, contradictory diagnoses and prescriptions have emerged on the basis of interests and political outlook of the analysis. On other issues, such as patent-system reform, similar suggestions have emerged from several different reports. The approaches suggested include the following:

Market, Regulatory, and Legal Environment

- Establish a public-private body to assess the impact of new regulations on innovation.
- Reduce the costs of tort litigation for the economy.
- Reform Section 404 of the Sarbanes–Oxley Act.
- Drop current efforts to expense stock options.
- Create best practices for collaborative standard-setting.
- Undertake market and regulatory reforms in the telecommunications industry with the goal of accelerating the speed and accessibility of networks.

Trade

- Increase focus on enforcement of the prevailing global rules for intellectual-property protection, particularly in China and in other countries where significant problems remain.
- Make completion of the Doha Round of world-trade talks a priority.

Intellectual Property

- Harmonize the US, European, and Japanese patent systems.
- Institute a postgrant open-review procedure for US patents.
- Stop diverting patent application fees to general revenue to provide the US Patent and Trademark Office (USPTO) with sufficient resources to modernize and improve performance.

- Shield some research uses of patented inventions from liability for infringement.
- Leverage the patent database as an innovation tool.

Tax Policy

- Make the R&D tax credit permanent, and extend coverage to research conducted in university–industry consortia.
- Provide new tax incentives for early-stage investments in innovative startups.
- Provide more favorable tax treatment (expensing and accelerated depreciation) for the purchase of high-technology manufacturing equipment to encourage industry to keep manufacturing in the United States.

Human Capital

- Create incentives for investments by employers and employees in lifelong learning, including the creation of tax-protected accounts.
- Restructure and expand worker-assistance programs like the Trade Adjustment Assistance program so that they are more flexible and cover workers displaced by reasons other than trade.
- Expedite the immigration process, including issuance of permanent residence status (green cards) to all master's and doctoral graduates of US institutions in science and engineering.
- Make H1-B visas "portable" to reduce the possibility of visa holder's being exploited and to reduce the negative impacts on US workers in those fields.
- Fund new programs that promote entrepreneurship at all levels of education.
- Reform policies toward health and pension benefits.
- Require companies operating in the United States to be transparent in reporting offshoring decisions.
- Use procurement policies to discourage government contractors from offshoring by requiring that certain tasks be performed by US workers.

New "Apollo"

- Gain presidential-level commitment to the proposition that sustaining and enhancing US ability to innovate is a key national priority.
- Have the President issue a major challenge encompassing federal research and all aspects of the innovation process to mobilize resources in pursuit of a critical national goal. The candidate fields for such a challenge include energy, space, and healthcare.

Support for Regional Innovation

- Establish a program of national innovation centers, or "hot spots," with matching funds from states and educational institutions.
- Designate a lead agency to coordinate regional economic-development programs to ensure that there is a common focus on innovation-based growth.

INNOVATION AND THE ECONOMY

Wm. A. Wulf points out that "there is no simple formula for innovation. There is, instead, a multi-component 'environment' that collectively encourages, or discourages, innovation."[2] This environment includes research funding, an educated workforce, a culture that encourages risk-taking, a financial system that provides patient capital for entrepreneurial activity, intellectual-property protection, and other elements.

The significance of this innovation environment has long been a subject of study. As far back as Adam Smith, economists have been interested in technologic innovation and its impact on economic growth.[3] Early in the 20th century, Joseph Schumpeter argued that innovation was the most important feature of the capitalist economy. Starting in the 1950s, Robert Solow and others developed methods of accounting for the sources of growth, leading to the observation that technologic change is responsible for over half the observed growth in labor productivity and national income. These methods are subject to continued debate and refinement. For example, over long periods the contributions of technologic change and other causes of growth—such as worker skills, capital deepening, and institutional change—are highly interactive and difficult to separate.

Other economists have focused on a more qualitative study of the institutions and practices underlying innovation in individual industries and entire economies. The effort to understand "national innovations systems" has been one focus of recent studies.[4] Others have examined the performance of particular industries.[5] The Sloan Foundation has given understanding innovation a high priority in its funding.[6]

---

[2]Wm. A. Wulf. 2005. "Review and Renewal of the Environment for Innovation." Unpublished Paper.

[3]J. Mokyr. Innovation in an Historical Perspective: Tales of Technology and Evolution. In B. Steil, D. G. Victor, and R. R. Nelson, eds. *Technological Innovation and Economic Performance*. Princeton, NJ: Princeton University Press, 2002.

[4]R. R. Nelson, ed. *National Innovation Systems: A Comparative Analysis*. New York: Oxford University Press, 1993.

[5]National Research Council. *US Industry in 2000: Studies in Competitive Performance*. Washington, DC: National Academy Press, 1999.

[6]See the Alfred P. Sloan Foundation Web site. Available at: http://www.sloan.org.

This literature underscores the importance of the environment for innovation and points to several lessons from recent history. Japan's growth trajectory in various S&E inputs and outputs (such as R&D investments, S&E personnel, and patents) since the early 1990s has been similar to what it was before.[7] Yet the Japanese economy's ability to reap the rewards of innovation in the form of higher productivity and incomes was much higher in the earlier period. This can be explained partly by the dual nature of the Japanese economy, where world-class manufacturing industries serving a global market exist side by side with inefficient industries, such as construction.[8] Economic mismanagement and a lack of flexibility in factor markets (labor and capital) also have played an important role.

In contrast, in the mid-1990s the United States saw a jump in productivity growth from the levels that had prevailed since the first oil shock of the early 1970s.[9] In addition to gains in information technology (IT) manufacturing productivity, productivity gains from IT use and the creation of new business methods that take advantage of IT were widespread throughout the economy (see Figure EI-1).

It is important to note that science and technology and the innovation process are not zero-sum games in the international context.[10] The United States has proved adept in the past at taking advantage of breakthroughs and inventions from abroad, such as the jet engine and monoclonal antibodies.[11]

Groups and individuals have made numerous recommendations for change in the US environment for innovation.

MARKET, REGULATORY, AND LEGAL ENVIRONMENT

Many analyses of innovation focus on the supply side of the equation, such as the size and composition of R&D spending, the number of S&E graduates, and so forth. The importance of the demand side is sometimes

---

[7]A. S. Posen. Japan. In R. Nelson, B. Steil, and D. Victor, eds. *Technological Innovation and Economic Performance*. Princeton, NJ: Princeton University Press, 2002. Pp. 74-111.

[8]D. W. Jorgenson and M. Kuroda. Technology, Productivity, and the Competitiveness of US and Japanese Industries. In T. Arrison, C. F. Bergsten, E. M. Graham, and M. C. Harris, eds. *Japan's Growing Technological Capability: Implications for the US Economy*. Washington, DC: National Academy Press, 1992.

[9]W. Norhaus. *The Source of the Productivity Rebound and the Manufacturing Employment Puzzle*. NBER Working Paper 11354. Cambridge, MA: National Bureau of Economic Research, 2005.

[10]Wm. A. Wulf. Observations on Science and Technology Trends: Their Potential Impact on Our Future. In A. G. K. Solomon, ed. *Technology Futures and Global Wealth, Power and Conflict*. Washington, DC: Center for Strategic and International Studies, 2005.

[11]NAS/NAE/IOM. *Capitalizing on Investments in Science and Technology*. Washington, DC: National Academy Press, 1999.

	Average share	Contribution to aggregate productivity growth					
	1998-2002	1999	2000	2001	2002	2003	Average
Gross domestic product	100.00	2.07	2.05	1.71	2.90	3.72	**2.49**
Private industries	87.50	1.96	2.08	2.00	2.97	3.43	**2.49**
Finance, insurance, real estate, etc.	19.74	0.60	1.36	0.92	0.33	0.38	**0.72**
Computer and electronic products	1.66	0.47	0.64	0.09	0.37	0.58	**0.43**
Retail trade	6.91	0.12	0.31	0.45	0.37	0.43	**0.33**
Information	4.62	0.22	0.14	0.21	0.34	0.56	**0.30**
Professional and business services	11.53	0.13	-0.18	0.41	0.42	0.47	**0.25**
Durable goods other than computers	6.78	0.01	0.30	0.01	0.49	0.26	**0.21**
Wholesale trade	6.08	0.31	-0.19	0.52	0.27	-0.07	**0.17**
Nondurable goods	5.75	0.12	0.04	0.00	0.51	0.39	**0.21**
Transportation and warehousing	3.01	0.06	0.15	0.02	0.16	0.19	**0.11**
Utilities	1.97	0.07	0.17	-0.06	0.18	0.17	**0.10**
Arts, entertainment, recreation, etc.	3.54	0.03	0.00	-0.03	0.02	0.05	**0.01**
Agriculture, forestry, etc.	1.03	0.00	0.10	-0.12	0.06	0.09	**0.03**
Government	12.44	0.06	0.04	-0.07	-0.04	0.14	**0.03**
Mining	1.04	0.12	-0.04	-0.09	0.08	-0.07	**0.00**
Educational services, health care, etc.	7.09	-0.01	0.01	-0.05	0.01	0.03	**0.00**
Other services, except government	2.40	-0.07	-0.05	-0.07	-0.08	0.04	**-0.05**
Construction	4.42	-0.18	-0.13	0.03	-0.03	0.00	**-0.06**

FIGURE EI-1 Contribution of different industries to the productivity rebound, by broad industry group, 1998-2003.
SOURCE: W. Nordhaus. *The Source of the Productivity Rebound and the Manufacturing Employment Puzzle.* NBER Working Paper 11354. Cambridge, MA: National Bureau of Economic Research, 2005. Table 4, p. 24. Available at: http://www.nber.org/papers/w11354.

neglected. The imperative of meeting the needs of demanding buyers and consumers plays a key role in driving the creation and diffusion of innovations. An open dynamic market is the source of US competitive strength in a range of industries. Even under the "Dell model"—in which development, manufacturing, and other functions are sourced and performed around the globe—contact with customers and knowledge of their needs is a critical capability that Dell keeps inhouse.[12]

In contrast, industries and economies where markets are closed, competition is limited, or consumer rights are not protected tend to act as a drag on innovation and growth. McKinsey and Company's international studies on sector productivity during the 1990s showed that competitive markets were the key factor separating successes and failures.[13]

A wide variety of policies and practices influence the market, regulatory, and legal environment for innovation. These include financial regulations,

[12]T. L. Friedman. *The World Is Flat: A Brief History of the 21st Century.* New York: Farrar, Straus, and Giroux, 2005. Pp. 414-419.

[13]W. W. Lewis. *The Power of Productivity: Wealth, Poverty, and the Threat to Global Stability.* Chicago: University of Chicago Press, 2004.

where the Sarbanes–Oxley Act has produced a number of changes in recent years. In addition, the costs of US approaches to litigation affecting product liability and securities fraud are a perennial target of industry groups.

Given the fact that the United States has lagged behind a number of other countries in broadband access (see Figure EI-2) and the potential positive impact of better and cheaper network access for the economy and the research enterprise in particular, the complex regulations governing telecommunications, the broadcast spectrum, and related areas would seem a promising target of reform.

Possible federal actions include the following:

- "The impact of new regulations on market investments in innovation should be more carefully and collaboratively assessed by a public-private Financial Markets Intermediary Committee, where periodic meetings can score existing and proposed legislation. This committee would follow the model of the Foreign Exchange Committee and Treasury Borrowing Committee."[14]
- "The country should set a goal to reduce the costs of tort litigation from the current level of two percent of GDP [gross domestic product]—some $200 billion—down to one percent."[15]
- Reform Section 404 of the Sarbanes–Oxley Act, which requires an internal control report in the company's annual report. "Many small and medium-sized companies have serious concern with Section 404 and the expense of the internal control reporting requirements. Small and medium-sized companies are disproportionately burdened by Section 404, and these provisions need to be examined to ensure a proper balance between accountability and bureaucracy."
- Drop efforts to expense stock options. "No industry has benefited more than the high-tech industry from the use of stock options. Stock options provide employees with a direct link to the growth and profitability of companies. They also are an essential tool for attracting and retaining the best workforce, especially for small businesses and start-ups who do not always have the capital to compete on salary alone. Already China and India have learned from the successful use of stock options in Silicon Valley and are using it to attract and retain businesses and employees."
- "The Federal government, through the Internal Revenue Service or Treasury Department, should establish clear guidelines in the Internal Revenue Code on the acceptability of investment of foundation assets in start-up ventures."[16]

[14]Council on Competitiveness, 2004, p. 65.
[15]Council on Competitiveness, 2004, p. 65
[16]Council on Competitiveness, 2004, p. 62.

Rank	Country	Broadband Subscribers per 100 Inhabitants
1	South Korea	24.9
2	Hong Kong	20.9
3	The Netherlands	19.4
4	Denmark	19.3
5	Canada	17.6
16	United States	11.4

FIGURE EI-2 Ranking of select countries by broadband subscribers per capita.
SOURCE: M. Calabrese, Vice President and Director, Wireless Future Program, New America Foundation. "Broadcast to Broadband: Completing the Digital Television Transition Can Jumpstart Affordable Wireless Broadband." US Senate Testimony, July 12, 2005.

- "The Federal government should encourage best practices and processes for standards bodies to align incentives for collaborative standard setting, and to encourage broad participation."[17]
- Congress should "use the DTV transition to encourage *both* licensed and unlicensed wireless broadband networks as competitive alternatives to wireline cable and DSL offerings."[18]
- "Provide industry the incentives to promote broadband and cellular penetration. Countries like South Korea and Italy have realized enormous competitive advantages by investing heavily in broadband and cellular deployment. Just as the interstate highway system dramatically increased the efficiency and productivity of the US economy half a century ago, so too can efficient communications networks have the same positive effect today. Broadband and cellular diffusion also foster competitive advantages by creating demand for cutting edge products and services."[19]

TRADE

Multilateral trade liberalization has been a goal of US policy-makers of both political parties since the end of World War II. The renewal of large US trade deficits in recent years has spurred debate over how to correct it and other global imbalances. The very large US deficit with China has pro-

[17]Council on Competitiveness, 2004, p. 70.

[18]M. Calabrese, Vice President and Director, Wireless Future Program, New America Foundation. Testimony to the Committee on Commerce, Science and Transportation, US Senate. Hearing on "Broadcast to Broadband: Completing the Digital Television Transition Can Jumpstart Affordable Wireless Broadband." July 12, 2005.

[19]American Electronics Association, 2005, p. 26.

duced calls for exchange-rate adjustment and other measures. In many important respects, China's industrial-development strategy has followed the export-led "playbook" developed by Japan, Korea, and other high-growth Asian economies during the 1960s, 1970s, and 1980s.[20]

Improving the protection of intellectual property worldwide, and especially in such large countries as China where piracy rates are high, has been a policy focus of industry groups (see Figure EI-3). It is important to note that China's laws and policies have come into line with international standards as a result of its accession to the World Trade Organization, so the main issue is enforcement.

Possible federal actions include the following:

- "Promote stronger enforcement of intellectual property protection worldwide. Intellectual property is typically the core asset of any high-tech company. From patents and copyrights to software and trade secrets, intellectual property forms the basis of the knowledge economy. Far too often, foreign legal systems do not adequately protect the owner of these valuable creations, resulting in the loss of literally billions of dollars. The Business Software Alliance estimated that 36 percent of software worldwide was illegally pirated in 2003. This translates to a $29 billion loss in revenue. In China, this figure is 92 percent and the revenue loss is estimated at $3.8 billion. Digital technology has made intellectual property theft that much easier on a wide scale. When foreign companies and consumers can steal this hard-earned property, the profitability and, ultimately, the competitiveness of US companies suffer."
- Make conclusions of the Doha Round a top priority. "The United States economy has gained greatly from liberalization of trade worldwide and from the rules-based system facilitated by the World Trade Organization (WTO). The Doha round of trade talks broke down in the summer of 2003 as negotiations on agriculture and certain service sectors reached an impasse. As a result, the United States risks losing momentum in further opening global markets to US products and services."[21]

INTELLECTUAL PROPERTY

With the rise of knowledge-based industries and a number of legislative, judicial, and administrative actions, intellectual-property protection in the United States has been significantly strengthened over the last 25 years.[22]

---

[20]R. Samuelson. *China's Devalued Concession.* The Washington Post, July 26, 2005. P. A19.

[21]American Electronics Association, 2005, p. 25.

[22]W. M. Cohen and S. A. Merrill, eds. *Patents in the Knowledge-Based Economy.* Washington, DC: The National Academies Press, 2003.

Piracy of $100 Million or More

	$M		$M
United States	$6,645	Sweden	$ 304
China	3,565	Denmark	226
France	2,928	South Africa	196
Germany	2,286	Norway	184
United Kingdom	1,963	Indonesia	183
Japan	1,787	Thailand	183
Italy	1,500	Turkey	182
Russia	1,362	Finland	177
Canada	889	Taiwan	161
Brazil	659	Malaysia	134
Spain	634	Czech Republic	132
Netherlands	628	Austria	128
India	519	Hungary	126
Korea	506	Saudi Arabia	125
Australia	409	Hong Kong	116
Mexico	407	Argentina	108
Poland	379	Ukraine	107
Belgium	309	Greece	106
Switzerland	309		

FIGURE EI-3 Ranking of 2004 piracy loses.
SOURCE: Business Software Alliance and IDC. *Second Annual BSA and IDC Global Software Piracy Study*. Washington, DC: Business Software Alliance. Available at: http://www.bsa.org/globalstudy/upload/2005-Global-Study-English.pdf.

With the increase in the value of a US patent have come an increase in patenting and greater focus by companies and other inventors on the management of intellectual property as an asset. In this environment, debate continues on how to tweak US intellectual-property policies so that they maximize incentives for the generation and broad diffusion of innovations.

Possible federal actions include the following:

• "Reduce redundancies and inconsistencies among national patent systems. The United States, Europe, and Japan should further harmonize patent examination procedures and standards to reduce redundancy in search and examination and eventually achieve mutual recognition of results. Differences that need reconciling include application priority (first-to-invent versus first-inventor-to-file), the grace period for filing an application after publication, the best mode requirement of US law, and the US exception to the rule of publication of patent applications after 18 months. This objective should continue to be pursued on a trilateral or even bilateral basis if multilateral negotiations are not progressing."[23]

---

[23]National Research Council. *A Patent System for the 21st Century*. Washington, DC: The National Academies Press, 2004. P. 8.

• "Strengthen USPTO capabilities. To improve its performance the USPTO needs additional resources to hire and train additional examiners and fully implement a robust electronic processing capability. Further, the USPTO should create a strong multidisciplinary analytical capability to assess management practices and proposed changes, provide an early warning of new technologies being proposed for patenting, and conduct reliable, consistent, reputable quality reviews that address office-wide and individual examiner performance. The current USPTO budget is not adequate to accomplish these objectives."[24]

• "Institute an Open Review procedure. Congress should seriously consider legislation creating a procedure for third parties to challenge patents after their issuance in a proceeding before administrative patent judges of the USPTO. The grounds for a challenge could be any of the statutory standards—novelty, utility, non-obviousness, disclosure, or enablement—or even the case law proscription in patenting abstract ideas and natural phenomena. The time, cost, and other characteristics of this proceeding should make it an attractive alternative to litigation to resolve patent validity questions both for private disputants and for federal district courts. The courts could more productively focus their attention on patent infringement issues if they were able to refer validity questions to an Open Review proceeding."[25]

• "Leverage the patent database as an innovation tool. Develop pilot projects (jointly funded by industry, universities and government) to highlight techniques for leveraging patent data for discovery."[26]

TAX POLICY

Tax policy is another element of the environment for innovation. The research and experimentation tax credit (popularly known as the R&D tax credit) is a longstanding feature of the tax code, although it is generally renewed year to year. The tax treatment of investments in startup companies and purchases of high-technology manufacturing equipment have also been the focus of recent recommendations.

Possible federal actions include the following:

• "The federal government should provide a 25 percent tax credit for early stage investments when made through qualified angel funds. The indi-

---

[24]Ibid., p. 7. Similar recommendations appear in Council on Competitiveness, 2004, and American Electronics Association, 2005. The latter two reports recommend stopping diversion of patent-application fees to general revenue.

[25]National Research Council, 2004, p. 6. A similar recommendation appears in Council on Competitiveness, 2004.

[26]Council on Competitiveness, 2004, p. 70.

viduals participating in these funds would need to make a minimum investment of $50,000 each year in order to receive the tax credit. Acceptable investments would be restricted to those that meet requirements for revenue size and age of firm."[27]

• "Enact a permanent, restructured R&E tax credit and extend the credit to research conducted in university-industry consortia."[28]

• Allow more favorable tax treatment of purchases of high-technology manufacturing equipment. "Accelerated depreciation or expensing of high technology equipment would have a particularly positive investment impact. Many of our economic competitors—who actively seek to lure investment in semiconductor manufacturing overseas—offer far more favorable tax treatment than that offered in the United States. As part of the discussion of fundamental reforms of the tax code to promote investment and manufacturing in the US, the Congress should consider allowing companies to expense high technology equipment."[29]

• "Use the required repeal of the Foreign Sales Corporation exemption to fund a revenue-neutral tax credit for investment in information-processing equipment, software, and industrial equipment. In response to WTO rulings, Congress passed a reduction of the corporate tax rate, which really does little to encourage companies to be more competitive and innovative. An investment tax credit would help companies increase investment which would in turn boost productivity. Moreover, it would make US companies more likely to invest in equipment in the United States and not overseas."[30]

HUMAN CAPITAL

A highly skilled, flexible labor force is an essential component of this nation's ability to reap the benefits of innovation. Recent debates over workforce issues have revolved around several issues.

The first trend is that growing numbers of service industries and their labor forces are becoming subject to global competition, a condition with which manufacturing industries have long familiarity. Offshore outsourcing of business process and IT jobs, or "offshoring," is growing rapidly (see

---

[27]Ibid., p. 62.

[28]Ibid., p. 59. There are similar recommendations in numerous other reports, including National Academy of Engineering. *Mastering a New Role: Prospering in a Global Economy*. Washington, DC: National Academy Press, 1993; and American Electronics Association, 2005.

[29]Semiconductor Industry Association Web site. Available at: http://www.sia-online.org/backgrounders_ tax.cfm.

[30]R. Atkinson. *Meeting the Offshoring Challenge*. Washington, DC: Progressive Policy Institute, 2004.

Figure EI-4). Aspects of research and education are included. There are strong disagreements about what outsourcing means, the ultimate impacts, and policy prescriptions.[31] In any case, the trend reinforces the imperative for the promotion of lifelong learning in the United States. As illustrated by Figure EI-5, working adults and other nontraditional students are of growing importance in fields like computer science. Calls to rethink approaches to incentives for continuing education and trade-displacement assistance programs have come from several quarters.

A second element focuses on the immigration of scientists, engineers, and other skilled professionals who contribute to the innovation process. Several recent reports have suggested ways to encourage skilled foreigners to continue immigrating. US openness to people and ideas from around the world is a longstanding strength of the American environment for innovation.[32] In particular, immigrant scientist-engineer-entrepreneurs from Alexander Graham Bell and Andrew Carnegie to Andrew Grove have played key roles in the creation of leading US companies and entire industries.

A third human-capital issue is the reform of health insurance, pensions, and other public and private benefits infrastructures. The goals here are to make these systems sustainable from a long-term cost perspective and to help them support a workforce that is increasingly mobile and less likely to be employed by large organizations for extended periods.

A fourth issue is the promotion of education about entrepreneurship at various educational levels, including S&E education. Among the recommendations that have been suggested are these:

- "Create the human capital investment tax credit to promote continuous education. Companies often lack incentives to invest in educating and retraining workers as they risk losing that return on investment if the worker subsequently leaves the firm. By providing human-capital investment tax credits, the US government can encourage companies to retrain workers by reducing or eliminating out-of-pocket costs. At the forefront of technology innovation, companies are often the best predictor of what skills will be most valuable in the future. Continuous retraining, education, and skills acquisition ensure that fewer technology workers will find themselves suddenly displaced with no skills to participate in the constantly shifting

[31]For a point-counterpoint see R. Hira and A. Hira. *Outsourcing America: What's Behind Our National Crisis and How We Can Reclaim American Jobs*. Washington, DC: AMACOM Books, 2005; D. Farrell, M. Laboissière, R. Pascal, J. Rosenfeld, C. de Segundo, S. Stürze, and F. Umezawa. *The Emerging Global Labor Market*. New York: McKinsey Global Institute, 2005.

[32]NAS/NAE/IOM, 1999.

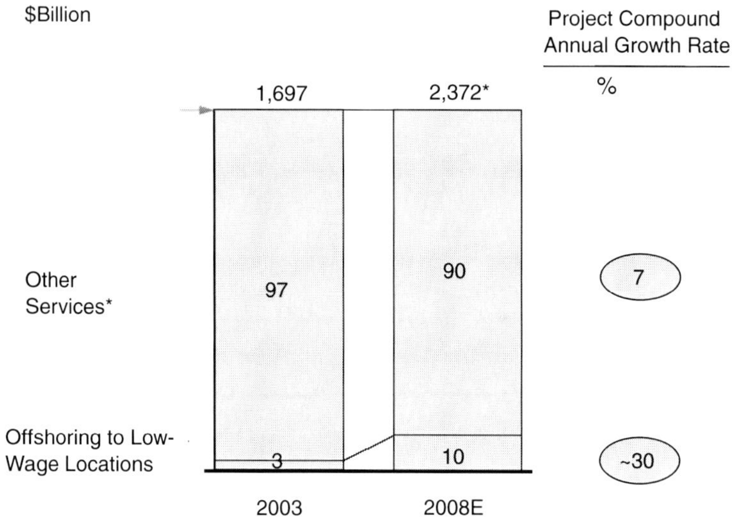

FIGURE EI-4 Business Process Outsourcing/IT offshore to low-wage locations as a percentage of total global services exports, 2003 and 2008.
NOTE: *Estimated at 6 percent annual growth from 2002 figure.
SOURCE: McKinsey and Company. "The Emerging Global Labor Market." June 2005. Executive Summary, p. 19.

high-tech industry. Furthermore, society would benefit from the continuous education of workers, which also increases productivity and decreases downtime between jobs."[33]

- Create lifelong learning accounts for employees that allow tax-exempt contributions by workers and tax credits for employer contributions.[34]
- "Reform and rename the Trade Adjustment Assistance Program to cover workers displaced for reasons other than trade, including service sector workers."[35]
- "Offer more flexibility and focus under federal-state employment and training programs. States and the federal government should have more discretion to devote employment and training resources toward high-performance programs, high-growth skills and skills in demand by local firms."[36]

---

[33]American Electronics Association, 2005, p. 26.
[34]Council on Competitiveness, 2004, p. 54.
[35]Council on Competitiveness, 2004, p. 56.
[36]Ibid.

Academic Institution	Number of 2001 Bachelor's Degrees Awarded
1. Strayer University	840
2. DeVry Institute of Technology (Addison, IL)	477
3. CUNY Bernard Baruch College	465
4. University of Maryland Baltimore County	463
5. DeVry Institute of Technology (Phoenix, AZ)	440
6. DeVry Institute of Technology (Cty of Industry, CA)	349
7. Rutgers the State University of New Jersey	336
8. DeVry Institute of Technology (Kansas City, MO)	316
9. DeVry Institute of Technology (Long Beach, CA)	301
10. James Madison University	393

FIGURE EI-5 Top producers of computer science bachelor's degree, 2001.
SOURCE: American Association for the Advancement of Science. *Preparing Women and Minorities for the IT Workforce: The Role of Nontraditional Educational Pathways.* Washington, DC: American Association for the Advancement of Science, 2005. Available at: http://www.aaas.org/publications/books_reports/ITW/PDFs/Complete_book.pdf.

- "Expand temporary wage supplements that help move workers more quickly off unemployment insurance and into new jobs and on-the-job training. The Alternative Trade Adjustment for Older Workers Program should be expanded to include younger workers and should not be linked exclusively to trade dislocation."[37]
- "Re-institute H1-B training grants to ensure that Americans are trained in the skills and fields for which companies now bring in foreign nationals."[38]
- "Establish an expedited immigration process, including automatic work permits and residency status for foreign students who: a) hold graduate degrees in S&E from American universities, b) have been offered jobs by US-based employers and who have passed security screening tests."[39]

[37]Ibid.
[38]Ibid.
[39]Council on Competitiveness, 2004, p. 51.

• "Give green cards to all US trained master and doctoral students. Accredited US colleges and universities award 8,000 doctoral and 56,000 master's degrees in S&E to foreign nationals per year. Instead of sending these people back to their countries, they should be given a Green Card to stay in the United States. These people will make significant contributions to the economy and workforce. The United States benefits by keeping them here."[40]

• "H1-B visas should be made 'portable' so that a foreign temporary nonimmigrant worker can more easily change jobs in the United States."[41]

• The National Science Foundation should take a significant role in funding pilot efforts to create innovation-oriented learning environments in K–12 and higher education. It also should sponsor research into the processes involved in teaching creativity, inventiveness, and commercialization in technical environments.[42]

• The federal government should create legal certainty for cash-balance pension plans to ensure that employers can continue to offer them. These plans are popular with many employees and have significant advantages over many defined-contribution plans.[43]

• Have the states and the federal government encourage the widespread availability of Health Savings Accounts, including affordable options for low-income workers, as a health-insurance option that provides portability for employees.[44]

• "States and the federal government should define a role for government re-insurance of higher-cost healthcare expenses, so as to reduce the cost of employer-provided coverage and reduce the cost of healthcare to employees."[45]

• "Government procurement rules should favor work done in the United States and should restrict the offshoring of work in any instance where there is not a clear long-term economic benefit to the nation or where the work supports technologies that are critical to our national economic or military security."[46]

---

[40]American Electronics Association, 2005, p. 25. A similar recommendation appears in Council on Competitiveness, 2004.

[41]National Research Council. *Building a Workforce for the Information Economy*. Washington, DC: National Academy Press, 2001.

[42]Council on Competitiveness, 2004, p. 53.

[43]Council on Competitiveness, 2004, p. 55.

[44]Ibid.

[45]Ibid.

[46]Institute of Electrical and Electronics Engineers. *Position Statement on Offshore Outsourcing*. Washington, DC: Institute of Electrical and Electronics Engineers, 2004. Available at: www.ieeeusa.org/policy/positions/offshoring.asp. A similar recommendation appears on the Economic Policy Institute Web site.

• Require transparent disclosure of offshoring. "The publicly owned firms that engage in offshoring ought to at least be transparent in their business dealings, offering layoff notices and providing clear accounting of the employment in their various units, both domestic and abroad."[47]

SUPPORTING CLUSTERS AND REGIONS

The tendency of innovative capabilities (such as research, manufacturing, educational institutions, and the workforce) to conglomerate in specific regions has been a subject of economic inquiry for some time.[48] The Council on Competitiveness sponsored a multiyear initiative to study the phenomenon in the US context.[49] One recent analysis postulates that regions need to draw a "creative class" human-resource base to compete effectively in knowledge-intensive industries.[50] Although many of the policy levers to promote regional innovation are in the hands of state and local governments, the federal government could play a larger role through such actions as the following:

• "The federal government should create at least ten Innovation Hot Spots over the next five years. State and local economic development entities and educational institutions should raise matching funds and develop proposals to operate these pilot national innovation centers."[51]

• "Innovation Partnerships need to be created to bridge the traditional gap that has existed between the long-term discovery process and commercialization. These new partnerships would involve academia, business and government, and they would be tailored to capture regional interests and economic clusters."[52]

• "The federal government should establish a lead agency for economic development programs to coordinate regional efforts and ensure that a common focus on innovation-based growth is being implemented."[53]

---

[47]Economic Policy Institute. *EPI Issue Guide: Offshoring*. Washington, DC: Economic Policy Institute, 2004.

[48]M. J. Piore and C. F. Sabel. *The Second Industrial Divide: Possibilities for Prosperity*. New York: Basic Books, 1984.

[49]Council on Competitiveness. *Clusters of Innovation: Regional Foundations of US Competitiveness*. Washington, DC: Council on Competitiveness, 2001.

[50]R. Florida. *The Rise of the Creative Class . . . and How It's Transforming Work, Leisure, Community, & Everyday Life*. New York: Basic Books, 2002.

[51]Council on Competitiveness, 2004, p. 62.

[52]Ibid., p. 53.

[53]Ibid., p. 62.

NEW "APOLLO," "SPUTNIK," OR "MANHATTAN PROJECT"

As part of the 2004-2005 debate over the sustainability of US S&E leadership, some individuals and groups have called for a presidential-level challenge to mobilize resources and national imagination in an effort that also would grow the S&E enterprise. Somewhat related is the call for the President to identify innovation as having a major national priority. Specific recommendations include the following:

- Launch an explicit national innovation strategy and agenda led by the President. "Innovation is the critical pathway to building prosperity and competitive advantage for advanced economies. Yet no single institution in government or the private sector has the horizontal responsibility for strengthening the innovation ecosystem at the national level—it is and always will be a shared responsibility. The United States should establish an explicit national innovation strategy and agenda, including an aggressive public policy strategy that energizes the environment for national innovation."[54]
- "Establish a focal point within the Executive Office of the President to frame, assess and coordinate strategically the future direction of the nation's innovation policies. This could be either a Cabinet-level interagency group, or a new, distinct mission assigned to the National Economic Council."[55]
- "Establish an explicit innovation agenda. Direct the President's economic advisors to analyze the impact of current economic policies on US innovation capabilities and identify opportunities for immediate improvement."[56]
- "Direct the Cabinet officers to undertake a policy, program and budget review and propose initiatives designed to foster innovation within and across departments. This is an opportunity to break down 'stovepipes' and foster closer collaboration among the agencies to meet clear national needs."[57]
- "The United States should build an integrated healthcare capability by the end of the decade."[58]
- Apply information technology, research, and systems-engineering tools to US healthcare delivery.[59]
- Launch a US-China crash program to develop alternative energies.[60]

---

[54]Ibid., p. 66.

[55]Ibid.

[56]Ibid.

[57]Ibid.

[58]Ibid., p. 74.

[59]National Academy of Engineering and Institute of Medicine. *Building a Better Delivery System: A New Engineering/Health Care Partnership*. Washington, DC: The National Academies Press, 2005.

[60]Friedman, 2005, p. 413.

Scientific Communication and Security

SUMMARY

Among the fundamental tenets of science is openness—minimizing restrictions on communication among scientists is considered essential to progress. The United States has achieved and maintained its pre eminence in science and technology (S&T) in part by embracing the values of scientific openness. And this openness has no natural, and certainly no national, boundaries in an increasingly international scientific enterprise.

Openness may pose risks, however. Adversaries may take advantage of ready access to information to acquire knowledge with which to do harm. Economic competitors may use open communication to pursue their own interests at the expense of the United States.

The United States has sought to limit these potential negative consequences by setting some limits on scientific communication. A system to protect intellectual property seeks to ensure that the applications of discoveries initially benefit those who make the breakthroughs. In the realm of national and homeland security, the US government carries out some research and development in secret and restricts access to certain types of information to keep it away from those who may have hostile intent.

The scientific and technical community recognizes that it has a responsibility to help protect the United States, as it has in the past, by harnessing

This paper summarizes findings and recommendations from a variety of recently published reports and papers as input to the deliberations of the Committee on Prospering in the Global Economy of the 21st Century. Statements in this paper should not be seen as the conclusions of the National Academies or the committee.

the best S&T to help counter terrorism and other national-security threats, even though this may mean accepting some limitations on its work. However, there is concern that some of the policies on scientific communication enacted in the wake of the September 11 terrorist attacks and the anthrax mailings and others under consideration will undermine the strength of science in the United States without genuinely advancing security. Various organizations, including the National Academies, have offered recommendations to address these concerns:

- Continue to support the principle set forth in National Security Decision Directive 189 that federally funded fundamental research, such as that conducted in universities and laboratories, should "to the maximum extent possible" be unrestricted.
- Create a clearly defined regulatory "safe harbor" for fundamental research so that universities in particular can have confidence that activities within the safe harbor are in compliance, thus permitting a focus on whatever occurred outside the safe harbor.
- Regularly review and update the lists of information and technologies subject to controls maintained by federal agencies with the goal of restricting the focus of the controls and removing controls on readily available technologies. Carry out the process across as well as within agencies, and include input from the S&T community.
- With regard to the specific issue of "deemed exports," do not change the current system of license requirements for use of export-controlled equipment in university basic research until the following steps have been implemented:
  - Greatly narrow the scope of controlled technologies requiring deemed-export licenses, and ensure that the list remains narrow going forward.
  - Delete all controlled technology from the list whose manuals are available in the public domain, in libraries, on the Internet, or from the manufacturers.
  - Delete all equipment from the list that is available for purchase on the open market overseas from foreign or US companies.
  - Clear international students and postdoctoral fellows for access to controlled equipment when their visas are issued or shortly thereafter so that their admission to a university academic program is coupled with their access to use of export-controlled equipment.
- Undertake a systematic review to determine the number and provisions of all existing types of "sensitive but unclassified" information in the federal government. Using that baseline, require a further review and justification for the maintenance of any category. Tie remaining categories to an explicit statutory or regulatory framework that includes procedures to request access to information and appeal decisions.

- In implementing federal security policies for S&T personnel:
  - Engage S&T personnel in the development and implementation plans for security measures.
  - Continue to accept non-US citizens as visitors and in some cases staff, expedite security reviews for visitors, and more generally work to avoid prejudice against foreigners.
  - Focus and limit security efforts to address the most important security situations.
- Create new or expand existing mechanisms to engage the S&T community in advisory capacities and to improve communication channels.
  - Encourage communication among the diverse communities involved in security issues—policy, S&T, national and homeland security, law enforcement, and intelligence—so that policies regarding scientific communication are both effective and broadly accepted.
  - Build bridges among these communities, particularly in areas of S&T, such as the life sciences, where there is little history of working with the government on security issues.

SECRET RESEARCH AND CLASSIFICATION OF INFORMATION

The US government handles issues of secrecy through a complex mix of statutes, regulations, and procedures that govern the control of classified information, public access to government information, and the maintenance of government records. With two exceptions, the government has no authority to designate information produced outside this legal framework as classified.[1] In the wake of September 11, President Bush extended classification authority to several departments and agencies that had not previously been involved in such matters, such as the Department of Agriculture, the Environmental Protection Agency, and the Department of Health and Human Services.

Controversies over whether areas of scientific research should be restricted in the name of national security recurred throughout the Cold War. During the early 1980s, the Reagan administration sought to restrict scientific communication in a number of fields. That controversy eventually led to a presidential directive in 1985, influenced in part by a report from the National Academy of Sciences.[2] National Security Decision Directive 189

---

[1]The first exception is through the Atomic Energy Act; information related to nuclear weapons may be "born classified" without any prior involvement of the government in its generation. The second exception, under the Invention Secrecy Act of 1951, permits information received as part of the patent-application process to be classified.

[2]National Research Council. *Scientific Communication and National Security*. Washington, DC: National Academy Press, 1982.

(NSDD-189) states that federally funded fundamental research, such as that conducted in universities and laboratories, should "to the maximum extent possible" be unrestricted.[3] Where restriction is deemed necessary, the control mechanism is formal classification. "No restrictions may be placed upon the conduct or reporting of federally-funded fundamental research that has not received national security classification, except as provided in applicable US statutes." The policy set out in NSDD-189 is still in force and has been reaffirmed by several senior George W. Bush administration officials.[4]

Over the years, reports and statements from the National Academies and other organizations have strongly supported the principle set forth in NSDD-189 as essential to maintaining the vitality of fundamental research in the United States.[5] Some have suggested that President Bush should reissue the directive as a signal of its continuing importance and his administration's commitment to scientific openness. Others are concerned that, given current controversies and security concerns, the interagency process necessary for such an action could result in a weaker presidential statement. At a minimum, the federal government could:

- Continue to support the principle set forth in National Security Decision Directive 189 that federally funded fundamental research, such as that conducted in universities and laboratories, should "to the maximum extent possible" be unrestricted.

"SENSITIVE" RESEARCH AND CONTROLS ON INFORMATION

Serious concerns can arise over whether information is properly classified, whether too much information is classified, and how such decisions are made, but these debates over the classification of scientific research take place within a system of reasonably well-specified and understood rules. Far more problematic is the interest in designating certain areas of research

[3]"Fundamental" research is defined as "basic and applied research in science and engineering, the results of which ordinarily are published and shared broadly within the scientific community, as distinguished from proprietary research and from industrial development, design, production and product utilization, the results of which ordinarily are restricted for proprietary or national security reasons." National Security Decision Directive 189, September 21, 1985.

[4]Letter to Dr. Harold Brown from Condoleeza Rice, Assistant to the President for National Security Affairs, November 1, 2001. John Marburger, Director of the Office of S&T Policy, Executive Office of the President, reaffirmed NSDD-189 in a speech to a workshop on "Scientific Openness and National Security" at the National Academies on January 9, 2003.

[5]Recent examples include National Research Council. *Assessment of Department of Defense Basic Research*. Washington, DC: The National Academies Press, 2005. P. 6; Center for Strategic and International Studies. *Security Controls on Scientific Information and the Conduct of Scientific Research*. Washington, DC: Center for Strategic and International Studies, June 2005.

and certain types of knowledge—wherever they are produced and however they are funded—as "sensitive but unclassified" (SBU).

The problem of "sensitive information" is not new. Classification is only one of the ways in which the US government controls public access to information. Across the federal government, there are dozens of categories that apply narrowly or broadly to specific types of information (see Figure SCS-1).[6] Some of the categories are defined in statute, some through regulation, and some only through administrative practices. In addition, different agencies may assign a variety of civil and even criminal penalties for violation of their restrictions.[7]

Here, the fundamental issue is the scope of restrictions—that is, how much should the government try to control? When the primary US opponent was another technologically sophisticated state, the Soviet Union, the case could be made that one should focus on S&T areas that could truly make a difference in terms of adding to Soviet capabilities or undermining those of the United States. With the fall of the Soviet Union, some argue that the range of less technologically sophisticated opponents, including terrorists, now confronting the United States means that the government should try to deny access to the much wider range of information and technologies that could be useful to them.

While recognizing the legitimate concerns that others may take advantage of open access to information, technologies, and materials for malicious purposes, past examinations of the potential tradeoffs between openness and security have concluded that the United States is best served by focusing its efforts on protecting fewer, very-high-value areas of S&T.[8] This is particularly true in fields where knowledge is advancing quickly and diffusing rapidly; otherwise, the United States may expend its efforts in attempts to control knowledge and technology that are readily available elsewhere. In addition, many of the existing and proposed lists of "sensitive"

---

[6]The CSIS Commission on Science and Security in the 21st Century identified at least 20 types of information that could be considered "sensitive" within the Department of Energy, most without consistent, departmentwide definitions or application. Center for Strategic and International Studies. *Science and Security in the 21st Century: A Report to the Secretary of Energy on the Department of Energy Laboratories*. Washington, DC: Center for Strategic and International Studies, 2002. P. 55.

[7]G. J. Knezo. *"Sensitive But Unclassified" and Other Federal Security Controls on Scientific and Technical Information: History and Current Controversy*. Washington, DC: Congressional Research Service, April 2, 2003. P. 10.

[8]This is a fundamental conclusion of the Corson report and is echoed in other reports, such as National Research Council. *A Review of the Department of Energy Classification Policy and Practice*. Washington, DC: National Academy Press, 1995; Commission on Protecting and Reducing Government Secrecy (the Moynihan Commission). *Secrecy*. Washington, DC: US Government Printing Office, 1997; Center for Strategic and International Studies. *Security Controls on Scientific Information and the Conduct of Scientific Research*. Washington, DC: Center for Strategic and International Studies, June 2005.

Data Category	**Description**
FOIA Exempted	Any information that is exempted from mandatory disclosure under the Freedom of Information Act.
Intelligence Activities	Information that involves or is related in intelligence activities, including collection methods, personnel, and unclassified information.
Cryptologic Activities	Information that involves encryption/decryption of information; communications security equipment, keys, algorithms, processes; information involving the methods and internal workings of cryptologic equipment.
Command and Control	Information involving the command and control of forces, troop movements.
Weapon and Weapon Systems	Information that deals with the design, functionality, and capabilities of weapons and weapon systems both fielded and un-fielded.
RD&E	Research, development, and engineering data on un-fielded products, projects, systems, and programs that are in the development or acquisition phase.
Logistics	Information dealing with logistics, supplies, materials, parts and parts requisitions, including quantities and numbers.
Medical Care/HIPAA	Information dealing with personal medical care, patient treatment, prescriptions, physician notes, patient charts, x-rays, diagnosis, etc.
Personnel Management	Information dealing with personnel, including evaluations, individual salaries, assignments, and internal personnel management.
Privacy Act Data	Information covered by the Privacy Act of 1974 (5 U.S.C. § 552A)
Contractual Data	Information and records pertaining to contracts, bids, proposals, and other data involving government contracts.
Investigative Data	Information and data pertaining to official criminal and civil investigations such as investigator notes and attorney-client privileged information.

FIGURE SC&S-1 Examples of "sensitive but unclassified" and other controlled information.
SOURCE: Congressional Research Service. *"Sensitive But Unclassified" and Other Federal Security Controls on Scientific and Technical Information: History and Current Controversy*. CRS Report for Congress. Order Code RL31845. February 20, 2004.

information and materials tend to consist of broad and general categories, making it potentially difficult for researchers to know whether their activities are in or out of bounds.

These considerations suggest two general principles and a number of specific recommendations:

- *Principle 1:* Construct "high fences" around narrow areas—that is, maintain stringent security around sharply defined and narrowly circumscribed areas, but reduce or eliminate controls over less sensitive material.
  - Regularly review and update the lists maintained by federal agencies of information and technologies subject to controls with the goal of restricting their focus and removing controls on readily available technologies.

– Carry out the process across as well as within agencies, and include input from the S&T community.

- *Principle 2:* Avoid the creation of categories of SBU information and consolidate existing ones.

– Undertake a systematic review to determine the number and provisions of all existing types of SBU in the federal government.

– Using that baseline, require a further review and justification for the maintenance of any category. Tie remaining categories to an explicit statutory or regulatory framework that includes procedures to request access to information and appeal decisions.

"DEEMED EXPORTS": A SPECIAL CURRENT CASE

The controls governed by the Export Administration Act and its implementing regulations extend to the transfer of "technology." *Technology* is considered "*specific information* necessary for the 'development,' 'production,' or 'use' of a product," and providing such information to a foreign national within the United States may be considered a "deemed export" whose transfer requires an export license[9] [italics added]. The primary responsibility for administering deemed exports lies with the Department of Commerce (DOC), but other agencies may have regulations to address the issue. Deemed exports are currently the subject of significant controversy.

In 2000, Congress mandated annual reports by agency offices of inspector general (IG) on the transfer of militarily sensitive technology to countries and entities of concern; the 2004 reports focused on deemed exports. The individual agency IG reports and a joint interagency report concluded that enforcement of deemed-export regulations had been ineffective; most of the agency reports recommended particular regulatory remedies.[10]

---

[9]"Generally, technologies subject to the Export Administration Regulations (EAR) are those which are in the United States or of US origin, in whole or in part. Most are proprietary. Technologies which tend to require licensing for transfer to foreign nationals are also dual-use (i.e., have both civil and military applications) and are subject to one or more control regimes, such as National Security, Nuclear Proliferation, Missile Technology, or Chemical and Biological Warfare." *"Deemed Exports" Questions and Answers*, Bureau of Industry and Security, Department of Commerce.

The International Traffic in Arms Regulations (ITAR), administered by the Department of State, control the export of technology, including technical information, related to items on the US Munitions List. Unlike the EAR, however, "publicly available scientific and technical information and academic exchanges and information presented at scientific meetings are not treated as controlled technical data."

[10]Reports were produced by the DOC, DOD, the Department of Energy (DOE), the Department of State, the Department of Homeland Security, and the Central Intelligence Agency. Only the interagency report and the reports from DOC, DOD, and DOE are publicly available.

The DOC sought comments from the public about the recommendations from its IG before proposing any changes. The department earned praise for this effort to reach out to potentially affected groups and is currently reviewing the 300 plus comments it received, including those from the leaders of the National Academies.[11]

On July 12, 2005, the Department of Defense (DOD) issued a notice in the *Federal Register* seeking comments on a proposal to amend the Defense Federal Acquisition Regulation Supplement (DFARS) to address requirements for preventing unauthorized disclosure of export-controlled information and technology under DOD contracts that follow the recommendations in its IG report. The proposed regulation includes a requirement for access-control plans covering unique badging requirements for foreign workers and segregated work areas for export-controlled information and technology, and it makes no mention of the fundamental-research exemption.[12] Comments are due by September 12, 2005.

Many of the comments in response to the DOC expressed concern that the proposed changes were not based on systematic data or analysis and could have a significant negative impact on the conduct of research in both universities and the private sector, especially in companies with a substantial number of employees who are not US citizens. Similar comments are expected in response to the DOD proposals. Among the recommendations that have been offered to date to address these concerns are the following:

• Create a clearly defined regulatory "safe harbor" for fundamental research so that universities can have confidence that activities within the safe harbor are in compliance with security restrictions, thus permitting a focus on whatever occurred outside the safe harbor.[13]

• Do not change the current system of license requirements for use of export-controlled equipment in university basic research until the following steps have been implemented:

– Greatly narrow the scope of controlled technologies requiring deemed-export licenses, and ensure that the list remains narrow going forward.

– Delete all controlled technology from the list whose manuals are available in the public domain, in libraries, on the Internet, or from the manufacturers.

– Delete all equipment from the list that is available for purchase on the open market overseas from foreign or US companies.

---

[11]The letter from the presidents of the National Academies may be found at http://www7.nationalacademies.org/rscans/Academy_Presidents_ Comments_to_DOC.PDF.

[12]*Federal Register* 70(132)(July 2005):39976-39978. Available at: http://a257.g.akamaitech.net/7/257/2422/01jan20051800/edocket.access.gpo.gov/2005/05-13305.htm.

[13]See footnote 11.

– Clear international students and postdoctoral fellows for access to controlled equipment when their visas are issued or shortly thereafter so that their admission to a university academic program is coupled with their access to use of export-controlled equipment.[14]

ENGAGING THE S&T COMMUNITY IN THE CHALLENGES OF ACHIEVING SECURITY

In the wake of September 11 and the anthrax mailings, the S&T community, as in past times of crisis and along with other Americans, responded to the new challenges to US security. This response has occurred on many levels, from helping to analyze current and potential threats to working on ways in which advances in S&T can improve national and homeland security.[15] This has required active engagement by the S&T community with policy-makers, particularly in national and homeland security, in law enforcement, and in intelligence, where many of the parties at the table are likely to lack experience dealing with one another. It also involves continuing efforts to ensure that highly qualified S&T personnel are attracted to working on problems related to national and homeland security.

Press reports since September 11 have suggested that officials in the DOD and DHS are concerned about attracting eligible workers, especially those with specialties in demand in open parts of the private sector. Since a significant portion of the work may be restricted or classified, this issue is largely a subset of the wider problem addressed in other background papers of ensuring that sufficient qualified US citizens are available to do the work. It also involves ensuring that restrictions on non-US citizens as employees are appropriate.

In addition, attracting personnel requires the creation of a work environment that will enable R&D in particular to be "cutting-edge." For example, scientists working in a restricted or classified environment, especially at federal laboratories, still need to interact with the wider scientific community, including foreign visitors and collaborators, where much of the innovation most relevant to their work is taking place. In the wake of a series of scandals over alleged security lapses in the DOE nuclear-weapons complex in the late 1990s, the department imposed a number of new and

[14]These recommendations were made by Dan Mote, president of the University of Maryland, at a May 6, 2005, workshop at the National Academies and cited in the letter from the National Academies' presidents.

[15]For a comprehensive examination of the potential contributions of S&T, see National Research Council. *Making the Nation Safer: The Role of Science and Technology in Countering Terrorism.* Washington, DC: The National Academies Press, 2002. Guides to additional reports and current projects of the National Academies related to homeland security may be found at: http://www.nationalacademies.org/subjectindex/sec.html.

expanded security restrictions. This sparked substantial concern about ensuring that the scientific quality of the laboratories could be sustained, and several organizations made proposals they believed would provide an appropriate balance between openness and security, these including[16]

- Engage S&T personnel in the development and implementation plans for security measures.
- Continue to accept non-US citizens as visitors and in some cases staff, expedite security reviews for visitors, and more generally work to avoid prejudice against foreigners.
- As with recommendations for other situations, focus and limit security efforts to address the most important security situations.

Beyond attracting S&T personnel, it is essential to engage the broader S&T community in efforts to bring the latest S&T to bear on security problems. Much of the relevant research and many of the best ideas seem likely to come from outside the government and its own network of laboratories. Tapping these resources involves meeting several needs. One is ensuring an attractive climate for undertaking security-related R&D in universities and the private sector. Another is engaging the S&T community in a variety of advisory capacities and communication channels. Some observers have recommended a variety of new mechanisms or expanded and revised roles for existing mechanisms, including the following:

- Encourage communication among the diverse communities involved in security issues—policy, S&T, national and homeland security, law enforcement, and intelligence—so that policies regarding scientific communication are both effective and broadly accepted.
- Build bridges among these communities, particularly in areas of S&T, such as the life sciences, where there is little history of working with the government on security issues.[17]

---

[16]National Research Council. *Balancing Scientific Openness and National Security Controls at the Nuclear Weapons Laboratories*. Washington, DC: National Academy Press, 1999; Center for Strategic and International Studies. *Science and Security in the 21st Century: A Report to the Secretary of Energy on the Department of Energy Laboratories*. Washington, DC: Center for Strategic and International Studies, 2002.

[17]See the recommendations, for example, in National Research Council. *Biotechnology Research in an Age of Terrorism*. Washington, DC: The National Academies Press, 2004.

Science and Technology Issues in National and Homeland Security

SUMMARY

Keeping a technological edge over adversaries of the United States has long been a key component of our national security strategy. US preeminence in science and technology (S&T) is considered essential to achieving that goal, so throughout the Cold War the United States generously funded research and development, including basic research, that could contribute to national security. Since 1950, "defense" funding has been the largest component of the overall federal R&D budget, and it has been a majority of that funding since fiscal year (FY) 1981 (see Figure NHS-1). That investment has provided substantial spinoffs to the private sector, adding to the knowledge base and innovation that have fueled US productivity and prosperity.

In the wake of the September 11 attacks and the anthrax mailings, the nation has looked to S&T to help meet the new challenges of homeland security. Meanwhile, the US military is in the midst of a "transformation" that depends on taking advantage of new and emerging technologies to respond to the diffuse and uncertain threats that characterize the 21st century.

The current pursuit of national and homeland security is taking place in a profoundly different environment, however. The end of the Cold War and

---

This paper summarizes findings and recommendations from a variety of recently published reports and papers as input to the deliberations of the Committee on Prospering in the Global Economy of the 21st Century. Statements in this paper should not be seen as the conclusions of the National Academies or the committee.

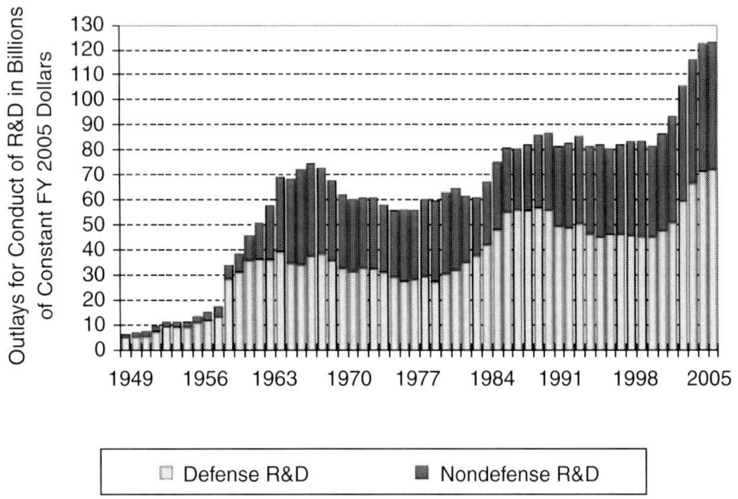

FIGURE NHS-1 Federal spending in defense and nondefense R&D, 1949-2005. SOURCE: American Association for the Advancement of Science. *Chart: Federal Spending on Defense and Nondefense R&D*. Washington, DC: AAAS, February 2005. Available at: http://www.aaas.org/spp/rd/histde06.pdf.

the increasing commercialization and globalization of the traditional sources of S&T innovation for security have produced significant challenges for US national and homeland security policies. Many proposals to ensure continuing US S&T leadership see defense funding as essential to supporting this goal, requiring policies that would be able to serve both economic and national and homeland security objectives.

Federal actions that have been proposed include the following:

• Raise the level of S&T spending to 3% of Department of Defense (DOD) spending and restore DOD's historical commitment to basic research by directing 20% of its S&T budget to long-term research.

– Increase the budget for mathematics, the physical sciences, and engineering research by 12% a year for the next 7 years within the research accounts of the Department of Energy (DOE), the National Science Foundation (NSF), the National Institute of Standards and Technology (NIST), and the DOD.

– Within the DOD, set the balance of support for 6.1 basic research more in favor of unfettered exploration than of research related to short-term needs.

- For homeland security R&D:

  – Commit to increase the portion of support that the Department of Homeland Security (DHS) devotes to basic research, perhaps by setting targets to be achieved within 5-10 years as the most immediate needs are satisfied.

  – Undertake a comprehensive review to identify opportunities across the entire federal homeland security R&D budget to support increased investments in basic and applied research.

  – On the applied R&D side, search for technologies that can reduce costs or provide ancillary benefits to civil society to ensure a sustainable effort against terrorist threats.

- Conduct a review of the current military and dual-use export-control systems to identify policies that narrowly target exports of concern without needlessly burdening peaceful commerce; strengthen the multilateral cooperation essential to any effective export-control regime; streamline export classification, licensing, and reporting processes; and afford the President the authority and flexibility needed to advance US interests.
- Establish a new framework for coordinating multilateral export controls based on harmonized export-control policies and enhanced defense cooperation with close allies and friends.
- Assess whether the current system of the national laboratories that carry out defense-related research has the structure, personnel, and resources to provide the cutting-edge work and innovation to support national and homeland security R&D needs.
- Create a new National Defense Education Act (NDEA) for the 21st century. The new NDEA would include portable graduate fellowships, institutional traineeships, incentives to create professional science and engineering (S&E) master's programs, undergraduate loan forgiveness, grants to support new and innovative undergraduate curricula, grants to expand K–12 education outreach, summer training and research opportunities for K–12 teachers, employer S&E and foreign-language educational tax breaks, national laboratory and federal service professional incentives, and additional funds for program evaluation.

THE NATIONAL AND HOMELAND SECURITY R&D PORTFOLIO

With the end of the Cold War, US defense investment, already declining in the wake of the Reagan Administration's massive buildup, entered the longest period of sustained decline since the end of World War II, with deep cuts in funding for weapons procurement and R&D. September 11 and the wars in Afghanistan and Iraq have more than restored overall funding levels, but serious concerns remain about the size and even more the mix of the R&D portfolio. In recent years, more and more emphasis has gone to devel-

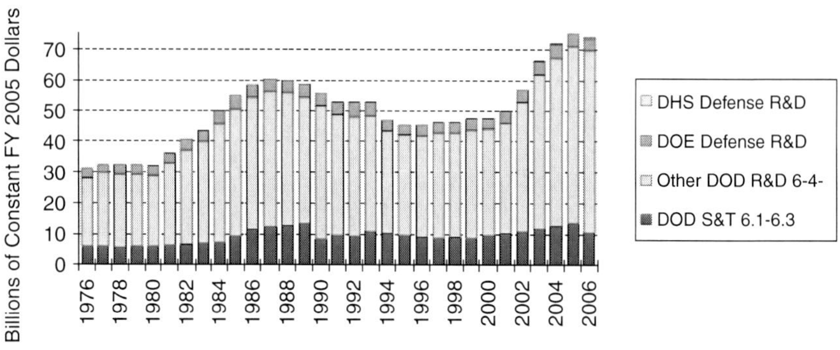

FIGURE NHS-2 Trends in defense R&D, FY 1976-FY 2006.
SOURCE: American Association for the Advancement of Science. *Chart: Trends in Defense R&D: FY 1976-2006*. Washington, DC: American Association for the Advancement of Science, February 2005. Available at: http://www.aaas.org/spp/rd/trdef06c.pdf.

opment as opposed to research (see Figure NHS-2). The portion of the DOD R&D budget devoted to basic research (the "6.1" account) has declined in constant dollars from 3.3% in FY 1994 to an estimated 1.9% in FY 2005 (see Figure NHS-3).[1] In addition, within that account there has been increasing emphasis on research that appears more likely to yield short-term payoffs rather than the more open exploration that has been so important to past advances. The President's budget request for FY 2006 called for a 13% cut in the 6.1 account, which by July 2005 the House of Representatives had partially restored to a 4% decrease. The House also called for a 4.2% gain in applied research (the "6.2" account) rather than the 15% reduction called for by the President's budget request, although the gain would come largely in the form of earmarks.[2]

Beyond meeting the immediate perceived R&D needs of the US military, broad service policy documents, such as *Joint Vision 2010* and *2020,* look toward substantial expansions in the breadth and depth of S&T to support US strategy.[3] The transformation goals set forth in DOD's 2001

[1]Funding for the 6.2 "applied research" account has gone up and down but now is 5.5% in FY 2005 compared with 7.6% in FY 1994. Constant dollar and percentage calculations by the Council on Competitiveness based on American Association for the Advancement of Science, "Historical Table: Trends in DOD 'S&T,' 1994-2005."

[2]American Association for the Advancement of Science. "Update on R&D in FY 2006 DOD House Appropriations." July 2005.

[3]National Research Council. *Assessment of Department of Defense Basic Research*. Washington, DC: The National Academies Press, 2005.

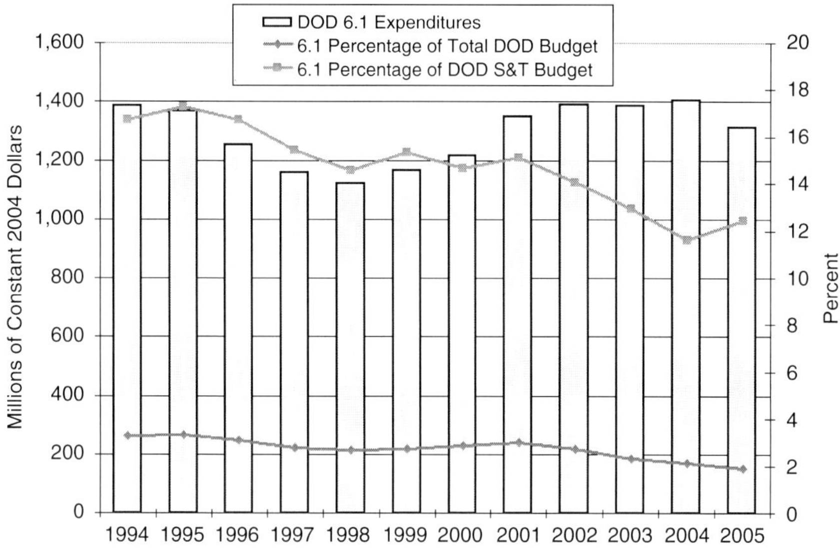

FIGURE NHS-3 Department of Defense (DOD) 6.1 expenditures, 1994-2005, in millions of constant 2004 dollars.
SOURCE: National Science Board. *Science and Engineering Indicators 2004*. NSB 04-01. Arlington, VA: National Science Foundation, 2004.

Quadrennial Defense Review (QDR) also depend on continuing to exploit the enhanced capabilities that can emerge from advances in S&T; the report called for significantly increasing S&T spending within the DOD budget.[4]

Achieving these goals will require a return to the traditional strong support for basic and applied research, in particular in the physical sciences and engineering. These goals also will demand initiatives in new and emerging areas of S&T, such as those called for by the QDR and a recent Defense Science Board study.[5] In addition, these changes are considered essential to sustaining the role that defense research has played in improving the broader health of the US S&T enterprise.

Among the actions that have been proposed for the federal government are these:

---

[4]Department of Defense. *Quadrennial Defense Review Report*. Washington, DC: Department of Defense, 2001.

[5]Defense Science Board. *The Defense Science Board 2001 Summer Study on Defense Science and Technology*. Washington, DC: Defense Science Board, 2001.

• Raise the level of S&T spending to 3% of DOD spending[6] and restore DOD's historical commitment to basic research by directing 20% of its S&T budget to long-term research.[7]

• Increase the budget for mathematics, the physical sciences, and engineering research by 12% a year for the next 7 years within the research accounts of DOE, NSF, NIST, DOD.

• Within DOD, set the balance of support for 6.1 basic research more in favor of unfettered exploration than of research related to short-term needs.

Funding for R&D for homeland security is a much more recent enterprise. The majority of US homeland security R&D funding actually occurs outside DHS (see Table NHS-1).[8] After annual increases of more than $200 million in each of its first 3 years, the FY 2006 budget request for DHS R&D slowed to a 3.6% increase, or $44 million, for a total of $1.3 billion. To date, both the House and the Senate have essentially retained the requested levels, but each has made changes in how the funds would be allocated. Efforts to consolidate all DHS R&D programs into the department's Directorate for S&T are scheduled to be completed in FY 2006.[9]

Basic research is at present a relatively small portion of the federal homeland security R&D portfolio. The priority is instead on efforts to use S&T to develop and field new methods and measures to increase security as quickly as possible.[10] The primary exception is the biodefense program, in particular the very large National Institutes of Health research program.

The question of the balance across the homeland security R&D portfolio is an open issue. If more funding for basic research is a goal, options for the federal government include the following:

• Commit to increase the portion of support that DHS devotes to basic research, perhaps by setting targets to be achieved within 5-10 years as the most immediate needs are satisfied.

• Undertake a comprehensive review to identify opportunities across

---

[6]Ibid., p. 41.

[7]Council on Competitiveness. *Innovate America*. Washington, DC: Council on Competitiveness, 2004.

[8]American Association for the Advancement of Science. "Table 4: Federal Homeland Security-Related R&D by Agency." March 2005.

[9]American Association for the Advancement of Science. "R&D Funding Update on R&D in the FY 2006 DHS Budget." 2005.

[10]For a comprehensive examination of the potential contributions of science and technology, see National Research Council. *Making the Nation Safer: The Role of Science and Technology in Countering Terrorism*. Washington, DC: The National Academies Press, 2002. Guides to the additional reports and current projects of the National Academies related to homeland security may be found at http://www.nationalacademies.org/subjectindex/sec.html.

TABLE NHS-1 US Homeland Security R&D Funding, by Agency, FY 2002 to FY 2006

	FY 2002 Actual	FY 2003 Actual	FY 2004 Actual	FY 2005 Estimate	**FY 2006 Budget**	Change Amount	FY 05-06 %
Agriculture	175	155	40	161	**172**	11	6.8%
Commerce	20	16	23	73	**82**	9	11.9%
Department of Defense	259	212	267	362	**394**	32	8.7%
Department of Energy	50	48	47	92	**81**	–12	–12.5%
Department of Homeland Security	266	737	1,028	1,243	**1,287**	44	3.6%
Environmental Protection Agency	95	70	52	33	**94**	61	185.1%
Health and Human Services	177	1,653	1,724	1,796	**1,802**	6	0.4%
- *National Institutes of Health*	*162*	*1,633*	*1,703*	*1,774*	***1,781***	*6*	*0.4%*
National Aeronautics and Space Adm.	73	73	88	88	**92**	4	4.5%
National Science Foundation	229	271	321	326	**329**	3	1.0%
Transportation	106	7	3	0	**0**	0	—
All Other	48	47	32	42	**92**	50	118.8%
Total Homeland Security R&D	1,499	3,290	3,626	4,216	**4,425**	208	4.9%
(Total Homeland Security Spending)	*32,881*	*42,447*	*40,834*	*46,015*	***49,943***	*3,928*	*8.5%*

NOTE: American Association for the Advancement of Science, based on Office of Management and Budget (OMB) data from OMB's *2003 Report to Congress on Combating Terrorism* and *Budget of the US Government FY 2006*. Figures adjusted from OMB data by AAAS to include conduct of R&D and R&D facilities, and revised estimates of DHS R&D. Figures do not include non-R&D homeland security activities, nor do they include DOD R&D investments in overseas combating terrorism. Funding for all years includes regular appropriations and emergency supplemental appropriations.

SOURCE: American Association for the Advancement of Science. *Guide to R&D Funding Data: Historical Data*. Washington, DC: American Association for the Advancement of Science, 2005. Available at: http://www.aaas.org/spp/rd/guihist.htm.

the entire federal homeland security R&D budget to support increased investments in basic and applied research.

- On the applied R&D side, search for technologies that can reduce costs or provide ancillary benefits to civil society to ensure a sustainable effort against terrorist threats.

NEW SOURCES OF INNOVATION FOR SECURITY: THE TECHNOLOGY TRANSFER DILEMMA

Traditionally, US government programs were the primary driver for research into new defense-related technologies. DOD relied on a dedicated domestic industrial base, supported largely by the results of generous DOD-funded R&D in the commercial sector and universities.

That Cold War model no longer exists because of the deep cuts in US defense research investment already discussed and the dramatic increases in private-sector R&D investment, particularly in the high-technology areas such as information and communications technologies essential to transformation. The US government has attempted to come to terms with this new situation through a variety of initiatives to enable it to take advantage of innovation from the commercial sector that could "spin on" to enhance military capabilities.

The dramatic consolidation and increasing globalization of many sectors of the traditional defense industrial base also have encouraged US efforts to find ways to enhance technology cooperation with close friends and allies. In the decade following the end of the Cold War, the 15 major US defense contractors shrank to four huge firms (see Figure NHS-4).[11] Many US defense firms have embraced a global business model, and non-US firms, primarily from Europe, have gained access to the US defense market on their own or in cooperation with US companies.[12]

These fundamental changes in the sources and structures of innovation for national security have also made it easier for US adversaries to gain access to knowledge and technology that could improve their capabilities.[13] Policies to draw on innovation from firms in the commercial sector with global mar-

---

[11] A. R. Markusen and S. S. Costigan. The Military Industrial Challenge. In A. R. Markusen and S. S. Costigan, eds. *Arming the Future: A Defense Industry for the 21st Century*. New York: Council on Foreign Relations, 1999. P. 8.

[12] "Transformed? A Survey of the Defence Industry." *The Economist*, July 20, 2002; K. Hayward. "The Globalization of Defence Industries." *Survival* (Summer 2001).

[13] See, for example, National Intelligence Council. *Mapping the Global Future: Report of the National Intelligence Council's 2020 Project*. Washington, DC: National Intelligence Council, December 2004; Defense Science Board Task Force on Globalization and Security. *Final Report*. Washington, DC: Office of the Under Secretary of Defense for Acquisition and Technology, 1999.

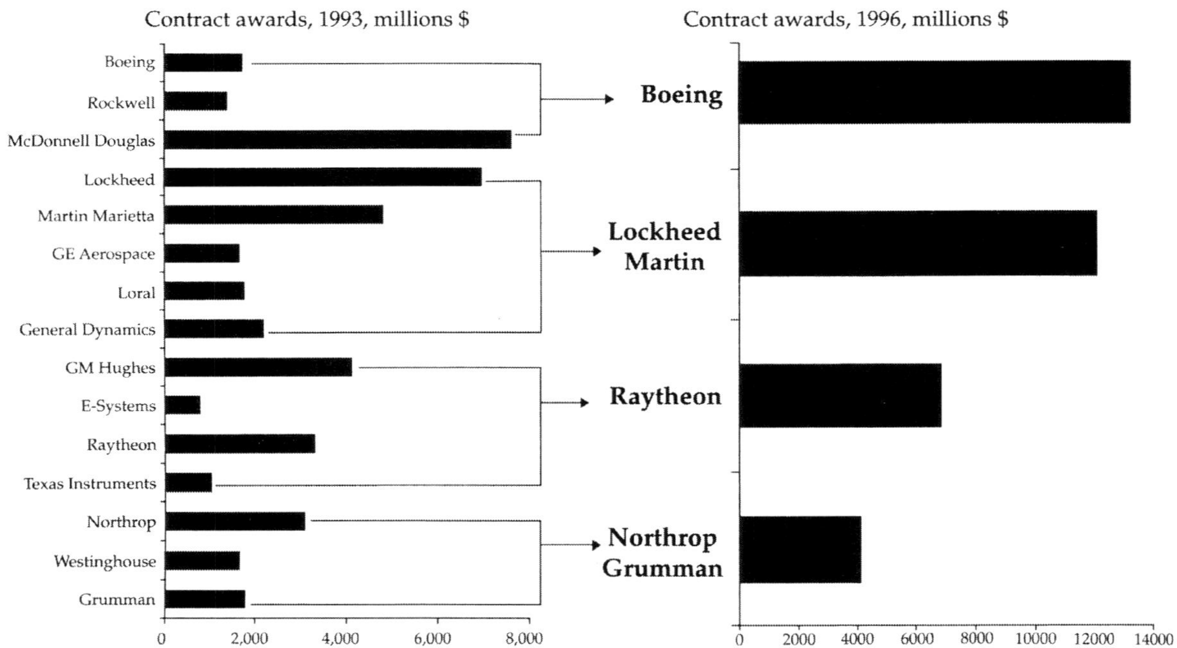

FIGURE NHS-4 US defense mergers in the 1990s.
SOURCE: A. R. Markusen and S. S. Costigan. *Arming The Future*. New York: Council on Foreign Relations Press, 1999. P. 8. Figure 1-1.

kets and international workforces or to enhance international technology cooperation potentially clash with longstanding US efforts to control the leakage of technology. September 11 and increasing concerns for terrorism—especially using nuclear, chemical, or biologic agents—have exacerbated these tensions. Faced with adversaries who are far less technologically sophisticated or who are relying on technology to make rapid advances in their capabilities—and for whom a much broader range of US technologies is thus potentially relevant than for a technologically advanced opponent like the Soviet Union—there is a natural inclination to broaden the scope of US control efforts to cover as much as possible that could be of use.

There is increasing concern that current policy initiatives serve neither technology transfer and cooperation on the one hand nor proliferation prevention on the other.[14] In part, this is because technology-transfer policy is being pursued largely through a policy apparatus constructed during the Cold War that critics from many quarters charge has never genuinely adjusted to the new threats facing the United States. According to critics, continued reliance on this apparatus—in particular, the current export-control regime for military and so-called dual-use goods and technologies—might do relatively little to prevent others from gaining access to US products and know-how while damaging the capacity of the United States to draw on innovation in the commercial sector for both economic and national- and homeland-security objectives.

While critics generally share profound dissatisfaction with the current system, there is little consensus within or among the federal government, Congress, and the affected communities about remedies for the situation. These disputes are not new, but they take on particular force now because of the depth and extent of the disputes and because of their potential impact on efforts to promote the health and capacity of the US S&T enterprise.

For the federal government, there are a number of possible options, including these:

- Conduct a review of the current US military and dual-use export-control systems to identify policies that narrowly target exports of concern without needlessly burdening peaceful commerce; strengthen the multilat-

---

[14]See, for example, Defense Science Board Task Force on Globalization and Security, 1999; Center for Strategic and International Studies. *Technology and Security in the 21st Century: US Military Export Control Reform*. Washington, DC: Center for Strategic and International Studies, 2001; Government Accountability Office. *Defense Trade: Arms Export Control System in the Post-9/11 Environment*. GAO-05-234. Washington, DC: Defense Science Board, February 16, 2005; Government Accountability Office. *Defense Trade: Arms Export Control Vulnerabilities and Inefficiencies in the Post-9/11 Security Environment*. GAO-05-468R. Washington, DC: Defense Science Board, April 7, 2005.

eral cooperation essential to any effective export-control regime; streamline export classification, licensing, and reporting processes; and afford the President the authority and flexibility needed to advance US interests.[15]

- Establish a new framework for coordinating multilateral export controls based on harmonized export-control policies and enhanced defense cooperation with close allies and friends.[16]

THE ROLE OF THE NATIONAL LABORATORIES IN NATIONAL AND HOMELAND SECURITY

Over the course of the Cold War, the United States created a system of national and federal laboratories, some devoted exclusively to research related to national security and some serving multiple roles. The DOE, for example, maintains 10 national laboratories that are managed through contracts with universities and private firms.[17] The DOD maintains a much larger system. Other laboratories maintained by such agencies as National Aeronautics and Space Administration may also conduct defense-related work. DHS has turned to some of the existing DOE laboratories to support its new R&D enterprise;[18] it also is creating the National Bioterrorism Analysis and Countermeasures Center to handle its large biodefense-research portfolio. Some of these laboratories do a mix of classified and unclassified research, and others carry out only unclassified work, in some cases to ensure the maximal openness for their basic-research programs.

Since the end of the Cold War, questions have arisen periodically about the continuing relevance of the national laboratory system. Periodic reviews of the DOE laboratories, for example, have proposed substantial changes, including consolidation of the laboratories and significant changes in management structures.[19] More general concerns include how to ensure the quality of scientific personnel in the laboratories and whether measures should

---

[15]Center for Strategic and International Studies. *Technology and Security in the 21st Century: US Military Export Control Reform*. Washington, DC: Center for Strategic and International Studies, 2001.

[16]Henry L. Stimson Center and Center for Strategic and International Studies. *Enhancing Multilateral Export Controls for US National Security*. Washington, DC: The Henry L. Stimson Center, 2001.

[17]See, for example, http://www.energy.gov/engine/content.do?BT_CODE=ST_SS16.

[18]See http://www.dhs.gov/dhspublic/display?theme=27&content=3000/.

[19]See, for example, Department of Energy. *Task Force on Alternatives Futures for the Department of Energy National Laboratories* (the "Galvin Commission"). Washington, DC: Secretary of Energy Advisory Board, 1995; General Accounting Office. *Department of Energy National Laboratories Need Clearer Vision and Better Management*. GAO/RCED-95-10. Washington, DC: General Accounting Office, January 1995; National Research Council. *Maintaining High Scientific Quality at Los Alamos and Livermore National Laboratories*. Washington, DC: The National Academies Press, 2004.

introduce greater competition to increase the incentives for the laboratories to draw on the best personnel and ideas in the private sector.[20]

Options for the federal government to address these issues include an initial effort to:

- Assess whether the current system of the national laboratories that carry out defense-related research has the structure, personnel, and resources to provide the cutting-edge work and innovation to support national and homeland security R&D needs.

NATIONAL DEFENSE EDUCATION ACT

Adopted by Congress in 1958, the original NDEA was intended to boost education and training in security and national-defense-related fields. NDEA was a response to the launch of Sputnik and the emerging threat to the United States posed by the Soviet Union. NDEA was funded with approximately $400 million to $500 million (in constant 2004 dollars). NDEA provided funding to enhance research facilities; fellowships to thousands of graduate students pursuing degrees in science, mathematics, engineering, and foreign languages; and low-interest loans for undergraduates in these areas.

By the 1970s, the act had been largely superseded by other programs, but its legacy remains in the form of several federal student-loan programs.[21] The legislation ultimately benefited all of higher education as the notion of defense was expanded to include most disciplines and fields of study.[22]

The DOD workforce is critical to our nation's security planning. This workforce, however, has experienced a real attrition of more than 13,000 personnel over the last 10 years. At the same time, DOD projects that its workforce demands will increase by more than 10% (by 2010). Indeed, several major studies[23] since 1999 argue that the number of US graduates in

---

[20]See, for example, National Research Council. *National Laboratories: Building New Ways to Work Together—Report of a Workshop*. Washington, DC: The National Academies Press, 2005; and the suggestions about personnel in Defense Science Board. *The Defense Science Board 2001 Summer Study on Defense Science and Technology*. Washington, DC: Defense Science Board, 2001.

[21]Association of American Universities. *A National Defense Education Act for the 21st Century. Renewing Our Commitment to US Students, Science, Scholarship, and Society.* White Paper. Washington, DC: Association of American Universities, 2005. Available at: http://www.aau.edu/education/NDEAOP.pdf.

[22]M. Parsons. "Higher Education Is Just Another Special Interest" *The Chronicle of Higher Education* 51(22)(2005):B20. Available at: http://chronicle.com/prm/weekly/v51/i22/22b02001.htm.

[23]See, for example, the National Science Board's companion paper to *Science and Engineering Indicators 2004*. Arlington, VA: National Science Foundation, 2004.

critical areas is not meeting national, homeland, and economic security needs (see Figure NHS-5). Science, engineering, and language skills continue to have very high priority across government and industrial sectors.

Many positions in critical-skills areas require security clearances, meaning that only US citizens may apply. While over 95% of undergraduates are US citizens, in many of the S&E fields less than 50% of those earning PhDs are US citizens. Retirements also loom on the horizon: over 60% of the federal S&E workforce is over 45, a large proportion of whom are employed by DOD (see Table NHS-2). DOD and other federal agencies face increased competition from domestic and global commercial interests for top-of-their-class, security-clearance-eligible scientists and engineers.

To ensure adequate human resources in fields important for homeland security, the National Research Council in the report *Making the Nation Safer* recommended that there be a human-resource development program similar to the NDEA.[24] National weapons laboratories have instituted specific programs to recruit and hire critically skilled people to staff nuclear-stockpile stewardship programs—for which US citizenship is a primary consideration—including graduate and postdoctoral internship programs, programs involving local high schools and universities, and support for current employees to gain additional training (see Table NHS-3). Human-resources offices are attempting to solve workforce problems through a number of independent actions. Many agencies now have direct-hire authorities and can offer significant signing bonuses in special cases. A recent Government Accountability Office report indicates these multi-approach programs are a major reason that DOD laboratories currently do not have significant problems locating the necessary people to fill critical-skills positions.[25]

DOD has proposed, as part of the department's 2006 appropriations,[26] to create and fund NDEA 2005 (see Figure NHS-6). This program would extend a 2004 pilot SMART program and, as with the original NDEA, would provide scholarships and fellowships to students in critical fields of science, mathematics, engineering, and foreign languages. It would expand

---

[24]National Research Council, 2002.

[25]Government Accountability Office. *National Nuclear Security Administration: Contractors' Strategies to Recruit and Retain and Critically Skilled Workforce Are Generally Effective.* GAO-05-164. Washington, DC: Government Accountability Office, 2005.

[26]See H.R. 1815, National Defense Authorization Act for Fiscal Year 2006 § Sec. 1105. Science, Mathematics, and Research for Transformation (SMART) Defense Education Program—National Defense Education Act (NDEA), Phase I. Introduced to the House on April 26, 2005; on June 6, 2005, referred to Senate committee; status as of July 26, 2005: received in the Senate and read twice and referred to the Committee on Armed Services.

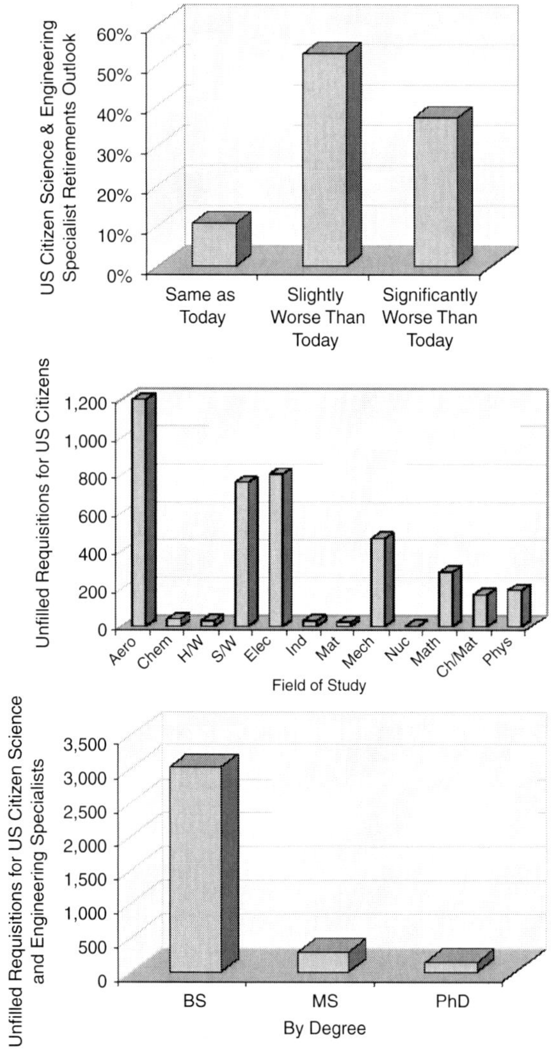

FIGURE NHS-5 Unfilled requisitions for US citizen scientists and engineers, by degree and field.
NOTE: Aero = Aerospace, Chem = Chemical, H/W = Hardware, S/W = Software, Elec = Electrical, Ind = Industrial, Mat = Materials, Mech = Mechanical, Nuc = Nuclear, Ch/Mat = Chemical/Materials, Phys = Physical.
SOURCE: E. Swallow. Chair, National Defense Industry Association Space Division and Chair, Industry Study on Critical Workforce Issues. Presentation at the National Defense Industry Association meeting, April 2005. Available at: http://proceedings.ndia.org/4340/swallow.pdf.

TABLE NHS-2 Percentage of Federal S&E Workforce Over the Age of 45, 1999-2002

	1999 (percent)	2000 (percent)	2001 (percent)	2002 (percent)
Total S&Es	44.2	43.5	43.1	43.4
All sci	26.1	25.4	25.6	26.9
Comp/Math	45.5	43.9	44.0	45.3
Life sci	11.4	11.2	11.0	10.9
Physical sci	26.7	26.2	26.1	26.2
Social sci	20.4	20.4	19.7	19.6
All eng	66.7	66.4	66.2	66.7
Aerospace	44.7	43.6	43.0	42.8
Chemical	62.3	63.6	65.7	67.6
Civil	61.8	61.3	60.6	60.1
EE&Comp	79.3	79.1	78.5	79.1
Industrial	81.1	80.2	79.4	79.4
Mechancial	88.2	88.2	88.4	89.2
Other eng	54.6	55.1	55.5	55.9

SOURCE: Based on National Science Foundation. *Federal Scientists and Engineers: 1998-2002*. NSF 05-304. Arlington, VA: National Science Foundation, 2005. Table 11.

on the original act in providing scholarships to undergraduates, including those pursuing associate degrees. The program would cover tuition, room and board, internships, tutors, and travel for all students. DOD requires a service commitment on completion of studies.

DOD has requested $10.3 million in its FY 2006 budget request for this program. SMART was initiated in 2005 as a pilot program and funded at $2.5 million. The program has generated considerable interest among students: SMART currently funds 25 students, but DOD vetted over 600 applications.[27]

Possible actions include:

- Create a new NDEA for the 21st century to promote the education and training of students in science, technology, engineering, mathematics, and foreign languages. The new NDEA would include portable graduate fellowships, institutional traineeships, incentives to create professional S&E

[27]J. Brainard. "Defense Department Hopes to Revive Sputnik-Era Science-Education Programs" *The Chronicle of Higher Education* 51(36)(2005):A18. Available at: http://chronicle.com/prm/weekly/v51/i36/36a01802.htm.

TABLE NHS-3 National Weapon Laboratories Personnel Recruitment Programs

	Program	Sponsor
Pre-College (K–12)	Materials World Modules	Army
	STARBASE	OSD-RA
	eCybermission	Army
Undergraduate	Awards to Stimulate and Support Undergraduate Research Education (ASSURE)	AFOSR with NSF
	Research Assistantships in Microelectronics	DARPA with Semiconductor Industries Association
	Science, Mathematics, and Research for Transformation (SMART)	AFOSR
Graduate	National Defense Science and Engineering Graduate Fellowships	NDSEG
	Naval Research—S&T for Americas Readiness (N-STAR)	Navy with NSF
	SMART	AFOSR

NOTE: OSD-RA = Office of the Assistant Secretary of Defense for Reserve Affairs, AFOSR = Air Force Office of Scientific Research, DARPA = Defense Advanced Research Projects Agency.
SOURCE: B. Berry, Acting Deputy Undersecretary for Laboratories and Basic Science. "STEM Education Act" presentation at STARBASE Directors' Conference, April 7, 2005. Available at: http://www.starbasedod.com/resources/SME%20Briefing-STARBASE%20Directors%20Conf%204-7-05v5%20wo%20Backup.ppt.

master's programs, undergraduate loan forgiveness, grants to support innovative undergraduate curricula, grants to expand K–12 education outreach, summer training and research opportunities for K–12 teachers, employer S&E and foreign-language educational tax breaks, national-laboratory and federal service professional incentives, and additional funds for program evaluation.[28]

[28]National Research Council, 2002; R. M. Sega, Director of Defense Research and Engineering, DOD. Testimony Before the Subcommittee on Emerging Threats and Capabilities of the Senate Armed Services Committee, March 9, 2005. Available at: http://armed-services.senate.gov/statemnt/2005/March/Sega%2003-09-05.pdf; Association of American Universities, 2005. White Paper.

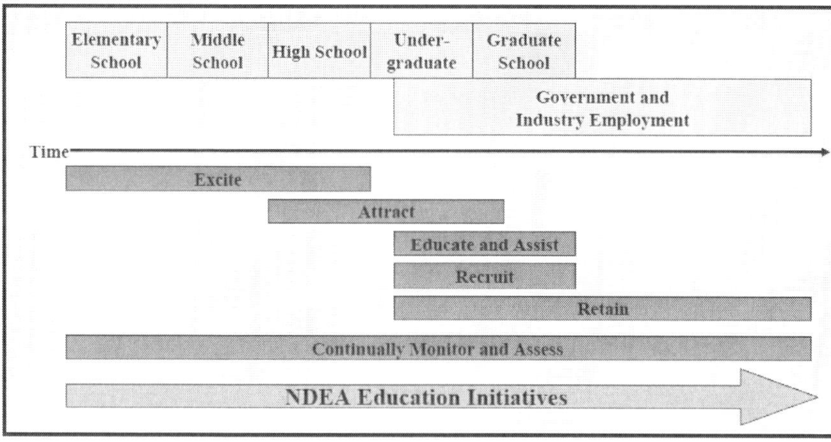

FIGURE NHS-6 DOD strategy for proposed National Defense Education Act (NDEA) within its current portfolio of workforce programs.
SOURCE: E. Swallow. Chair, National Defense Industry Association Space Division and Chair, Industry Study on Critical Workforce Issues. Presentation at the National Defense Industry Association meeting, April 2005. Available at: http://proceedings.ndia.org/4340/swallow.pdf.

Appendix E

Estimated Recommendation Cost Tables

The following tables provide each implementation action and a "back of the envelope" estimate of the incremental annual cost to the federal government for each. These cost estimates are to illustrate the size of the program the committee is proposing and are not definitive. To obtain a more definitive estimate, economic modeling would be necessary.

Since the prepublication version of the report was released on October 12, 2005, policy-makers have requested a 10-year annual cost estimate for the committee's recommendations. That estimate is what is presented here. In developing this estimate, the committee collected more detailed information than it had at that time to develop these cost estimates. In addition, we have attempted to provide sufficient information so that the cost estimates could be re-created.

In many cases, programs are phased in over a number of years. In addition, some of the scholarship and fellowship programs take 3-4 years to reach steady state.

The aggregate cost estimate of all the recommendations remain in the same range as the committee developed in October 2005. The cost estimate at that time was \$9.2 to \$23.8 billion. The current aggregate cost estimate is \$8.6 to \$19.8 billion.

Gathering Storm Annual Cost Estimate, by Recommendation, 2007-2016 (millions of dollars)

Recommendation	2007	2008	2009	2010	2011
K–12	376	747	1219	1689	1976
Research	1670	2790	4010	5430	6850
Higher Education	1100	1700	2300	2675	2675
Innovation	5423	5423	5423	5423	5423
TOTAL	**8569**	**10660**	**12952**	**15217**	**16924**

Gathering Storm Annual Cost Estimate, by Action Item, 2007-2016 (millions of dollars)

Action Item	**Recommendation**	**2007**	**2008**	**2009**	**2010**
A-1	Recruit 10,000 Teachers	110	295	550	815
A-2a	Summer Institutes	40	80	120	120
A-2b	Master's Program	46	92	158	224
A-2c	AP/IB Teacher Incentives	100	179	265	363
A-2d	Science and Math Curriculum	20	20	20	20
A-3	AP/IB Student Incentives	60	81	106	147
Total (K–12)		*376*	*747*	*1219*	*1689*
B-1	Long-Term Basic Research	800	1700	2700	3900
B-2	Early-Career Research Award	20	40	60	80
B-3	Research Infrastructure	500	500	500	500
B-4	High-Risk Research	0	0	0	0
B-5	ARPA-Energy	300	500	700	900
B-6	Presidential Innovation Award	50	50	50	50
Total (Research)		*1670*	*2790*	*4010*	*5430*
C-1	Undergraduate Scholarships	375	750	1125	1500
C-2	Graduate Portable Fellowships	225	450	675	675
C-3	Continuing Educ. Tax Credit	500	500	500	500
C-4	Expedite Visa Processing	0	0	0	0
C-5	Intl Student Work Permits	0	0	0	0
C-6	Skills-Based Immigration Policy	0	0	0	0
Total (Higher Education)		*1100*	*1700*	*2300*	*2675*
D-1	Intellectual Property Protection	323	323	323	323
D-2	R&D Tax Credit	5100	5100	5100	5100
D-3	Innovation Tax Incentives Study	0	0	0	0
D-4	Ubiquitous Broadband Internet	0	0	0	0
Total (Incentives for Innovation)		*5423*	*5423*	*5423*	*5423*
		2007	**2008**	**2009**	**2010**
Grand Total (Billions of Dollars)		8.6	10.7	13	15.2

5-Year Total	2012	2013	2014	2015	2016	% of Total in 2016
6007	2071	2091	2091	2091	2091	**10.5**
20750	8250	9650	9650	9650	9650	**48.6**
10450	2675	2675	2675	2675	2675	**13.5**
27115	5423	5423	5423	5423	5423	**27.3**
64322	**18419**	**19839**	**19839**	**19839**	**19839**	**99.9**

2011	**5-Year Total**	**2012**	**2013**	**2014**	**2015**	**2016**
965	2735	1040	1040	1040	1040	1040
120	480	120	120	120	120	120
290	810	310	330	330	330	330
400	1307	400	400	400	400	400
20	100	20	20	20	20	20
181	575	181	181	181	181	181
1976	6007	*2071*	*2091*	*2091*	*2091*	*2091*
5200	14300	6600	8000	8000	8000	8000
100	300	100	100	100	100	100
500	2500	500	500	500	500	500
0	0	0	0	0	0	0
1000	3400	1000	1000	1000	1000	1000
50	250	50	50	50	50	50
6850	20750	*8250*	*9650*	*9650*	*9650*	*9650*
1500	5250	1500	1500	1500	1500	1500
675	2700	675	675	675	675	675
500	2500	500	500	500	500	500
0	0	0	0	0	0	0
0	0	0	0	0	0	0
0	0	0	0	0	0	0
2675	10450	*2675*	*2675*	*2675*	*2675*	*2675*
323	1615	323	323	323	323	323
5100	25500	5100	5100	5100	5100	5100
0	0	0	0	0	0	0
0	0	0	0	0	0	0
5423	27115	*5423*	*5423*	*5423*	*5423*	*5423*
2011	10045	**2012**	**2013**	**2014**	**2015**	**2016**
16.9	64.4	18.4	19.8	19.8	19.8	19.8

Action A-1 (Recruit 10,000 Teachers) Detailed Analysis

Year	2007	2008	2009	2010
Millions of Dollars				
Scholarships	75	225	450	675
Bonuses	0	0	0	40
Institutional Awards	35	70	100	100
Total	110	295	550	815
Scholarship if $15,000 per Year				
Number of Old Students	5000	5000	15000	30000
Number of New Students		10000	15000	15000
Scholarship $	15000	15000	15000	15000
Total $M	75	225	450	675
Bonuses				
Number of Teachers	0	0	0	4000
Bonus $	0	0	0	10000
Total $M	0	0	0	40
Institutional Matching Grants				
Number of Institutions	35	70	100	100
Grant $M	1	1	1	1
Total $M	35	70	100	100

Action A-2a (Summer Institutes) Detailed Analysis

Year	2007	2008	2009	2010
Number of Teachers	16500	33500	50000	50000
Cost per Teacher ($)	2400	2400	2400	2400
Total ($M)	**40**	**80**	**120**	**120**

Action A-2b (Master's Program) Detailed Analysis

Year		2007	2008	2009
Number of Programs	100	200	300	400
Students per Program	20	20	20	20
Number of Students/Program	2000	4000	6000	8000
Cost per Student	23000	23000	23000	23000
Cost per Program ($M)	46	92	138	184
Number of Graduates			2000	4000
Cost per Bonus			10000	10000
Bonus Cost ($M)			20	40
Total ($M)	46	92	158	224

	Breakdown of Program Cost ($) (100 students over a 5-year period)
Course Development	100,000
Administrative Staff Cost	500,000
Tuition	1,500,000
Transportation Reimbursement	150,000
Equipment	50,000
Total $/Student	23,000

NOTE: Program cost estimate based on email from Hai-Lung Dai of the University of Pennsylvania to Deborah Stine dated December 20, 2005.

2011	2012	2013	Steady State
825	900	900	900
40	40	40	40
100	100	100	100
965	1040	1040	1040
45000	55000	60000	60000
10000	5000		
15000	15000	15000	15000
825	900	900	900
4000	4000	4000	4000
10000	10000	10000	10000
40	40	40	40
100	100	100	100
1	1	1	1
100	100	100	100

2011	2012	2013	Steady State
50000	50000	50000	50000
2400	2400	2400	2400
120	**120**	**120**	**120**

2010	2011	2012	2013	Steady State
500	500	500	500	
20	20	20	20	
10000	10000	10000	10000	
23000	23000	23000	23000	
230	230	230	230	
6000	8000	10000	10000	
10000	10000	10000	10000	
60	80	100	100	
290	310	330	330	

Action A-2c (AP/IB Teacher Incentive Program) Detailed Analysis

Year	2007	2008	2009
Number of Math and Science AP/IB Teachers	14000	28000	42000
Number Laying the Foundation (LTF) (Pre-AP/IB) Teachers	16000	32000	48000
Training Fee per Teacher	800	800	800
Total Training Cost ($M)	24	48	72
Stipend Fee/AP/IB Teacher	1800	1800	1800
Total Stipends for AP/IB Teachers ($M)	25	50	76
Stipend Fee/LTF Teacher	1000	1000	1000
Total Stipends for LTF Teachers ($M)	16	32	48
Estimated Number of Passing AP/IB Scores	313500	385000	495000
Teacher Bonus for Passing AP/IB Scores	100	100	100
Total AP/IB Teacher Bonuses ($M)	31	39	50
Estimated Number of Passing LTF Scores	160000	384000	768000
Teacher Bonus for Passing LTF Scores	25	25	25
Total LTF Teacher Bonuses ($M)	4	10	19
Total ($M)	100	179	265

Action A-2d (Curriculum Standards)
Cost: $100 million over 5 years

Action A-3 (AP/IB Student Incentive Program) Detailed Analysis

Year	2007	2008
Number of Math and Science AP/IB Students	570000	700000
Exam Fee (50% of $82 fee)	41	41
Total AP-IB Exam Fee Cost	23	29
Estimated Number of Passing AP/IB Scores	313500	385000
Student Mini-Scholarship for Passing AP/IB Scores	100	100
Total Mini-Scholarships	31	39
Number of Pre-AP/IB Students Taking Test	640000	1280000
Exam Fee (50% of $20 fee)	10	10
Total Pre-AP/IB Exam Fee Cost	6	13
Total ($M)	60	81

2010	2011	2012	2013	Steady State
56000	56000	56000	56000	56000
64000	64000	64000	64000	64000
800	800	800	800	800
96	96	96	96	96
1800	1800	1800	1800	1800
101	101	101	101	101
1000	1000	1000	1000	1000
64	64	64	64	64
700000	870000	870000	870000	870000
100	100	100	100	100
70	87	87	87	87
1280000	2080000	2080000	2080000	2080000
25	25	25	25	25
32	52	52	52	52
363	400	400	400	400

2009	2010	2011	2012	Steady State
900000	1250000	1500000	1500000	1500000
41	41	41	41	41
37	51	62	62	62
495000	700000	870000	870000	870000
100	100	100	100	100
50	70	87	87	87
1920000	2560000	3200000	3200000	3200000
10	10	10	10	10
19	26	32	32	32
106	147	181	181	181

Action B-1 (Research Funding) Detailed Analysis: The base for this cost estimate was determined using information from the following National Science Foundation (NSF) publication: Federal Funds for Research and Development: Fiscal Years 2002, 2003, and 2004 (Publication No: NSF 05-307), which identifies the amount of basic research funding, by agency, in the physical sciences, mathematics, computer sciences, and engineering. (FY 2004 is the most recent year this information is available.) This is summarized in the table below:

Preliminary federal obligations for basic research in the physical sciences, mathematics, computer sciences, and engineering, by agency, FY 2004[a]

AGENCY	Funding (Billions)	Percent of Total
All agencies	$7.4	
Department of Defense	$1.1	15%
Department of Energy	$2.1	28%
NASA	$1.5	21%
NSF	$1.9	26%

[a]Other agencies which fund these areas are Department of Commerce, Department of Health and Human Services (HHS), US Department of Agriculture, Department of Homeland Security, Department of Interior, Veteran's Administration, and Environmental Protection Agency. HHS is largest at 6%; remainder at 1% or less for a total of 10%.
SOURCE: National Science Foundation. *Federal Funds for Research and Development: Fiscal Years 2002, 2003, and 2004*. NSF 05-307. Arlington, VA: National Science Foundation, 2005.

The committee also proposed that all of DOD basic research funding be increased. The following number provides the amount of DOD basic research funding included and not included in the table above (i.e., not in the research fields identified):

DOD Basic Research Funding (including above): $1.6 billion
(not including above): $0.5 billion

This provides a total base: $7.9 billion ($7.4B + $0.5B), which is that increased at a rate of 10% per year over the next 7 years as indicated in the recommendation.

Year	2007	2008	2009	2010	2011	2012	2013
Millions of Dollars	800	1700	2700	3900	5200	6600	8000

Action B-2 (Early-Career Research Grants) Detailed Analysis: As with the scholarship programs, this estimate gradually increases over time as the $500,000 research grants are payable over 5 years.

Year	2007	2008	2009	2010	Steady State
#Previous Researchers		200	400	600	800
#New Researchers	200	200	200	200	200
Research $	100000	100000	100000	100000	100000
Total $M	20	40	60	80	100

Action B-3 (Advanced Research Instrumentation and Facilities) Detailed Analysis: This proposed action indicates that the funding will be $500 million per year for the next 5 years.

Year	2007	2008	2009	2010	2011
Millions of Dollars	500	500	500	500	500

Action B-4 (High-Risk Research) Detailed Analysis: No new funds required for this action—just a specification that 8% of existing funds will be directed toward high-risk research.

Action B-5 (ARPA-E) Detailed Analysis: This recommendation indicates that ARPA-E will be funded at $300 million per year, gradually increasing over 5-6 years to $1 billion per year at which time it would be evaluated.

Year	2007	2008	2009	2010	2011	2012	2013	2014	2015	2016
Millions of Dollars	300	500	700	900	1000	1000	1000	1000	1000	1000

Action B-6 (Presidential Innovation Award) Detailed Analysis: This program for a Presidential Innovation award would not require a monetary award, but one may be provided as indicated here.

Year	2007	2008	2009	2010	2011	2012	2013	2014	2015	2016
Millions of Dollars	50	50	50	50	50	50	50	50	50	50

Action C-1 (Undergraduate Scholarships) Detailed Analysis: The committee recommends that up to $20,000/year be provided for scholarships. This analysis focuses on the mid-range scholarship of $15,000/year. This allows the amount of the scholarship to vary relative to an institution's tuition. Because scholarship programs provide funding for a number of years, this program will take several years to reach steady state (when the number of new students entering is equal to the number of old students graduating and no longer receiving a scholarship).

Year	2007	2008	2009	2010	Steady State
#Old Students		25000	50000	75000	100000
#New Students	25000	25000	25000	25000	
Scholarship $	15000	15000	15000	15000	15000
Total $M	375	750	1125	1500	1500

Action C-2 (Graduate Fellowships) Detailed Analysis: As with the undergraduate scholarships, the committee recommends that up to $20,000/year be provided for fellowships. This analysis focuses on the mid-range fellowship of $15,000/year. This allows the amount of the fellowship to vary relative to an institution's tuition. Because fellowship programs provide funding for a number of years, this program will take several years to reach steady state. This program also provides a stipend for the graduate student at $30,000/year.

Year	2007	2008	2009	Steady State
#Old Students		5000	10000	15000
#New students	5000	5000	5000	
Fellowship $	15000	15000	15000	15000
Stipend $	30000	30000	30000	30000
Total $M	225	450	675	675

Action C-3 (Continuing Education) Detailed Analysis: Based on a similar existing credit, the budget for this action is assumed to be $500 million per year. Education tax credits are available to individual families to help defray the cost of college. According to the College Board, the federal government provided $81.5 billion in student aid during 2003-2004. This can be roughly disaggregated as follows: 70% loans, 21% grants, 8% tax benefits. The tax benefits (Hope and Lifetime Learning Credits) amounted to approximately $6.3 billion. The $500 million the committee proposes is a modest amount in this context designed to spur corporate sponsorship of continuing education. The cap on tax credits could be increased over time or even lifted altogether. The following Web page provides an overview of existing tax credits: http://www.ed.gov/offices/OPE/PPI/HOPE/index.html.

Actions C-4 to C-7 (International Students and Scholars) Detailed Analysis: These are assumed to have no major cost as they involve only a policy change.

Action D-1 (Enhance Intellectual Property Protection) Detailed Analysis: The base amount for this estimate is the FY 2006 request for the US Patent and Trademark Office which was $1.7 billion, a 10% increase from FY 2005. The proposed $340 million per year, reflects a 20% increase in FY 2007 in order to institute a post-grant review system and enhance other capabilities aimed at increasing the quality and timeliness of patent examinations. The other proposed changes do not require funding.

Year	2007	2008	2009	2010	2011	2012	2013	2014
Millions of Dollars	340	340	340	340	340	340	340	340

Action D-2 (Strengthen the R&D Tax Credit) Detailed Analysis: Tax credits vary from year to year. In FY 2005 the R&D tax credit was forecast to cost $5.1 billion. Since the amount of the R&D tax credit in any given year depends on overall corporate profitability and tax liability, as well as on R&D spending decisions, the overall cost cannot be determined precisely in advance. Maintaining the existing R&D tax credit, therefore, is assumed to result in no incremental cost. The proposed doubling of the existing R&D tax credit is assumed to cost about the same as the existing R&D tax credit, but the actual increment could be larger or smaller. In addition, the proposed expansion of the number of companies eligible for the R&D tax credit is not reflected in the $5.1 billion figure as the number of companies who might potentially be involved is unknown.

Year	2007	2008	2009	2010	2011	2012	2013	2014
Millions of Dollars	5100	5100	5100	5100	5100	5100	5100	5100

Action D-3 (Innovation Tax Incentive Study) Detailed Analysis: The proposed studies should be able to be performed within existing budgets.

Action D-4 (Ubiquitous Broadband Internet) Detailed Analysis: The most important steps would not necessarily entail federal outlays.

Appendix F

K–12 Education Recommendations Supplementary Information

JUSTIFICATION FOR NUMBERS OF TEACHERS AND STUDENTS IN THE AP/IB AND PRE-AP/IB PROGRAMS RECOMMENDED IN *ACTION A-2*

Students

The goal is to have 1,500,000 high school students taking at least one Advanced Placement (AP) or International Baccalaureate (IB) mathematics or science exam by 2010, an increase to 23% from 6.5% of US high school juniors and seniors who took at least one AP math or science exam in 2004, with 700,000 passing the exam[1] (see Exhibit 1). AP/IB classes must be open to all students.

---

[1]AP passing score is 3-5; note that some colleges do not allow credit for AP coursework unless a score of 5 is achieved. IB scores on a 7-point scale, and 5 or higher is considered passing.

Exhibit 1 US Public School Enrollment and AP Participation

	Projected 2004[a]	Projected 2010[b]
Total Grade 9–12 Enrollment	14,700,000	14,600,000
Total Grade 11–12 Enrollment		6,500,000
	Actual 2004[c]	Projected 2010
Number of High School Jr./Sr. Taking at Least One AP Mathematics or Science Exam	380,000	1,500,000
Percent of Jr./Sr. Taking at Least One AP Mathematics or Science Exam	6.5%	23%
AP Mathematics or Science Teachers	33,000	100,000
Students per AP Teacher	11.5	15

[a]The College Board.
[b]Statistical Abstract of the United States: 2004-2005. Table 202.
[c]The College Board.

The proposed AP incentive program (APIP) has increased the number of students taking AP exams. To measure AP participation in a school, district, state, or nation, we calculated the number of students taking AP exams per 1,000 juniors and seniors. In 2005, the number of students taking AP exams in all math, science, or English in the Dallas 10 districts was 2.3 times that of the national level (see Exhibit 2).

Exhibit 2 Students Taking AP Math, Science, and English Exams per 1,000[2] Juniors and Seniors Enrolled

Dallas 10 APIP Schools	245 students
Texas Public Schools	131 students
US Public Schools	105 students

Teachers—AP/IB

The AP and pre-AP programs as proposed would provide professional development for 150,000 teachers now in the classroom to teach rigorous math and science courses in middle and high schools. Of these, 70,000 will

[2]"Per 1,000" is calculated on the best enrollment data available at the time.

teach Advanced Placement or International Baccalaureate courses in mathematics and science.[3] In addition, 80,000 teachers in grades 6–11 who are now in the classroom will receive training, teachers guides, and assessments instruments, such as those available in the Laying the Foundation program, to prepare them to teach pre-AP mathematics and science courses that lead up to AP or IB courses. The proposed professional development program for AP/IB teachers is 7 days a year for 4 years; for Laying the Foundation teachers it is 8 days a year for 4 years.

Assuming 10% attrition among the current 33,000 AP mathematics and science teachers and by training an additional 70,000 teachers, public high schools would have an estimated 100,000 mathematics and science teachers capable of teaching AP or IB courses in place by 2010. This number is based on a realistic goal with the capacity to provide quality professional training for teachers on a large scale. As they become more productive and confident as teachers, they will recruit more students into demanding mathematics and science courses. We then realistically can expect steady increases in the numbers of junior and senior students who will take AP/IB mathematics and science exams to 1.5 million students by 2010, with increases well beyond 2010.

Teachers—Pre-AP/IB

This proposal will provide pre-AP math and science training in content and pedagogy for 80,000 teachers who are currently in grades 6–11 classrooms. The 4-year training program includes 8 days of training each year for 4 years and the classroom materials (vertically aligned curriculum, lesson plans, laboratory exercises, and diagnostics) needed to teach the more demanding math and science courses. By 2010, these teachers will help an estimated 5 million students each year develop critical thinking and problem-solving skills in order to enlarge the AP pipeline in math and science. This represents an estimated 20% of US students who will be enrolled in grades 6–11 in 2010 (see Exhibit 3).

[3]Including AP calculus, computer science, statistics, biology, chemistry, physics, and environmental science.

Exhibit 3 K–12 Students, Teachers, and Salaries[a]

	# Students	# Teachers	Average Salary	# Science and Math Teachers
K–5	29,627,634	1,781,900	$46,408	
6–8				350,702[b] (191K in science, 160K in mathematics)
9–12	18,504,864	1,264,723	$47,120	
High School Grads (2003-2004)	2,771,781			
Total (Fall 2003)	48,132,518	3,046,623	$46,752	(1,700,000)[c]

[a]Unless otherwise noted, figures, excerpts, and charts are for the 2003-2004 school year, as reported by National Education Association. *Rankings and Estimates*. Atlanta, GA: NEA Research, 2005. Available at: http://www.nea.org/edstats/images/05rankings.pdf.
[b]For the 1999-2000 school year.
[c]From Glenn Commission report, 2000. Includes ALL primary school teachers, as well as specialty teachers in middle and upper grades.
NOTE: In 2003, there were 15,397 US school districts, and the average amount spent per K–12 student from all revenue sources was $8,248.

Appendix G

Bibliography

Adelman, C. 1998. Women and Men on the Engineering Path: A Model for Analysis of Undergraduate Careers. Washington, DC: US Department of Education. Available at http://www.nae.edu/nae/diversitycom.nsf/98b72da8aad70f1785256da20053deaf/85256cfb00484b5c85256da000002f83/$FILE/Adelman_Women_and_Men_of_the_Engineering_Path.pdf.

Allen, M. 2003. Eight Questions on Teacher Preparation: What Does the Research Say? Washington, DC: Education Commission of the States. Available at: http://www.ecs.org/tpreport/.

Alliance for Science and Technology Research in America. 2004. Basic Research: Investing in America's Innovation Future, a presentation for the House Republican High-Tech Working Group, March 31.

Alphonso, C. 2005. Facing security hurdles, top students flock to Canada. The Globe and Mail, February 22.

American Association for the Advancement of Science. 2004. Trends in Federal Research by Discipline, FY 1976-2004. October. Available at: http://www.aaas.org/spp/rd/disc04tb.pdf and http://www.aaas.org/spp/rd/discip04c.pdf.

American Association for the Advancement of Science. 2005. Historical Data on Federal R&D, FY 1976-2006. March 22. Available at: http://www.aaas.org/spp/rd/hist06p2.pdf.

American Electronics Association. 2004. Losing the Competitive Advantage? The Challenge for Science and Technology in the United States. Washington, DC: American Electronics Association.

American Electronics Association. 2004. Offshore Outsourcing in an Increasingly Competitive and Rapidly Changing World: A High-Tech Perspective. Washington, DC: American Electronics Association, March.

Andrews, E. L. 2005. The doctrine was not to have one; Greenspan will leave no road map to his successor. New York Times, August 26, p. C1.

Arndt, M. 2005. No longer the lab of the world: U.S. chemical plants are closing in droves as production heads abroad. BusinessWeek, May 2. Available at: http://www.businessweek.com/magazine/content/05_18/b3931106.htm and http://www.usnews.com/usnews/biztech/articles/051010/10energy.htm.

Aslanbeigui, N., and V. Montecinos. 1998. Foreign students in US doctoral Programs. Journal of Economic Perspectives 12:171-182.

Association of American Universities. 2005. A National Defense Education Act for the 21st Century. Renewing Our Commitment to US Students, Science, Scholarship, and Society (White paper). Washington, DC: AAU. Available at: http://www.aau.edu/education/NDEAOP.pdf.

Association of American Universities, Committee on Graduate Education. 1998. Graduate Education. Washington, DC: AAU.

Athreye, S. S. 2003. The Indian Software Industry. Carnegie Mellon Software Industry Center Working Paper 03-04. Pittsburgh, PA: Carnegie Mellon University, October.

Atkinson, R. 2004. Meeting the Offshoring Challenge. Washington, DC: Progressive Policy Institute.

Atkinson, R. D. 2004. The Past and Future of America's Economy: Long Waves of Innovation That Power Cycles of Growth. Northampton, MA: E. Elgar.

Attewell, P. 2001. The winner take-all high school: Organizational adaptations to educational stratification. Sociology of Education 74(4):267-296.

Attiyeh, G., and R. Attiyeh. 1997. Testing for bias in graduate school admissions. Journal of Human Resources 32:524-548.

Auriol, L. Why do we need indicators on careers of doctorate holders? Workshop on User Needs for Indicators on Careers of Doctorate Holders. Paris: OECD, September 27. Available at: http://www.olis.oecd.org/olis/2004doc.nsf.

Austin, C., L. Brady, T. Insel, and F. Collins. 2004. NIH molecular libraries initiative. Science 306:1138-1139.

Autor, D., L. Katz, and M. Kearney. 2005. Trends in U.S. Wage Inequality: Re-Assessing the Revisionists. Working Paper 11627. Cambridge, MA: National Bureau of Economic Research.

Ayers, W. M. 2002. MIT: The Impact of Innovation. Boston, MA: Bank Boston. Available at: http://web.mit.edu/newsoffice/founders/Founders2.pdf.

Babco, E. 2002. Trends in African American and Native American Participation in STEM Higher Education. Washington, DC: Commission on Professionals in Science and Technology.

Bar Shalom, A., and R. Cook-Deegan. 2002. Patents and innovation in cancer therapeutics: Lessons from CellPro. The Milbank Quarterly 80(December):iii-iv, 637-676.

Bardhan, A., and C. Kroll. 2003. The New Wave of Outsourcing. Fisher Center Research Reports #1103. Berkeley, CA: Fisher Center for Real Estate and Urban Economics, November 2.

Bauer, P. W. 1999. Are We in a Productivity Boom? Evidence from Multifactor Productivity Growth. Cleveland, OH: Federal Reserve Bank of Cleveland, October 15. Available at: www.clevelandfed.org/research/Com99/1015.pdf, Table 1.

Berkner, L. K., S. Cuccaro-Alamink, and A. C. McCormick. 1996. Descriptive Summary of 1989-90 Beginning Postsecondary Students: 5 Years Later with an Essay on Postsecondary Persistence and Attainment. NCES 96155. Washington, DC: National Center for Education Statistics.

Berliner, D. C., and B. J. Biddle. 1995. The Manufactured Crisis: Myths, Fraud, and the Attack on America's Public Schools. New York: Addison-Wesley.

Bhagwati, J., A. Panagariya, and T. N. Srinivasan. 2004. The muddles over outsourcing. Journal of Economic Perspectives 18(Summer):93-114.

Blasie, C., and G. Palladino. 2005. Implementing the professional development standards: A research department's innovative masters degree program for high school chemistry teachers. Journal of Chemical Education 82(4):567-570.

Bogumil, J. 2003. The Brain Drain in an Era of Liberalism. Ottawa, ON: Canadian Bureau for International Education.
Bonvillian, W. B. 2004. Meeting the new challenge to US economic competitiveness. Issues in Science and Technology 21(1)(Fall):75-82.
Boskin, M. J., and L. J. Lau. 1992. Capital, Technology, and Economic Growth. In Technology and the Wealth of Nations, N. Rosenberg, R. Landau, and D. C. Mowery, eds. Stanford, CA: Stanford University Press.
Boylan, M. 2004. Assessing Changes in Student Interest in Engineering Careers Over the Last Decade. CASEE, National Academy of Engineering. Available at: http://www.nae.edu/NAE/caseecomnew.nsf/weblinks/NFOY-6GHJ7B/$file/Engineering%20Interest%20-%20HS%20through%20College_V21.pdf.
Braxton, J. M. 2002. Reworking the Student Departure Puzzle. Nashville, TN: Vanderbilt University Press.
Brown, H. 2004. Council of Graduate Schools Finds Declines in New International Graduate Student Enrollment for Third Consecutive Year. Washington, DC: Council of Graduate Schools, November 4.
Bubnoff, A. von. 2005. Asia squeezes Europe's lead in science. Nature 436(7049)(July 21):314.
Budget of the United States Government, Fiscal Year 2006. 2005. Analytical Perspectives. Washington, DC: US Government Printing Office. Available at: http://a255.g.akamaitech.net/7/255/2422/07feb20051415/www.gpoaccess.gov/usbudget/fy06/pdf/spec.pdf.
Building Engineering and Science Talent. 2004. The Talent Imperative. San Diego, CA: BEST.
Building Engineering and Science Talent. 2004. A Bridge for All: Higher Education Design Principles in Science, Technology, Engineering and Mathematics. San Diego, CA: BEST. Available at: http://www.bestworkforce.com.
Bureau of Industry and Security. 2004. Deemed Export Controls May Not Stop the Transfer of Sensitive Technology to Foreign Nationals in the U.S. Final Inspection Report No. IPE-16176. Washington, DC: Office of Inspections and Program Evaluations.
Bush, V. 1945. Science: The Endless Frontier. Washington, DC: US Government Printing Office.
Bush pushes ubiquitous broadband by 2007. Reuters, March 26, 2004.
Business Roundtable. 2005. Tapping America's Potential. Washington, DC: Business Roundtable.
Business Roundtable. 2006. Innovation and U.S. Competitiveness: Addressing the Talent Gap. Public Opinion Research. Washington, DC: Business Roundtable, January 12. Available at: http://www.businessroundtable.org/pdf/20060112Two-pager.pdf.
Business-Higher Education Forum. 2005. A Commitment to America's Future: Responding to the Crisis in Mathematics and Science Education. Washington, DC: American Council on Education.
Busquin, P. Investing in people. Science 303:145.
Butts, S., and R. Killoran. 2003. Industry-University Research in Our Times: A White Paper. Available at: http://www7.nationalacademies.org/guirr/IP_background.html.
Bybee, R. W., and E. Stage. 2005. No country left behind. Issues in Science and Technology. (Winter):69-75.
Calabrese, M. 2005. Testimony to the Committee on Commerce, Science and Transportation, United States Senate, Hearing on Broadcast to Broadband: Completing the Digital Television Transition Can Jumpstart Affordable Wireless Broadband, July 12.
Callan, B., S. Costigan, and K. Keller. 1997. Exporting U.S. High Tech: Facts and Fiction About the Globalization of Industrial R&D. New York: Council on Foreign Relations.
Center for Strategic and International Studies. 1996. Global Innovation/National Competitiveness. Washington, DC: CSIS.

Center for Strategic and International Studies. 2001. Technology and Security in the 21st Century: US Military Export Control Reform. Washington, DC: CSIS.
Center for Strategic and International Studies. 2005. Security Controls on Scientific Information and the Conduct of Scientific Research. Washington, DC: CSIS, June.
Center for Strategic and International Studies. 2005. Technology Futures and Global Power, Wealth and Conflict. Washington, DC: CSIS, May, p. viii.
Center for Sustainable Energy Systems. 2005. US Energy System Factsheet. Ann Arbor, MI: University of Michigan, August.
Centers for Medicare and Medicaid Services. 2005. National Heath Expenditures. Available at: http://www.cms.hhs.gov/NationalHealthExpendData/downloads/tables.pdf.
Central Intelligence Agency. 2001. Long-Term Global Demographic Trends: Reshaping the Geopolitical Landscape. Langley, VA: CIA, July, p. 25.
Chelleraj, G., K. E. Maskus, and A. Mattoo. 2004. The Contribution of Skilled Immigration and International Graduate Students to US Innovation. Working Paper 04-10. Boulder, CO: University of Colorado.
Clinton, W. J. 1997. Commencement Address at Morgan State University in Baltimore, Maryland. Public Papers of the Presidents of the United States, Books I and II. Washington, DC: Government Printing Office, May 18. Available at: http://www.gpoaccess.gov/pubpapers/wjclinton.html.
Clotfeltner, C. T., R. G. Ehrenberg, M. Getz, and J. J. Siegfried. 1991. Economic Challenges in Higher Education. Chicago, IL: University of Chicago Press.
Cochran-Smith, M., and K. M. Zeichner. 2005. Studying Teacher Education. Washington, DC: American Educational Research Association.
Cohen, D. K., and H. C. Hill. 2000. Instructional policy and classroom performance: The mathematics reform in California. Teachers College Record 102(2):294-343.
Cohen, W. 2001. East Asia at the Center: Four Thousand Years of Engagement with the World. New York: Columbia University Press.
Cohen, W. M., and S. A. Merrill, eds. 2003. Patents in the Knowledge-Based Economy. Washington, DC: The National Academies Press.
College Board. 2004. Trends in Student Aid 2004. Washington, DC: College Board.
Colvin, G. 2005. Can Americans compete? Is America the world's 97-lb. weakling? Fortune, July 25.
Commission on Protecting and Reducing Government Secrecy (the Moynihan Commission). 1997. Secrecy. Washington, DC: US Government Printing Office.
Commission on the Future of the United States Aerospace Industry. November 2002. Final Report. Arlington, VA: Commission on the Future of the United States Aerospace Industry. Available at: www.ita.doc.gov/td/aerospace/aerospacecommission/AeroCommission FinalReport.pdf.
Committee for Economic Development, Research and Policy Committee, Learning for the Future. 2003. Changing the Culture of Math and Science Education to Ensure a Competitive Workforce. New York: Committee for Economic Development.
Conference Board of Canada. 1999. The Economic Implications of International Education for Canada and Nine Comparator Countries: A Comparison of International Education Activities and Economic Performance. Ottawa: Department of Foreign Affairs and International Trade.
Congressional Commission on the Advancement of Women and Minorities in Science, Engineering, and Technology Development. 2000. Land of Plenty: Diversity as America's Competitive Edge in Science, Engineering, and Technology. Arlington, VA: National Science Foundation.
Consortium for Policy Research in Education. 2002. A Report on the Eighth Year of the Merck Institute for Science Education. Philadelphia, PA: CPRE, University of Pennsylvania. Available at: http://www.mise.org/pdf/cpre2000_2001.pdf.

Constable, G., and B. Somerville. 2003. A Century of Innovation: Twenty Engineering Achievements That Transformed Our Lives. Washington, DC: Joseph Henry Press.

Corporate R&D scorecard. 2005. Technology Review (September):56-61.

Council of Economic Advisers. 1995. Economic Report of the President. Washington, DC: US Government Printing Office.

Council of Economic Advisors. 1995. Supporting Research and Development to Promote Economic Growth: The Federal Government's Role. Washington, DC: White House, October.

Council of Graduate Schools. 2004. Ph.D. Completion and Attrition: Policy, Numbers, Leadership, and Next Steps. Washington, DC: Council of Graduate Schools.

Council on Competitiveness. 1991. Gaining New Ground: Technology Priorities for America's Future. Washington, DC: Council on Competitiveness.

Council on Competitiveness. 2001. Clusters of Innovation: Regional Foundations of US Competitiveness. Washington, DC: Council on Competitiveness.

Council on Competitiveness. 2004. Innovate America: National Innovation Initiative Summit and Report: Thriving in a World of Challenge and Change. Washington, DC: Council on Competitiveness.

Council on Government Relations. 1999. The Bayh-Dole Act: A Guide to the Law and Implementing Regulations. Washington, DC: CGR. Available at: http://www.ucop.edu/ott/bayh.html.

Council on Governmental Relations. 2003. Report of the Working Group on the Cost of Doing Business. Washington, DC: CGR, June 2.

Crow, M., and B. Bozeman. 1998. Limited by Design: R&D Laboratories and the U.S. National Innovation System. New York: Columbia University Press, pp. 5-6.

Dalton, D. H., M. G. Serapio, Jr., and P. G. Yoshida. 1999. Globalizing Industrial Research and Development. Washington, DC: US Department of Commerce, Technology Administration, Office of Technology Policy.

Davis, T. M. 2003. Atlas of Student Mobility. New York: Institute of International Education.

Defense Science Board. 2001. The Defense Science Board 2001 Summer Study on Defense Science and Technology. Washington, DC: DSB.

Department of Commerce. 2002. Commission on the Future of the United States Aerospace Industry, Final Report. Arlington, VA: Department of Commerce. Available at: http://www.ita.doc.gov/td/aerospace/aerospacecommission/AeroCommissionFinalReport.pdf.

Department of Defense. 2000. Quadrennial Defense Review Report. Washington, DC: Department of Defense.

Department of Defense. 2001. Quadrennial Defense Review Report. Washington, DC: Department of Defense.

Department of Energy. 1995. Task Force on Alternatives Futures for the Department of Energy National Laboratories (Galvin Commission). Washington, DC: Department of Energy.

Department of State. 2004. Revision to Visas Mantis Clearance Procedure. State 153587, No. 22. Available at: http://travel.state.gov/visa/state153587.html.

Dertouzos, M., R. Lester, and R. Solow. 1989. Made in America: Regaining the Productive Edge. Cambridge, MA: MIT Press.

Dickeson, R. C. 2004. Collision Course: Rising College Costs Threaten America's Future and Require Shared Solutions. Indianapolis, IN: Lumina Foundation for Education, Inc.

Dresselhaus, M. S., and I. L. Thomas. 2001. Alternative energy technologies. Nature 414:332-337.

Dubois, L. H. 2003. DARPA's Approach to Innovation and Its Reflection in Industry. In Reducing the Time from Basic Research to Innovation in the Chemical Sciences: A Workshop Report to the Chemical Sciences Roundtable. Washington, DC: National Academy Press, pp. 37-48.

Dupree, A. H. 1986. Science in the Federal Government: A History of Policies and Activities, 2nd ed. Baltimore, MD: Johns Hopkins University Press.
Economic Policy Institute. 2004. EPI Issue Guide: Offshoring. Washington, DC: Economic Policy Institute.
Ehrenberg, R. G., and L. Zhang. 2004. The Changing Nature of Faculty Employment. Working Paper 44. Ithaca, NY: Cornell Higher Education Research Institute.
Eisenberg, R. 2003. Science and the law: Patent swords and shields. Science 299(5609):1018-1019.
European Commission. 2005. Enterprise and Industry Directorate-General. New and Emerging Science and Technology (NEST) Programme. Available at: http://www.cordis.lu/nest/home.html.
Evenson, R. E. 2001. Economic impacts of agricultural research and extension. In Handbook of Agricultural Economics Vol. 1, B. L. Gardner and G. C. Rausser, eds. Rotterdam: Elsevier, pp. 573-628.
Finn, M. G. 2003. Stay Rates of Foreign Doctorate Recipients from US Universities, 2001. Oak Ridge, TN: Oak Ridge Institute for Science and Education.
Florida, R. 2002. The Rise of the Creative Class . . . and How It's Transforming Work, Leisure, Community, & Everyday Life. New York: Basic Books.
Florida, R. 2005. The Flight of the Creative Class: The New Global Competition for Talent. New York: Harper Business.
Forrester Research. 2004. Near-Term Growth of Offshoring Accelerating. Cambridge, MA: Forrester Research, May 14.
Fox, M. F., and P. Stephan. 2001. Careers of young scientists: Preferences, prospects, and reality by gender and field. Social Studies of Science 31:109-122.
Freeman, R. 2005. It's a flat world, after all. New York Times, April 3, Section 6, Column 1, Magazine Desk, p. 33.
Freeman, R. B. 2005. Does Globalization of the Scientific/Engineering Workforce Threaten US Economic Leadership? Working Paper 11457. Cambridge, MA: National Bureau of Economic Research.
Freeman, R., E. Weinstein, E. Marincola, J. Rosenbaum, and F. Solomon. 2001. CAREERS: Competition and careers in biosciences. Science 294(5550):2293-2294.
Freeman, R. B., E. Jin, and C.-Y. Shen. 2004. Where Do New US-Trained Science-Engineering PhDs Come From? Working Paper 10544. Cambridge, MA: National Bureau of Economic Research.
Friedman, T. L. 2005. The World Is Flat: A Brief History of the 21st Century. New York: Farrar, Straus, and Giroux, pp. 414-419.
Friedman, T. 2005. The end of the rainbow. New York Times, June 29.
Galvin Panel report. 1995. Task Force on Alternative Futures for the Department of Energy National Laboratories, Secretary of Energy Advisory Board. Washington, DC: US Department of Energy.
Geoffrey, C. 2005. America isn't ready. Fortune, July 25.
Gereffi, G., and V. Wadhwa. 2005. Framing the Engineering Outsourcing Debate: Placing the United States on a Level Playing Field with China and India. Available at: http://memp.pratt.duke.edu/downloads/duke_outsourcing_2005.pdf.
Gerstner, L. V., Jr. 2004. Teaching at Risk: A Call to Action. New York: City University of New York. Available at: http://www.theteachingcommission.org.
Glenn Commission. 2000. Before It's Too Late: A Report to the Nation from the National Commission on Mathematics and Science Teaching for the 21st Century. Washington, DC: US Department of Education.

Golde, C. M., and T. M. Dore. 2000. At Cross Purposes: What the Experiences of Doctoral Students Reveal About Doctoral Education. Philadelphia, PA: A Survey Initiated by The Pew Charitable Trusts.

Gomory, R., and W. Baumol. 2001. Global Trade and Conflicting National Interests. Cambridge, MA: MIT Press.

Gomory, R. E., and H. T. Shapiro. 2003. Globalization: Causes and Effects. Issues in Science and Technology. Washington, DC: The National Academies Press.

Goo, S. K. 2004. Two-way traffic in airplane repair. Washington Post, June 1.

Goo, S. K. 2005. Airlines outsource upkeep. Washington Post, August 21. Available at: http://www.washingtonpost.com/wp-dyn/content/article/2005/08/20/AR2005082000979.html.

Gordon, R. J. 2002. The United States. In Technological Innovation and Economic Performance, B. Steil, D. G. Victor, and R. R. Nelson, eds. Princeton, NJ: Princeton University Press, pp. 49-73.

Gordon, R. J. 2002. Technology and Economic Performance in the American Economy. Working Paper 8771. Cambridge, MA: National Bureau of Economic Research.

Gordon, R. J. 2004. Why Was Europe Left at the Station When America's Productivity Locomotive Departed? Working Paper 10661. Cambridge, MA: National Bureau of Economic Research. Available at: http://www.nber.org/papers/w10661.

Government Accountability Office. 2004. Border Security: Improvements Needed to Reduce Time Taken to Adjudicate Visas for Science Students and Scholars. GAO-04-371. Washington, DC: GAO.

Government Accountability Office. 2005. National Nuclear Security Administration: Contractors' Strategies to Recruit and Retain and Critically Skilled Workforce Are Generally Effective. GAO-05-164. Washington, DC: GAO.

Government Accountability Office. 2005. Border Security: Streamlined Visas Mantis Program Has Lowered Burden on Science Students and Scholars, but Further Refinements Needed. GAO-05-198. Washington, DC: GAO.

Government Accounting Office. 1998. Best Practices: Elements Critical to Successfully Reducing Unneeded RDT&E Infrastructure. USGAO Report to Congressional Requesters. Washington, DC: GAO.

Grabowski, H., J. Vernon, and J. DiMasi. 2002. Returns on research and development for 1990s new drug introductions. Pharmacoeconomics 20(Supplement 3):11-29.

Gralla, P. 2004. U.S. lags in broadband adoption despite VoIP demand, says report. EE Times Online. Available at: http://www.eet.com/showArticle.jhtml?articleID=55800449.

Gross, G. 2003. CEOs defend moving jobs offshore at tech summit. InfoWorld, October 9.

Hall, B. H., and J. van Reenen. 1999. How Effective Are Fiscal Incentives for R&D? A Review of the Evidence. Working Paper 7098. Cambridge, MA: National Bureau of Economic Research.

Hamilton, C. Winter 2003. University Technology Transfer and Economic Development: Proposed Cooperative Economic Development Agreements Under the Bayh-Dole Act. John Marshall Law Review.

Henry L. Stimson Center and Center for Strategic and International Studies. 2001. Enhancing Multilateral Export Controls for US National Security. Washington, DC: Henry L. Stimson Center.

Heyman, G. D., B. Martyna, and S. Bhatia. 2002. Gender and achievement-related beliefs among engineering students. Journal of Women and Minorities in S&E 8:33-45.

Hicks, D. 2004. Asian countries strengthen their research. Issues in Science and Technology 20(4)(Summer 2004):75-78.

Hira, R. 2004. Rochester Institute of Technology, presentation to Committee on Science, Engineering, and Public Policy, Workshop on International Students and Postdoctoral Scholars, National Academies, July.

Hira, R., and A. Hira. 2005. Outsourcing America: What's Behind Our National Crisis and How We Can Reclaim American Jobs. Washington, DC: AMACOM Books.

Hobbs, F., and N. Stoops. 2004. Demographic Trends in the 20th Century. US Census Bureau, CENSR-4. Washington, DC: US Census Bureau, November.

Holm-Nielsen, L. B. 2002. Promoting science and technology for development: The World Bank's Millennium Science Initiative. Paper delivered on April 30, 2002, to the First International Senior Fellows meeting, The Wellcome Trust, London, UK.

Holmstrom, E. I., C. D. Gaddy, V. V. Van Horne, and C. M. Zimmerman. 1997. Best and Brightest: Education and Career Paths of Top S&E Students. Washington, DC: Commission on Professionals in Science and Technology.

Hughes, K. H. 2005. Building the Next US Century: The Past and Future of US Economic Competitiveness. Washington, DC: Woodrow Wilson Center Press.

Hughes, K. H. 2005. Facing the global competitiveness challenge. Issues in Science and Technology 21(4):72-78.

Hundt, R. 2003. Why is government subsidizing the old networks when "Big Broadband" convergence is inevitable and optimal? New America Foundation Issue Brief December.

Hunter, K. 2005. Education key to jobs, Microsoft CEO says. August 17. Available at: Stateline.org.

IMD International. 2005. World Competitiveness Yearbook. London: Thompson Learning.

Information Technology Association of America. 2004. The Impact of Offshore IT Software and Services Outsourcing on the US Economy and the IT Industry. Lexington, MA: Global Insight (USA), March.

Institute for International Education. 2004. Open Doors Report on International Educational Exchange. New York: Institute for International Education.

Institute of Electrical and Electronics Engineers. 2004. Position Statement on Offshore Outsourcing. Washington, DC. Available at: www.ieeeusa.org/policy/positions/offshoring.asp.

International Association of Pharmaceutical Manufacturers & Associations. 2004. A Review of Existing Data Exclusivity Legislation in Selected Countries. January. Available at: http://www.who.int/intellectualproperty/topics/ip/en/Data.exclusivity.review.doc.

Jackson, R., and N. Howe. 2003. The 2003 Aging Vulnerability Index. Washington, DC: CSIS and Watson Wyatt Worldwide, p. 43.

Joint Chiefs of Staff. 2001. Joint Vision 2020. Washington, DC: Department of Defense.

Jones, B. 2005. Age and Great Innovation. Working Paper 11359. Cambridge, MA: National Bureau of Economic Research. Available at: http://www.nber.org/papers/w11359.

Kalita, S. M. Virtual secretary puts new face on Pakistan. Washington Post, May 10, 2005, p. A01.

Kane, T. J., and P. R. Orszag. 2003. Higher Education Spending: The Role of Medicaid and the Business Cycle. Policy Brief #124. Washington, DC: Brookings Institution.

Kanellos, M. 2004. IBM sells PC group to Lenovo. News.com, December 8.

Kapur, D., and J. McHale. 2005. Sojourns and Software: Internationally Mobile Human Capital and High-Tech Industry Development in India, Ireland, and Israel. Oxford, UK: Oxford University Press.

Kerr, W. 2004. Ethnic Scientific Communities and International Technology Diffusion. Working Paper. Available at: http://econ-www.mit.edu/faculty/download_pdf.php?id=994.

King, A. F. 2005. Policy Implications of Changes in Higher Education Finance, presentation to the National Academies' Board on Higher Education and Workforce, Washington, DC, April 21-22.

King, D. A. 2004. The scientific impact of nations. Nature 430(6997)(July 15):311-316.
King, J. L. 2003. Patent examination procedures and patent quality. In Patents in the Knowledge-Based Economy, W. M. Cohen and S. A. Merrill, eds. Washington, DC: The National Academies Press, pp. 54-73.
Kissler, J. 2005. Why It Is in the Interest to Address the Growing Gap Between Public and Private Universities. Oakland, CA: University of California.
Knezo, G. J. 2003. Sensitive but Unclassified and Other Federal Security Controls on Scientific and Technical Information: History and Current Controversy. Washington, DC: Congressional Research Service.
Korean Ministry of Science and Technology (MOST). Available at: http://www.most.go.kr/most/english/link_2.jsp.
Lane, N., and T. Kalil. 2005. The National Nanotechnology Initiative: Present at the Creation. Issues in Science and Technology 21(Summer):49-54.
Laudicina, P. A. 2005. World Out of Balance: Navigating Global Risks to Seize Competitive Advantage. New York: McGraw-Hill.
Leonard, J. A. 2003. How Structural Costs Imposed on U.S. Manufacturers Harm Workers and Threaten Competitiveness. Washington, DC: National Association of Manufacturers. Available at: http://www.nam.org/s_nam/bin.asp?CID=216&DID=227525&DOC=FILE.PDF.
Lewin, T. 2005. Many going to college are not ready, report says. New York Times, August 17.
Lewis, W. W. 2004. The Power of Productivity: Wealth, Poverty, and the Threat to Global Stability. Chicago: University of Chicago Press.
Lim, P. J. 2006. Looking ahead means looking abroad. New York Times, January 8.
Lu, A. 2002. The Decision Cycle for People Going to Graduate School. Stamford, CT: Peterson's Thomson Learning.
Madey v. Duke Univ. 307 F.3d 1351. Available at: 2002 U.S. App. LEXIS 20823, 64 U.S.P.Q.2d. (BNA) 1737 (Fed. Cir. 2002).
Mandel, M. J. 2004. Rational Exuberance: Silencing the Enemies of Growth and Why the Future Is Better Than You Think. New York: Harper Business, p. 27.
Mann, W. C. 2003. Globalization of IT Services and White Collar Jobs. Washington, DC: Institute for International Economics.
Mansfield, E. 1991. Academic research and industrial innovation. Research Policy 20:1-12.
Markusen, A. R., and Costigan, S. S. 1999. The Military Industrial Challenge. In Arming the Future: A Defense Industry for the 21st Century, A. R. Markusen and S. S. Costigan, eds. New York: Council on Foreign Relations, p. 8.
Mashelkar, R. A. 2005. India's R&D: Reaching for the top. Science 307:1415-1417.
Math and Science Expert Panel. 2004. Exemplary Promising Mathematics Programs. Washington, DC: US Department of Education.
May, R. M. 2004. Raising Europe's game. Nature 430:831.
McKinsey and Company. 2003. Offshoring: Is It a Win-Win Game? New York: McKinsey and Company, August.
McKinsey and Company. 2005. The Emerging Global Labor Market: Part II—The Supply of Offshore Talent in Services. New York: McKinsey and Company, June, p. 23.
Mehlman, B. 2003. Offshore Outsourcing and the Future of American Competitiveness. Business Roundtable Working Group
Merck Institute for Science Education (MISE). Available at: http://www.mise.org/mise/index.jsp.
Mervis, J. 2003. Down for the count. Science 300(5622)(May 16):1070-1074.
Ministry of Science and Technology. 2004. Chinese Statistical Yearbook 2004. People's Republic of China, Chapter 21, Table 21-11. Available at: http://www.stats.gov.cn/english/statisticaldata/yearlydata/yb2004-e/indexeh.htm.

Moore, S., and Simon, J. L. 1999. The Greatest Century That Ever Was: 25 Miraculous Trends of the Last 100 Years. Policy Analysis No. 364. Washington, DC: Cato Institute, December 15.

Nadiri, M. I. 1993. Innovations and Technical Spillovers. Working Paper 4423. Cambridge, MA: National Bureau of Economic Research.

NASSCOM. 2005. Strategic Review 2005. India: National Association of Software and Service Companies. Chapter 6, Sustaining the India Advantage. Available at: http://www.nasscom.org/strategic2005.asp.

The National Academies. 2005. Policy Implications of International Graduate Students and Postdoctoral Scholars in the United States. Washington, DC: The National Academies Press.

National Academy of Engineering. 1999. Concerning Federally Sponsored Inducement Prizes in Engineering and Science. Washington, DC: National Academy Press.

National Academy of Engineering. 2003. The Impact of Academic Research on Industrial Performance. Washington, DC: The National Academies Press.

National Academy of Engineering and Institute of Medicine. 2005. Building a Better Delivery System: A New Engineering/Health Care Partnership. Washington, DC: The National Academies Press.

National Academy of Engineering and National Research Council. 2005. Enhancing the Community College Pathway to Engineering Careers. Washington, DC: The National Academies Press.

National Academy of Sciences. 1996. Ozone Depletion, Beyond Discovery Series. Washington, DC: National Academy Press, April.

National Academy of Sciences, National Academy of Engineering, and Institute of Medicine. 1993. Science, Technology, and the Federal Government: National Goals for a New Era. Washington, DC: National Academy Press.

National Academy of Sciences, National Academy of Engineering, and Institute of Medicine. 1995. Reshaping the Graduate Education of Scientists and Engineers. Washington, DC: National Academy Press.

National Academy of Sciences, National Academy of Engineering, and Institute of Medicine. 1999. Capitalizing on Investments in Science and Technology. Washington, DC: National Academy Press.

National Academy of Sciences, National Academy of Engineering, and Institute of Medicine. 2000. Enhancing the Postdoctoral Experience for Scientists and Engineers. Washington, DC: National Academy Press.

National Academy of Sciences, National Academy of Engineering, and Institute of Medicine. 2002. Observations on the President's Fiscal Year 2003 Federal Science and Technology Budget. Washington, DC: The National Academies Press, pp. 14-16.

National Center for Education Statistics. 1999. Highlights from TIMSS. Available at: http://nces.ed.gov/pubs99/1999081.pdf.

National Center for Education Statistics. 2004. Schools and Staffing Survey: Qualifications of the Public School Teacher Workforce: Prevalence of Out-of-Field Teaching 1987-88 to 1999-2000 (Revised), p. 10. Available at: http://nces.ed.gov/pubs2002/2002603.pdf.

National Center for Education Statistics. 2004. Digest of Education Statistics 2004. Washington, DC: Institute of Education Sciences, Department of Education, Table 250. Available at: http://nces.ed.gov/programs/digest/d04/tables/dt04_250.asp.

National Center for Education Statistics. 2005. International Outcomes of Learning in Mathematics Literacy and Problem Solving: PISA 2003 Results from the U.S. Perspective. Washington, DC: US Department of Education, pp. 15, 29. Available at: http://nces.ed.gov/pubs2005/2005003.pdf.

National Center for Education Statistic. 2006. Public Elementary and Secondary Students, Staff, Schools, and School Districts: School Year 2003–04. Available at: http://nces.ed.gov/pubs2006/2006307.pdf.
National Center for Education Statistics. 2006. The Nation's Report Card: Mathematics 2005. Available at: http://nces.ed.gov/nationsreportcard/pdf/main2005/2006453.pdf.
National Center for Teaching and America's Future. 1996. Doing What Matters Most: Teaching for America's Future. New York: NCTAF.
National Council of Teachers of Mathematics. 2000. Principles and Standards for School Mathematics. Washington, DC: NCTM. Available at: http://standards.nctm.org.
National Critical Technologies Panel. 1991. Report of the National Critical Technologies Panel. Washington, DC: US Government Printing Office.
National Critical Technologies Panel. 1993. The Second Biennial Report of the National Critical Technologies Panel. Washington, DC: US Government Printing Office.
National Critical Technologies Panel. 1995. The National Critical Technologies Report. Washington, DC: US Government Printing Office.
National Energy Policy Development Group. 2001. National Energy Policy. Washington, DC: US Government Printing Office, May.
National Institutes of Health. 2001. Working Group on Construction of Research Facilities. A Report to the Advisory Committee of the Director, National Institutes of Health. Bethesda, MD: NIH.
National Institutes of Health. 2005. NIH Roadmap, High Risk Research. Available at: http://nihroadmap.nih.gov/highrisk/.
National Intelligence Council. 2004. Mapping the Global Future: Report of the National Intelligence Council's 2020 Project. Pittsburgh, PA: Government Printing Office.
National Research Council. 1982. Scientific Communication and National Security. Washington, DC: National Academy Press.
National Research Council. 1984. High-Technology Ceramics in Japan. Washington, DC: National Academy Press.
National Research Council. 1995. Allocating Federal Funds for Science and Technology. Washington, DC: National Academy Press.
National Research Council. 1995. Evolving the High Performance Computing and Communications Initiative to Support the Nation's Information Infrastructure. Washington, DC: National Academy Press.
National Research Council. 1996. National Science Education Standards. Washington, DC: National Academy Press.
National Research Council. 1997. Industrial Research and Innovation Indicators. Washington, DC: National Academy Press.
National Research Council. 1997. Improving Teacher Preparation and Credentialing Consistent with the National Science Education Standards: Report of a Symposium. Washington, DC: National Academy Press.
National Research Council. 1997. Science for All Children: A Guide to Improving Elementary Science Education in Your School District. Washington, DC: National Academy Press.
National Research Council. 1998. Trends in the Early Careers of Life Scientists. Washington, DC: National Academy Press.
National Research Council. 1999. Balancing Scientific Openness and National Security Controls at the Nuclear Weapons Laboratories. Washington, DC: National Academy Press.
National Research Council. 1999. A Strategic Plan for Education Research and Its Utilization. Washington, DC: National Academy Press.
National Research Council. 1999. Harnessing Science and Technology for America's Economic Future: National and Regional Priorities. Washington, DC: National Academy Press.

National Research Council. 1999. Challenges and Opportunities. Washington, DC: National Academy Press.

National Research Council. 1999. The Advanced Technology Program: Challenges and Opportunities, Washington, DC: National Academy Press.

National Research Council. 1999. The Small Business Innovation Research Program: Challenges and Opportunities. Washington, DC: National Academy Press.

National Research Council. 1999. Transforming Undergraduate Education in Science, Mathematics, Engineering, and Technology. Washington, DC: National Academy Press.

National Research Council. 1999. US Industry in 2000: Studies in Competitive Performance. Washington, DC: National Academy Press.

National Research Council. 1999. How People Learn: Brain, Mind, Experience, and School. Washington, DC: National Academy Press. Available at: http://books.nap.edu/catalog/6160.html.

National Research Council. 1999. Harnessing Science and Technology for America's Economic Future. Washington, DC: National Academy Press.

National Research Council. 1999. U.S. Industry in 2000: Studies in Competitive Renewal. Washington, DC: National Academy Press.

National Research Council. 1999. On Evaluating Curricular Effectiveness: Judging the Quality of K–12 Mathematics Evaluations. Washington, DC: National Academy Press.

National Research Council. 2000. New Practices for the New Millennium. Washington, DC: National Academy Press.

National Research Council. 2000. An Assessment of the Department of Defense's Fast Track Initiative. Washington, DC: National Academy Press.

National Research Council. 2000. Attracting Science and Mathematics Ph.D.s to Secondary School Education. Washington, DC: National Academy Press. Available at: http://www.nap.edu/catalog/9955.html.

National Research Council. 2000. The Small Business Innovation Research Program: An Assessment of the Department of Defense Fast Track Initiative. Washington, DC: National Academy Press.

National Research Council. 2000. How People Learn: Brain, Mind, Experience, and School: Expanded Edition. Washington, DC: National Academy Press.

National Research Council. 2001. Trends in Federal Support of Research and Graduate Education. Washington, DC: National Academy Press.

National Research Council. 2001. Building a Workforce for the Information Economy. Washington, DC: National Academy Press.

National Research Council. 2001. Educating Teachers of Science, Mathematics, and Technology: New Practices for a New Millennium. Washington, DC: National Academy Press.

National Research Council. 2001. Review of the Research Program of the Partnership for a New Generation of Vehicles. Washington, DC: National Academy Press.

National Research Council. 2001. The Advanced Technology Program: Assessing Outcomes. Washington, DC: National Academy Press.

National Research Council. 2001. Building a Workforce for the Information Economy. Washington, DC: National Academy Press.

National Research Council. 2002. Attracting PhDs to K–12 Education: A Demonstration Program for Science, Mathematics, and Technology. Washington, DC: National Academy Press.

National Research Council. 2002. Learning and Understanding: Improving Advanced Study of Mathematics and Science in U.S. Schools. Washington, DC: National Academy Press.

National Research Council. 2002. Making the Nation Safer: The Role of Science and Technology in Countering Terrorism. Washington, DC: The National Academies Press.

National Research Council. 2003. Securing the Future: Regional and National Programs to Support the Semiconductor Industry. Washington, DC: The National Academies Press.
National Research Council. 2003. Frontiers in Agricultural Research: Food, Health, Environment, and Communities. Washington, DC: The National Academies Press.
National Research Council. 2004. A Patent System for the 21st Century. Washington, DC: The National Academies Press.
National Research Council. 2004. An Assessment of the Small Business Innovation Research Program: Project Methodology. Washington, DC: The National Academies Press.
National Research Council. 2004. Biotechnology Research in an Age of Terrorism. Washington, DC: The National Academies Press.
National Research Council. 2004. Measuring Research and Development Expenditures in the U.S. Economy. Washington, DC: The National Academies Press.
National Research Council. 2004. Evaluation of the National Aerospace Initiative. Washington, DC: The National Academies Press.
National Research Council. 2004. Program Diversity and Assessment Challenges, Report of a Symposium. Washington, DC: The National Academies Press.
National Research Council. 2004. Engaging Schools: Fostering High-School Students' Motivation to Learn. Washington, DC: The National Academies Press.
National Research Council. 2005. Assessment of Department of Defense Basic Research. Washington, DC: The National Academies Press.
National Research Council. 2005. Bridges to Independence: Fostering the Independence of New Investigators in Biomedical Research. Washington, DC: The National Academies Press.
National Research Council. 2005. Avoiding Surprise in an Era of Global Technology Advances. Washington, DC: The National Academies Press.
National Research Council. 2005. National Laboratories: Building New Ways to Work Together. Washington, DC: The National Academies Press.
National Research Council. 2005. Advancing the Nation's Health Needs. Washington, DC: The National Academies Press.
National Science and Technology Council. 1995. Final Report on Academic Research Infrastructure: A Federal Plan for Renewal. Washington, DC: White House Office of Science and Technology Policy.
National Science and Technology Council. 2000. Ensuring a Strong US Scientific, Technical, and Engineering Workforce in the 21st Century. Washington, DC: Executive Office of the President of the United States.
National Science Board. 2003. Science and Engineering Infrastructure for the 21st Century: The Role of the National Science Foundation. Arlington, VA: NSF.
National Science Board. 2004. Committee on Programs and Plans, Charge to the Task Force on Transformative Research. Available at: http://www.nsf.gov/nsb/committees/cpptrcharge.htm.
National Science Board. 2004. Science and Engineering Indicators 2004. Arlington, VA: NSF. Available at: http://www.nsf.gov/sbe/srs/seind04/c4/c4h.htm.
National Science Board. 2004. Report to the National Science Board on the National Science Foundation's Merit Review Process Fiscal Year 2004. NSB-05-12. Arlington, VA: NSB.
National Science Foundation. 1999. Preparing Our Children: Math and Science Education in the National Interest. Arlington, VA: NSF.
National Science Foundation. 2001. Survey of Doctorate Recipients 2001, the NSF Survey of Graduate Students and Postdocs. Arlington, VA: NSF.
National Science Foundation. 2003. Blue-Ribbon Advisory Panel on Cyberinfrastructure, Revolutionizing Science and Engineering Through Cyberinfrastructure. Arlington, VA: NSF.

National Science Foundation. 2003. The Science and Engineering Workforce: Realizing America's Potential. Arlington, VA: NSF.
National Science Foundation. 2003. Graduate Enrollment Increases in Science and Engineering Fields, Especially in Engineering and Computer Sciences. NSF 03-315. Arlington, VA: NSF.
National Science Foundation. 2004. Graduate Enrollment in Science and Engineering Fields Reaches New Peak; First-Time Enrollment of Foreign Students Declines. NSF 04-326. Arlington, VA: NSF.
National Science Foundation. 2004. Survey of Graduate Students and Postdoctorates in Science and Engineering 2002. Arlington, VA: NSF.
National Science Foundation. 2005. Survey of Earned Doctorates, 2003. Arlington, VA: NSF.
National Science Foundation. 2005. Graduate Enrollment in Science and Engineering Programs Up in 2003, but Declines for First-Time Foreign Students: Info Brief. NSF 05-317. Arlington, VA: NSF.
National Science Foundation, Advisory Panel on Cyberinfrastructure. 2003. Revolutionizing Science and Engineering Through Cyberinfrastructure. Arlington, VA: NSF.
National Science Foundation, Division of Science Resources Statistics. 2000. Science and Engineering Research Facilities at Colleges and Universities, 1998. NSF-01-301. Arlington, VA: NSF.
Nelson, R. R., ed. 1993. National Innovation Systems: A Comparative Analysis. New York: Oxford University Press.
Noon, C. 2005. Starbuck's Schultz bemoans health care costs. Forbes Magazine, September 19.
Nordhaus, W. 1999. The Contribution of Improved Health and Living Standards. Working Paper 8818. Cambridge, MA: National Bureau of Economic Research.
Nordhaus, W. 2005. The Sources of the Productivity Rebound and the Manufacturing Employment Puzzle. Working Paper 11354. Cambridge, MA: National Bureau of Economic Research.
Office of Management and Budget. 2005. Budget of the United States Government, Fiscal Year 2006. Washington, DC: US Government Printing Office.
Office of Science and Technology of Canada. 2005. Immigration Policy Change Widens Door for Foreign Students and Scholars. Bridges 6. July 13. Available at: http://bridges.ostina.org.
Office of Science and Technology Policy. 2000. Analysis of Facilities and Administrative Costs at Universities. Washington, DC: Executive Office of the President.
Office of Scientific Research and Development. 1945. Science—The Endless Frontier. Washington, DC: US Government Printing Office.
Ohland, M. W., G. Zhang, B. Thorndyke, and T. J. Anderson. 2004. Grade-Point Average, Changes of Major, and Majors Selected by Students Leaving Engineering. 34th ASEE/IEEE Frontiers in Education Conference. Session T1G:12-17.
Organisation for Economic Co-operation and Development. 1998. Technology, Productivity, and Job Creation: Best Policy Practices. Paris: OECD Publications.
Organisation for Economic Co-operation and Development. 2002. International Mobility of the Highly Skilled. Policy Brief 92 2002 01 1P4. Paris: OECD Publications. Available at: http://www.oecd.org/dataoecd/9/20/1950028.pdf.
Organisation for Economic Co-operation and Development. 2003. Science, Technology and Industry Scoreboard, 2003 R&D Database. Paris: OECD Publications. Available at: http://www1oecd.org/publications/e-book/92-2003-04-1-7294/.
Organisation for Economic Co-operation and Development. 2003. Tax Incentive for Research and Development: Trends and Issues. Paris: OECD Publications. Available at: http://www.oecd.org/dataoecd/12/27/2498389.pdf.

Organisation for Economic Co-operation and Development. 2004. Science, Technology and Industry Outlook. Paris: OECD Publications, December, pp. 25, 67, 190. Available at: http://www.oecd.org/document/63/0,2340,en_2649_33703_33995839_1_1_1_1,00.html.

Organisation for Economic Co-operation and Development. 2004. The Economic Impact of ICT. Paris: OECD Publications.

Organisation for Economic Co-operation and Development. 2005. China Overtakes U.S. as World's Leading Exporter of Information Technology Goods. Paris: OECD Publications, December 12.

Organisation for Economic Co-operation and Development. 2005. OECD Broadband Statistics. Paris: OECD Publications, June. Update October 20, 2005. Available at: http://www.oecd.org/document/16/0,2340,en_2649_201185_35526608_1_1_1_1,00.html#data2004.

Organisation for Economic Co-operation and Development. 2005. Trends in International Migration: 2004 Annual Report. Paris: OECD Publications. See http://www.workpermit.com for more information on immigration policies in English-speaking countries and the European Union.

Organisation for Economic Co-operation and Development. 2005. Main Science & Technology Indicators. Paris: OECD Publications. Available at: http://www.oecd.org/document/26/0,2340,en_2649_34451_1901082_1_1_1_1,00.html.

Organisation for Economic Co-operation and Development. 2005. Education at a Glance 2005. Paris: OECD Publications. Available at: http://www.oecd.org/dataoecd/41/13/35341210.pdf.

Organisation for Economic Co-operation and Development. 2005. Program for International Student Assessment. Paris: OECD Publications. Available at: http://www.pisa.oecd.org.

Organization for International Investment. The Facts About Insourcing. Available at: http://www.ofii.org/insourcing/.

Parker, L. 1997. The Engineering Research Centers (ERC) Program: An Assessment of Benefits and Outcomes. Arlington, VA: NSF. Available at: www.nsf.gov/pubs/1998/nsf9840/nsf9840.htm.

Parsons, M. 2005. Higher education is just another special interest. The Chronicle of Higher Education 51(22):B20. Available at: http://chronicle.com/prm/weekly/v51/i22/22b02001.htm.

Peterson, P. G. 2002. The shape of things to come: Global aging in the 21st century. Journal of International Affairs 56(1)(Fall). New York: Columbia University Press.

Pew Research Center. 2005. U.S. Image Up Slightly, But Still Negative, American Character Gets Mixed Reviews. Pew Global Attitudes Project. Washington, DC. Available at: http://pewglobal.org/reports/display.php?ReportID=247.

Piore, M. J., and C. F. Sabel. 1984. The Second Industrial Divide: Possibilities for Prosperity. New York: Basic Books.

Popper, S. W., and C. S. Wagner. 2002. New Foundations for Growth: The US Innovation System Today and Tomorrow. Arlington, VA: RAND.

Popper, S. W., C. S. Wagner, and E. V. Larson. 1998. New Forces at Work: Industry Views Critical Technologies. Santa Monica, CA: RAND.

Powell, K. 2005. Hot house high. Nature 435:874-875.

Preeg, E. H. 2003. The Emerging Chinese Advanced Technology Superstate. Arlington, VA: Manufacturers Alliance/MAPI and Hudson Institute.

President's Council of Advisors on Science and Technology. 1997. Federal Energy Research and Development for the Challenges of the Twenty-First Century-Report of the Energy Research and Development Panel. Washington, DC: The President's Committee of Advisors on Science and Technology.

President's Council of Advisors on Science and Technology. 2004. Sustaining the Nation's Innovation Ecosystems, Information Technology Manufacturing and Competitiveness. Washington, DC: White House Office of Science and Technology Policy.

President's Information Technology Advisory Committee. 2005. Computational Science: Ensuring America's Competitiveness. Washington, DC: National Coordination Office for Information Technology Research and Development (NCO/ITR&D).

Prestowitz, C. 2005. Three Billion New Capitalists: The Great Shift of Wealth and Power to the East. New York: Basic Books.

Raven, P. 2005. Biodiversity and our common future. Bulletin of the American Academy of Arts & Sciences 58:20-24.

Rizzo, M. 2005. State Preferences for Higher Education Spending: A Panel Data Analysis, 1977-2001. Paper presented at Cornell Higher Education Research Institute's Annual Conference, "Assessing Public Higher Education at the Start of the 21st Century," Ithaca, NY, May 22-23.

Roach, S. 2004. More jobs, worse work. New York Times, July 22.

Romer, P. 1993. Implementing a National Technology Strategy with Self-Organizing Industry Investment Boards. In Brookings Papers on Economic Activity: Microeconomics (2), M. N. Baily, P. C. Reiss, and C. Winston, eds. Washington, DC: Brookings Institution.

Romer, P. M. 2000. Should the Government Subsidize Supply or Demand in the Market for Scientists and Engineers? Working Paper 7723. Cambridge, MA: National Bureau for Economic Research. Available at: http://www.nber.org/papers/w7723/.

Rutherford, F. J., and A. Ahlgren. 1990. Science for all Americans. Washington, DC: American Association for the Advancement of Science, October. Available at: http://www.project2061.org/default_flash.htm.

Samuelson, P. A. 2004. Where Ricardo and Mill rebut and confirm arguments of mainstream economists supporting globalization. Journal of Economic Perspectives 18(Summer):135-146.

Samuelson, R. 2005. China's devalued Concession. Washington Post, July 26.

Samuelson, R. 2005. The world is still round. Newsweek, July 25.

Saxenian, A. L. 1999. Silicon Valley's New Immigrant Entrepreneurs. San Francisco: Public Policy Institute. Available at: http://www.ccis-ucsd.org/PUBLICATIONS/wrkg15.PDF.

Saxenian, A. L. 2002. Brian circulation: How high-skill immigration makes everyone better off. The Brookings Review 20(1)(Winter 2002). Washington, DC: Brookings Institution.

Schmidt, W. H., C. McKnight, R. T. Houang, and D. E. Wiley. 2005. The Heinz 57 Curriculum: When More May Be Less. Paper presented at the 2005 annual meeting of the American Education Research Association, Montreal, Quebec.

Scott, A., G. Steyn, A. Geuna, S. Brusoni, and W. E. Steinmeuller. 2001. The Economic Returns of Basic Research and the Benefits of University-Industry Relationships. Report for the UK Government Office of Science and Technology. Brighton: SPRU (Science and Technology Policy Research), University of Sussex. Available at: http://www.sussex.ac.uk/spru/documents/review_for_ost_final.pdf.

Secretary of Energy's Advisory Board, Task Force on the Future of Science Programs at the Department of Energy. 2003. Critical Choices: Science, Energy and Security. Final Report. Washington, DC: US Department of Energy, October 13.

Semiconductor Industry Association. 2005. Choosing to Compete. December 12. Available at: www.choosetocompete.org/.

Seymour, E., and N. Hewitt. 1997. Talking About Leaving: Why Undergraduates Leave the Sciences. Boulder, CO: Westview Press.

Shanghai's Jiao Tong University Institute of Higher Education. 2004. Academic Ranking of World Universities. Available at: http://ed.sjtu.edu.cn/rank/2004/2004Main.htm.

Sharma, D. C., and M. Yamamoto. 2004. How India is handling international backlash. CNET news.com, May 6.
Sigma Xi. 2004. Sigma Xi National Postdoctoral Survey. Available at: http://postdoc.sigmaxi.org.
Smith, T. 2001. The Retention and Graduation Rates of 1993-1999 Entering Science, Mathematics, Engineering, and Technology Majors in 175 Colleges and Universities. Norman, OK: Center for Institutional Data Exchange and Analysis (C-IDEA), the University of Oklahoma.
Solow, R. M. 1957. Technical change and the aggregate production function. The Review of Economics and Statistics 39:312-320.
Solow, R. M. 1960. Investment and Technical Progress. Available at: http://nobelprize.org/economics/laureates/1987/index.html.
Segal, A. 2004. Is America losing its edge? Innovation in a globalized world. Foreign Affairs November/December.
Spencer, S. 2005. 2004. CEO Study: A Statistical Snapshot of Leading CEOs. Available at: http://content.spencerstuart.com/sswebsite/pdf/lib/Statistical_Snapshot_of_Leading_CEOs_relB3.pdf#search='ceo%20educational%20background'.
Spotts, P. N. 2005. Pulling the plug on science? Christian Science Monitor, April 14.
Steil, B., D. G. Victor, and R. R. Nelson. 2002. Technological Innovation and Economic Performance. Princeton, NJ: Princeton University Press.
Stephan, P. E., and S. G. Levin, eds. 2005. Foreign Scholars in US Science: Contributions and Costs in Science and the University. Madison, WI: University of Wisconsin Press.
Stokes, D. E. 1997. Pasteur's Quadrant: Basic Science and Technological Innovation. Washington, DC: Brookings Institution.
Subotnik, R. F., K. M. Stone, and C. Steiner. 2001. Lost generation of elite talent in science. Journal of Secondary Gifted Education 13:33-43.
Task Force on the Future of American Innovation. 2005. The Knowledge Economy: Is the United States Losing Its Competitive Edge, Benchmarks for Our Innovation Future. Washington, DC: The Task Force on the Future of US Innovation, February.
Teitelbaum, M. S. 2003. Do we need more scientists? The Public Interest 153:40-53.
Tobias, S. 1990. They're Not Dumb, They're Different. Stalking the Second Tier. Tucson, AZ: Research Corporation.
Tufts Center for the Study of Drug Development. 2001. Backgrounder: How New Drugs Move Through the Development and Approval Process. November 1. Available at: http://csdd.tufts.edu/NewsEvents/RecentNews.asp?newsid=4.
United Nations Conference on Trade and Development. 2004. World Investment Report 2004: The Shift Towards Services. New York and Geneva: United Nations.
United Nations, Department of Economic and Social Affairs, Population Division. 1999. The World at Six Billion. October 12. Available at: http://www.un.org/esa/population/publications/sixbillion/sixbillion.htm.
United States Bureau of Labor Statistics. 2005. International Comparisons of Hourly Compensation Costs for Production Workers in Manufacturing, 2004. November 18. Available at: ftp://ftp.bls.gov/pub/news.release/History/ichcc.11182005.news.
United States Commission on National Security. 2001. Road Map for National Security: Imperative for Change. Washington, DC: US Commission on National Security.
United States of America's Embassy. 1996. China's Science and Technology Policy for the Twenty-First Century—A View from the Top. Report from the US Embassy, Beijing, November.
US Bureau of Labor Statistics. 2005. Business Employment Dynamics: First Quarter 2005. November 18. Available at: http://www.bls.gov/rofod/3640.pdf.
US Bureau of the Census. 1975. Historical Statistics of the United States, Colonial Times to 1970. Part 1, Series B 107-15. Washington, DC: US Census Bureau.

US Census Bureau. 2000. Statistical Abstract of the United States. Washington, DC: US Census Bureau.
US Census Bureau Database. Total Mid-Year Population, 2004-2050. Available at: http://www.census.gov/ipc/www/idbsprd.html.
US Congress House of Representatives Committee on Science. 1998. Unlocking Our Future: Toward a New National Science Policy (the "Ehlers Report"). September 24. Washington, DC: US Congress. Available at: http://www.house.gov/science/science_policy_report.htm.
US Commission on National Security. 2001. Road Map for National Security: Imperative for Change. Washington, DC: US Commission on National Security.
US Department of Commerce. 2004. Spectrum Policy for the 21st Century: The President's Spectrum Policy Initiative. Report 1. Washington, DC: US Department of Commerce, June. Available at: http://www.ntia.doc.gov/reports/specpolini/presspecpolini_report1_06242004.htm.
US Department of Education. 1994. Prisoners of Time. National Education Commission on Time and Learning. Washington, DC: US Department of Education.
US Department of Education. 1998. Pursuing Excellence: A Study of Twelfth-Grade Mathematics and Science Achievement in International Context. National Center for Education Statistics 98-049. Washington, DC: US Government Printing Office.
US Department of Education. 2000. NAEP 1999 Trends in Academic Progress: Three Decades of Academic Performance. National Center for Education Statistics 2000-469. Washington, DC: US Department of Education.
US Department of Energy. 2001. Infrastructure Frontier: A Quick Look Survey of the Office of Science Laboratory Infrastructure. Washington, DC: DOE, April.
US Department of Energy. 2005. Laboratory Science Teacher Professional Development Program: About LSTPD. Washington, DC: Office of Science, Office of Workforce Development for Teachers and Scientists.
US Department of Labor. 2001. Report on the American Workforce. Washington, DC: Department of Labor. Available at: http://www.bls.gov/opub/rtaw/pdf/rtaw2001.pdf.
US General Accounting Office. 2001. Contributions to and Results of the Small Business Technology Transfer Program. Statement by Jim Wells. GAO-01-867T. Washington, DC: GAO.
US General Accounting Office. 2002. Lessons Learned from Previous Research Could Benefit FreedomCAR Initiative. Statement of Jim Wells. GAO-02-810T. Washington, DC: GAO.
US Geological Survey. 1998. Building Safer Structures. Fact Sheet 167-95. Reston, VA: USGS, June. Available at: http://quake.wr.usgs.gov/prepare/factsheets/SaferStructures/SaferStructures.pdf.
US Geological Survey. 1998. Speeding Earthquake Disaster Relief. Fact Sheet 097-95. Reston, VA: USGS, June. Available at: http://quake.wr.usgs.gov/prepare/factsheets/Mitigation/Mitigation.pdf.
US Patent and Trademark Office. 2006. USPTO Annual List of Top 10 Organizations Receiving Most U.S. Patents. January 10. Available at: http://www.uspto.gov/web/offices/com/speeches/06-03.htm.
Vedder, R. 2004. Growing Broke by Degree: Why College Costs Too Much. Washington, DC: AEI Press.
Venezia, A., M. W. Kirst, and A. L. Antonio. 2003. Betraying the College Dream: How Disconnected K–12 and Postsecondary Education Systems Undermine Student Aspirations. Stanford, CA: The Bridge Project, Stanford University. Available at: http://www.stanford.edu/group/bridgeproject/betrayingthecollegedream.pdf.
Wagner, C. S., and S. W. Popper. 2003. Identifying critical technologies in the USA. Journal of Forecasting 22:113-128.

Walsh, K. 2005. Foreign High-Tech R&D in China: Risks, Rewards, and Implications for US-China Relations. Washington, DC: Henry L. Stimson Center.

Wilson, D. J. 2002. Is embodied technological change the result of upstream R&D? Industry-level evidence. Review of Economic Dynamics 5(2):342-362.

Wulf, Wm. A. 2005. Review and Renewal of the Environment for Innovation. Unpublished paper.

Wulf, Wm. A. 2005. Observations on Science and Technology Trends: Their Potential Impact on Our Future. In Technology Futures and Global Wealth, Power and Conflict, A. G. K. Solomon, ed. Washington, DC: CSIS.

Zumeta, W., and J. S. Raveling. 2001. The Best and the Brightest for Science: Is There a Problem Here? In Innovation Policy in the Knowledge-Based Economy, M. P. Feldman and A. N. Link, eds. Boston: Kluwer Academic Publishers, pp. 121-161.

Zumeta, W., and J. S. Raveling. 2002. Attracting the best and the brightest. Issues in Science and Technology (Winter).

Zumeta, W., and J. S. Raveling. 2004. The Market for Ph.D. Scientists: Discouraging the Best and Brightest? Discouraging All? AAAS Symposium, February 16. Available at: http://www.eurekalert.org/pub_releases/2004-02/uow-rsl021304.php.

Index

B

G

H

L

N

O

P

S

T

X

Y